Applications of Metal–Organic Frameworks and Their Derived Materials

Scrivener Publishing
100 Cummings Center, Suite 541J
Beverly, MA 01915-6106

Publishers at Scrivener
Martin Scrivener (martin@scrivenerpublishing.com)
Phillip Carmical (pcarmical@scrivenerpublishing.com)

Applications of Metal–Organic Frameworks and Their Derived Materials

Edited by

Inamuddin, Rajender Boddula, Mohd Imran Ahamed and Abdullah M. Asiri

Scrivener Publishing

WILEY

This edition first published 2020 by John Wiley & Sons, Inc., 111 River Street, Hoboken, NJ 07030, USA and Scrivener Publishing LLC, 100 Cummings Center, Suite 541J, Beverly, MA 01915, USA

For more information about Scrivener publications please visit www.scrivenerpublishing.com.

Wiley Global Headquarters
111 River Street, Hoboken, NJ 07030, USA

For details of our global editorial offices, customer services, and more information about Wiley products visit us at www.wiley.com.

Library of Congress Cataloging-in-Publication Data

Names: Inamuddin, 1980– editor. | Boddula, Rajender, editor. | Ahamed, Mohd Imran, editor. | Asiri, Abdullah M., editor.
Title: Applications of metal-organic frameworks and their derived materials / edited by Inamuddin, Rajender Boddula, Mohd Imran Ahamed, and Abdullah M. Asiri.
Description: Hoboken, NJ : Wiley-Scrivener, 2020. | Includes bibliographical references and index.
Identifiers: LCCN 2020015462 (print) | LCCN 2020015463 (ebook) | ISBN 9781119650980 (cloth) | ISBN 9781119651161 (adobe pdf) | ISBN 9781119650959 (epub)
Subjects: MESH: Metal-Organic Frameworks–chemistry | Nanostructures–chemistry | Biosensing Techniques
Classification: LCC QD411 (print) | LCC QD411 (ebook) | NLM QD 411 | DDC 547/.05–dc23
LC record available at https://lccn.loc.gov/2020015462
LC ebook record available at https://lccn.loc.gov/2020015463

Cover image: Kris Hackerott
Cover design by Russell Richardson

Set in size of 11pt and Minion Pro by Manila Typesetting Company, Makati, Philippines

MIX
Paper from responsible sources
FSC® C013604

Contents

Preface

Metal–organic frameworks (MOFs) are porous crystalline polymers constructed by metal sites and organic building blocks. Since the discovery of MOFs in the 1990s, they have received tremendous research attention for various applications due to their high surface area, controllable morphology, tunable chemical properties, and multifunctionalities. These applications include MOFs as precursors and self-sacrificing templates for synthesizing metal oxides, heteroatom-doped carbons, metal-atoms encapsulated carbons, etc. These nanomaterials present new opportunities for versatile applications such as biomedical, energy conversion and storage, catalysis, and environmental remediation, etc. Hence, awareness and knowledge about MOFs and their derived nanomaterials with conceptual understanding are essential for the advanced material community.

Applications of Metal–Organic Frameworks and Their Derived Materials aim to explore down-to-earth applications in fields such as biomedical, environmental, energy, and electronics. This book provides an overview of the structural and fundamental properties, synthesis strategies, and versatile applications of MOFs and their derived nanomaterials. It gives an updated and comprehensive account of the research in the field of MOFs and their derived nanomaterials. This book will be beneficial for graduate and postgraduate students, faculty members, and research and development specialists working in the area of inorganic chemistry and material science, and chemical engineering, as well as industry professionals and nanotechnologists. Based on thematic topics, the book contains the following 14 chapters:

Chapter 1 presents some recent progress in the sensing field of MOFs and their derived materials. Different types of sensors are outlined based on signal transduction mechanism. The present problems and future development of MOFs and their derived materials in sensing field are also reported.

Chapter 2 discusses briefly the principle of piezo/ferroelectrity and the historical developments of MOF materials applied in piezo/ferroelectric applications.

Chapter 3 briefly surveys the fabrication and functionalization strategies of MOFs and their derived materials. The effects of the construction agents, synthesis techniques, synthesis conditions, and synthesis constitutions are discussed.

Chapter 4 focuses on the use of MOFs in membrane transport having immense potential in clinical and theronaustic applications. It also focuses on the recent developments of MOF materials used in clinical and theronaustic applications.

Chapter 5 discusses the introductory idea about the structure, classification, and properties of MOFs. The major part of the chapter discusses the role of MOF as an electrocatalyst for electrochemical sensing as well as in (electro)organic reactions for organic group transformations.

Chapter 6 focuses on the development of MOFs incorporated with ionic liquid (IL) for their potential application as ion conducting composite polymer electrolyte membranes for rechargeable batteries. Interaction of IL ions with metal nodes and organic linkers of MOFs and ion transport dynamics are discussed through XPS, scanning EXAFS, XANES, and dielectric spectroscopy studies.

Chapter 7 elaborates the use of MOF-based catalysts for the synthesis of fine chemicals. It explains the catalytic effect of MOFs and MOF composites in some of the most common reactions, such as oxidation, cycloaddition, esterification, and C–C bond formation, used in the synthesis of fine chemicals.

Chapter 8 discusses unique physiochemical properties of MOF and derived materials and their application as catalysts for various types of hydrogenation reactions. It also covers the exceptional variation and intensity of MOF-based catalyst structures and their selective application in hydrogenation reaction of various compounds.

Chapter 9 discusses the use of MOFs and their derived materials as sorbents in solid phase extraction applications, including miniaturized approaches. These novel materials constitute a powerful alternative to commercial sorbents due to several outstanding properties, thus providing more selective and sensitive analysis of complex samples.

Chapter 10 discusses the medical applications of MOFs as drug carriers of cancer drugs and as photosensitizers in photodynamic therapy. Additionally, the uses of MOFs as antibacterial and antifungal agents are discussed. The mechanisms of antimicrobial action of MOFs are also presented in addition to the advantages of bioMOFs.

Chapter 11 focuses on the general synthesis of MOFs and the adsorptive removal of ibuprofen, diclofenac, macroxen, and oxybenzone using MIL-100 by adopting *in silico* process.

Chapter 12 discusses the adsorption performance of some MOFs against volatile organic compounds (VOCs), the effect of some key features of MOF to the adsorption performance, the development of MOF composite for improvement of adsorption performance, an analytical method for modeling the adsorption, and factors influencing the adsorption performance.

Chapter 13 discusses the MOF-based materials as an electrocatalyst in diverse applications. Recent developments of MOF in energy and environmental application such as water splitting, hydrogen evolution reaction, oxygen evolution reaction, carbon dioxide reduction, and electrochemical sensing are summarized.

Chapter 14 reviews recent investigations of MOF pristine and MOF composites as photocatalysts for applications of energy (hydrogen production, CO_2 conversion), and environment (degradation of organic/inorganic pollutants). MOF composites, including metals, semiconductors, and multicomponent systems are analyzed in terms of preparation methods, properties, and reaction mechanisms involved in the selected photocatalytic reactions.

Key Features

- Overviews on MOFs and their derived nanomaterials/composites
- Addresses a wide range of applications in organo/photo/electro catalysis, sensors, adsorption, energy conversion, and storage

- Provides an understanding of MOFs and derived materials
- Discusses the fundamental and solutions to the applied problems of MOFs

Editors
Inamuddin,
Rajender Boddula,
Mohd Imran Ahamed
Abdullah M. Asiri

1

Application of MOFs and Their Derived Materials in Sensors

Yong Wang[1,2*], Chang Yin[1] and Qianfen Zhuang[1]

[1]*College of Chemistry, Nanchang University, Nanchang, China*
[2]*Jiangxi Province Key Laboratory of Modern Analytical Science, Nanchang University, Nanchang, China*

Abstract
In the past years, the application of metal organic framework (MOFs) and their derived materials in sensors has attracted wide attention due to their outstanding physical and chemical properties such as large specific surface area, tunable pore size, easy design/functionalization, high stability, good catalytic ability, and so on. In this chapter, we present some recent progress in the sensing field of MOFs and their derived materials. Depending on the signal transduction mechanism, different types of sensors are outlined. Moreover, the present problems and future development of MOFs and their derived materials are also presented.

Keywords: Metal organic framework, derived materials, sensor

1.1 Introduction

Sensor is a material or device that measures a physical or chemical quantity and converts it into an observable signal for detection of specific chemicals at trace levels [1–3]. Generally, sensing-based detection methods are superior to those traditional detection methods such as titration, chromatography, mass spectrometry and so on, because of its rapidity, simplicity, low cost, and suitability for large-scale sample screening [1–3]. On the basis of these advantages, the sensing-based approaches have been widely used in fields of environmental and industrial monitoring, drug

**Corresponding author*: wangyong@ncu.edu.cn

Inamuddin, Rajender Boddula, Mohd Imran Ahamed and Abdullah M. Asiri (eds.) Applications of Metal–Organic Frameworks and Their Derived Materials, (1–32) © 2020 Scrivener Publishing LLC

quality monitoring, forensic analysis, food safety, medical diagnostics, and national security [1–3]. However, at present, sensors suffer from some disadvantages like poor sensitivity, limited selectivity, slow response time, low lifetime, and stability.

To over these limitations, many advanced materials have been developed to construct various robust sensors [4–10]. Among them, metal–organic frameworks (MOFs) are especially attractive as a novel sensing material [11–18]. This material is a kind of crystalline material possessing nanopore network structure, which are formed by self-assembly of coordination between transition-metal cations and oxygen or nitrogen-containing polydentate organic linkers (see Figure 1.1) [19].

Generally, the large specific surface area of MOFs improves sensor sensitivity [11–18]. The metal ions or polydentate ligands of MOFs can be rationally designed to modulate the pore size of MOF, the number and orientation of catalytically active sites of MOFs, and the different interaction force between the analyte and MOF receptor, which enhance the sensor selectivity and sensitivity [11–18]. In addition, the interaction force and the structural matching between the analyte and MOF receptors can be modulated to increase the reversibility and response time of the sensor toward analyte, leading to the regeneration and real-time monitoring of the sensor [11–18]. On the other hand, MOF-derived materials are usually obtained using MOFs and/or other materials as precursors by various

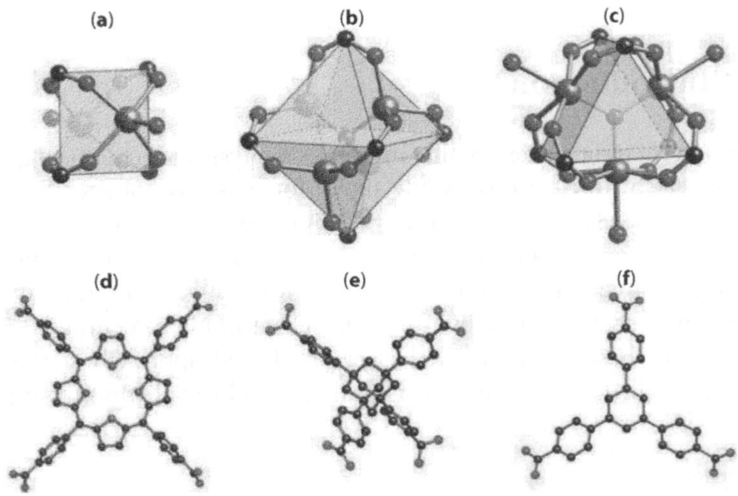

Figure 1.1 (a)–(c) Inorganic secondary building units (SBUs) of MOFs. (d)–(f) Organic ligands of MOFs. Reproduced with permission from Ref. [19]. Copyright 2004 Elsevier.

strategies such as high-temperature calcinations, hydrothermal synthesis, solvothermal synthesis, and so on [20–25]. These derived materials can not only retain the original structure of MOFs, but also improve some performances of these materials like electric conductivity, stability, water-solubility, catalytic activity, and mass-transfer ability [20–25]. These merits from the MOF-derived materials lead to the enhancement of the sensor's performances.

In the chapter, four different types of sensor, namely, optical sensor, electrochemical sensor, field-effect transistor-based sensor, and mass-sensitive sensor, is respectively described on the basis of the signal transduction mechanism. Particularly, we focus on the use of MOFs and their derived materials in the construction of sensors for detection of analyte, and summarize some representative investigations on the use of MOFs and their derived materials in the above-mentioned four types of sensor.

1.2 Application of MOFs and Their Derived Materials in Sensors

1.2.1 Optical Sensor

Optical sensor has recently attracted wide attention owing to its operation simplicity, time efficiency, and good reproducibility. MOFs and their derived materials can be easily designed to introduce optical probes, facilitating the construction of optical sensor. In addition, MOFs and their derived materials are demonstrated to possess nanozyme activity, which can catalyze various substrates into optical substances. As a result, the nanomaterials with artificial enzymes can be conveniently combined with different optical substances to construct various optical sensors. On the basis of optical transduction mechanism, optical sensors are usually classified into three types: colorimetric sensors, fluorescent sensors, chemiluminescent sensors. Therefore, the three different types of optical sensors will be introduced in the following section.

1.2.1.1 Colorimetric Sensor

In 2013, Li's group synthesized a Fe-MIL-88NH$_2$ MOF using 2-aminoterephthalic acid and FeCl$_3$ as precursors in a medium of acetic acid, and found for the first time that the MOF acted as a peroxidase, and could catalyze the oxidation of 3,3′,5,5′-tetramethylbenzidine (TMB) by H$_2$O$_2$ to produce a blue product [26]. Subsequently, they combined the MOF materials with glucose oxidase to construct a sensitive colorimetric sensor for glucose detection (see Figure 1.2) [26]. Following the work, many researchers exploited the peroxidase-like activity of the MOFs and their derived materials to construct various colorimetric sensor. Tan *et al.* [27] prepared a nanocomposite (CuNPs@C) composed of copper nanoparticles dispersed in a carbon matrix by one-pot thermal decomposition of a copper-based MOF precursor. The CuNPs@C can also possess peroxidase-like activity, which catalyze the reaction between H_2O_2 and 3,3,5,5-tetramethylbenzidine (TMB). Because this CuNPs surface does not contain a stabilizer, a higher affinity of CuNPs@C toward H_2O_2 can be obtained. Depending on the inhibition of TMB oxidation by ascorbic acid (AA), the material can be used to construct a colorimetric quenching sensor for detecting AA. Hou *et al.* used magnetic zeolitic imidazolate framework 8 to pack glucose oxidase, and then constructed the nanocomposite-based colorimetric sensor for glucose [28]. Dong *et al.* [29] encapsulated cobalt nanoparticles into Fe MOF-derived magnetic carbon to produce a nanocomposite, and found that the nanocomposite had much stronger peroxidase-like activity than pure CoNPs and magnetic carbon. Therefore, they combined glucose oxidase with the CoNPs/MC to construct a robust glucose sensor.

Metal ions are present in the ecosystem, and have important influence on the ecosystem. Hence, the construction of sensor for detection of metal ions is necessary for industrial processes, medical diagnosis, and environmental monitoring. In 2015, Gao *et al.* [30] synthesized a thermostable magnesium metal–organic framework (Mg-MOF), and found that many nanoholes containing non-coordinating nitrogen atoms were present in the material, which is suitable for hosting Eu^{3+} ions. On the basis of the energy level matching and energy transfer between the Eu^{3+} and the

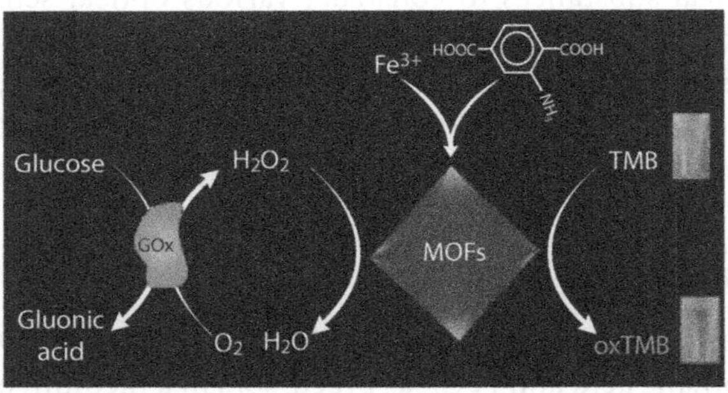

Figure 1.2 Schematic illustration of the peroxidase-like activity of Fe-MIL-88NH_2 MOFs using TMB and H_2O_2 as reactants and their applications for glucose sensing. Reproduced from Ref. [26] with permission. Copyright 2013 Royal Society of Chemistry.

Mg-MOF, they constructed a sensitive sensor for detection of Eu^{3+} ions. Khalil *et al.* [31] used UiO-66 metal–organic frameworks to accommodate diethyldithiocarbamate (DDTC) chromophore, and the obtained DDTC/UiO-66 was used for the construction of digital image-based colorimetric sensor for Cu^{2+} detection. Zeng *et al.* [32] synthesized bimetallic (Eu–Tb) lanthanide (Ln) metal–organic frameworks (MOFs) using Tb^{3+}/Eu^{3+} and 1,4-benzenedicarboxylate (BDC) as precursors for on-site sensitive and selective detection of Pb^{2+} in environmental waters. Li *et al.* [33] synthesized a composite containing Pt nanoparticle and UiO-66-NH_2 with permanent porosity, strong thermal and high chemical stability, and found that the material displayed high peroxidase-like activity. However, the peroxidase-like behavior of the material was suppressed in the presence of Hg^{2+}, due to the Hg^{2+}/Pt nanoparticle specific interaction. Therefore, they followed the principle to realize the construction of Hg^{2+} sensor (see Figure 1.3).

Recently, Wang *et al.* [34] has exploited the partial oxidation of cerium(III) to prepare the mixed-valence state cerium-based metal–organic framework (MVC-MOF) with the oxidase activity, and demonstrated that the oxidase activity of the MVC-MOF could be suppressed by single-stranded DNA (ssDNA). However, the oxidase activity of the material can be prevented in the presence of double-stranded DNA (dsDNA). Following the principle, the authors designed a Hg^{2+}-specific thymine-rich ssDNA (T-ssDNA) together with MVC-MOF for colorimetric detection of Hg^{2+}. Li *et al.* [35] used porous MOFs with highly stable hierarchical structure to immobilize enzymes, and found that the materials improved the adsorption capacity, and possessed strong pH-resisting ability. Therefore, they combined three enzymes like glucose oxidase, uricase, and horseradish

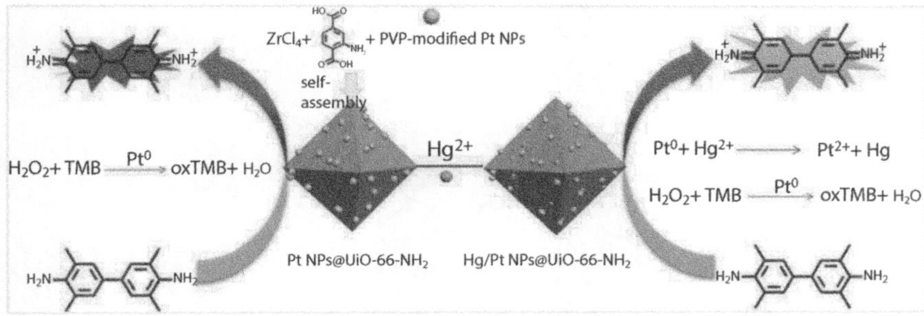

Figure 1.3 Synthesis of Pt NP@UiO-66-NH_2 nanocomposites with peroxidase-like activity for Hg^{2+} detection. Reproduced from Ref. [33] with permission. Copyright 2017 American Chemical Society.

peroxidase (HRP) with the porous MOFs to fabricate multi-enzyme sensors for detection of glucose and uric acid. In addition, this constructed sensor displayed good sensitivity, high selectivity, and recyclability. Zhang *et al.* [36] employed the dsDNA and G-quadruplex/hemin labeled on a metal organic framework of type MIL-101(Fe) as a peroxidase mimetic, and the clenbuterol-specific aptamer-labeled magnetic beads and complementary DNA (MB/Apt-cDNA) as capture probes for colorimetric sensing of clenbuterol. Dalapati *et al.* [37] prepared Ce MOFs with oxidase-like catalytic properties, and found that cysteine could reduce the oxidase-like catalytic properties of the MOFs, thus facilitating the construction of cysteine colorimetric sensor.

Volatile organic compounds (VOCs) usually correspond to the formation process of ozone and secondary aerosols in atmospheric environment, and play an important role in environmental pollutants and human health. Razavi *et al.* [38] employed 3,6-di(pyridin-4-yl)-1,4-dihydro-1,2,4,5-tetrazine (H_2DPT) and 4,4′-oxybis(benzoic acid) (H_2OBA) as ligands to prepare dihydrotetrazine-functionalized pillared metal–organic framework, and found that the MOFs displayed a high and selective response toward chloroform, and underwent a remarkable yellow-pink color change. Based on the principle, they constructed a colorimetric sensor for detection of chloroform (see Figure 1.4). Zhang's group [39] prepared a [Zn(N-(4-carboxybenzyl)-(3,5-dicarboxyl) pyridinium)(4,4′-bipy)$_{0.5}$]·$2H_2O$ MOF, and found that the exposure to NH_3, ethylamine (EA), and n-propylamine (PA) vapor led to

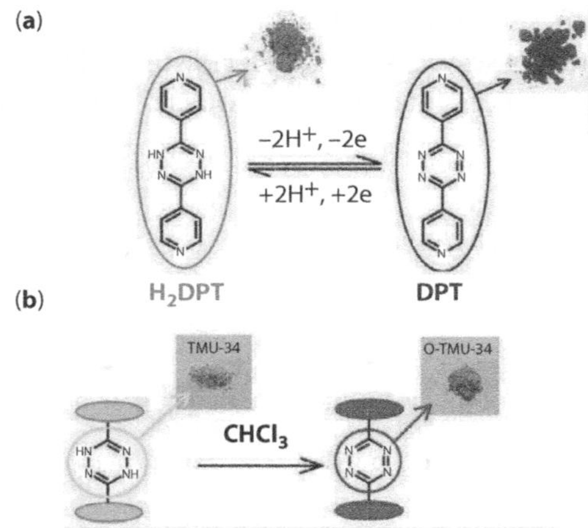

Figure 1.4 (a) Reversible transformation of H_2DPT into DPT. (b) Conversion of a dihydrotetrazine to a tetrazine moiety inside the TMU-34 framework upon the addition of chloroform. Reproduced with permission from Ref. [38]. Copyright 2017 John Wiley and Sons.

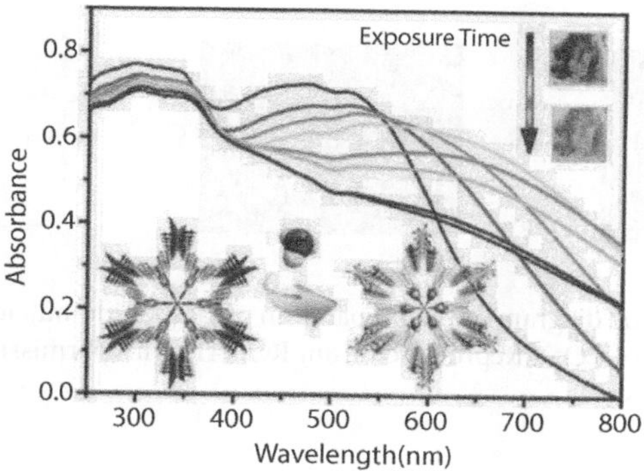

Figure 1.5 Application of FJU-56 MOFs for colorimetric sensing of NH_3. Reproduced from Ref. [40] with permission. Copyright 2018 American Chemical Society.

the rapid color change from colorless to yellow. They demonstrated that the photochromic behaviors of the material were ascribed to the photo-induced electron transfer and the formation of bipyridinium radicals, and the material could be used for the fabrication of photochromic and vapochromic colorimetric sensors. Zhang *et al.* [40] used tris-(4-tetrazolylphenyl)amine (H_3L) ligand to prepare three-dimensional mixed-valence cobalt(II/III) MOF (FJU-56) for sensitive, selective, and recyclable colorimetric sensing of ammonia (see Figure 1.5). The addition of ammonia causes the noticeable color change from red to brown, thus facilitating the visual detection by naked eyes.

1.2.1.2 Fluorescence Sensor

Among the optical sensors, fluorescence sensors attract the attention from most scientists because of its high sensitivity, relatively good selectivity, and rapidity. In 2006, Wong *et al.* [41] designed and synthesized lanthanide-mucicate MOFs using mucic acid, $TbCl_3$, and triethylamine as precursors, and found that the MOFs had recognition ability for CO_3^{2-} anion, and the addition of CO_3^{2-} caused the fluorescence enhancement of MOFs. Based on the principle, they constructed a fluorescence sensor for CO_3^{2-} detection (see Figure 1.6).

In 2009, Li's group [42] synthesized a highly bright MOFs, [Zn_2(4,4′-biphenyldicarboxylate)$_2$-(1,2-bipyridylethene)], and applied the materials for fast, reversible, sensitive sensing of 2,4-dinitrotoluene (DNT), and 2,3-dimethyl-2,3-dinitrobutane (DMNB) on the basis of Fluorescence redox quenching mechanism. After that, Pramanik *et al.* [43] synthesized

Figure 1.6 Schematic diagram of the preparation of lanthanide-mucicate MOFs and florescence sensing of CO_3^{2-}. Reproduced from Ref. [41] with permission. Copyright 2006 John Wiley and Sons.

a highly fluorescent $[Zn_2(oba)_2(bpy)]_3$ DMA MOF, and found that the aromatic compounds with electron-withdrawing groups quenched its fluorescence, but those with electron-donating groups enhanced its fluorescence. Different aromatic compounds were then determined by the $[Zn_2(oba)_2(bpy)]_3$ DMA MOF-based fluorescence sensor. The sensing mechanism results from electron transfer by molecular orbital and electronic band structure calculations as well as electrochemical measurements. Yang *et al.* [44] prepared fluorescent MIL-53(Al) MOF using Al^{3+} and 1,4-benzenedicarboxylic acid as precursors, and discovered that when adding Fe^{3+}, the MIL-53(Al) transformed into MIL-53(Fe) with low fluorescence. Then, they successfully fabricated a fluorescence sensor for Fe^{3+} detection. Li *et al.* [45] designed and prepared two clusters containing coordination-unsaturated metal sites from $\{M3\}_x$ SBUs. The two clusters could selectively adsorb CO_2, and the addition of Ba^{2+} and Cu^{2+} ions could significantly enhance or quench their fluorescence. Subsequently, they constructed fluorescent sensors for detection of Ba^{2+} and Cu^{2+}. They found that the fluorescence quenching by Cu^{2+} originated from the cation exchange of Cu^{2+} with $[NH_2(CH_4)_2]+$ in the pores of the framework, and the fluorescence enhancement by Ba^{2+} was attributed to the strong binding of Ba^{2+} and the uncoordinated carbonyl oxygen in BTC, the suitable Ba^{2+} ion radius, and the appropriate coordination ability of the specific solvent. Gole *et al.* [46] selected the ligands with aromatic tags for the synthesis of electron-rich MOFs, and then found that the fluorescent MOFs can selectively and sensitively quench the explosive nitroaromatic compounds, which is the basis of the construction of fluorescent sensor for detection of explosive nitroaromatic compounds. Liu *et al.* [47] constructed fluorescent MOF sensor for detection of the explosives, and used quantum theoretical calculations, periodic crystal models, cluster models, and fluorescence spectra to investigate the sensing mechanism in detail. They noted that explosives bound to MOFs via π–π stacking and hydrogen bonding, which contributed to the intermolecular electron transfer from the conduction bands of the MOFs to the LUMO of explosives, resulting in the remarkable fluorescence quenching. Yu *et al.* [48] realized the detection of nitroaromatic compounds and metal ions using two Zn(II) MOFs, namely, $Zn_3L_3(DMA)_2(H_2O)_3$ and $Zn_3L_3(DMF)_2$ (L represents 4,4′-stilbenedicarboxylic acid). They found that the two MOFs exhibited a highly sensitive and selective fluorescence quenching toward nitroaniline and Fe^{3+}, and the first MOF also displayed a fluorescence quenching toward Al^{3+}. As shown in Figure 1.7, Yang *et al.* [49] synthesized Cu MOFs using the polar tritopic quaternized carboxylate ligands together with 4,4′-dipyridyl sulfide ligands, and found that the MOF could interact with two carboxyfluorescein-labeled ssDNA via electrostatic, π–π stacking, and/or hydrogen-bonding interactions, which leads to the photoinduced electron-transfer fluorescence quenching.

The sensing principle was then used for detection of longer HIV ds-DNA or SUDV RNA sequences. After that, Tan *et al.* [50] used thermal decomposition approach to treat Fe MOFs (MIL-88A) to obtain a derived porous material containing magnetic carbon, and followed Yang's principle to construct a fluorescence sensor for ssDNA detection (see Figure 1.8). It was demonstrated that the proposed sensor had low background signal, high sensitivity, good selectivity, and even can distinguish single nucleotide mismatched nucleotides. Sun *et al.* [51] exploited Fe-MIL-88 MOFs and H_2O_2 to construct a "turn-on" fluorescence sensing platform for detection of biothiols. They found that Fe-MIL-88 with H_2O_2 exhibited weak fluorescence, but the addition of biothiols noticeably enhanced the fluorescence. To explain this fluorescence sensing mechanism, the authors demonstrated by various analytical technique that biothiols bound to Fe-MIL-88 through hydrogen bonding and electrostatic force, and then the biothiols reduced Fe^{3+} in the Fe-MIL-88 to Fe^{2+}, which catalyze the decomposition of H_2O_2 via Fenton reaction. The •OH radicals produced by Fenton reaction further

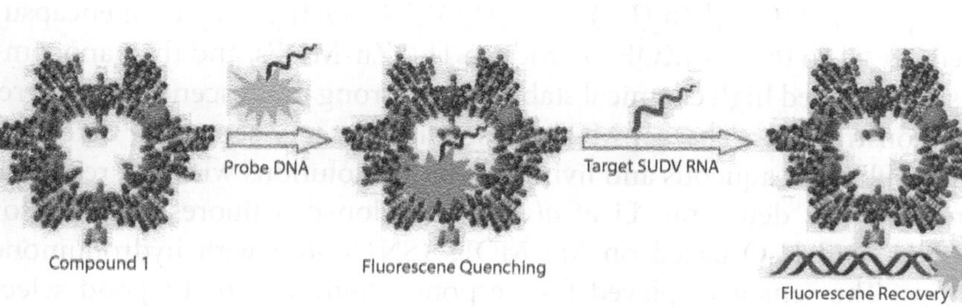

Figure 1.7 Schematic diagram of detection of Sudan virus RNA by carboxyfluorescein-labeled single-stranded DNA together with Cu MOFs. Reproduced from Ref. [49] with permission. Copyright 2015 American Chemical Society.

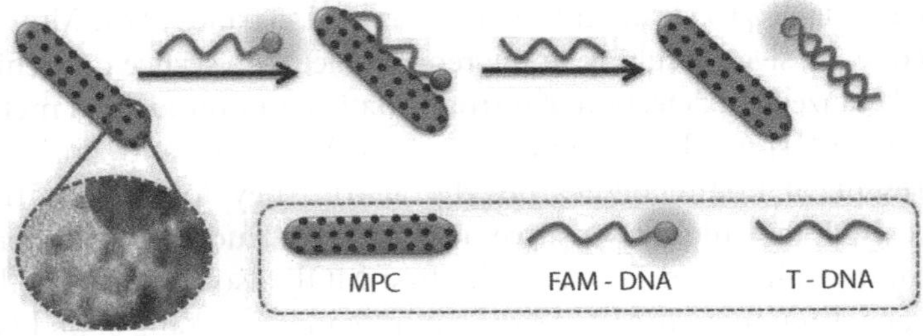

Figure 1.8 Fluorescence detection of target T-DNA based on magnetic porous carbon nanocomposite. Reproduced from Ref. [50] with permission. Copyright 2016 Elsevier Inc.

oxidize the terephthalic acid ligand to highly fluorescent product, leading to the fluorescence enhancement. Wu *et al.* [52] reported a hybrid of T-rich FAM-labeled ssDNA and UiO-66-NH_2 MOFs, which can be employed to fluorescence sensing of Hg^{2+}. The T-rich FAM-labeled ssDNA can associate with the UiO-66-NH_2 MOFs via π–π stacking, and hydrogen-bonding interactions between the DNA bases and the aromatic organic linker in the UiO-66-NH_2 MOFs. The interaction makes the FAM moiety close to the surface of the UiO-66-NH_2 MOFs, and as a result, the fluorescence of FAM undergoes a photoinduced-energy-transfer quenching process. However, the addition of Hg^{2+} caused the conformational change of T-rich FAM-labeled ssDNA from random-coil structure to hairpin structure, leading to the recovery of the fluorescence. Yang *et al.* [53] solvothermally synthesized Zn MOF and Cd MOF, and characterized their structure by single-crystal X-ray diffraction, IR and elemental analysis. They found that the MOFs possessed high thermal stability, good hydrolytic robustness, and strong fluorescence, and could be used to construct a fluorescence quenching sensor for H_2S. Ji *et al.* [54] self-assembled Zn^{2+} with ligand 4,4′,4″-[(1,3,5-triazine-2,4,6-triyl)tris(sulfanediyl)]tribenzoic acid into a microporous MOF, $\{[Zn_4(L^{3-})_2(O^{2-})(H_2O)_2]\cdot 4EtOH\}_n$. They then encapsulated Tb^{3+} into the Zn-MOF to produce Tb@Zn-MOFs, and the nanocomposite displayed high chemical stability and strong fluorescence. Therefore, they constructed a Tb@Zn-MOFs-based fluorescence sensor for detection of phosphate in aqueous and living cell buffer solutions with fast response time and low detection. Li *et al.* [55] developed a fluorescence sensor for NH_3 and H_2O based on Mg MOFs (SNNU-88) with hydroquinone groups. The sensor displayed fast response, high sensitivity, good selectivity, and old reusability due to the combination of the redox property of the 2,5-dihydroxyterephthalic acid (DHBDC) ligand and the pore size. The sensing mechanism was ascribed to the excited state intramolecular

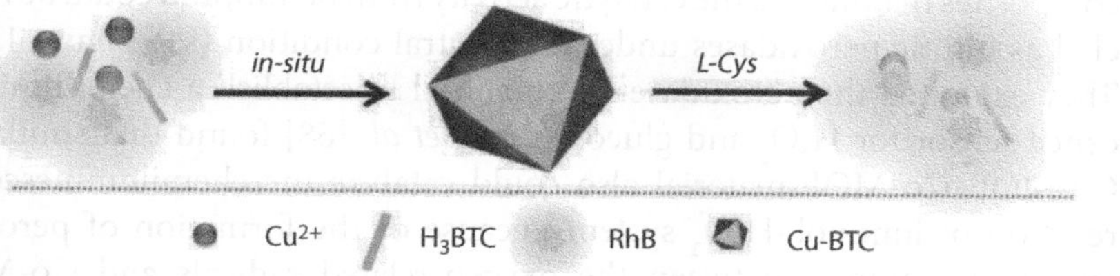

Figure 1.9 Fluorescence sensor for L-Cys using *in situ* synthesized RhB@Cu-BTC. Reproduced with permission from Ref. [56]. Copyright 2018 Elsevier Inc.

proton transfer process in DHBDC. In 2018, Zhao *et al.* [56] incorporated rhodamine B (RhB) into Cu-BTC MOFs to form RhB@Cu-BTC through *in situ* synthesis strategy, and found that thiol-containing amino acid, L-cysteine, could induce the collapse of RhB@Cu-BTC framework, and released the fluorescent RhB (see Figure 1.9). Based on the principle, a sensitive turn-on fluorescent sensor for thiol-containing amino acid was successfully developed.

Ratiometric sensors with dual signals are usually superior to the single signal sensor because the ratio of the two signals can give a self-reference correction to the intrinsic or extrinsic factors, improving the sensing accuracy and sensitivity. Yan *et al.* [57] used a postsynthetic method to prepare the Eu^{3+}-containing MOF-253 with two different color-emitting, and exploited the ratio of the two signals to construct a ratiometric fluorescent pH sensor. Subsequently, many researchers [58–65] constructed various ratiometric fluorescence sensor for detection of phosphate, Hg^{2+}, 2,4,6-trinitrophenol, H_2S, F^-, D_2O, 6-mercaptopurine, anthrax spore biomarker, and temperature.

1.2.1.3 Chemiluminescent Sensor

In addition to colorimetric sensors and fluorescence sensors, the chemiluminescence sensors based on MOFs and their derived materials have been recently investigated by many researchers [66–70]. In 2015, Zhu *et al.* [66] found for the first time that the chemiluminescence reaction of luminol–H_2O_2 system could be catalyzed in the presence of Cu-BTC (HKUST-1) owing to the radical generation and electron-transfer processes on the surface of Cu-BTC. However, the addition of dopamine quenched the chemiluminescence of the system, and thus they constructed a chemiluminescence sensor for detection of dopamine. After that, Luo *et al.* [67] embedded hemin into the HKUST-1 MOF materials, and found that the composites maintained the catalytic activity from hemin, and could be recycled as mimic peroxidases under the neutral condition (see Figure 1.10). They exploited the peroxidase-like material to establish a chemiluminescence sensor for H_2O_2 and glucose. Yang *et al.* [68] found that similar to Cu-MOF, Co-MOF material also could catalyze the chemiluminescence reaction of luminol–H_2O_2 system because of the formation of peroxide analogous complex between the oxygen-related radicals and Co-MOF material. Depending on the chemiluminescence system, they established a sensor for detection of L-cysteine. Chen's group [69] noticed that the composites containing graphene oxide (GO) and the copper-based metal organic frameworks (HKUST-1) could noticeably improve catalytic chemiluminescence between luminol and H_2O_2, and the addition of ascorbic acid decreased the chemiluminescence of the system. The quenching of chemiluminescence can be used to detect ascorbic acid in commercial food and fruit juices. Zhou *et al.* [70] used Fe-MIL-88B-NH_2 MOF nanoparticles to label the antibody, and employed the labeled antibody to capture the analyte alpha-fetoprotein (AFP). Subsequently, the Fe-MIL-88B-NH_2 NPs labeled antibody was dissolved in hydrochloric acid to release Fe^{3+}, which can be sensitively and selectively detected by a luminol–H_2O_2 chemiluminescence system.

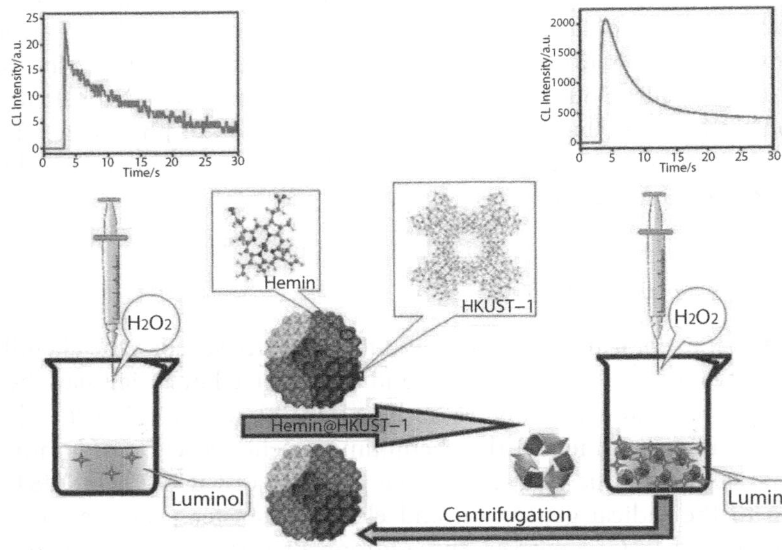

Figure 1.10 Construction of Hemin@HKUST-1-based chemiluminescence sensor for detection of H_2O_2. Reproduced with permission from Ref. [67]. Copyright 2015 American Chemical Society.

1.2.2 Electrochemical Sensor

Electrochemical technique is one of the most promising methods because of its rapidity and simplicity, low cost, easy miniaturization, high sensitivity and high selectivity. The MOFs and/or their derived materials have high specific area, strong conductivity, good stability, excellent catalytic activity, and fast response time, and thus are suitable for the construction of robust electrochemical sensors. Hence, MOFs and their derived materials have recently drawn wide interest from electroanalytical chemists. Current sensors based on MOFs and their derived materials can be divided into amperometric sensors, impedance sensors, electrochemiluminescence sensors, and photoelectrochemical sensors.

1.2.2.1 Amperometric Sensor

Among the electrochemical sensors, amperometric sensor is one of the most widely used sensors due to its high sensitivity and rapidity. In 2012, Hu's group [71] employed ZIF-8 MOFs to synthesize Co_3O_4 nanoparticles, and found that the template-based method can control the size of Co_3O_4 nanoparticle to improve the electrocatalytic activity. Therefore, they constructed a amperometric sensor for glucose and H_2O_2 based on the Co_3O_4 nanoparticles. Wang *et al.* [72] prepared Zn_4O(1,4-benzenedicarboxylate)$_3$ MOFs and used the materials to modify the carbon paste electrode. Then, they preadsorbed lead ion at the modified electrode surface and used differential pulse stripping voltammetry technique for sensing of lead ion. Xiao *et al.* [73] prepared the MOF derived materials, namely nitrogen doped microporous carbon (NMC) composite via a carbonization process of zeolitic imidazolate framework-8, and then used nafion-bismuth/NMC composite to construct a voltammetric sensor for Cd^{2+} and Pb^{2+}. It was demonstrated that the high sensitivity of the sensor originated from the excellent physicochemical properties and the synergistic effects of the NMC materials. Wang's group [74] modified benzoic acid groups onto reduced graphene oxide (rGO), and then immersed the material in Co^{2+} and 1,3,5-benzentricarboxylic acid solution to give a nanocomposite. Subsequently, the nanocomposite was pyrolyzed to obtain the MOF derived materials. It was demonstrated that the MOF derived materials displayed a good response toward glucose, and the good response was attributed to the good performances of the derived materials. Wu *et al.* [75] designed and prepared an electro-conducting water-stable Cu MOF, $\{[Cu_2(HL)_2(\mu_2\text{-}OH)_2(H_2O)_5]\cdot H_2O\}_n$ (H_2L = 2,5- dicarboxylic acid-3,4-ethylene dioxythiophene), and directly used the MOFs to constructed a electrochemical sensor for simultaneous detection of simultaneous detection of ascorbic acid and L-tryptophan. Wang *et al.* [76] prepared three kinds of Fe-MOFs (MIL-88A) as templates, and pyrolyzed them to obtain different hierarchical Fe_3O_4/carbon superstructures (see Figure 1.11). Then they performed a detailed investigation on the thermal decomposition process of different MOF templates on the products. In addition, they realized the electrochemical sensing of N-acetyl cysteine using the derived materials.

Wang's group [77] prepared the composites containing carbon spheres and Al-MIL-53-$(OH)_2$ MOFs, and further used the C/Al-MIL-53-$(OH)_2$ and the nafion polymer to modify the glassy carbon electrode for the construction of dopamine sensor. It was found that the dopamine current signals noticeably enhanced owing to the good electrical conductivity and the large surface area of the MOF nanocomposite, and the film-forming ability of nafion. Cui *et al.* [78] designed Fe porphyrinic MOF-labeled DNA probe. In the probe, Fe porphyrinic MOFs possessed high mimic peroxidase performance, and there was a Pb-dependent RNAzyme, GR-5. The addition of Pb^{2+} can specifically cleaved the RNAzyme to release the short Fe porphyrinic MOF-linked DNA fragment, which hybridize with hairpin DNA on the screen-printed carbon electrode surface. The immobilized

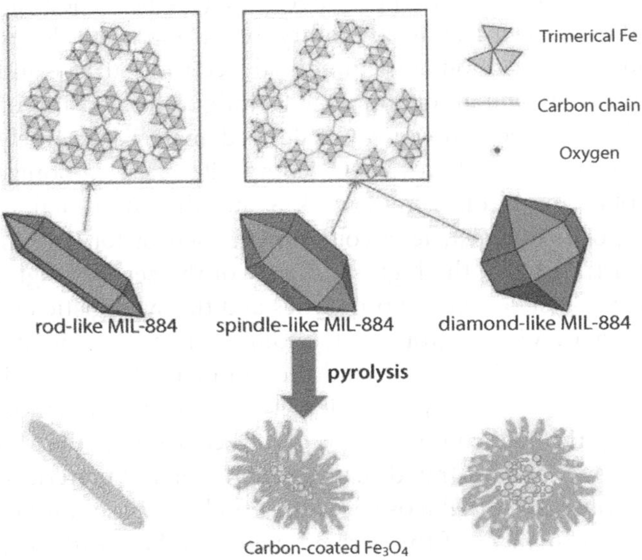

Figure 1.11 Schematic diagram of the formation process of Fe_3O_4/carbon superstructures derived from MIL-88A. Reproduced with permission from Ref. [76]. Copyright 2015 Springer Nature.

Fe porphyrinic MOF on the screen-printed carbon electrode surface can electrochemically catalyze the oxidation of TMB by H_2O_2 to produce enzymatically amplified electrochemical signals, which is suitable for sensitive detection of Pb^{2+} (see Figure 1.12).

Liu *et al.* [79] prepared Ni/NiO/C composites by pyrolysis of Ni MOFs in nitrogen and air atmosphere, and found that the Ni/NiO/C can be combined with the matrix of myoglobin (Mb)/hemoglobin (Hb) to construct an enzymatic electrochemical sensor for nitrite detection. The sensor displayed good electrochemical properties because the synergistic effect of MOFs-derived Ni, NiO, and carbon hybrids can accelerate the electron transfer process. Jia *et al.* [80] pyrolyzed a bimetallic NiCo MOF into a MOF-derived nanocomposite containing $NiCo_2O_4$, CoO, and Co/Ni nanoparticles packed with carbon nanotubes, and used the composite as an effective platform for immobilizing the probe DNA of human immune deficiency to electrochemical analysis of HIV-1 DNA in human serum samples. The electrochemical sensor exhibited high sensitivity, high selectivity, and good stability. Our group has recently reported the electrodepositing of Cu-BTC MOFs onto glassy carbon electrode (GCE) with electroreduced graphene oxide (ERGO) modification to construct a electrochemical sensing platform for 2,4,6-trinitrophenol (TNP). Importantly, the sensor can detect TNP when other similarly structured compounds (2-nitrophenol, 4-nitrophenol, 2,4-dinitrophenol) coexist [81]. The high sensitivity of the sensor was demonstrated to be ascribed to the high electrical conductivity and high electrocatalytic activity of the nanocomposite.

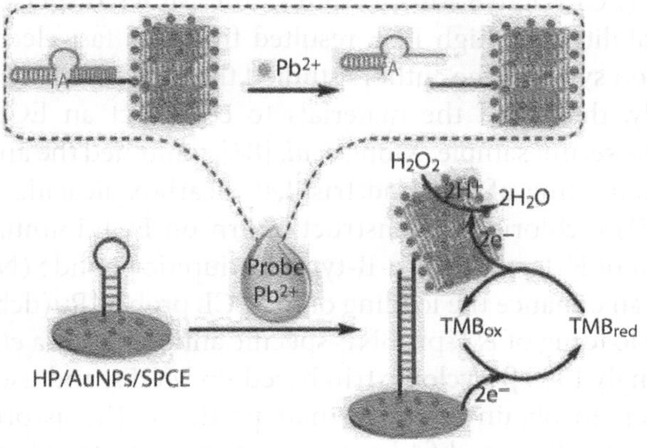

Figure 1.12 Fabrication of screen-printed carbon electrode modified with DNA functionalized Fe porphyrin MOFs for Pb^{2+} sensing. Reproduced with permission from Ref. [78]. Copyright 2015 American Chemical Society.

1.2.2.2 *Impedimetric, Electrochemiluminescence, and Photoelectrochemical Sensor*

In addition to amperometric electrochemical sensor, other types of MOFs and their derived materials-based electrochemical sensors such as impedance, electrochemiluminescence, and photoelectrochemical sensors have recently been investigated by many researchers. In 2015, Deep *et al.* [82] synthesized the Cd MOFs $[Cd(aminoterephthalic\ acid)(H_2O)_2]_n$, and conjugated this Cd MOFs film with anti-parathion antibody to construct a sensor for parathion based on the change of electrochemical impedance spectroscopy (EIS). Peterson *et al.* [83] reported UiO-66-NH_2 MOF-based EIS sensor for the explosive simulant 2,6-dinitrotoluene. They found that the interaction of the functional groups of the MOFs and analyte could cause the change of the dielectric properties, which leads to the EIS change. Zhou *et al.* [84] constructed a sensitive EIS aptasensor for carcinoembryonic antigen (CEA). They prepared the peroxidase-like nanocomposite consisting of Cu MOFs and Pt nanoparticles, CEA aptamer, hemin, and GOx. The nanocomposite can be then immobilized onto the modified electrode via the reaction between CEA and its aptamers. GOx initiated the cascade reaction of 3,3-diaminobenzidine (DAB) and glucose to produce H_2O_2 and precipitates with no conductivity. In the case, the electron transfer rate of this electrochemical probe, $[Fe(CN)_6]^{4-/3-}$, was hindered reduced, and the related EIS signal was enhanced. The signal enhancement resulted from the catalysis of the nanocomposite, and cascade reaction (see Figure 1.13).

In 2015, Xu *et al.* [85] reported the first example for electrochemiluminescence (ECL)-active Ru/Zn MOFs. The MOFs displayed high ECL and high stability. The high ECL resulted from the fast electron transfer of the reaction system. The authors studied the ECL mechanism in details. Subsequently, they used the materials to construct an ECL sensor for cocaine in the serum sample. Xiong *et al.* [86] combined the antibodies and the MOFs consisting of zinc and tris(4,4′-dicarboxylicacid2,2′-bipyridyl) ruthenium(II) dichloride to construct a turn-on ECL immunosensor for the detection of N-terminal pro-B-type natriuretic peptide (NT-proBNP). The MOFs can enhance the loading of the ECL probe, $[Ru(dcbpy)_3]^{2+}$, and improve the loading of NT-proBNP-specific antibodies. Ma *et al.* [87] prepared a strongly ELC β-cyclodextrin based Pb MOFs, and used the MOFs to reduce Ag^+ to obtain many Ag nanoparticles. The as-prepared Ag@Pb(II)-β-CD was then modified onto glass carbon electrodes, and the antibody of PSA was immobilized into the Ag@Pb(II)-β-CD via Ag nanoparticles. The sensing platform can realize PSA detection. Yuan's group [88]

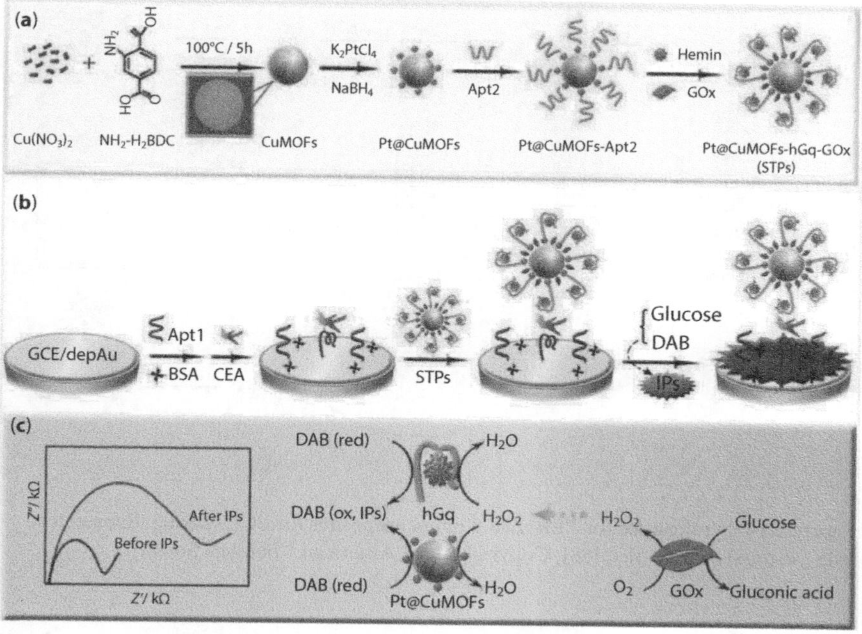

Figure 1.13 (a) Preparation of Pt@CuMOFs-hGq-GOx nanocomposite, (b) fabrication of the EIS sensor for CEA, and (c) cascade catalysis EIS amplification. Reproduced from Ref. [84] with permission. Copyright 2017 Elsevier Inc.

synthesized Ru-PEI@ZIF-8 nanocomposites with high ECL and excellent stability using a self-enhanced ruthenium polyethylenimine (Ru-PEI) complex doped zeolitic imidazolate framework-8, and its application for sensing of telomerase activity (see Figure 1.14). To further enhance the sensitivity of the constructed sensor, the authors exploited an enzyme-assisted DNA recycle-amplification technique to amplify the telomerase activity signal, and successfully applied this sensor to detect telomerase activity from cancer cells. Dong *et al.* [89] encapsulated $Ru(bpy)_3^{2+}$ in UiO-67 MOFs to form $Ru(bpy)_3^{2+}$/UiO-67, and used the ECL probe to construct a competitive-based immunosensor for diethylstilbestrol (DES) detection. The ECL sensor displayed high anti-interference ability and high stability for DES detection. Yan *et al.* [90] prepared a nanocomposite containing Au/Pt nanoparticles and UiO-66 MOFs to construct a sensitive and selective ECL sensor for protein kinase A (PKA) activity detection and corresponding inhibitor screening. They exploited the dual catalysis and recognition of Au/Pt nanoparticles for PKA. In the presence of ATP, PKA on the electrode was phosphorylated, and the phosphorylated PKA can bind to the

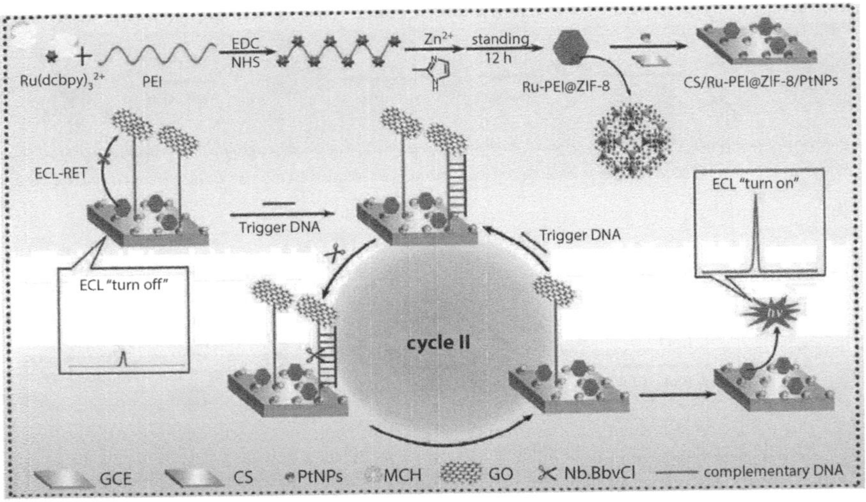

Figure 1.14 Synthesis of Ru-PEI@ZIF-8 nanocomposites for ECL sensing. Reproduced with permission from Ref. [88]. Copyright 2017 American Chemical Society.

UiO-66 defects of the nanocomposite via Zr–O–P bonds. The immobilized probe can enhance the ECL signals due to the strong synergistic catalysis of the nanocomposite, and inhibited these nanoparticles from aggregating.

In 2013, Kuang's group [91] adopted a self-template method to prepare nanocomposite containing ZnO and ZIF-8. The nanocomposite displayed vertically standing arrays like nanorod arrays and nanotube arrays. They found that the ZnO@ZIF-8 nanorod arrays exhibited different photoelectrochemical response to the hole scavengers. The photoelectrochemical signals in the presence of H_2O_2 were much more intense than those with ascorbic acid, which may be attributed to the pore size of the ZIF-8 shell and the different sizes of the two molecules. The ZnO@ZIF-8 nanorod arrays were then used to construct a photoelectrochemical H_2O_2 sensor. Jin *et al.* [92] solvothermally synthesized a TiO_2-modified MOF, and immobilized it on GCE. The results showed that the modified electrode had high photoelectrocatalytic activity, and could be used for developing a photoelectrochemical sensor for clethodim. The authors also studied the photoelectrochemical sensing mechanism in details. They found that the glassy carbon electrode delivered excited electrons to make positively charged holes leave on the TiO_2-modified MOF surface, and these holes reacted in the presence of H_2O to give hydroxy radicals, which were rapidly captured by clethodim to increase efficiency of charge separation. As a result, the photocurrent was noticeably enhanced. In 2016, Zhang *et al.*

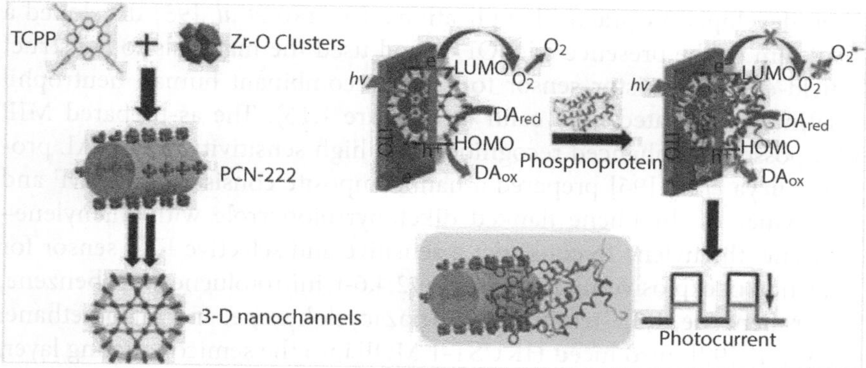

Figure 1.15 Preparation of PCN-222 and its application for photoelectrochemical sensing of phosphoprotein. Reproduced with permission from Ref. [93]. Copyright 2016 American Chemical Society.

[93] prepared Zr porphyrinic MOFs using tetrakis(4-carboxyphenyl) porphyrin ligand, and the ligand played a role as a light-harvesting substance. Oxygen and dopamine around these ligands can be enriched because of its particular porosity and tunable structures. As a result, these led to the improvement of the photoelectric conversion efficiency. Therefore, the Zr porphyrinic MOFs were used for constructing a label-free, sensitive, selective, turn-off photoelectrochemical sensor for detection of a phosphoprotein, α-casein (see Figure 1.15). Wang *et al.* [94] synthesized a Zr-based UiO-66 MOF with $[Ru(bpy)_3]^{2+}$, and then immobilized this material to the TiO_2/ITO electrode modified with phosphorylated kemptide. Subsequently, they found that the excited electrons from $[Ru(bpy)_3]^{2+}$ could enter into the TiO_2 conduction band to give photocurrent under visible light irradiation, and thus the observation could be exploited to construct a photoelectrochemical sensor for protein kinase activities. The sensor possessed a high sensitivity and high selectivity, which was attributed to the MOF defect recognition, and large specific surface area and high porosity of MOF.

1.2.3 Field-Effect Transistor Sensor

Field-effect transistor (FET) sensor is made up of source and drain electrodes. Each of the two electrodes keeps contact with a semiconductor layer, and the charge density is controlled by an electric field used between the semiconductor and a gate electrode. In the past few years, many field effect transistor sensors based on MOFs and their derived materials have been developed in practical applications. Iskierko *et al.* [95] developed a MIP film in the presence of MOF-5, and used the materials to construct a field-effect transistor sensor to detect recombinant human neutrophil gelatinase-associated lipid calin (see Figure 1.16). The as-prepared MIP film possessed enhanced recognition and high sensitivity for NGAL protein. Surya *et al.* [96] prepared a nanocomposite consisting of MOF and copolymer of thiophene flanked diketopyrrolopyrrole with thienylene–vinylene–thienylene to construct a sensitive and selective FET sensor for detection of explosive analytes, such as 2,4,6-trinitrotoluene, nitrobenzene, dinitrobenzene, 1,3,5-trinitro-1,3,5-triazacyclohexane, and nitromethane. Jang *et al.* [97] introduced HKUST-1 MOF into the semiconducting layer, and combined the material with poly(3-hexylthiophene-2,5-diyl) (P3HT) to construct a FET humidity sensor for detecting water. The HKUST-1/P3HT composites displayed a high sensitivity because of the good gas capture ability and the porosity of HKUST-1. In addition, the sensor showed fast response and can be recycled. Gardner *et al.* [98] demonstrated the use of MOFs to tune the selectivity of chemical adsorption and work function shift. They respectively used three different kinds of MOFs for FET-based sensing of water, NO_2, and NH_3. Wang *et al.* [99] developed an effective method for constructing Ni-MOF FET through *in situ* growing $Ni_3(HITP)_2$ membrane as the FET channel materials. The as-prepared film was large-area, dense and uniform, and controlling reaction time can modulate the film thickness and density of MOF-FET. The Ni-MOF FET can be successfully used for developing a sensor for detection of gluconic acid.

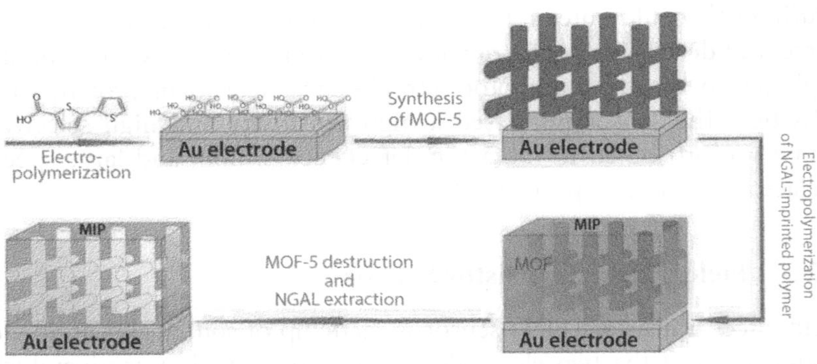

Figure 1.16 Preparation of MIP film in the presence of MOF-5 for the construction of FET-based NGAL sensor. Reproduced with permission from Ref. [95]. Copyright 2016 American Chemical Society.

1.2.4 Mass-Sensitive Sensor

Mass is one of the most basic properties of any analyte. On the basis of the advantage, many researchers have recently focused on the development of the sensors. Mass sensitive sensors based on MOFs and their derived materials can be divided into two types: quartz crystal microbalance (QCM) sensors and piezoelectric sensors. In 2011, Si *et al.* [100] used a simple method to synthesize amine-decorated microporous MOF, CAU-1, and found that the material could strongly adsorb methanol. Subsequently, they constructed a QCM sensor for methanol. Meilikhov *et al.* [101] synthesized heterostructured non-centrosymmetric binary Janus MOF $[Cu_2(dicarboxylate)_2(dabco)]_n$ with the dicarboxylate ligand, and found that the surface functionality of MOF strongly effect the analyte affinity, and the MOF could selectively adsorb the small and polar methanol. Therefore, they believed that the heterogeneous MOFs could be applied to design the coatings for the modulation of sensing performance. After that, Hou *et al.* [102] synthesized a new porous metal–organic framework (MOF) $\{[Cu_4(OH)_2(tci)_2(bpy)_2]\cdot 11H_2O\}$, to construct a sensitive and selective QCM sensor for small molecules, like methanol, ethanol, acetone and acetonitrile. Wannapaiboon *et al.* [103] exploited the liquid-phase epitaxial growth strategy to prepare Zn MOFs with hierarchical structure, and used the materials to construct a selective QCM-based sensor for alcohols and methanol. Jiao *et al.* [104] solvothermally synthesized a mixed-valent Cu^{I}/Cu^{II} MOF and used QCM technique to construct a sensor for water. Qian *et al.* [105] employed MIL-101 MOF to synthesize metolcarb-specific molecularly imprinted nanoparticles, and combined QCM technique to construct a metolcarb sensor. Zhou *et al.* [106] synthesized a Cu MOF, $[Cu_3L_2(H_2O)_{2.75}]\cdot 0.75H_2O\cdot 1.75DMA$ (H_3L = 4-(2-carboxyphenoxy)-isophthalic acid, DMA = dimethylacetamide), and exploited QCM technique to construct a humidity sensor. Sun's group [107] modified the Al(OH)(1,4-NDC) nanoscale MOF onto the surface of QCM and used the modified sensing platform for detection of pyridine. They demonstrated by density functional theory calculation that the selectivity of the pyridine sensor results from the larger binding energy of the MOF to pyridine and large mass of pyridine than water. Tchalala *et al.* [108] used fluorinated MOFs to selectively remove and sense SO_2, and found that the KAUST-7 (NbOFFIVE-1-Ni) and KAUST-8 (AlFFIVE-1-Ni) MOFs had a high affinity for SO_2. Then, they used QCM to realize the SO_2 sensing. Abuzalat *et al.* [109] used Cu-BTC/polyaniline (PANI) nanocomposite to construct a QCM-based hydrogen sensor. Zeinali's group [110] employed MIL-101(Cr) MOF to construct a QCM-based sensor for pyridine detection.

In 2012, Wen *et al.* [111] solvothermally synthesized a new MOF, namely, $[Mn_5(NH_2bdc)_5(bimb)_5 \cdot (H_2O)_{0.5}]_n$ (bimb =4,4′-bis(1-imidazolyl)biphenyl), and found that the MOF material had a typical ferroelectric behavior, suggesting that the MOF materials can be potentially applied for the construction of piezoelectric sensor.

1.3 Conclusion

In the chapter, the applications of MOFs and their derived materials in sensors are summarized, focusing mainly on the applications of these materials in optical sensors, electrochemical sensors, field-effect transistor-based sensors, and mass-sensitive sensors. Although the sensing behavior of MOFs and their derived materials has been investigated comprehensively, some challenges remain, which hinder the real application of these materials.

First of all, the relationship between physico-chemical properties (pore size, electrical conductivity, luminescence, catalytic active site, chemical stability, interaction force, etc.) and structure/composition of MOFs and their derived materials deserves to be studied in details because the understanding of the underlying physics and the important structure–property relationship is very helpful for the modulation and improvement of the sensor's performance, like sensitivity, selectivity, response time, stability, and regeneration. In this aspect, synthesis approaches, analytical characterization techniques, as well as theoretical calculations are strongly needed to solve the problem.

Second, how can be the sensor's performance (sensitivity, selectivity, response time, stability, and regeneration) enhanced? After understanding the aforementioned structure–property relationship, some important factors should be discovered, and some modulation routes can be naturally created to improve the sensor's performance. In addition, some other materials and techniques will also aid the improvement of the sensor's performance. For examples, the other different materials (DNA aptamers, antibodies, enzymes) can be introduced to enhance the sensor's selectivity; different amplification techniques like enzyme-amplification, polymer-amplification, DNA-amplification can be introduced to increase the sensor's sensitivity.

Finally, most sensors based on the MOFs and their derived materials cannot be applied in real samples at present because of the presence of severe matrix effect and background interferences in real samples. To solve the problems, it is appealing to combine the standard addition or ratiometic techniques with chemometrics tools (multivariate curve resolution by alternating least-squares, alternating trilinear decomposition, parallel factor analysis, etc.) with second-order advantages.

With the aforementioned efforts, it is anticipated that major progress and exciting researching into the sensing application of the MOFs and their derived materials will take place in the near future.

Acknowledgments

This book chapter was supported by the National Natural Science Foundation of China (NSFC-21864017 and NSFC-31960495), and the Science and Technology Innovation Platform Project of Jiangxi Province (20192BCD40001).

References

1. Howes, P.D., Chandrawati, R., Stevens, M.M., Colloidal nanoparticles as advanced biological sensors. *Science*, 346, 53, 2014.
2. Swager, T.M. and Mirica, K.A., Introduction: Chemical sensors. *Chem. Rev.*, 119, 1–2, 2019.
3. Ricco, A.J. and Crooks, R.M., Chemical sensors. *Acc. Chem. Res.*, 31, 200, 1998.
4. Haupt, K. and Mosbach, K., Molecularly imprinted polymers and their use in biomimetic sensors. *Chem. Rev.*, 100, 2495–2504, 2000.
5. Willner, I. and Zayats, M., Electronic aptamer-based sensors. *Angew. Chem. Int. Ed.*, 46, 6408–6418, 2007.
6. Farka, Z., Jurik, T., Kovar, D., Trnkova, L., Skladal, P., Nanoparticle-based immunochemical biosensors and assays: Recent advances and challenges. *Chem. Rev.*, 117, 9973–10042, 2017.
7. Hatchett, D.W. and Josowicz, M., Composites of intrinsically conducting polymers as sensing nanomaterials. *Chem. Rev.*, 108, 746–769, 2008.
8. Stewart, M.E., Anderton, C.R., Thompson, L.B., Maria, J., Gray, S.K., Rogers, J.A., Nuzzo, R.G., Nanostructured plasmonic sensors. *Chem. Rev.*, 108, 494–521, 2008.
9. Yetisen, A.K., Naydenova, I., Vasconcellos, F.D.C., Blyth, J., Lowe, C.R., Holographic sensors: Three-dimensional analyte-sensitive nanostructures and their applications. *Chem. Rev.*, 114, 10654–10696, 2014.
10. Schroeder, V., Savagatrup, S., He, M., Lin, S.B., Swager, T.M., Carbon nanotube chemical sensors. *Chem. Rev.*, 119, 599–663, 2019.
11. Kreno, L.E., Leong, K., Farha, O.K., Allendof, M., Duyne, R.P.V., Hupp, J.T., Metal–organic framework materials as chemical sensors. *Chem. Rev.*, 112, 1105–1125, 2012.

12. Lei, J.P., Qian, R.C., Ling, P.H., Cui, L., Ju, H.X., Design and sensing applications of metal–organic framework composites. *Trends Anal. Chem.*, 58, 71–78, 2014.
13. Hu, Z.C., Deibert, B.J., Li, J., Luminescent metal–organic frameworks for chemical sensing and explosive detection. *Chem. Soc. Rev.*, 43, 5815–5840, 2014.
14. Kumar, P., Deep, A., Kim, K.H., Metal organic frameworks for sensing applications. *Trends Anal. Chem.*, 73, 39–53, 2015.
15. Yi, F.Y., Chen, D.X., Wu, M.K., Han, L., Jiang, H.L., Chemical sensors based on metal–organic frameworks. *ChemPlusChem*, 81, 675–690, 2016.
16. Lustig, W.P., Mukherjee, S., Rudd, N.D., Desai, A.V., Li, J., Ghosh, S.K., Metal–organic frameworks: functional luminescent and photonic materials for sensing applications. *Chem. Soc. Rev.*, 46, 3242–3285, 2017.
17. Liu, L.T., Zhou, Y.L., Liu, S., Xu, M.T., The applications of metal–organic frameworks in electrochemical sensors. *ChemElectroChem*, 5, 6–19, 2018.
18. Koo, W.T., Jang, J.S., Kim, I.D., Metal–organic frameworks for chemiresistive sensors. *Chem.*, 5, 1938–1963, 2019.
19. Rowsell, J.L.C. and Yaghi, O.M., Metal–organic frameworks: A new class of porous materials. *Microporous Mesoporous Mater.*, 73, 3–14, 2004.
20. Song, Y.H., Li, X., Sun, L.L., Wang, L., Metal/metal oxide nanostructures derived from metal–organic frameworks. *RSC Adv.*, 5, 7267–7279, 2015.
21. Xia, W., Mahmood, A., Zou, R.Q., Xu, Q., Metal–organic frameworks and their derived nanostructures for electrochemical energy storage and conversion. *Energy Environ. Sci.*, 8, 1837–1866, 2015.
22. Zhao, S.N., Song, X.Z., Song, S.Y., Zhang, H.J., Highly efficient heterogeneous catalytic materials derived from metal–organic framework supports/precursors. *Coord. Chem. Rev.*, 337, 80–96, 2017.
23. Mai, H.D., Rafiq, K., Yoo, H., Nano metal–organic framework-derived inorganic hybrid nanomaterials: Synthetic strategies and applications. *Chem. Eur. J.*, 23, 5631–5651, 2017.
24. Wu, H.B. and Lou, X.W., Metal–organic frameworks and their derived materials for electrochemical energy storage and conversion: Promises and challenges. *Sci. Adv.*, 3, eaap9252, 2017.
25. Dang, S., Zhu, Q.L., Xu, Q., Nanomaterials derived from metal–organic frameworks. *Nat. Rev. Mater.*, 3, 17075, 2018.
26. Liu, Y.L., Zhao, X.J., Yang, X.X., Li, Y.F., A nanosized metal–organic framework of Fe-MIL-88NH_2 as a novel peroxidase mimic used for colorimetric detection of glucose. *Analyst*, 138, 4526–4531, 2013.
27. Tan, H.L., Ma, C.J., Gao, L., Li, Q., Song, Y.H., Xu, F.G., Wang, T., Wang, L., Metal–Organic Framework-Derived Copper Nanoparticle@Carbon Nanocomposites as Peroxidase Mimics for Colorimetric Sensing of Ascorbic Acid. *Chem. Eur. J.*, 20, 16377–16383, 2014.
28. Hou, C., Wang, Y., Ding, Q.H., Jiang, D., Li, M., Zhu, W.W., Pan, D., Zhu, H., Liu, M.Z., Facile synthesis of enzyme-embedded magnetic metal–organic

frameworks as a reusable mimic multi-enzyme system: Mimetic peroxidase properties and colorimetric sensor. *Nanoscale*, 7, 18770–18779, 2015.
29. Dong, W.F., Zhuang, Y.X., Li, S.Q., Zhang, X.D., Chai, H.X., Huang, Y.M., High peroxidase-like activity of metallic cobalt nanoparticles encapsulated in metal–organic frameworks derived carbon for biosensing. *Sens. Actuators B Chem.*, 255, 2050–2057, 2018.
30. Gao, Y.F., Zhang, X.Q., Sun, W., Liu, Z.L., A robust microporous metal–organic framework as a highly selective and sensitive, instantaneous and colorimetric sensor for Eu^{3+} ions. *Dalton Trans.*, 44, 1845–1849, 2015.
31. Khalil, M.M.H., Shahat, A., Radwan, A., El-Shahat, M.F., Colorimetric determination of Cu(II) ions in biological samples using metal–organic framework as scaffold. *Sens. Actuators B Chem.*, 233, 272–280, 2016.
32. Zeng, X.L., Zhang, Y.J., Zhang, J.Y., Hu, H., Wu, X., Long, Z., Hou, X.D., Facile colorimetric sensing of Pb^{2+} using bimetallic lanthanide metal–organic frameworks as luminescent probe for field screen analysis of lead-polluted environmental water. *Microchem. J.*, 134, 140–0145, 2017.
33. Li, H.P., Liu, H.F., Zhang, J.D., Cheng, Y.X., Zhang, C.L., Fei, X.Y., Xian, Y.Z., Platinum nanoparticle encapsulated metal–organic frameworks for colorimetric measurement and facile removal of mercury(II). *ACS Appl. Mater. Interfaces*, 9, 46, 40716–40725, 2017.
34. Wang, C.H., Tang, G.E., Tan, H.L., Colorimetric determination of mercury(II) via the inhibition by ssDNA of the oxidase-like activity of a mixed valence state cerium-based metal–organic framework. *Microchim. Acta*, 185, 475, 2018.
35. Liu, X., Qi, W., Wang, Y.F., Su, R.X., He, Z.M., A facile strategy for enzyme immobilization with highly stable hierarchically porous metal–organic frameworks. *Nanoscale*, 9, 17561–17570, 2017.
36. Zhang, Y.D., Ren, H.X., Miao, Y.B., Visualization and colorimetric determination of clenbuterol in pork by using magnetic beads modified with aptamer and complementary DNA as capture probes, and G-quadruplex/hemin and DNA antibody on the metal–organic framework MIL-101(Fe) acting as a peroxidase mimic. *Microchim. Acta*, 186, 515, 2019.
37. Dalapati, R., Sakthivel, B., Ghosalya, M.K., Dhakshinamoorthy, A., Biswas, S., A cerium-based metal–organic framework having inherent oxidase-like activity applicable for colorimetric sensing of biothiols and aerobic oxidation of thiols. *CrystEngComm*, 19, 5915–5925, 2017.
38. Razavi, S.A.A., Masoomi, M.Y., Morsali, A., Stimuli-responsive metal–organic framework (MOF) with chemo-switchable properties for colorimetric detection of $CHCl_3$. *Chem. Eur. J.*, 23, 12559–12564, 2017.
39. Li, L., Hua, Y., Guo, Y., Wang, H.Y., Li, X.N., Zhang, H., Bifunctional photo- and vapochromic behaviors of a novel porous zwitterionic metal–organic framework. *New J. Chem.*, 43, 3428–3431, 2019.
40. Zhang, J.D., Ouyang, J., Ye, Y.X., Li, Z.Y., Lin, Q.J., Chen, T., Zhang, Z.J., Xiang, S.C., Mixed-valence cobalt(II/III) metal–organic framework for ammonia

sensing with naked-eye color switching. *ACS Appl. Mater. Interfaces*, 10, 27465–27471, 2018.
41. Wong, K.L., Law, G.L., Yang, Y.Y., Wong, W.T., A highly porous luminescent terbium-organic framework for reversible anion sensing. *Adv. Mater.*, 18, 1051–1054, 2006.
42. Lan, A.J., Li, K.H., Wu, H.H., Olson, D.H., Emge, T.J., Ki, W., Hong, M.C., Li, J., A luminescent microporous metal–organic framework for the fast and reversible detection of high explosives. *Angew. Chem. Int. Ed.*, 48, 2334–2338, 2009.
43. Pramanik, S., Zheng, C., Zhang, X., Emge, T.J., Li, J., New microporous metal–organic framework demonstrating unique selectivity for detection of high explosives and aromatic compounds. *J. Am. Chem. Soc.*, 133, 4153–4155, 2011.
44. Yang, C.X., Ren, H.B., Yan, X.P., Fluorescent metal–organic framework MIL-53(Al) for highly selective and sensitive detection of Fe^{3+} in aqueous solution. *Anal. Chem.*, 85, 7441–7446, 2013.
45. Li, Y.W., Li, J.R., Wang, L.F., Zhou, B.Y., Chen, Q., Bu, X.H., Microporous metal–organic frameworks with open metal sites as sorbents for selective gas adsorption and fluorescence sensors for metal ions. *J. Mater. Chem. A*, 1, 495–499, 2013.
46. Gole, B., Bar, A.K., Mukherjee, P.S., Modification of Extended open frameworks with fluorescent tags for sensing explosives: competition between size selectivity and electron deficiency. *Chem. Eur. J.*, 20, 2276–2291, 2014.
47. Liu, L., Hao, J.Y., Shi, Y.T., Qiu, J.S., Hao, C., Roles of hydrogen bonds and p–p stacking in the optical detection of nitro-explosives with a luminescent metal–organic framework as the sensor. *RSC Adv.*, 5, 3045, 2015.
48. Yu, Z.C., Wang, F.Q., Lin, X.Y., Wang, C.M., Fu, Y.Y., Wang, X.J., Zhao, Y.N., Li, G.D., Selective fluorescence sensors for detection of nitroaniline and metal Ions based on ligand-based luminescent metal–organic frameworks. *J. Solid State Chem.*, 232, 96–101, 2015.
49. Yang, S.P., Chen, S.R., Liu, S.W., Tang, X.Y., Qin, L., Qiu, G.H., Chen, J.X., Chen, W.H., Platforms formed from a three-dimensional Cu-based zwitterionic metal–organic framework and probe ss-DNA: Selective fluorescent biosensors for human immunodeficiency virus 1 ds-DNA and Sudan virus RNA sequences. *Anal. Chem.*, 87, 12206–12214, 2015.
50. Tan, H.L., Tang, G., Wang, Z.X., Li, Q., Gao, J., Wu, S.M., Magnetic porous carbon nanocomposites derived from metal–organic frameworks as a sensing platform for DNA fluorescent detection. *Anal. Chim. Acta*, 940, 136–142, 2016.
51. Sun, Z.J., Jiang, J.Z., Li, Y.F., A sensitive and selective sensor for biothiols based on the turn-on fluorescence of the Fe-MIL-88 metal–organic frameworks–hydrogen peroxide system. *Analyst*, 140, 8201, 2015.
52. Wu, L.L., Wang, Z., Zhao, S.N., Meng, X., Song, X.Z., Feng, J., Song, S.Y., Zhang, H.J., A metal–organic framework/DNA hybrid system as a novel

fluorescent biosensor for mercury(II) ion detection. *Chem. Eur. J.*, 22, 477–480, 2016.

53. Yang, X.F., Zhu, H.B., Liu, M., Transition-metal-based (Zn^{2+} and Cd^{2+}) metal–organic frameworks as fluorescence "turn-off" sensors for highly sensitive and selective detection of hydrogen sulfide. *Inorganica Chim. Acta*, 466, 410–416, 2017.
54. Ji, G.F., Gao, X.C., Zheng, T.X., Guan, W.H., Liu, H.T., Liu, Z.L., Postsynthetic metalation metal–organic framework as a fluorescent probe for the ultrasensitive and reversible detection of PO_4^{3-} Ions. *Inorg. Chem.*, 57, 17, 10525–10532, 2018.
55. Li, Y.P., Li, S.N., Jiang, Y.C., Hu, M.C., Zhai, Q.G., A semiconductor and fluorescence dual-mode room-temperature ammonia sensor achieved by decorating hydroquinone into a metal–organic framework. *Chem. Commun.*, 54, 9789–9792, 2018.
56. Zhao, X.D., Zhang, Y.Z., Han, J.M., Jing, H.M., Gao, Z.Q., Huang, H.L., Wang, Y.Y., Zhong, C.L., Design of "turn-on" fluorescence sensor for L-cysteine based on the instability of metal–organic frameworks. *Microporous Mesoporous Mater.*, 268, 88–92, 2018.
57. Lu, Y. and Yan, B., A ratiometric fluorescent pH sensor based on nanoscale metal–organic frameworks (MOFs) modified by europium(III) complexes. *Chem. Commun.*, 50, 13323–13326, 2014.
58. Dai, C., Yang, C.X., Yan, X.P., Ratiometric fluorescent detection of phosphate in aqueous solution based on near infrared fluorescent silver nanoclusters/metal–organic shell composite. *Anal. Chem.*, 87, 11455–11459, 2015.
59. Wu, P.Y., Liu, Y.H., Liu, Y., Wang, J.R., Li, Y., Liu, W., Wang, J., Cadmium-based metal–organic framework as a highly selective and sensitive ratiometric luminescent sensor for mercury(II). *Inorg. Chem.*, 54, 11046–11048, 2015.
60. Fu, R.B., Hu, S.M., Wu, X.T., Luminescent zinc phosphonates for ratiometric sensing of 2,4,6-trinitrophenol and temperature. *Cryst. Growth Des.*, 16, 5074–5083, 2016.
61. Zhang, X., Hu, Q., Xia, T.F., Zhang, J., Yang, Y., Cui, Y.J., Chen, B.L., Qian, G.D., Turn-on and ratiometric luminescent sensing of hydrogen sulfide based on metal–organic frameworks. *ACS Appl. Mater. Interfaces*, 8, 32259–32265, 2016.
62. Yang, Z.R., Wang, M.M., Wang, X.S., Yin, X.B., Boric-acid-functional lanthanide metal organic frameworks for selective ratiometric fluorescence detection of fluoride ions. *Anal. Chem.*, 89, 1930–1936, 2017.
63. Hawes, C.S. and Gunnlaugsson, T., Multichannel luminescent lanthanide polymers as ratiometric sensors for D_2O. *Chem.*, 2, 463–465, 2017.
64. Jin, M., Mou, Z.L., Zhang, R.L., Liang, S.S., Zhang, Z.Q., An efficient ratiometric fluorescence sensor based on metal–organic frameworks and quantum dots for highly selective detection of 6-mercaptopurine. *Biosens. Bioelectron.*, 91, 162–168, 2017.

65. Luan, K., Meng, R.Q., Shan, C.F., Cao, J., Jia, J.G., Liu, W.S., Tang, Y., Terbium functionalized micelle nanoprobe for ratiometric fluorescence detection of anthrax spore biomarker. *Anal. Chem.*, 90, 3600–3607, 2018.
66. Zhu, Q., Chen, Y.L., Wang, W.F., Zhang, H.G., Ren, C.L., Chen, H.L., Chen, X.G., A sensitive biosensor for dopamine determination based on the unique catalytic chemiluminescence of metal–organic framework HKUST-1. *Sens. Actuators B Chem.*, 210, 500–507, 2015.
67. Luo, F.Q., Lin, Y.L., Zheng, L.Y., Lin, X.M., Chi, Y.W., Encapsulation of hemin in metal–organic frameworks for catalyzing the chemiluminescence reaction of the H_2O_2–luminol system and detecting glucose in the neutral condition. *ACS Appl. Mater. Interfaces*, 7, 11322–11329, 2015.
68. Yang, N., Song, H.J., Wan, X.Y., Fan, X.Q., Su, Y.Y., Lv, Y., A metal (Co)-organic framework-based chemiluminescence system for selective detection of L-cysteine. *Analyst*, 140, 2656–266, 2015.
69. Zhu, Q., Dong, D., Zheng, X.J., Song, H.Q., Zhao, X.R., Chen, H.L., Chen, X.G., Chemiluminescence determination of ascorbic acid using graphene oxide@copper-based metal–organic frameworks as a catalyst. *RSC Adv.*, 6, 25047–25055, 2016.
70. Zhou, J.R., Long, Z., Tian, Y.F., Ding, X.D., Wu, L., Hou, X.D., A chemiluminescence metalloimmunoassay for sensitive detection of alpha-fetoprotein in human serum using Fe-MIL-88B-NH_2 as a label. *Appl. Spectrosc. Rev.*, 51, 517–526, 2016.
71. Hou, C.T., Xu, Q., Yin, L.N., Hu, X.Y., Metal–organic framework templated synthesis of Co_3O_4 nanoparticles for direct glucose and H_2O_2 detection. *Analyst*, 137, 5803–5808, 2012.
72. Wang, Y., Wu, Y.C., Xie, J., Hua, X.Y., Metal–organic framework modified carbon paste electrode for lead sensor. *Sens. Actuators B Chem.*, 177, 1161–1166, 2013.
73. Xiao, L.L., Xua, H.B., Zhou, S.H., Song, T., Wang, H.H., Li, S.Z., Gan, W., Yuan, Q.H., Simultaneous detection of Cd(II) and Pb(II) by differential pulseanodic stripping voltammetry at a nitrogen-doped microporous carbon/nafion/bismuth-film electrode. *Electrochim. Acta*, 143, 143–151, 2014.
74. Song, Y.H., Wei, C.T., He, J., Li, X., Lu, X.P., Wang, L., Porous Co nanobeads/rGO nanocomposites derived from rGO/Co-metal organic frameworks for glucose sensing. *Sens. Actuators B Chem.*, 220, 1056–1063, 2015.
75. Wu, X.Q., Ma, J.G., Li, H., Chen, D.M., Gu, W., Yang, G.M., Cheng, P., Metal–organic framework biosensor with high stability and selectivity in a biomimic environment. *Chem. Commun.*, 51, 9161–9164, 2015.
76. Wang, L., Zhang, Y.Y., Li, X., Xie, Y.Z., He, J., Yu, J., Song, Y.H., The MIL-88A-derived Fe_3O_4-carbon hierarchical nanocomposites for electrochemical sensing. *Sci. Rep.*, 5, 14341, 2015.
77. Wang, Y., Ge, H.L., Ye, G.Q., Chen, H.H., Hu, X.Y., Carbon functionalized metal organic framework/Nafion composites as novel electrode materials for ultrasensitive determination of dopamine. *J. Mater. Chem. B*, 3, 3747–3753, 2015.

78. Cui, L., Wu, J., Li, J., Ju, H.X., Electrochemical sensor for lead cation sensitized with a DNA functionalized porphyrinic metal–organic framework. *Anal. Chem.*, 87, 20, 10635–10641, 2015.
79. Liu, D., Dong, S.Y., Wei, W.B., Fu, Y.L., Huang, T.L., Ni/NiO/C composites derived from nickel based metal–organic frameworks for improved enzyme-based biosensor. *J. Electrochem. Soc.*, 164, B495–B501, 2017.
80. Jia, Z.K., Ma, Y.S., Yang, L.Y., Guo, C.P., Zhou, N., Wang, M.H., He, L.H., Zhang, Z.H., $NiCo_2O_4$ spinel embedded with carbon nanotubes derived from bimetallic NiCo metal–organic framework for the ultrasensitive detection of human immune deficiency virus-1 gene. *Biosens. Bioelectron.*, 133, 55–63, 2019.
81. Wang, Y., Cao, W., Wang, L.Y., Zhuang, Q.F., Ni, Y.N., Electrochemical determination of 2,4,6-trinitrophenol using a hybrid film composed of a copper-based metal organic framework and electroreduced graphene oxide. *Microchim. Acta*, 185, 315, 2018.
82. Deep, A., Bhardwaj, S.K., Paul, A.K., Kim, K.-H., Kumar, P., Surface assembly of nano-metal organic framework on amine functionalized indium tin oxide substrate for impedimetric sensing of parathion. *Biosens. Bioelectron.*, 65, 226–231, 2015.
83. Peterson, G.W., McEntee, M., Harris, C.R., Klevitch, A.D., Fountain, A.W., Soliz, J.R., Balboa, A., Hauser, A.J., Detection of an explosive simulant via electrical impedance spectroscopy utilizing the UiO-66-NH_2 metal–organic framework. *Dalton Trans.*, 45, 17113–17116, 2016.
84. Zhou, X.X., Guo, S.J., Gao, J.X., Zhao, J.M., Xue, S.Y., Xu, W.J., Glucose oxidase-initiated cascade catalysis for sensitive impedimetric aptasensor based on metal–organic frameworks functionalized with Pt nanoparticles and hemin/G-quadruplex as mimicking peroxidases. *Biosens. Bioelectron.*, 98, 83–90, 2017.
85. Xu, Y., Yin, X.B., He, X.W., Zhang, Y.K., Electrochemistry and electrochemiluminescence from a redox-active metal–organic framework. *Biosens. Bioelectron.*, 68, 197–203, 2015.
86. Xiong, C.Y., Wang, H.J., Liang, W.B., Yuan, Y.L., Yuan, R., Chai, Y.Q., Luminescence-functionalized metal–organic frameworks based on a ruthenium(II) complex: A signal amplification strategy for electrogenerated chemiluminescence immunosensors. *Chem. Eur. J.*, 21, 9825–9832, 2015.
87. Ma, H.M., Li, X.J., Yan, T., Li, Y., Zhang, Y., Wu, D., Wei, Q., Du, B., Electrochemiluminescent immunosensing of prostate-specific antigen based on silver nanoparticles-doped Pb (II) metal–organic framework. *Biosens. Bioelectron.*, 79, 379–385, 2016.
88. Xiong, C.Y., Liang, W.B., Zheng, Y.N., Zhuo, Y., Chai, Y.Q., Yuan, R., Ultrasensitive assay for telomerase activity via self-enhanced electrochemiluminescence ruthenium complex doped metal–organic frameworks with high emission efficiency. *Anal. Chem.*, 89, 3222–3227, 2017.
89. Dong, X., Zhao, G.H., Liu, L., Li, X., Wei, Q., Cao, W., Ultrasensitive competitive method-based electrochemiluminescence immunosensor for

diethylstilbestrol detection based on $Ru(bpy)_3^{2+}$ as luminophor encapsulated in metal–organic frameworks UiO-67. *Biosens. Bioelectron.*, 110, 201–206, 2018.
90. Yan, Z.Y., Wang, F., Deng, P.Y., Wang, Y., Cai, K., Chen, Y.H., Wang, Z.H., Liu, Y., Sensitive electrogenerated chemiluminescence biosensors for protein kinase activity analysis based on bimetallic catalysis signal amplification and recognition of Au and Pt loaded metal–organic frameworks nanocomposites. *Biosens. Bioelectron.*, 109, 132–138, 2018.
91. Zhan, W.W., Kuang, Q., Zhou, J.Z., Kong, X.J., Xie, Z.X., Zheng, L.S., Semiconductor@metal–organic framework core-shell heterostructures: A case of ZnO@ZIF8 nanorods with selective photoelectrochemical response. *J. Am. Chem. Soc.*, 135, 1926–1933, 2013.
92. Jin, D.Q., Xu, Q., Yu, L.Y., Hu, X.Y., Photoelectrochemical detection of the herbicide clethodim by using the modified metal–organic framework amino-MIL-125(Ti)/TiO_2. *Microchim. Acta*, 182, 1885–1892, 2015.
93. Zhang, G.Y., Zhuang, Y.H., Shan, D., Su, G.F., Cosnier, S., Zhang, X.J., Zirconium-based porphyrinic metal–organic framework (PCN-222): Enhanced photoelectrochemical response and its application for label-free phosphoprotein detection. *Anal. Chem.*, 88, 11207–11212, 2016.
94. Wang, Z.H., Yan, Z.Y., Wang, F., Cai, J.B., Guo, L., Su, J.K., Liu, Y., Highly sensitive photoelectrochemical biosensor for kinase activity detection and inhibition based on the surface defect recognition and multiple signal amplification of metal–organic frameworks. *Biosens. Bioelectron.*, 97, 107–114, 2017.
95. Iskierko, Z., Sharma, P.S., Prochowicz, D., Fronc, K., D'Souza, F., Toczydłowska, D., Stefaniak, F., Noworyta, K., Molecularly imprinted polymer (MIP) film with improved surface area developed by using metal–organic framework (MOF) for sensitive lipocalin (NGAL) determination. *ACS Appl. Mater. Interfaces*, 8, 19860–19865, 2016.
96. Surya, S.G., Nagarkar, S.S., Ghosh, S.K., Sonar, P., Rao, V.R., OFET based explosive sensors using diketopyrrolopyrrole and metal organic framework composite active channel material. *Sens. Actuators B Chem.*, 223, 114–122, 2016.
97. Jang, Y.J., Jung, Y.E., Kim, G.W., Lee, C.Y., Metal–organic frameworks in a blended polythiophene hybrid film with surface-mediated vertical phase separation for the fabrication of a humidity sensor. *RSC Adv.*, 9, 529–535, 2019.
98. Gardner, D.W., Gao, X., Fahad, H.M., Yang, A.T., He, S., Javey, A., Carraro, C., Maboudian, R., Transistor-based work function measurement of metal–organic frameworks for ultra-low-power, rationally designed chemical sensors. *Chem. Eur. J.*, 25, 2019. doi: 10.1002/chem.201902483.
99. Wang, B.F., Luo, Y.Y., Liu, B., Duan, G.T., Field-effect transistor based on *in-situ* grown metal–organic framework film as liquid-gated sensing device. *ACS Appl. Mater. Interfaces*, 2019. doi: 10.1021/acsami.9b14319.

100. Si, X.L., Jiao, C.L., Li, F., Zhang, J., Wang, S., Liu, S., Li, Z.B., High and selective CO_2 uptake, H_2 storage and methanol sensing on the amine-decorated 12-connected MOF CAU-1. *Energy Environ. Sci.*, 4, 4522–4527, 2011.
101. Meilikhov, M., Furukawa, S., Hirai, K., Fischer, R.A., Kitagawa, S., Binary Janus porous coordination polymer coatings for sensor devices with tunable analyte affinity. *Angew. Chem. Int. Ed.*, 52, 341–345, 2013.
102. Hou, C.Y., Bai, Y.L., Bao, X.L., Xu, L.Z., Lin, R.G., Zhu, S.R., Fang, J.H., Xu, J.Q., A metal–organic framework constructed using a flexible tripodal ligand and tetranuclear copper cluster for sensing small molecules. *Dalton Trans.*, 44, 7770–7773, 2015.
103. Wannapaiboon, S., Tu, M., Sumida, K., Khaletskaya, K., Furukawa, S., Kitagawa, S., Fischer, R.A., Hierarchical structuring of metal–organic framework thin-films on quartz crystal microbalance (QCM) substrates for selective adsorption applications. *J. Mater. Chem. A*, 3, 23385–23394, 2015.
104. Jiao, C.L., Jiang, X., Chu, H.L., Jiang, H.Q., Sun, L.X., A mixed-valent Cu^{I}/Cu^{II} metal–organic framework with selective chemical sensing properties. *CrystEngComm*, 18, 8683–8687, 2016.
105. Qian, K., Deng, Q.L., Fang, G.Z., Wang, J.P., Pan, M.F., Wang, S., Pu, Y.H., Metal–organic frameworks supported surface-imprinted nanoparticles for the sensitive detection of metolcarb. *Biosens. Bioelectron.*, 79, 359–363, 2016.
106. Zhou, Z.Q., Li, M.X., Wang, L.Y., He, X., Chi, T., Wang, Z.X., Antiferromagnetic copper(II) metal–organic framework based quartz crystal microbalance sensor for humidity. *Cryst. Growth Des.*, 17, 6719–6724, 2017.
107. Xua, F., Sun, L.X., Huang, P.R., Sun, Y.J., Zheng, Q., Zou, Y.J., Chu, H.L., Yan, E.H., Zhang, H.Z., Wang, J.H., Du, Y., A pyridine vapor sensor based on metal–organic framework-modified quartz crystal microbalance. *Sens. Actuators B Chem.*, 254, 872–877, 2018.
108. Tchalala, M.R., Bhatt, P.M., Chappanda, K.N., Tavares, S.R., Adil, K., Belmabkhout, Y., Shkurenko, A., Cadiau, A., Heymans, N., Weireld, G.D., Maurin, G., Salama, K.N., Eddaoudi, M., Fluorinated MOF platform for selective removal and sensing of SO2 from flue gas and air. *Nat. Commun.*, 10, 1328, 2019.
109. Abuzalat, O., Wong, D., Park, S.S., Kim, S., High-performance, room temperature hydrogen sensing with a Cu-BTC/polyaniline nanocomposite film on a quartz crystal microbalance. *IEEE Sens. J.*, 13, 4789–4795, 2019.
110. Haghighi, E. and Zeinali, S., Nanoporous MIL-101(Cr) as a sensing layer coated on a quartz crystal microbalance (QCM) nanosensor to detect volatile organic compounds (VOCs). *RSC Adv.*, 9, 24460–24470, 2019.
111. Wen, L.L., Zhou, L., Zhang, B.G., Meng, X.G., Qu, H., Li, D.F., Multifunctional amino-decorated metal–organic frameworks: nonlinear-optic, ferroelectric, fluorescence sensing and photocatalytic properties. *J. Mater. Chem.*, 22, 22603–22609, 2012.

2

Applications of Metal–Organic Frameworks (MOFs) and Their Derivatives in Piezo/Ferroelectrics

H. Manjunatha[1*], K. Chandra Babu Naidu[2†], N. Suresh Kumar[3], Ramyakrishna Pothu[4] and Rajender Boddula[5]

[1]*Department of Chemistry, GITAM Deemed to be University, Bangalore, India*
[2]*Department of Physics, GITAM Deemed to be University, Bangalore, India*
[3]*Department of Physics, JNTUA, Anantapuramu, A.P, India*
[4]*College of Chemistry and Chemical Engineering, Hunan University, Changsha, China*
[5]*CAS Center for Excellence in Nanoscience, National Center for Nanoscience and Technology, Beijing, China*

Abstract

Metal–organic frameworks (MOFs) are the important class of materials. In this chapter, we specifically focused on the MOFs for applications in sensors actuators, microelectromechanical devices, ultrasonic resonators and energy harvesting technology. In addition, the MOFs and their derivatives like the barium titanate, lead titanate and lead zirconium titanate based organic materials are discussed for piezo/ferroelectric properties. In view of this, the corresponding parameters like saturation polarization (P_s), dielectric constant, and piezoelectric coefficient (d_{33}) are elucidated for MOFs. Moreover, the importance of porous structure of MOFs is discussed for sensors actuators, microelectromechanical devices, ultrasonic resonators device applications along with piezo/ferroelectric parameters.

Keywords: Metal–organic frameworks, piezoelectricity, ferroelectricity, dielectric constant, spontaneous polarization, piezoelectric coefficient

**Corresponding author*: hanumanjunath80@gmail.com
†*Corresponding author*: chandrababu954@gmail.com

Inamuddin, Rajender Boddula, Mohd Imran Ahamed and Abdullah M. Asiri (eds.) Applications of Metal–Organic Frameworks and Their Derived Materials, (33–62) © 2020 Scrivener Publishing LLC

2.1 Introduction

2.1.1 Brief Introduction to Piezo/Ferroelectricity

Most of the research done all over the world these days is on finding alternative energy sources to the traditional and much-exploited energy sources like fossil fuels. As these energy sources are fast depleting, researchers are looking at other options of generating electricity in order to address the increasing global energy demands. Piezoelectric materials are one such option. Piezoelectric materials area class of dielectric materials which produce electricity when subjected to a mechanical stress or pressure. The word "piezo" in Greek refers to "to press or squeeze". Thus piezoelectric materials are the materials which convert mechanical energy into electrical energy (direct piezoelectric effect) [1]. The converse is also possible with these materials, i.e., when an electric potential is applied to piezoelectric materials, they develop a mechanical stress in their crystal structure and in turn a strain (Inverse piezoelectric effect). Such devices which convert any other form of energy into electrical energy are called as energy harvesters and can be used to generate energy when direct electricity is not available and batteries are not a possible option.

On the other hand ferroelectrics are a special type of dielectrics and subclass of piezoelectrics and pyroelectrics which exhibit a spontaneous electric polarization whose direction can be reversed by applying external electric fields [2]. This switchability of spontaneous polarization results in a hysteresis loop between polarization and the electric field. All ferroelectric materials exhibit piezoelectricity. Many crystalline materials, ceramic materials, polymers, liquid crystals, organic matter such as bones, proteins, DNA, etc., have been investigated as piezo/ferroelectric materials. Piez/ferroelectric materials have various potential applications. Starting from the first application of piezoelectric quartz crystal as ultrasonic transducer (technically called sonar used in submarines during world war I) to communication circuit components, sensors and actuators, filters, resonators, ultrasonic motors, energy harvester, microelectromechanical systems (MEMs), data storage, optical devices for instance ferroelectric random access memories (FeRAM), and field effective transistors [3–7].

2.2 Fundamentals of Piezo/Ferroelectricity

The applications of piezo/ferroelectric materials depends on various piezo/ferroelectric characteristics, important to mention are the piezoelectric constants (*d*, *g*, *h*, and *e*), Curie temperature (T_c), which represents a phase transition, an electromechanical coupling coefficient (*k*), a parameter used to compare diverse piezoelectric materials and is a measure of interchange of electrical and mechanical energy. Ferroelectricity is described by parameters such as spontaneous polarization (P_s), ramanent polarization (P_r), and coercive field (E_c). The following discussion will focus on the theoretical aspects of piezo/ferroelectric principle.

A dielectric is a solid, electrically insulating material that can be polarized by an applied electric field. A dielectric material when placed in an external electric field, *E*, undergoes polarization (*P*) and results in separation of positive and negative charges and develops a dipole moment, which is given by the product of positive and negative charges and the separation between them. This charge separation in the material is owing to the arrangement of small molecular-dipoles along the direction of the functional electric field. Piezoelectricity is observed in dielectric materials without center of symmetry. Non-centrosymmetry or chirality is one of the most essential crystallographic features of piezoelectric materials. Among the 32 crystallographic point groups, 20 are non-centrosymmetric (piezoelectric) and out of these 20, 10 point groups posses a unique spontaneously polarized polar axis (pyroelectric). If such polarization is reversed by applying external electric field then these pyroelectrics can be ferroelectric [7]. The piezoelectric effect was found to be due to displacement of electric charges (electrical dipoles in the form of ions or molecules in the crystal structure) under applied mechanical stress. Mathematically it is given by the equation,

$$P = d \times \sigma \qquad \text{(2.1) Direct Piezoelectric effect and}$$

$$\varepsilon = d \times E \qquad \text{(2.2) Inverse Piezoelectric effect}$$

where P = Polarization or electric charge density (pC/m^2)

σ = Mechanical stress (N/m^2)

ε = Strain and

d = piezoelectric coefficient (pC/N or m/V).

According to thermodynamic principles [8], the piezoelectric coefficient, d, is same for both direct piezoelectric effect and converse piezoelectric effect. In matrix method, it is given by

$$\begin{pmatrix} P1 \\ P2 \\ P3 \end{pmatrix} = \begin{pmatrix} 0 & 0 & 0 & 0 & d15 & 0 \\ 0 & 0 & 0 & d24 & 0 & 0 \\ d31 & d32 & d33 & 0 & 0 & 0 \end{pmatrix} \begin{pmatrix} T1 \\ T2 \\ T3 \\ T4 \\ T5 \\ T6 \end{pmatrix} \& \begin{pmatrix} S1 \\ S2 \\ S3 \\ S4 \\ S5 \\ S6 \end{pmatrix}$$

$$= \begin{pmatrix} 0 & 0 & d31 \\ 0 & 0 & d32 \\ 0 & 0 & d33 \\ 0 & d24 & 0 \\ d15 & 0 & 0 \\ 0 & 0 & 0 \end{pmatrix} \begin{pmatrix} E1 \\ E2 \\ E3 \end{pmatrix}$$

Piezoelectric coefficient of a material can be measured along the direction of the deformation of the crystal structure upon applying electric field. Piezoelectric effect is always orientation-dependent(1-D to 3-D) and one can measure the longitudinal piezoelectric coefficient, which is designated as d_{zz} and is given by the following equation for a tetragonal symmetry group [8, 9].

$$d_{zz} = (d_{31} + d_{15}) \, \mathrm{Sin}2\theta \mathrm{Cos}\theta + d_{33}\mathrm{Cos}3\theta,$$

where θ is the angle between (001) piezoelectric crystal axis and measurement direction. If the direction of (001) crystal axis is perpendicular to the sample. That is, the θ becomes equal to zero. Hence, the d_{zz} and d_{33} become equal.

Ferroelectric materials are characterized by spontaneous polarization (P_s) along one or more polar axis that can be switched or reversed by the applications of an external electric field (E) below their Curie temperature (T_c). Ferroelectrics have a permanent electric-dipole in their crystal unit cell which may be reversed by applying external fields. This results in a polar-field hysteresis loop (*P–E*) as shown in the Figure 2.1. The hysteresis loop macroscopically reflects the motion of electric domains (regions of spontaneous polarization in crystals). The net spontaneous polarization that exists in the crystal even subsequently the external electric field is detached is called as remanent polarization (P_r) and the coercive field E_c

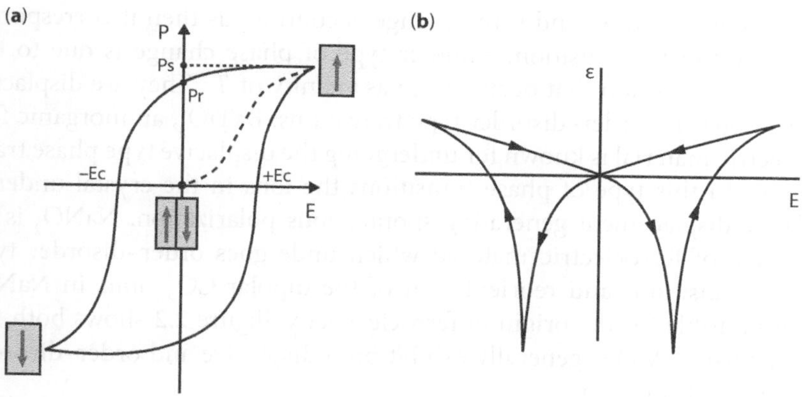

Figure 2.1 Ferroelectric hysteresis (a) and butterfly loops (b).

is defined as the field required to set the polarization back to zero. On the other hand, E_c can be described as a threshold field at which the polarization switches sign. From P–E hysteresis loops one can obtain the values of P_s, P_r, and E_c of a ferroelectric material. There are some ferroelectric materials which show additional ferroelectric–ferroelectric transitions below Curie temperature (T_c) and above Curie temperature (T_c) they said to be in paraelectric phase where the material has higher symmetry and behaves like a normal dielectric with no hysteresis.

It is well known that ferroelectric behavior of compounds has direct link to origin and motion of electric polarization in the structure of materials. During the emergence of charge or polarity the crystal structure undergoes a phase transition. A phase transition is the transformation of a thermodynamic system from one phase of the material to another. Ferroelectric materials undergo phase transition from paraelectric phase to ferroelectric phase. Such transitions always lead to changes in the physical properties of the materials. In true ferroelectrics, spontaneous polarization (P_s) occurs due to the alignment of intrinsic dipoles present in the solid compound. This alignment of dipole moments and origin of charged condition directs toward transitions in structure to ferroelectric phase from paraelectric phase. The compound becomes asymmetric due to loss of some symmetry elements under the critical temperature by cooling and the process is named as symmetry-breaking. At this stage, an order parameter is introduced to measure degree of order in the crystal and this parameter is called spontaneous polarization (P_s). The value of P_s is generally zero in the phase which occurs above T_c and is non-zero for the phase that occurs below T_c for a crystal. If the change in P_s value is discontinuous then it is called first-order transition and if the change is continuous then it corresponds to second-order transition. Another type of phase change is due to the nature of transition that occur at the as vicinity of T_c. They are displacive transitions and order–disorder type transitions. $BaTiO_3$, an inorganic ferroelectric material is known for undergoing the displacive type phase transitions. In this type of phase transitions the ions in the crystal undergo relative displacement generating spontaneous polarization. $NaNO_2$ is an example of ferroelectric material which undergoes order–disorder type phase transitions and reorientation of the dipolar CO_3^{2-} ions in $NaNO_3$ is responsible for the origin of ferroelectricity. Figure 2.2 shows both the mechanisms. MOFs generally exhibit both displacive and order–disorder type phase transitions.

There are several analytical methods for identifying ferroelectricity in a new material and each analytical technique will make use of a property of the material to identify its ferroelectricity. For example, XRD analysis can be used to assign the martial to which crystal class it belongs to. If it belongs to anyone of the 21 non-centrosymmetric crystals point groups then the material is considered to be ferroelectric. Other techniques used to identify ferroelectricity are thermal analysis (due to the fact that all ferroelectrics are pyroelectrics, pyroelectric coefficient $P = dP_s/dT$), dielectric spectroscopy (dielectric constant (ε) is affected by phase transition

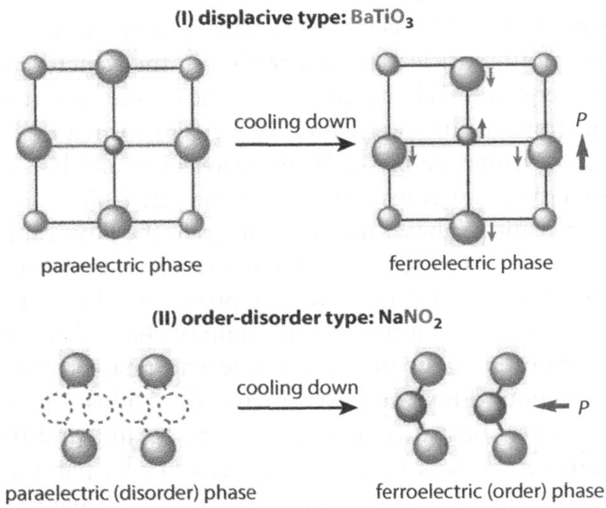

Figure 2.2 Schematic illustrations for the displacive-type and order–disorder type ferroelectric.

effects and it displays an irregularity at critical temperature), second harmonic generation response (SHG), which is based on the fact that all ferroelectrics are essentially non-linear optical materials, piezoelectric force microscopy (all ferroelectrics are piezo-electrics), differential scanning calorimetry, which gives information about the degree of order in ferroelectric phase transitions and electrical measurements (displacement loops, *D-E* plots, PUND technique, etc.) [10]. Figure 2.3 shows the analytical tools generally used to measure piezo/ferroelectricity of materials.

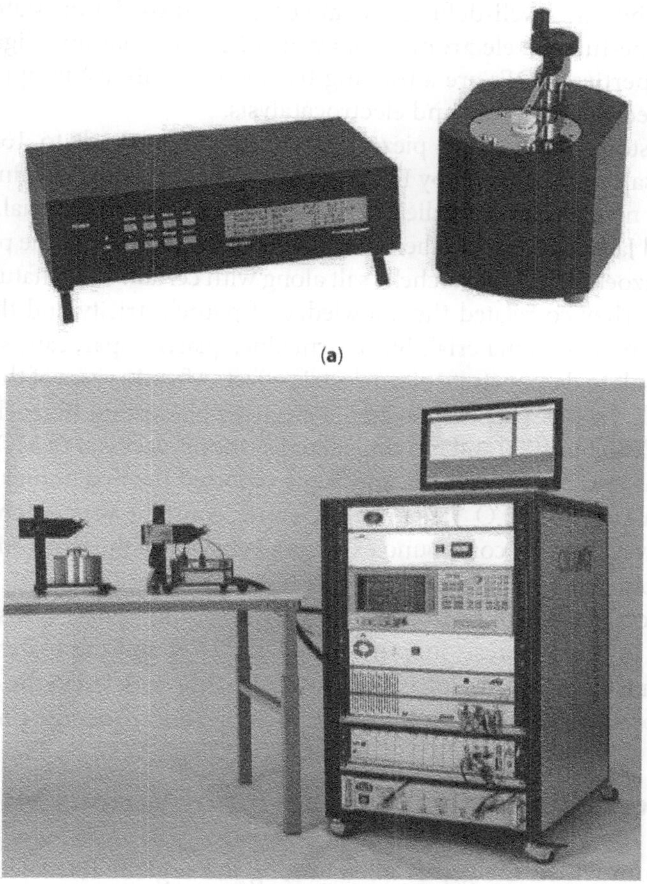

(a)

(b)

Figure 2.3 Photographs of (a) piezoelectric force microscopy and (b) TF analyzer 2000 used to measure ferroelectricity.

2.3 Metal–Organic Frameworks for Piezo/Ferroelectricity

Metal–organic frameworks are highly porous, three-dimensional, hybrid nanocrystalline structures composed of inorganic metals linked to organic ligands through coordinate bonds in an ordered manner. They are also called as metal coordination compounds which combine the properties of both organic (chromophore, chirality, and tailorable) and inorganic (electronic d and f orbital properties) compounds within a single molecular scale composite. These composites offer lot of flexibility not only in their synthesis but also well-defined metal centers and oxidation states which provide fine-tunable electronics behaviors of the surrounding ligands. All these properties MOFs are attracting tremendous interest in applications such as electronic devices and electrocatalysis.

The history of MOFs as piezo/ferroelectrics dates back to 1655 when Rochelle salt was separated by Elie Seignette in France. However, in the next 200 years no significant studies were carried out on this material. In 1880 Pierre and Jacques Curie brothers, for the first time, discovered the phenomenon of piezoelectricity in Rochelle salt along with certain other natural materials [11]. They correlated the knowledge of pyroelectricity and the crystal structures of natural materials like tourmaline, quartz, topaz, cane sugar, and Rochelle salt to demonstrate piezoelectric effect. After 40 years of this discovery, Joseph Valasek [12] demonstrated electric polarization (ferroelectricity) in Rochelle salt in 1920 and it is considered as the predecessor of MOFs. Later, toward the end of 1960, Okada and Sugie reported the first antiferroelectric MOF, $Cu(HCOO)_2.H_2O$ [13]. The dielectric constant versus temperature studies reveal that the compound exhibits a behavior of the first-order transition at around 235 K that shifts to higher temperature of 245 K with deuterate substitution in place of hydrogen in H_2O. An addition phase transition was observed in the hysteresis loop measured at 227 K and the material transformed into a ferroelectric phase below 227 K. This work led to the discovery and exploration of numerous MOFs for their piezo/ferroelectric properties.

2.4 Ferro/Piezoelectric Behavior of Various MOFs

MOFs from which the gap is filled among the organic ferroelectrics and pure-inorganic ferroelectrics are gaining more interest as these hybrids can be tuned by varying the metal ions and organic ligands to get desired electrical properties. Number of MOFs has been studied so far by many researchers for their electrical and mechanical properties. Flexibility in synthesis with the selection of organic ligands had lead to the synthesis of MOFs based on tartarates, formats, amino acids, propionates, sulfates, halogenometallate, UiO-66 series nanocrystals, NU-1000, NUS-6 series nanocrystals and other compounds. The summary of the ferroelectric properties of these MOFs has been listed in Table 2.1.

Rochelle salt (RS), which is potassium sodium tartrate tetrahydrate with the chemical formula $[KNa(C_4H_4O_6)]\ 4H_2O$ is the first typical ferroelectric MOFs [11, 12] and belongs to tartrate family. Two Curie points are found for RS at 255 K and 297 K. It exists in paraelectric phase below 255 K and above 297 K. However, between these temperature conditions it shows monoclinic P_{21} phase corresponds to ferroelectric-phase. Pepinsky *et al.* [14] proposed a phase transition mechanism called order–disorder mechanism for these changes in crystals and later Solans *et al.* [15] proved that the ferroelectricity in RS arises due to the formation of two non-equivalent organic chains along *a* axis. The dielectric constant, ε versus temperature (*T*) studies of the material revealed that it exhibits two anomalies at 255 K and 297 K as shown in Figure 2.4. The spontaneous polarization (P_s) for RS is found to be 0.25 μC.cm^{-2} at 273 K. When the hydrogen of RS salt was replaced by deuterium, $[KNa(C_4D_4O_6)]4D_2O$, the values of T_c shifted to 251 K and 308 K. The saturation polarization value of the deuterated RS was found to be 0.35 μC.cm^{-2}. This small change in the P_s value for non-deuterated and deuterated RS supports an order–disorder type mechanism for phase transition. Other tartrates studied for their ferroelectric properties are $[(NH_4)Na(C_4H_4O_6)]._4H_2O$ [16] and $[MLi(C_4H_4O_6)].H_2O(M = NH_4,Tl)$ [17], $[(NH_4)Na(C_4H_4O_6)]_4.HO$ shows a dielectric anomaly and crystallizes into $P2_12_12$ an orthorhombic, paraelectric phase above 109 K and it crystallizes into $P2_1$, a monoclinic, ferroelectric phase under 110K.$[MLi(C_4H_4O_6)].H_2O\left(M{=}NH_4^+,K^+\right)$ compounds exhibit a second ordered phase-transition to ferroelectric $P2_1$ (monoclinic) phase from paraelectric $P2_12_12$ (orthorhombic) phase at T_c values of 104 K (for M = NH_4, with a dielectric constant of 140) and 11 K (for M = Tl with a dielectric constant of 5000), respectively. This high dielectric constant of 5000 for $[LiTl(C_4H_4O_6)].H_2O$ was found to unusual to the usual behavior for ferroelectric and it was attributed to contributions of domain wall motions [18].

Formate based compounds impart variety of properties to MOFs starting from magnetism, porosity to ferroelectricity. Among $[Cu(HCOO)_2(H_2O)_2].2H_2O$ [19], $[Mn_3(HCOO)_6].C_2H_5OH$ [20], $(NH_4)[M(HCOO)_3]$ (M = Zn, Mg) [21], $[(CH_3)_2)NH_2][M(HCOO)_3]$ (M = Mn, Fe, Co, Ni, Zn) [22, 23] and $[C(NH_2)_3][M(HCOO)_3]$ (M = Cu, Cr) [24, 25] compounds, $[(CH_3)_2)NH_2][M(HCOO)_3]$ (M = Mn, Fe, Co, Ni, Zn) compounds studied by Cheetham *et al.* are the first lead-free hybrid frameworks with traditional

Table 2.1 The ferroelectric properties of various materials.

MOFs with chemical formula	Ferro-electric properties		
	P_s (μC. cm^{-2})	E_c (kV. cm^{-1})	Ref.
Tartarates $KNa(C_4H_4O_6).4H_2O$	0.25 (273 K)		[11]
$(NH_4)Na(C_4H_4O_6).4H_2O$	–		[12]
Formates $[Cu(HCOO)_2(H_2O)].2H_2O$			[19]
$[Mn_3(HCOO)_6]$.C2H5OH			[20]
$(NH_4)[Zn(HCOO)_3]$	1		[21]
$[(CH_3)NH_2][M(HCOO)_3]$ (M =, Mn, Fe, Co, Ni, Zn)	6		[26]
$[Sm(HCOO)_3]$	0.4		[101]
Amino acids $[Ag(NH_3CH_2COO)(NO_3)]$	0.60 (100 K)		
$(NH_3CH_2COO)_2.MnCl_2.2H_2O$	1.3 (273)	5.6	[29]
$[Ca(CH_3NH_2CH_2COO)_3Cl_2]$	0.27		[30]
$[Ca(CH_3NH_2CH_2COO)_3Br_2]$			[31]
$[Ca\{(CH_3)_3NCH_2COO\}(H_2O)Cl_2]$	2.5 (46 K)		[35]
$[Ln_2Cu_3\{NH(CH_2COO)_2\}_6\}].9H_2O$			[32, 33]
$[Cu^I_2Cu^{II}(CTDA)(4,4'\text{-bpy})_2].6H_2O$			[34]
Propionates $Ca_2Sr(CH_3CH_2COO)_6$	3.0 (228 K)		[36]
$Ca_2Ba(CH_3CH_2COO)_6$			[37]
$Ca_2Pb(CH_3CH_2COO)_6$			[38]
Sulfate Family $[C(NH_2)_3]M(XO_4)_2$ (M = Al, V, Cr, Ga, X = S, Se) M/R Al/S	0.35 (293 K)	1.7 (293 K)	[39, 41]

(Continued)

Table 2.1 The ferroelectric properties of various materials. (*Continued*)

MOFs with chemical formula	Ferro-electric properties		
	P_s (μC. cm^{-2})	E_c (kV. cm^{-1})	Ref.
V/S	0.38 (293 K)	6 (293 K)	[40]
Cr/S	0.37 (293 K)		[39]
Ga/S	0.36 (293 K)	3.6 (293 K)	[39, 41]
Al/Se	0.45 (293 K)		[39, 41]
Cr/Se	0.47 (293 K)		[39, 41]
Ga/Se	0.47 (293 K)		[39, 41]
$[(CH_3)_2NH_2][M(H_2O)_6](SO_4)_2$ (M = Al, Ga)			
$[CH_3NH_3][M(H_2O)_6](RO_4)_2.6H_2O$ M/R Al/S	1.0 (175 K)	6 (175 K)	[42, 45, 46]
V/S	0.9 (155 K)	6 (155 K)	[42]
Cr/S	1.0 (162 K)	6 (162 K)	[42]
Fe/S	1.3 (167 K)	6 (167 K)	[42]
Ga/S	0.4 (just below T_C)		[43]
In/S	1.2 (162 K)	6 (162 K)	[42]
Al/Se	1.2 (165 K)	1.0 (165 K)	[42, 43]
Cr/Se	0.43 (191 K)	3.3 (191 K)	[43]
Ga/Se	0.42 (191 K)	2.7 (191 K)	[43]
$[H_2dbco][Cu(X_2O)_6](SeO_4)_2$ (X = H or D)	1.51	1.5	[47]
Halogenometallate Family $A_2[MX_4]$ Family Mn/Cl			[49]
Fe/Cl			[48]

(*Continued*)

Table 2.1 The ferroelectric properties of various materials. (*Continued*)

MOFs with chemical formula	Ferro-electric properties		
	P_s (μC. cm^{-2})	E_c (kV. cm^{-1})	Ref.
Co/Cl	0.0025		[50]
Cu/Br	0.07 (2.39)		[51]
Zn/Cl	0.006		[52]
Zn/I	0.13 (150 K)		[53]
$A_3[M_2X_9]$ $(CH_3)_2NH_2$/Sb/Cl	0.66 (210 K)		[57]
$(CH_3)_2NH_2$/Sb/Br	0.07 (170 K)		[58, 59]
$(CH_3)_3NH$/Sb/Cl	2.0 (320)		[60]
CH_3NH_3/Sb/Br	0.13 (100 K)		[61]
CH_3NH_3/Bi/Br	0.06 (100 K)		[62]
$(CH_3)_4P$/Sb/Cl	0.050 (just below 135 K)		[63]
$(CH_3)_4P$/Bi/Cl	0.012 (just below 151 K)		[63]
$(CH_3)_4P$/Sb/Br	0.003 (just below 193 K)		[63]
$(CH_3)_4P$/Bi/Br	0.001 (just below 205 K)		[63]
$A_5[M_2X_{11}]$ CH_3NH_3/Bi/Cl	0.86 (285 K)		[64, 65]
CH_3NH_3/Bi/Br	0.70 (285 K)		[66]
Imidazolium/Sb/Br	0.18 (137 K)		[67]
Imidazolium/Bi/Cl	0.6 (100 K)		[68]
Imidazolium/Bi/Br	0.26 (130 K)		[69]

(*Continued*)

Table 2.1 The ferroelectric properties of various materials. (*Continued*)

	Ferro-electric properties		
MOFs with chemical formula	**P_s (μC. cm^{-2})**	**E_c (kV. cm^{-1})**	**Ref.**
Pyridinium/Bi/Br	0.3 (105 K)		[70]
Miscellaneous Compounds $[H_2dbco]_2[CuCl_3(H_2O)_2]Cl_3$. 3 H_2O	1.0 (153 K)	5	[71]
$[CN_4H_8][ZrF_6]$	0.45 (295)		[72]
$[CH_3NH_3]_2[Al(H_2O)_6]X_5$ (X = Cl, Br)	0.60 (64 K)		[73, 74]
[Co(titb)(L)].3H_2O with H2L=4,4′-(ethane-1,2-diyl) dibenzoic acid (H2L) and titb = 1,3,5-tris (imidazol-1-ylmethyl)-2,4, 6-trimethylbenzene	0.063	9.12	[75]
$[H_2N(CH_3)2][Ba(H_2O)(BTB)]$	–	–	[76]
$[C_2H_5NH_3][Na_{0.5}Fe_{0.5}(HCOO)_3]$	–	–	[77]
$[Mn(4\text{-tzba})(bpy)_2.2O]$ (bpy).3H_2O with tzba = 4-tetrazolbenzoic acid; bpy = 2,2′-blpyridine)	–	–	[78]
$[CuL_2(H_2O)_2].(NO_3)_2.(H_2O)_{1.5}.(CH_3OH)$ with L = $[PhPO(NH_4Py)_2]$	21.79 (18.35)	5.9 (9.7)	[79]
$[Zn_2(mtz)(nic)_2(OH)]$.0.5nH_2O with Hphtz = 5-phenyltetrazole, Hnic = nicotinic acid	6.26	2.57	[80]
[Zn(phtz)(nic)]with Hphtz = 5-phenyltetrazole, Hnic = nicotinic acid	5.27	1.42	[80]
$\{[Fe(2,2'\text{-bipyridine})(CN)_4]_2$ Co-(4,4′-bipyridine)}.4H_2O	0.5 (10 K)	15 (10 K)	[81]

(*Continued*)

Table 2.1 The ferroelectric properties of various materials. (*Continued*)

MOFs with chemical formula	Ferro-electric properties		
	P_s (μC. cm^{-2})	E_c (kV. cm^{-1})	Ref.
$[Cu_4O(Lvcz)_2Br_4].H_2O$ HLvcz = voriconazole	0.093	0.128	[82]
$[Cu_2(HLvcz)_2I_2].H_2O$ HLvcz = voriconazole	0.098	0.0838	[82]
$[[Ag(HLvcz)_2]CF_3SO_3$Lvcz = voriconazole	0.112	0.0657	[82]
$[Zn(s\text{-}nip)_2]$s-nip = (S)-2-(1,8-napthalimido)-3-(4-imidazole) propanoate	0.294	4.08	[83]
$[Co(s\text{-}nip)_2].(H_2O)_{0.5}$ with s-nip = (S)-2-(1,8-napthalimido)-3-(4-imidazole)propanoate	0.0332	3.68	[83]
$[Cd_3(S\text{-}L)_4].(ClO_4)_2$ with L = 2-(1-(2-pyridine)-ethylimino)-5-bromo-6-methoxy-pheno	0.08	1.47	[84]
$[Cd(L_2)_2]_3.4H_2O$ with H_2L = 2-aminoisonicotinic acid	0.024	8.45	[85]
$[Cd(tib)(p\text{-}BDC\text{-}OH)].H_2O$ with tib = 1,3,5-tris(1-imidazolyl) benzene, p-H2BDC-R = 2-R-1, 4-benzenedicarboxylic acid	11.65	~1.2	[86]
$[InC_{16}H_{11}N_2O_8].1.5H_2O$	3.81	1.65	[87]
$[Cu_2L_4(H_2O)_2].(CLO_4)_4.(H_2O)_5.(CH_3OH)$ with L = $PhPO(NH\text{-}3\text{-}pyridyl)_2$	1.8	16	[88]
$[Cu_3L_6(H_2O)_3].(CLO_4)_5.(NO_3).(H_2O)_{11}$ with L = $PhPO(NH\text{-}3\text{-}pyridyl)_2$	0.55	30	[88]
$[Fe(tib)_{2/3}(H_2O)_4[SO_4]$ with tib = 1,3,5-tris(1-imidazolyl) benzene	0.029	0.4	[89]

(*Continued*)

Table 2.1 The ferroelectric properties of various materials. (*Continued*)

MOFs with chemical formula	Ferro-electric properties		
	P_s (μC. cm^{-2})	E_c (kV. cm^{-1})	Ref.
$[Ce_2(H_2O)_3(D\text{-}tar)_3].3H_2O$ with tar = tartrate	0.233	0.579	[90]
$[Ce_2(H_2O)_3(L\text{-}tar)_3].3H_2O$ with tar = tartrate	0.171	0.505	[90]
$Cu_2(bpy)(H_2O)(Clma)_2$ with Hclma = R-2-chloromandelic acid and bpe = 4,4′-dipyridine	0.167	21.4	[91]
Cu(bpp)(Clma) with Hclma = R-2-chloromandelic acid and bpp = 1,3-di(4-pyridyl)propane	0.183	1.69	[91]
$[(CH_3)_2NH_2][Mn (HCOO)_3]$	–	–	[92]
$[Ag_2(HPIDC)]$ with H_3PIDC = (pyridine-4-yl)-1H-imidazole-4,5-dicarboxylaic acid	0.48	2.65	[93]
$Zn_3(titmb)(BTC)_2(H_2O)$ with titmb = 1,3,5-tris(1-imidazol-1-ylmethyl)-2,4,6-trimethyl benzene and H_3BTC = 1,3, 5-benzene-tricarboxylic acid	0.0486	0.68	[94]
$[Mn(H_2O)_2(bpe)(SO_4)].H_2O$ with bpe = trans-2,2-bis(4-pyridyl) ethene	0.4177	8.8	[95]
$\{Co_2(L)\text{-}(bpe)(H_2O)\}.5H_2O$ with H_4L = N-(1,3-dicarboxy-5-benzyl)-carboxymethylglycine	2.6	1	[96]
[Zn(Mitz)Cl] with Mitz = 3-tetrazolyl-6-methyl-5-(4-pyridyl)-2-pyridone	0.51	2.6	[97]
$[Mn(tib)_2(H_2O)_4]SO_4$ with tib = 1,3,5-tris(1-imidazolyl) benzene	0.208	2	[97]

(*Continued*)

Table 2.1 The ferroelectric properties of various materials. (*Continued*)

MOFs with chemical formula	Ferro-electric properties		
	P_s (μC. cm^{-2})	E_c (kV. cm^{-1})	Ref.
[Co(tib)$_2$(H$_2$O)$_4$]SO$_4$ with tib = 1,3,5-tris(1-imidazolyl) benzene	0.383	2.6	[97]
[Sr(μ-BDC)(DMF)] with BDC = benzene-1,4-dicarboxylate	0.025	7.1	[98]
[Sr(μ-BDC)(DMF)] with BDC = benzene-1,4-dicarboxylate	0.83	0.81	[98]
Co(SDBA)(BIMB) with H$_2$SDBA = 4,4′-icarboxybiphenylsulfone and bimb = 4,4′-bis(1-imidazolyl) biphenyl	0.238	3.38	[99]
Mg(int)$_2$.H$_2$O with int = isonicotinate	~0.016	~2	[100]
[Sm(HCOO)$_3$]	0.4	6.59	[101]
[Cd(BDAC)]$_2$.H$_2$O with HBDAC = (1′H-[2,2′]biimidazoly-1-yl)-acetic acid	1.14	1.493	[102]
[Zn$_2$(TPOM)(5-OH-bdc)$_2$].(DMF)(H$_2$O)$_2$ with TPOM = tetrakis (4-pyridyloxymethylene) methane	0.451	5.755	[103]
[Cd(pmida)H$_2$O].1.8H$_2$O] with pmida = N-(4-pyridylmethyl) iminodiacetate	~1.75	37.5	[104]
[Cd$_3$(BPT)$_2$(H$_2$O)$_9$].2H$_2$O with BPT = biphenyl-3,4′, 5-tricarboxylate	0.039	10.48	[105]
[Ni$_2$(bptc)(en)$_2$(m2-H$_2$O)].2H$_2$O with H$_4$bptc = biphenyl-2,5,20,50-tetracaboxylic acid, en=ethylenediamine	–	–	[106]

(*Continued*)

Table 2.1 The ferroelectric properties of various materials. (*Continued*)

MOFs with chemical formula	Ferro-electric properties		
	P_s (μC. cm^{-2})	E_c (kV. cm^{-1})	Ref.
$[Cd_6(L)_4(Cam)_4(H_2O)_4].2H_2O$ with H_2Cam=enantiopure camphoric acid	–	17.11	[107]
$[Co(BIPA)(titmb)].H_2O$ with H_2BIPA = 5-bromoisophthalic acid and titmb = 1,3,5-tris(imidazol-1-ylmethyl)-2,4,6-tri-methylbenzene	0.06	0.5	[108]
$[CX_3CH_2YH_3]Mn(HCOO)_3$ with X=H or F, Y=N or P	–	–	[109]
$Mn_5(NH_2bdc)_5(bimb)_5$	2.556	0.35	[110]

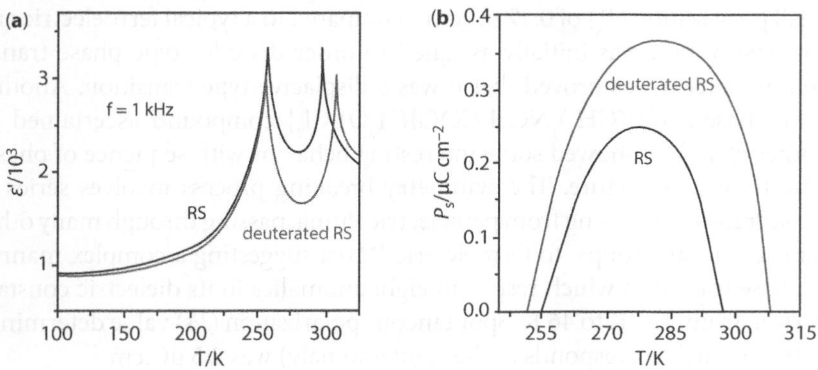

Figure 2.4 (a) Temperature dependence of dielectric constant, and (b) Ps versus T curve of Rochelle salt.

ABX_3 perovskite structure and show better ferroelectric properties. They exhibit multiferroic properties due to strong hydrogen bond ordering between 160 K and 185 K. Later, Sante *et al.* [26] have studied the ferroelectric behavior of $[(CH_3)_2NH_2][Mn(HCOO)_3]$ through density functional theory (DFT) calculations and by tuning organic cations they achieved a ferroelectric polarization of 6 μC.cm^{-2}. Apart from these $[C(NH_2)_3][M(HCOO)_3]$ (M = Cu, Cr) compounds exhibit switchable ferroelectric polarization due to the John–Teller distortion and the anti-ferro-distortion that transition metals like Cu and Cr could induce in the crystal structure. Ferroelectric properties exhibited by $(NH_4)[Zn(HCOO)_3]$ compound can be assigned to ammonium cation, which changes from a high temperature, disordered phase to a low temperature, ordered phase through the formation of hydrogen bonds between NH_4^+ of ammonium to oxygen atoms of metal formates and is responsible for ferroelectricity in the compound. The dielectric constant shows an anomaly along the *c* axis reflecting a phase transitionat T_c = 191 K and the spontaneous polarization (P_s) measured from the dielectric hysteresis loop was found to be 1.0 μC.cm^{-2}.

Among various amino acid-based MOFs, i.e., $[Ag(NH_3CH_2COO)(NO_3)]$ [27, 28], $(NH_3CH_2COO)_2.MnCl_2.2H_2O$ [29], $[Ca(CH_3NH_2CH_2COO)_3X_2]$ (X = Cl, Br) [30, 31], $[Ln_2Cu_3\{NH(CH_2COO)_2\}_6].9H_2O$(Ln = La, Gd, Ho, Nd, Sm, Er) [32, 33], $[CuI_2CuII(CDTA)(4,4'\text{-bpy})_2].6H_2O$ [34] compounds, $(NH_3CH_2COO)_2.MnCl_2.2H_2O$ shows spontaneous polarization, P_s of 1.3 μC.cm^{-2} at room temperature. It exhibits polarization up to 328 K and undergoes dehydration above this temperature losing polarization. It does not show any Curie temperature (T_c) as the compound undergoes decomposition above 328 K. $[Ca(CH_3NH_2CH_2COO)Cl_2]$ exhibits a peculiar, small polarization (P_s) of 0.27 μC.cm^{-2} compared to a typical ferroelectric and this observation was initially assigned to order-disorder type phase transition but later it was proved that it was a displacive type transition. Another amine-based $[Ca\{(CH_3)_3NCH_2COO\}(H_2O)_2Cl_2]$ compound ascertained by Rother *et al.* [35] showed some interesting behavior with sequence of phase-transitions in structure. The symmetry-breaking process involves series of phase transition starting from paraelectric Pnma, passing through many other paraelectric subgroups to ferroelectric P21ca suggesting a complex manner of phase transition which results in eight anomalies in its dielectric constant starting from 164 K to 46 K. Spontaneous polarization (*Ps*) value determined at 46 K (which corresponds to the eight anomaly) was 2.5 μC.cm^{-2}.

$[Ln_2Cu_3\{NH(CH_2COO)_2\}_6].9H_2O$ (Ln = La, Gd, Ho, Er, Nd, Sm) compounds reported by Kobayashi *et al.* have same structure belonging to trigonal space group P3c1 and the structural changes are due to guest water molecule occupying different sites in the channel. They exhibit large dielectric constants as high as 1300 and 350 (for Ln = Sm and La, respectively) and antiferroelectric behavior at high temperatures such as 400 K. These compounds are suitable for high-temperature applications of MOFs.

Long and coworkers reported $\left[Cu_2^{I}Cu^{II}(CDTA)(4,4'\text{-bpy})_2\right].6H_2O$ with water as guest molecule and it is confined to 1D as water wire in nano-channels exhibiting better dielectric transitions between 175 K and 277 K. The compound undergoes a phase transition from one-dimensional water (liquid) to ferroelectric ice (Solid) above 277 K. However, crystal structure analysis revealed that it transforms into *Fddd* space group which is centrosymmetric and violates symmetric requirement for ferroelectric. This could be due the fact that the correct sites of hydrogen-atoms cannot be located accurately. $Ca_2Sr(CH_3CH_2COO)_6$ [36], $Ca_2Ba(CH_3CH_2COO)_6$ [37], and $Ca_2Pb(CH_3CH_2COO)_6$ [38] are the MOFs studied for their dielectric behavior with propionate as organic molecule. In propionate-based MOFs, the variations in order and disorder involve the terminal $-CH_3$ set of propionic acid has significant part to play in transitions of ferroelectrics.

The ferroelectric performance of various sulfate-based MOFs has been listed in Table 2.1 [39–47]. Sulfate based MOFs with the general formula $[C(NH_2)_3][M(H_2O)_6](XO_4)_2$(M = Al, V, Cr, Ga, and X = S, Se) were the first to be reported as a family ferroelectrics Holden and coworkers [41, 43]. Similar to $(NH_3CH_2COO)_2.MnCl_2.2H_2O$ compound discussed earlier, this family of compounds does not show Curie temperature (T_c), which is assumed to be present behind their decomposition temperature. A typical compound with the formula, $[C(NH_2)_3][Al(H_2O)_6](SO_4)_2$, of the family (abbreviated as GASH (G = Guanidinium group, $C(NH_2)_3$) shows a temperature-independent dielectric constant of 6 up to 100°C and the polarization (*Ps*) value of 0.35 μC.cm^{-2}. However, higher P_s values have been reported for selenium-based compounds than sulfur based compounds of this family (see Table 2.1). Apart from this family of sulfate based compounds, other compounds have also been reported with the general formula $[(CH_3)_2NH_2][M(H_2O)_6](SO_4)_2$(M = Al, Ga) [44] $[CH_3NH_3][M(H_2O)_6](RO_4)_2.6H_2O$ [45, 46] and $[H_2dbco][Cu(X_2O)_6](SeO_4)_2$(X = HorD) [47] have also been reported for their ferroelectricity. $[(CH_3)_2NH_2][M(H_2O)_6](SO_4)_2$(M = Al, Ga) shows a spontaneous polarization (P_s) of 1.4 μC.cm^{-2} at 120 K. Among various $[CH_3NH_3][M(H_2O)_6](RO_4)_2.6H_2O$MOFs, when R = Se, the MOFs show higher spontaneous polarization (Table 2.1).

Halogenometallates with the general formula $A_y[M_mX_n]$ form another large family of MOFs here A = derivatives of protonated-amine, M = metal-ion and X = halide. The different subclass of the family with the general formulae, $A_2[MX_4]$ Family [48–53], $A[MX_3]$ Family [54–56], $A_m[M_nX_{3n+m}]$ family [57–70], exhibit maximum spontaneous polarization P_s, of 0.13 μC.cm^{-2}, 1–3 μC.cm^{-2} and 2.0 μC.cm^{-2}, respectively (see Table 2.1). The highest polarization of 2.0 μC/cm^2 reported $[(CH_3)_2NH_2]_3[Sb_2Cl_9]$ compound and the ferroelectric properties the compound was firmly associated with crystals consuming two-dimensional layers of polyanionic $M_2X_9(Sb_2Cl_9)$ units that are reserved in bulky-cations [57].

Recent research on peizo/ferroelectric MOFs other than the above discussed families of MOFs have resulted in number of new hybrid MOFs with variety of chemical formulae [71–110]. The electrical properties of all these MOFs have been summarized in Table 2.1. Among all these MOFs, $[Zn_2(phtz)(nic)_2(OH)].0.5nH_2O$(Hphtz = 5phenyltetrazole, Hnic = nicotinicacid), [Zn(phtz)(nic)](Hphtz = 5-phenyltetrazole, Hnic = nicotinicacid), [Cd(tib)(p-BDC-OH)].H2O (with tib = 1,3,5tris(1-imidazolyl)benzene, p-H2BDC-R = 2-R-1,4benzenedicarboxylicacid), $[InC_{16}H_{11}N_2O_8].1.5H_2O$, and $Mn_5(NH_2bdc)_5(bimb)_5$ show significant ferroelectric properties with the spontaneous polarization values P_s, of 6.26 μC.cm^{-2}, 5.27 μC.cm^{-2}, 11.65 μC.cm^{-2}, 3.81 μC.cm^{-2}, and 2.556 μC.cm^{-2}, respectively.

Studies of piezoelectric properties of MOFs have been very rarely reported despite of having well established electromechanical techniques such as piezoelectric force microscopy(PFM). Only few papers [111–113] discus piezoelectric properties of certain MOFs. A MOF based on Cd[Imazethapyr] [111] displays a d_{33} value of 60.10 pC N^{-1} when measured in a high accuracy PM200 piezometer. Though this value was smaller than practically used inorganic $BaTiO_3$ (300–2500 pC N^{-1}), the discovery was significant due to the fact that it was the first MOF with such piezo/ferroelectricity. Later two homochiral coordination polymers (MOFs) [112] with temperature-independent peizo/ferroelectricty were synthesized and studies for their electrical properties. The piezoelectric coefficient measured with a d_{22} value for one of the compounds (Complex 1) was 6.9 pC N^{-1}, which shows good temperature stability up to 100°C. Similarly, ZIF-8 [113] based MOFs also exhibited "soft" piezo/ferroelectricity but no classical peizo/ferroelectricity was observed in the lattice. Some water-soluble MOFs such as NUS-series, and UiO-series [114, 115] are often considered as piezoelectric and ferroelectric and studies have revealed that they exhibit a piezoelectric coefficient as d_{zz} vales of 2.0–3,5 pm V^{-1} (for NUS-6(Zr) and 1.5–2.5 pm V^{-1} (for NUS-6-(Hf). These values are less than that reported for typical inorganic, piezoelectric such as $BaTiO_3$ (18 pm V^{-1}) However, they are higher that reported for biopiezoelectrics materials like bones (0.7 pm V^{-1}). Studies exploring the piezoelectric properties of MOFs are of great importance owing to the fact that porous materials have attractive applications in the energy reaping. In addition, knowledge of piezoelectricity gives better understanding of ferroelectricity. We believe that future research on MOFs has to be done with more emphasis on their stress–strain or electromechanical behaviors.

2.5 Conclusion

We have discussed briefly the principle of piezo/ferroelectrity and the historical development of MOF materials for their piezo/ferroelectric applications. $[Zn_2(phtz)(nic)_2(OH)].0.5nH_2O$(Hphtz = 5-phenyltetrazole, Hnic = nicotinicacid), [Zn(phtz)(nic)](Hphtz = 5-phenyltetrazole, Hnic = nicotinic acid), [Cd(tib)(p-BDC-OH)].H2O (with tib = 1,3,5-tris(1-imidazolyl)benzene,p-H2BDC-R=2-R-1,4-benzenedicarboxylicacid), $[InC_{16}H_{11}N_2O_8].1.5H_2O$, and $Mn_5(NH_2bdc)_5(bimb)_5$ have shown highest ferroelectric property with the spontaneous polarization values P_s, of 6.26 $\mu C.cm^{-2}$, 5.27 $\mu C.cm^{-2}$, 11.65 $\mu C.cm^{-2}$, 3.81 $\mu C.cm^{-2}$, and 2.556 $\mu C.cm^{-2}$, respectively. A MOF based on Cd[Imazethapyr] shows highest piezoelectric coefficient value of 60.10 pC N^{-1}. Although these values are smaller than the practically used inorganic based $BaTiO_3$ or lead-based PZT material, they are higher than biopiezoelectricity. MOFs are have greater potential for Peizo/ferroelectricity based applications like sensors actuators, microelectromechanical devices, ultrasonic resonators and the most important energy harvesting due to their highly porous structure and flexibility to fine-tune their electrical properties by changing inorganic metals ions and organic ligands.

References

1. Vatansever, D., Siores, E., Shah, T., Alternative Resources for Renewable Energy: Piezoelectric and Photovoltaic Smart Structures, in: *Global Warming—Impacts and Future Perspective*, B.R. Singh (Ed.), InTech, Rijeka, 2012, p Ch. 10.
2. Sun, Y., Hu, Z., Zhao, D., Zeng, K., Probing nanoscale functionalities of metal–organic framework nanocrystals. *Nanoscale*, 9, 12163–12169, 2017.
3. Kaufmann, P., Rohring, S., Supancic, P., Deluca, M., Influence of ferroelectric domain Texture on the performance of multilayer piezoelectric actuators. *J. Eur. Ceram. Soc.*, 37, 2039–2046, 2017.
4. Bobić, J.D., Ivanov, M., Ilić, N.I., Dzunuzović, A.S., Vijatović Petrović, M.M., Banys, J., Ribic, A., Despotovic, Z., Stojanovic, B.D., PZT nickel ferrite and PZT-cobalt ferrite comparative study: Structural, dielectric, ferroelectric and magnetic properties of composite ceramics. *Ceram. Int.*, 44, 6551–6557, 2018.
5. Manotham, S., Butnoi, P., Jaita, P., Kumar, N., Chokethawai, K., Rujijangul, G., Cann, D.P., Large electric field-induced strain and large improvement in energy density of bismuth sodium potassium titanate-based piezoelectric ceramics. *J. Alloys Compd.*, 739, 457–467, 2018.
6. Ehterami, A., Kazemi, M., Nazari, B., Saraeian, P., Azami, M., Fabrication and characterization of highly porous barium titanate based scaffold coated by Gel/HA nanocomposite with high piezoelectric coefficient for Bone Tissue Engineering Applications. *J. Mech. Behav. Biomed. Mater.*, 79, 195–202, 2018.

7. Setter, N., Damjanovic, D., Eng, L., Fox, G., Gevorgian, S., Hong, S., Kingon, A., Kohlstedt, H., Park, N.Y., Stephenson, G.B., Stolitchnov, I., Taganstev, A.K., Taylor, D.V., Yamada, T., Streiffer, S., Ferroelectric thin films: Review of materials, properties, and applications. *J. Appl. Phys.*, 100, 051606, 2006.
8. Gabor, A.S. and Park, J.Y., Concepts, instrument, and model systems that enabled the rapid evolution of surface science. *Surf. Sci.*, 603, 1293–1300.
9. Broers, H., Schattenburg, M.L., Hector, S.D., Electron beam lithography-Resolution limits. *Microelectron. Eng.*, 32, 143–158, 1996.
10. Asadi, K. and Van der Veen, M.A., Ferroelectricity in Metal–Organic Frameworks: Characterization and Mechanisms. *Eur. J. Inorg. Chem.*, 2016, 27, 4332–4344, 2016.
11. Curie, P. and Curie, J., Dévelopment, par pression, de l'électricitépolairedans les cristaux émièdres à faces inclinées. *C. R. Acad. Sci.*, 91, 294–295, 1880.
12. Valasek, J., Piezo-electric and allied phenomena in Rochelle salt. *Phys. Rev.*, 17, 475–481, 1921.
13. Okada, K. and Sugie, H., Experimental Study of Antiferroelectric Copper FormateTetrahydrate and Its Deuterium Substitute. *J. Phys. Soc. Jpn.*, 25, 1128, 1968.
14. Frazer, B.C., Mc Keown, M., Pepinsky, R., Neutron diffraction studies of Rochelle-salt single crystals. *Phys. Rev.*, 94, 1435, 1954.
15. Solans, X., Gonzalez-Silgo, C., Ruiz-Perez, A., Structural Study on the Rochelle Salt C. *J. Solid State Chem.*, 131, 350–157, 1997.
16. Jona, F. and Pepinsky, R., Dielectric Properties of Some Double Tartrates. *Phys. Rev.*, 92, 1577, 1953.
17. Matthias, B.T. and Hulm, J.K., New Ferroelectric Tartrates. *Phys. Rev.*, 82, 108, 1951.
18. Fousek, J., Cross, L.E., Seely, K., Some properties of the ferroelectric lithium thallium tartrate. *Ferroelectrics*, 1, 63–70, 1970.
19. Okada, K., Antiferroelectric Phase Transition in Copper-FormateTetrahydrate. *Phys. Rev. Lett.*, 15, 252, 1965.
20. Cui, H., Wang, Z.-M., Takahashi, K., Okano, Y., Kobayashi, H., Kobayashi, A., Ferroelectric Porous Molecular Crystal, $[Mn_3(HCOO)_6](C_2H_5OH)$, Exhibiting Ferrimagnetic Transition. *J. Am. Chem. Soc.*, 128, 15074–15075, 2006.
21. Xu, G.-C., Ma, X.-M., Zhang, L., Wang, Z.-M., Gao, S., Disorder–Order Ferroelectric Transition in the Metal Formate Framework of $[NH_4][Zn(HCOO)_3]$. *J. Am. Chem. Soc.*, 132, 9588–9590, 2010.
22. Jain, P., Dalal, N.S., Toby, B.H., Kroto, H.W., Cheetham, A.K., Order–Disorder Antiferroelectric Phase Transition in a Hybrid Inorganic–Organic Framework with the Perovskite Architecture. *J. Am. Chem. Soc.*, 130, 10450–10451, 2008.
23. Jain, P., Ramachandran, V., Clark, R.J., Zhou, H.D., Toby, B.H., Dalal, N.S., Kroto, H.W., Cheetham, A.K., Multiferroic Behavior Associated

with an Order–Disorder Hydrogen Bonding Transition in Metal–Organic Frameworks (MOFs) with the Perovskite ABX_3 Architecture. *J. Am. Chem. Soc.*, 131, 13625–13627, 2009.
24. Stroppa, A., Jain, P., Barone, P., Marsman, M., Perez-Mato, J.M., Cheetham, A.K., Kroto, H.W., Picozzi, S., *Angew. Chem. Int. Ed.*, 50, 5847–5850, 2011.
25. Stroppa, A., Barone, P., Jain, P., Perez-Mato, J.M., Picozzi, S., Hybrid Improper Ferroelectricity in a Multiferroic and Magnetoelectric Metal-Organic Framework. *Adv. Mater.*, 25, 2284–2290, 2013.
26. Di Sante, D., Stroppa, A., Jain, P., Picozzi, S., Hybrid Improper Ferroelectricity in a Multiferroic and Magnetoelectric Metal-Organic Framework. *J. Am. Chem. Soc.*, 135, 18126–18130, 2013.
27. Pepinsky, R., Okaya, Y., Eastman, D.P., Mitsui, T., Ferroelectricity in glycine silver nitrate. *Phys. Rev.*, 107, 1538, 1957.
28. Choudhury, R.R., Panicker, L., Chitra, R., Sakuntala, T., Structural phase transition in ferroelectric glycine silver nitrate. *Solid State Commun.*, 145, 407, 2008.
29. Pepinsky, R., Vedam, K., Oakaya, Y., New room-temperature ferroelectric. *Phys. Rev.*, 110, 1309, 1958.
30. Makita, Y., Ferroelectricity in $(CH_3NHCH_2COOH)_3{\cdot}CaCl_2$. *J. Phys. Soc. Jpn.*, 20, 2073, 1965.
31. Haga, H., Onodera, A., Yamashita, H., Shiozaki, Y., New Phase Transition in Ferroelectric $(CH_3NHCH_2COOH)_3CaCl_2$ at Low Temperatures. *J. Phys. Soc. Jpn.*, 62, 1857, 1993.
32. Cui, H.B., Zhou, B., Long, L.S., Okano, Y., Kobayashi, H., Kobayashi, A., A Porous Coordination-Polymer Crystal Containing One-Dimensional Water Chains Exhibits Guest-Induced Lattice Distortion and a Dielectric Anomaly. *Angew. Chem. Int. Ed.*, 47, 3376, 2008.
33. Zhou, B., Kobayashi, A., Cui, H.B., Long, L.-S., Fujimori, H., Kobayashi, H., Anomalous Dielectric Behavior and Thermal Motion of Water Molecules Confined in Channels of Porous Coordination Polymer Crystals. *J. Am. Chem. Soc.*, 133, 5736–5739, 2011.
34. Zhao, H.-X., Kong, X.-J., Li, H., Jin, Y.-C., Long, L.-S., Zeng, X.C., Huang, R.-B., Zheng, L.-S., Transition from one-dimensional water to ferroelectric ice within a supramolecular architecture. *Proc. Natl. Acad. Sci. U.S.A.*, 108, 3481–3486, 2011.
35. Rother, H.J., Albers, J., Klöpperpieper, A., Phase transitions, critical dielectric phenomena and hysteresis effects in betaine calcium chloride dehydrate. *Ferroelectrics*, 54, 107–110, 1984.
36. Matthias, B.T. and Remeika, J.P., Ferroelectricity of Dicalcium Strontium Propionate. *Phys. Rev.*, 107, 1727, 1957.
37. Seki, S., Momotani, M., Nakatsu, K., Oshima, T., Polymorphic Phase Transition of Barium-dicalcium Propionate $BaCa_2(C_2H_5CO_2)_6$. *Bull. Chem. Soc. Jpn.*, 28, 411–416, 1955.

38. Itoh, K., Niwata, A., Abe, W., Kasatani, H., Nakamura, E., Structural Study of Ferroelectric Phase Transition in $Ca_2Pb(C_2H_5CO_2)_6$ Crystals. *J. Phys. Soc. Jpn.*, 61, 3593–3600, 1992.
39. Holden, A.N., Matthias, B.T., Merz, W.J., Remeika, J.P., New Class of Ferroelectrics. *Phys. Rev.*, 98, 546, 1955.
40. Remeika, J.P. and Merz, W., Guanidine Vanadium Sulfate Hexahydrate: A New Ferroelectric Material. *J. Phys. Rev.*, 102, 295, 1956.
41. Holden, A.N., Merz, W.J., Remeika, J.P., Matthias, B.T., Properties of Guanidine Aluminum Sulfate Hexahydrate and Some of Its Isomorphs. *Phys. Rev.*, 101, 962, 1956.
42. Jona, F. and Shirane, G., *Ferroelectric Crystals*, Pergamon Press, New York, 1962.
43. Shiozaki, Y., Nakamura, E., Mitsui, T. (Eds.), *Ferroelectrics and Related Substances*, Landolt-Börnstein New Series, Vol. III/36, Springer-Verlag, Berlin, 2006.
44. Kirpichnikova, L.F., Andreev, E.F., Ivanov, N.R., Shuvalov, L.A., Varikash, V.M., Some physical properties of new ferroelectric dimethylamin aluminium sulphate, Kristallografiya.33, 1437, 1988. *Sov. Phys. Crystallogr.* (English Transl.), 33, 855, 1988.
45. Pepinsky, R., Jona, F., Shirane, G., Ferroelectricity in the Alums. *Phys. Rev.*, 102, 1181, 1956.
46. Flecher, R.O. and Steeple, H., *Acta Crystallogr.*, 17, 290, 1964.
47. Zhang, W., Chen, L.-Z., Xiong, R.-G., Nakamura, T., Huang, S.D., New Ferroelectrics Based on Divalent Metal Ion Alum. *J. Am. Chem. Soc.*, 131, 12544, 2009.
48. Shimizu, H., Abe, N., Kokubo, N., Yasuda, N., Fujimoto, S., Yamaguchi, T., Sawada, S., Pressure-induced ferroelectricity in $\{N(CH_3)_4\}_2FeCl_4$. *Solid State Commun.*, 34, 363–368, 1980.
49. Mashiyama, H. and Tanisaki, S., Incommensurate-Commensurate Phase Transition of $\{N(CH_3)_4\}_2MnCl_4$. *J. Phys. Soc. Jpn.*, 50, 1413–1414, 1981.
50. Sawada, S., Shiroishi, Y., Yamamoto, A., Takashige, M., Matsuo, M., Ferroelectricity in $\{N(CH_3)_4\}_2$ $CoCl_4$. *Phys. Lett., A.*, 67, 56–58, 1978.
51. Wada, M., Suzuki, M., Sawada, A., Ishibashi, Y., Gesi, K., Ferroelectricity in $N(CH_3)_4]_2CuBr_4$ Crystal. *J. Phys. Soc. Jpn.*, 50, 1813–1814, 1981.
52. Sawada, S., Shiroishi, Y., Yamamoto, A., Takashige, M., Matsuo, M., Ferroelectricity in $\{N(CH_3)_4\}_2ZnCl_4$. *J. Phys. Soc. Jpn.*, 44, 687, 1978.
53. Gesi, K. and Perret, R., Ferroelectricity in $\{N(CH_3)_4\}_2ZnI_4$. *J. Phys. Soc. Jpn.*, 57, 3698, 1988.
54. Gesi, K., *J. Phys. Soc. Jpn.*, 59, 432v, 1990.
55. Fatuzzo, E., *Proc. Phys. Soc. (London)*, 76, 797, 1960.
56. Fatuzzo, E., Nitsche, R., Roetschi, H., Zingg, S., *Phys. Rev.*, 125, 514, 1962.
57. Jakubas, R., Ferroelectric phase transition in tris (dimethylammonium) nonachlorodiantimonate (III), $[NH_2(CH_3)_2]_3Sb_2Cl_9$. *Solid State Commun.*, 60, 389, 1986.

58. Jakubas, R., Sobczyk, L., Matuszewski, J., Ferroelectricity and phase transitions in tris (dimethylammonium) nonabromodiantimonate (III). *Ferroelectrics*, 74, 339–345, 1987.
59. Zaleski, J., Pawlaczyk, Cz., Jakubas, R., Unruh, H.-G., Structure and dynamic dielectric behaviour of ferroelectric $[NH_2(CH_3)_2]_3Sb_2Br_9$ (DMABA). *J. Phys.: Condens. Matter.*, 12, 7509, 2000.
60. Jakubas, R., Czapla, Z., Galewski, Z., Sobczyk, L., Ferroelectric phase transition in $[(CH_3)_3NH]_3Sb_2Cl_9$ (TMACA). *Ferroelectrics Lett.*, 5, 143–148, 1986.
61. Jakubas, R., Bator, G., Sobczyk, L., Mróz, J., Dielectric and pyroelectric properties of $(CH_3NH_3)_3Me_2Br_9$ (Me = Sb, Bi) crystals in the ferroelectric phase transition regions. *Ferroelectrics*, 158, 43–48, 1994.
62. Jakubas, R., Krzewska, U., Bator, G., Sobczyk, L., Structure and phase transition in $(CH_3NH_3)_3Bi_2Br_9$. A novel improper ferroelectrics. *Ferroelectrics*, 77, 129–135, 1988.
63. Wojtas, M. and Jakubas, R., Structure and properties of $[(CH_3)_4P]_3[Sb_2Cl_9]$ and $[(CH_3)_4P]_3[Bi_2Cl_9]$. *J. Phys.: Condens. Matter.*, 16, 7521, 2004.
64. Jakubas, R., Sobczyk, L., Lefebvre, J., A new ferroelectric crystal: $(CH_3NH_3)_5Bi_2Cl_{11}$. *Ferroelectrics*, 100, 143–149, 1989.
65. Szklarz, P., Gałązka, M., Zieliński, P., Bator, G., Phase transitions in the ferroelectric crystals [CH3NH3]5Bi2Cl11 and [CH3NH3]5Bi2Br11 studied by the nonlinear dielectric effect. *Phys. Rev., B*, 74, 184111, 2006.
66. Jakubas, R., A new ferroelectric compound: $(CH_3NH_3)_5Bi_2Br_{11}$. *Solid State Commun.*, 69, 267, 1989.
67. Piecha, A., Pietraszko, A., Bator, G., Jakubas, R., Structural characterization and ferroelectric ordering in $(C_3N_2H_5)_5Sb_2Br_{11}$. *J. Solid State Chem.*, 181, 1155–1166, 2008.
68. Jakubas, R., Piecha, A., Pietraszko, A., Bator, G., Structure and ferroelectric properties of $(C_3N_2H_5)_5Bi_2Cl_{11}$. *Phys. Rev., B*, 72, 104107, 2005.
69. Piecha, A., Białońska, A., Jakubas, R., Structure and ferroelectric properties of $[C_3N_2H_5]_5[Bi_2Br_{11}]$. *J. Phys.: Condens. Matter.*, 20, 325224, 2008.
70. Jozkow, J., Jakubas, R., Bator, G., Ferroelectric properties of $(C_5H_5NH)_5Bi_2Br_{11}$. *J. Chem. Phys.*, 114, 7239, 2001.
71. Zhang, W., Ye, H.-Y., Cai, H.-L., Ge, J.-Z., Xiong, R.-G., Huang, S.P.D., Discovery of New Ferroelectrics: $[H_2dbco]_2\cdot[Cl_3]\cdot[CuCl_3(H_2O)_2]\cdot H_2O$ (dbco = 1,4-Diaza-bicyclo[2.2.2]octane). *J. Am. Chem. Soc.*, 132, 7300–7302, 2010.
72. Bauer, M.R., Pugmire, D.L., Paulsen, B.L., Christie, R.J., Arbogast, D.J., Gallagher, C.S., Raveane, W.V., Nielson, R.M., Ross, C. R., II, Photinos, P., Abrahams, S.C., Aminoguanidinium hexafluorozirconate: A new ferroelectric. *J. Appl. Crystallogr.*, 34, 47–54, 2001.
73. Onoda-Yamamura, N., Ikeda, R., Yamamura, O., Matsuo, T., Methylammonium aluminum chloride hydrate: Its chemical composition and phase transitions. *Solid State Commun.*, 101, 647–651, 1997.

74. Gesi, K., Dielectric Study on the Phase Transitions in Ferroelectric $(CH_3NH_3)_2AlBr_5{\cdot}6H_2O$. *J. Phys. Soc. Jpn.*, 68, 3095–3099, 1999.
75. Wang, Y., Che, Y.X., Zheng, J.M., A 3D ferroelectric Co(II) polymer showing (3,5)-connected hms topology with 2-fold interpenetration. *Inorg. Chem. Commun.*, 21, 69–71, 2012.
76. Asha, K.S., Makkitaya, M., Sirohi, A., Yadav, L., Sheetb, G., Mandal, S., A series of s-block (Ca, Sr and Ba) metal–organic frameworks: Synthesis and structure–property correlation. *CrystEngComm*, 18, 1046–1053, 2016.
77. Ptak, M., Mączka, M., Gągor, A., Sieradzki, A., Stroppa, A., Di Sante, D., Perez-Mato, J.M., Macalik, L., Experimental and theoretical studies of structural phase transition in a novel polar perovskite-like $[C_2H_5NH_3][Na_{0.5}Fe_{0.5}(HCOO)_3]$ formate. *Dalton Trans.*, 45, 2574–2583, 2016.
78. Gao, J.X., Xiong, J.B., Xu, Q., Tan, Y.H., Liu, Y., Wen, H.R., Tang, Y.Z., Supramolecular interactions induced chirality transmission, second harmonic generation responses, and photoluminescent property of a pair of enantiomers from *situ* [2 + 3] cycloaddition synthesis. *Cryst. Growth Des.*, 16, 1559–1564, 2016.
79. Srivastava, A.K., Divya, P., Praveenkumar, B., Boomishankar, R., Potentially ferroelectric $\{Cu''L_2\}_n$ based two-dimensional framework exhibiting high polarization and guest-assisted dielectric anomaly. *Chem. Mater.*, 27, 5222–5229, 2015.
80. Liu, D.S., Sui, Y., Chen, W.T., Feng, P., Two new nonlinear optical and ferroelectric Zn (ll) compounds based on nicotinic acid and tetrazole derivative ligands. *Cryst. Growth Des.*, 15, 4020–4025, 2015.
81. Yang, J., Zhou, L., Cheng, J., Hu, Z., Kuo, C., Pao, C.W., Jang, L., Lee, J.F., Dai, J., Zhang, S., Feng, S., Kong, P., Yuan, Z., Yuan, J., Uwatoko, Y., Liu, T., Jin, C., Long, Y., Charge transfer induced multifunctional transitions with sensitive pressure manipulation in a metal – organic framework. *Inorg. Chem.*, 54, 6433–6438, 2015.
82. Li, Q., Wu, T., Lai, J.C., Fan, Z.L., Zhang, W.Q., Zhang, G.F., Cui, D., Gao, Z.W., Diversity of coordination modes, structures and properties of chiral metal–organic coordination complexes of the drug voriconazole. *Eur. J. Inorg. Chem.*, 5281–5290, 2015.
83. Yu, L., Hua, X.N., Jiang, X.J., Qin, L., Yan, X.Z., Luo, L.H., Han, L., Histidine-Controlled Homochiral and Ferroelectric Metal–Organic Frameworks. *Cryst. Growth Des.*, 15, 687–694, 2015.
84. Wen, H.R., Qi, T.T., Liu, S.J., Liu, C.M., Tang, Y.Z., Chen, J.L., Syntheses and structures of chiral tri- and tetranuclear Cd(II) clusters with luminescent and ferroelectric properties. *Polyhedron*, 85, 894–899, 2015.
85. Zhou, W.W., Wei, B., Wang, F.W., Fang, W.Y., Liu, D.F., Wei, Y.J., Xu, M., Zhao, X., Zhao, W., An acentric 3-D metal–organic framework with three-fold interpenetrated diamondoid network: Second-harmonic-generation response, potential ferroelectric property and photoluminescence. *RSC Adv.*, 5, 100956–100959, 2015.

86. Hua, J.A., Zhao, Y., Zhao, D., Kang, Y.S., Chen, K., Sun, W.Y., Functional group effects on structure and topology of cadmium(II) frameworks with mixed organic ligands. *RSC Adv.*, 5, 43268–43278, 2015.
87. Pan, L., Liu, G., Li, H., Meng, S., Han, L., Shang, J., Chen, B., Platero-Prats, A.E., Lu, W., Zou, X., Li, R.W., A Resistance-Switchable and Ferroelectric Metal–Organic Framework. *J. Am. Chem. Soc.*, 136, 17477–17483, 2014.
88. Srivastava, A.K., Praveenkumar, B., Mahawar, I.K., Divya, P., Shalini, S., Boomishankar, R., Anion Driven $[Cu^{II}L_2]_n$ Frameworks: Crystal Structures, Guest-Encapsulation, Dielectric, and Possible Ferroelectric Properties. *Chem. Mater.*, 26, 3811–3817, 2014.
89. Tan, Y.H., Yu, Y.M., Xiong, J.B., Gao, J.X., Xu, Q., Fu, C.W., Tang, Y.Z., Wen, H.R., Synthesis, structure and ferroelectric–dielectric properties of an acentric 2D framework with imidazole-containing tripodal ligands. *Polyhedron*, 70, 47–51, 2014.
90. Qi, J.L., Ni, S.L., Zheng, Y.Q., Xu, W., Syntheses, structural characterizations and ferroelectric properties of new Ce(III) coordination polymers via isomeric tartaric acid ligands. *Solid State Sci.*, 28, 61–66, 2014.
91. Qi, J.L., Ni, S.L., Xu, W., Zheng, Y.Q., Three Cu(II) (*R*)-2-chloromandelato complexes generated from dipyridyl-type ligands with different spacer lengths: Syntheses, crystal structures, and ferroelectric properties. *J. Coord. Chem.*, 67, 2287–2300, 2014.
92. Sanchez-Andujar, M., Gomez-Aguirre, L.C., PatoDoldan, B., Yanez-Vilar, S., Artiaga, R., Llamas-Saiz, A.L., Manna, R.S., Schnelle, F., Lang, M., Ritter, F., Haghighirad, A.A., Senaris-Rodriguez, M.A., First-order structural transition in the multiferroic perovskite-like formate $[(CH_3)_2NH_2][Mn(HCOO)_3]$. *CrystEngComm*, 16, 3558–3566, 2014.
93. Chen, L.Z., Huang, D.D., Ge, J.Z., Wang, F.M., A novel Ag(I) coordination polymers based on 2-(pyridin-4-yl)-1H-imidazole-4,5-dicarboxylic acid: Syntheses, structures, ferroelectric, dielectric and optical properties. *Inorg. Chim. Acta*, 406, 95–99, 2013.
94. Wang, X.F., Liu, G.X., Zhou, H., Syntheses, structures and physical properties of two zinc(II) coordination polymers with 1,3,5-tris(imidazol-1-ylmethyl)-2,4,6-trimethylbenzene and 1,3,5-benzenetricarboxylate. *Inorg. Chim. Acta*, 406, 223–229, 2013.
95. Xu, W. and Lin, J.L., A Mn(II) Coordination Polymer with Sulfate and trans-1,2-Bis(4-pyridyl)ethylene Bridges: Synthesis, Structure, Magnetic and Ferroelectric Properties. *Z. Naturforsch. B*, 68, 877–884, 2013.
96. Dong, X.Y., Li, B., Ma, B.B., Li, S.J., Dong, M.M., Zhu, Y.Y., Zang, S.Q., Song, Y., Hou, H.W., Mak, T.C.W., Ferroelectric Switchable Behavior through Fast Reversible De/adsorption of Water Spirals in a Chiral 3D Metal–Organic Framework. *J. Am. Chem. Soc.*, 135, 10214–10217, 2013.

97. Tan, Y.Z., Zhou, M., Huang, J., Tan, Y.H., Wu, J.S., Wen, H.R., *In Situ* Synthesis and Ferroelectric, SHG Response, and Luminescent Properties of a Novel 3D Acentric Zinc Coordination Polymer. *Inorg. Chem.*, 52, 1679–1681, 2013.
98. Guo, P.C., Chu, Z., Ren, X.M., Ning, W.H., Jin, W., Comparative study of structures, thermal stabilities and dielectric properties for a ferroelectric MOF $[Sr(\mu\text{-}BDC)(DMF)]_\infty$ with its solvent-free framework. *Dalton Trans.*, 42, 6603–6610, 2013.
99. Zhou, H., Liu, G.X., Wang, X.F., Wang, Y., Three cobalt(II) coordination polymers based on V-shaped aromatic polycarboxylates and rigid bis(imidazole) ligand: Syntheses, crystal structures, physical properties and theoretical studies. *CrystEngComm.*, 15, 1377–1388, 2013.
100. Liu, T., Luo, D., Xu, D., Zeng, H., Lin, Z., An open-framework rutile-type magnesium isonicotinate and its structural analogue with an anatase topology. *Dalton Trans.*, 42, 368–371, 2013.
101. Feng, D., Che, Y., Zheng, J., An acentric lanthanide–formate complex: Synthesis, structure, ferroelectric and magnetic properties. *J. Rare Earth.*, 30, 798–801, 2012.
102. Wang, J., Tao, J.Q., Xu, X.J., Tan, C.Y., Synthesis, Crystal Structure, and Properties of a Cadmium(II) Complex with the Flexible Ligand (1′H-[2, 2′] Biimidazoly-1-yl)-acetic Acid. *Z. Anorg. Allg. Chem.*, 9, 1261–1264, 2012.
103. Hu, J.-S., Yao, X.-Q., Zhang, M.-D., Qin, L., Li, Y.-Z., Guo, Z.-J., Zheng, H.-G., Xue, Z.-L., Syntheses, Structures, and Characteristics of Four New Metal–Organic Frameworks Based on Flexible Tetrapyridines and Aromatic Polycarboxylate Acids. *Cryst. Growth Des.*, 7, 3426–3435, 2012.
104. Hou, C., Liu, Q., Lu, Y., Okamura, T.A., Wang, P., Chen, M., Sun, W.Y., Metal–organic frameworks with *N*-(4-pyridylmethyl)iminodiacetate ligand: Synthesis, structure and sorption properties. *Microporous Mesoporous Mater.*, 152, 96–103, 2012.
105. Li, L., Ma, J., Song, C., Chen, T., Sun, Z., Wang, S., Luo, J., Hong, M., A 3D Polar Nanotubular Coordination Polymer with Dynamic Structural Transformation and Ferroelectric and Nonlinear-Optical Properties. *Inorg. Chem.*, 51, 2438–2442, 2012.
106. Guo, S.Q., Tian, D., Zheng, X., Zhang, H., A rare 4-connected neb topological metal–organic framework with ferromagnetic characteristics. *CrystEngComm*, 14, 3177–3182, 2012.
107. Su, Z., Lv, G.-C., Fan, J., Liu, G.X., Sun, W.Y., Homochiral ferroelectric three-dimensional cadmium(II) frameworks from racemic camphoric acid and 3,5-di(imidazol-1-yl)benzoic acid. *Inorg. Chem. Commun.*, 15, 317–320, 2012.
108. Liu, G.X., Xu, H., Zhou, H., Nishihara, S., Ren, X.M., Temperature-induced assembly of MOF polymorphs: Syntheses, structures and physical properties. *CrystEngComm*, 14, 1856–1864, 2012.

109. Di Sante, D., Stroppa, A., Jain, P., Picozzi, S., Tuning the Ferroelectric Polarization in a Multiferroic Metal–Organic Framework. *J. Am. Chem. Soc.*, 135, 18126–18130, 2013.
110. Wen, L., Zhou, L., Zhang, B., Meng, X., Qu, H., Li, D., Multifunctional amino-decorated metal–organic frameworks: Nonlinear-optic, ferroelectric, fluorescence sensing and photocatalytic properties. *J. Mater. Chem.*, 22, 22603–22609, 2012.
111. Fu, D.-W., Zhang, W., Xiong, R.-G., The first metal-organic framework(MOF) Imazethapyr and its SHG, piezoelectric and dielectric properties. *Dalton Trans.*, 3946–3948, 2008.
112. Yang, P., He, X., Li, M.-X., Ye, Q., Ge, J.-Z., Wang, Z.-X., Zhu, S.-R., Shao, M., Cai, H.-L., The first homochiral coordination polymer with temperature-independent piezoelectric and dielectric properties. *J. Mater. Chem.*, 22, 2398–2400, 2012.
113. Knebel, A., Geppert, B., Volgmann, K., Kolokolov, D.I., Stepanov, A.G., Twiefel, J., Heitjans, P., Volkmer, D., Caro, J., Defibrillation of soft porous metal–organic Frameworks with electric fields. *Science*, 358, 347–351, 2017.
114. Sun, Y., Hu, Z., Zhao, D., Zeng, K., Probing nanoscale functionalities of metal–organic framework nanocrystals. *Nanoscale*, 9, 12163–12169, 2017.
115. Sun, Y., Gao, J., Cheng, Y., Zhang, Y.-W., Zeng, K., Design of the Hybrid Metal–Organic Frameworks (MOFs) as Potential Supramolecular Piezo/Ferro-Electrics. *J. Phys. Chem. C.*, 123, 3122–3129, 2019.

3

Fabrication and Functionalization Strategies of MOFs and Their Derived Materials "MOF Architecture"

Demet Ozer

Hacettepe University, Department of Chemistry, Ankara, Turkey

Abstract

Metal–organic frameworks (MOFs) have attracted great interest because of their significant properties like a designable structure, controllable morphology, high porosity, high surface area, surface functionality, optical, electrical, and magnetic properties. These properties have achieved excellent interest in designing and fabricating functional MOFs for potential applications. This chapter briefly surveys the fabrication and functionalization strategies of MOFs and their derived materials. The effects of the construction agents (metal nodes, secondary building units, organic linkers), the synthesis techniques (solvothermal and hydrothermal method, microwave synthesis, electrochemical method, mechanochemical synthesis, sonochemical synthesis, diffusion method, and template method), the synthesis conditions (reaction time and temperature, cooling rate, and pressure) and the constitutions (metal source, concentration of reagents, solvents, additives, and pH) were summarized.

Keywords: Metal–organic frameworks, fabrication, functionalization, synthesis methods, synthesis strategies

3.1 Introduction

Metal–organic frameworks have received a great deal of interest and applied various intensive researches because of their changeable crystal shapes, designable framework, diverse porosity, tunable pore size,

Email: demetbaykan@hacettepe.edu.tr

Inamuddin, Rajender Boddula, Mohd Imran Ahamed and Abdullah M. Asiri (eds.) Applications of Metal–Organic Frameworks and Their Derived Materials, (63–100) © 2020 Scrivener Publishing LLC

high surface area, surface functionality, optical, electrical, and magnetic properties [1]. Using MOFs for the separation, adsorption, and storage of light gases, which have environmental and biological importance and also considered as an alternative energy source, has increased the interest in such materials [2–4]. The uniform pore size and shape of MOFs provide various materials separations like hydrocarbons [5], olefins [6], and toxic gases (CO, NH_3, etc.). The high surface area (500–7000 m^2/g) supports the enhancement of substrates around the active sites and contributes to catalytic activation. In different catalytic reactions including oxidation [7], reduction [8], hydrogenation [9], condensation [10], coupling [11], photocatalysis [12], electrocatalysis [13], biocatalysis [14], they possess both advantages of the heterogeneous and homogeneous catalysis as reusability, stability, high selectivity, high efficiency, well-defined active site and show excellent activity [15]. The flexible and porous structure of MOFs increases the adsorption capacity and MOF use as an excellent adsorbent for biological compounds [16], antibiotics [17] and toxic pollutants like arsenic [18], dye [19], etc. The highly ordered and tunable structure, give an opportunity to use MOFs in sensing applications for metals [20], biological molecules [21], environmental contaminants [22], and explosives [23]. As a delivery vehicle, MOF applies as a biodegradable therapeutic agent for the dispersion of bioactive gas molecules [24] and drugs [25]. In addition, they have used promising applications as thin-film [26], batteries [27], supercapacitors [28], fuel cells [29], ion-exchange [30], membrane production [31], magnetic resonance imaging [32], proton conduction [33], and luminescence [34].

Metal–organic frameworks, establishing a new topic between material science and coordination chemistry, originate from combinations of inorganic nodes (metal ions/cluster) and organic ligands. It enables the synthesis of different desired sizes frameworks with pores ranging from micropores (pore diameter < 2 nm) to mesopores (pore diameter 2–50 nm) by covalent bonds of extensive diversity of organic ligands with metal ions and secondary building units. The micropores MOFs are desirable for many applications [35]. Besides conventional methods, high temperature and pressure requiring methods as solvothermal and hydrothermal, microwave synthesis, electrochemical method, mechanochemical synthesis, sonochemical synthesis, diffusion method, and template method have been used in the synthesis of these materials [36]. Depending on the synthesis conditions and compositions, structural, thermal, and surface characteristics of the metal–organic framework has changed and gained new properties for various applications.

This chapter briefly surveys the fabrication and functionalization of MOFs and their derived materials according to expected properties and desired applications. The first question is which construction materials should be chosen and the effects of construction materials (metal nodes, ligand types, and secondary building units) were examined. The second question is which synthesis method is suitable for functionalization of MOFs and the effects of the various synthesis techniques (solvothermal and hydrothermal method, microwave synthesis, electrochemical method, mechanochemical synthesis, sonochemical synthesis, diffusion method, and template method) were investigated. The final question is which synthesis parameters and constitutions should be selected to enhance the functionality of MOFs. The effects of the synthesis conditions (reaction time and temperature, cooling rate, and pressure) and constitutions (metal source, the concentration of reagents, solvents, additives, and pH) were compared.

3.2 Fabrication and Functionalization of MOFs

The metal–organic frameworks (porous coordination polymers) are one-, two-, or three-dimensional crystalline porous structures designed by the bonding of metal nodes comprising the metal ion or ion clusters with organic linkers. They connect by coordination bonds and weak interactions, which promote the structural flexibility and diversity. For instance, 1,4-benzenedicarboxylate bridged zinc ion to obtain a 1D linear structure and then connected by π–π stacking and hydrogen bonds to create a 2D layered and 3D network structures [37]. Using various types of carboxylate ligands and nickel complexes, a robust 3D, a 2D square-grid and 3D diamondoid networks were formed (Figure 3.1). Because of a great number of metal nodes and organic ligands, designed new well-defined functional metal-organic frameworks with desired functional applications proceeds to be the charming scope of scientific research [38].

3.2.1 Metal Nodes

The metal nodes are the coordination centers of metal–organic frameworks and the coordination numbers of metal have changed the obtained crystal structure. The formation scheme of the MOFs according to metal coordination number was shown in Figure 3.2 [39]. The coordination numbers of metal (2–7) allows the complexes to form in various geometries such as linear, T-shape, tetrahedral, trigonal-bipyramidal and octahedral [40].

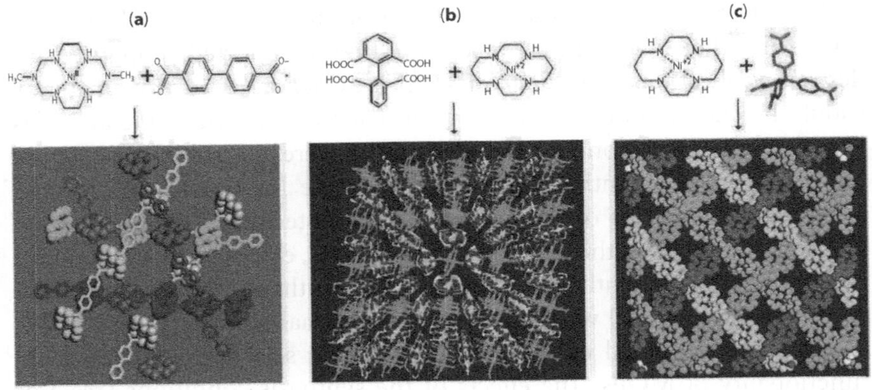

Figure 3.1 Structures of Ni-based MOFs, 3D network formed by the packing of 1D coordination polymers in a double network of threefold braids (a), 2D square-grid network (b), 3D diamondoid network (c). Reprinted with permission from Ref. [38], copyright (2012) ACS.

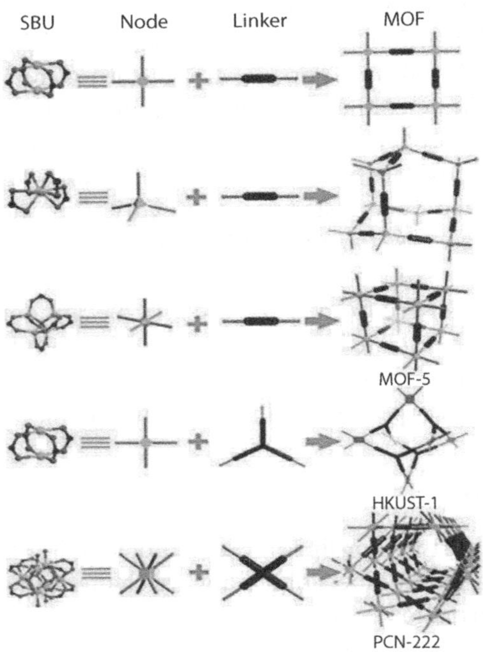

Figure 3.2 The formation scheme of MOFs. Reproduced from Ref. [39], with permission from RSC.

Two metal–organic frameworks were prepared using Co^{2+} and Zn^{2+} metal nodes and a bridging triazine ligand. Cobalt cluster has 1014 m^2/g BET surface area and applied as a heterogeneous catalyst for condensation reaction with high reusability when zinc cluster was used as a highly selective luminescence sensor with 1147 m^2/g BET surface area (Figure 3.3a) [41]. Polyhedral coordination polymers can be synthesized with lanthanides that have a coordination number of 7–10. A series of isostructural mesoporous MOFs were produced using different metal nodes and TMTB (4,4′,4″-(2,4,6-trimethylbenzene1,3,5-triyl) tribenzoic acid (Figure 3.3b). The eight-connected metal nodes and tritopic ligand generated M-NU-1200 framework including micropores and mesopores. The surface areas of the isostructural MOFs have changed according to metal nodes. Zr-NU-1200, Hf-NU-1200, Ce-NU-1200, and Th-NU-1200 have 2380, 1750, 1900, and 1300 m^2/g BET surface area. The catalytic performances of catalysts have also been influenced by the electronegativity and the oxidation states of metals. The Zr-NU-1200 framework showed the highest catalytic activity for alcohol oxidation [42].

In general, transition metals and their well-soluble nitrate, sulfate, chloride or acetate salts are used as a metal source. MIL-100(Fe) was produced from the reaction of ferric nitrate and trimesic acid on the HF-free conditions at 100 °C and used as an effective catalyst in the acetalization of aldehydes [43]. MIL-100 (Mg) was obtained using magnesium acetate and trimesic acid by one-pot self-assembly reaction at 150 °C for hydrogen storage applications [44]. MIL-53(Fe) was synthesized using iron (III) chloride and terephthalic acid (H_2BDC) by hydrothermal method at 150 °C [45]. Metal complexes are also used as nodes and have advantageous in changing the bond angle. Using a trimer building block $[In_3O(CO_2)_6]$ with a

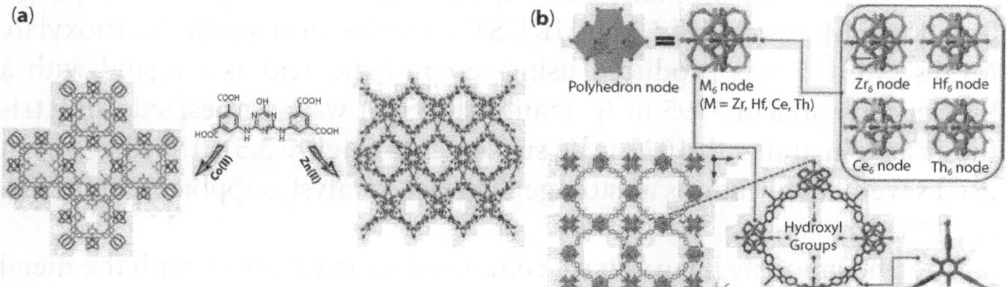

Figure 3.3 The obtained structures from Zn(II) and Co(II) clusters and a bridging organic ligand (a). Reprinted, permission from Ref. [41], copyright (2018) ACS. The formation of M-NU-1200 (M = Zr, Hf, Ce, and Th) (b). Reprinted with permission from Ref. [42], copyright (2019) ACS.

tetracarboxylic ligand obtain oxo-bridged trinuclear indium–carboxylate framework that has the high surface area and narrow pores and used as a hydrogen storage materials [46].

Mixing metal species are widely applied to create new MOFs and improve their properties. In MOF-5 [$Zn_4O(bdc)_3$], cobalt replaced zinc and the obtain CoZn-MOF-5 have higher H_2, CO_2 and CH_4, uptake capacity than MOF-5 because cobalt is incorporated into unexposed metal sites where gas molecules don't access [47]. The bimetallic CoFe-MOF was produced by treating stoichiometric amounts of iron(III) chloride and cobalt(II) chloride with 2-aminoterephthalic acid and the catalytic activity of the product was investigated in catalytic reduction of nitroarenes, catalytic dehydrogenation of sodium borohydride, and electrocatalytic water oxidation reactions. The porous CoFe-MOF demonstrated splendid activity, selectivity, and stability because of the synergic effect between the cobalt and iron metals [48].

3.2.2 Organic Linkers

Organic molecules including functional groups, mostly carboxylic acids, and nitriles, are used as ligands (also referred to as linkers or connectors) [49]. The neutral, anionic, and cationic forms of organic ligands provide several linking sites to construct functional metal-organic frameworks. The organic ligands possess electron donor atoms, such as oxygen and nitrogen are generally used to design MOFs. As an oxygen donor ligand, the carboxylate-containing linkers are the most opted linkers because of their high thermal and chemical stability, and flexibility. They have several coordination modes to combine metal ions and act as H-bond acceptors and donors. Some carboxylate ligands are given in Figure 3.4. From bidentate to polydentate aromatic carboxylates are extensively used to design functional MOFs. MOF-5 and HKUST-1 are the most studied carboxylate MOFs. MOF-5 was produced using terephthalic acid as a ligand with a high surface area of 3995 m^2/g, while HKUST-1 was synthesized using trimesic acid ligand with 1027 m^2/g surface area (Figure 3.5). They have been used several applications as storage material, catalyst, supporting material, etc. [50].

The phosphonate ligands have composed stronger bonds with the metal nodes than carboxylate ligands [51]. Using phosphonate ligands has been less common than the carboxylate ligands because the phosphonates have had solubility problem due to the complicated structure for deprotonation. Simple monophosphate ligands have formed a simple layer metal-organic framework. Di-, tri-, or monophosphate linkers have been used to

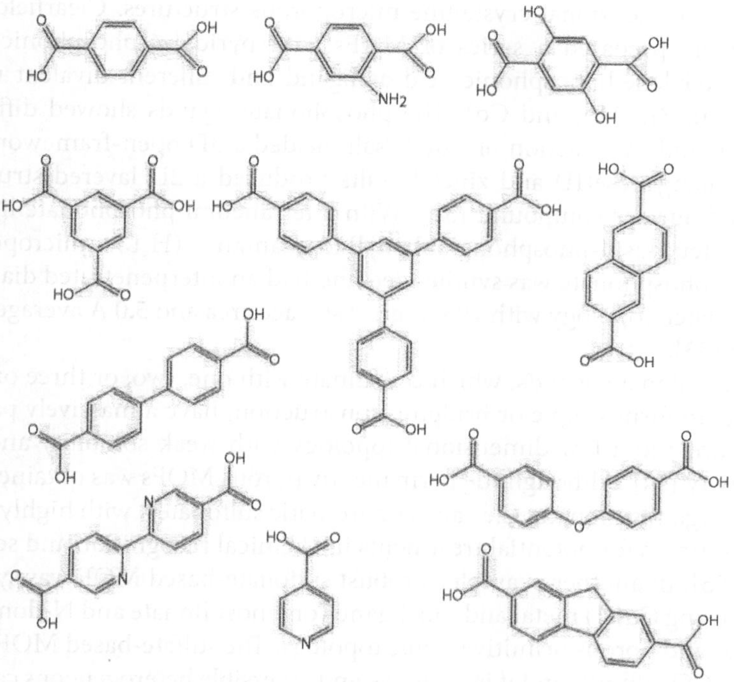

Figure 3.4 Various carboxylic acid ligands used for MOF synthesis.

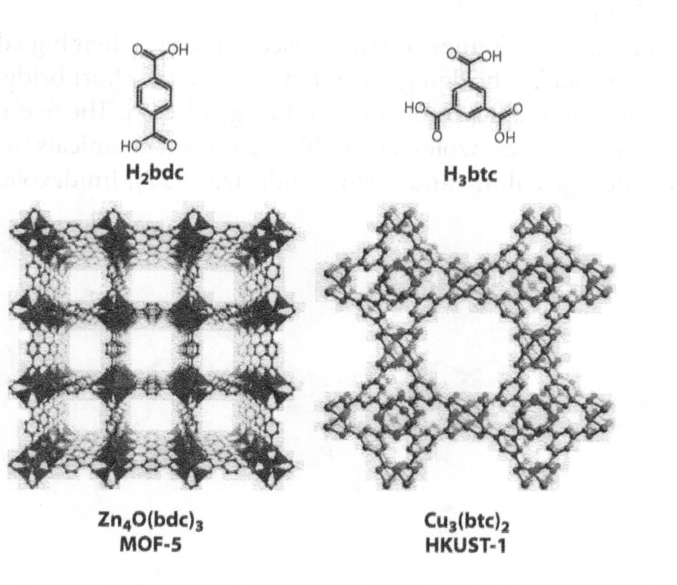

Figure 3.5 The MOF-5 and HKUST-1 structures. Reproduced with permission from Ref. [50] RSC.

get three-dimensional crystalline microporous structures. Clearfield and coworkers prepared a series of MOFs with pyridyl-4-phosphonic acid and p-xylylenediphosphonic acid as ligand and different divalent metal salts (Cu, Zn, Mn, and Co). The phosphonate ligands showed different ligation and the reaction of Cu(II) salt yielded a 3D open-framework. By using manganese(II) and zinc(II) salts produced a 2D layered structure and a dinuclear compound [52]. With a tetrahedral phosphonate ligand, 1,3,5,7-tetrakis(4-phosphonatophenyl)adamantane (H_8L), microporous copper phosphonate was synthesized and had an interpenetrated diamondoid related topology with 198 m^2gm^{-1} surface area and 5.0 Å average pore width [53].

The sulfonate ligands, which coordinate with one, two, or three oxygen donors to form simple or bridging construction, have a massively packed structure and a low-dimensional topology with weak solubility and low reactivity [54]. Although the permanently porous MOFs was obtained utilizing organosulfonates (i.e., alkyl or aromatic sulfonates) with highly polar pore surfaces for potential treatments in chemical recognition and separation [55]. In another example, a robust sulfonate-based MOF was synthesized using Cu(II) metal and mix ligand (organosulfonate and N-donor) as a polar and porous primitive-cubic topology. The sulfate-based MOF has a strong CO_2 affinity, and it is an active and reversible heterogeneous catalyst for CO_2 fixation [56]. Various phosphoric acid and sulfonic acid ligands are given in Figure 3.6.

The metal azolate frameworks have used because of their high directional coordination skill in bridging metal ions despite the short bridging length and complication of deprotonation of the ligand [57]. The five-membered nitrogen heterocycles, azoles, are highly significant chemicals that are used in medicine, agriculture and various industries [57]. Imidazole, pyrazole,

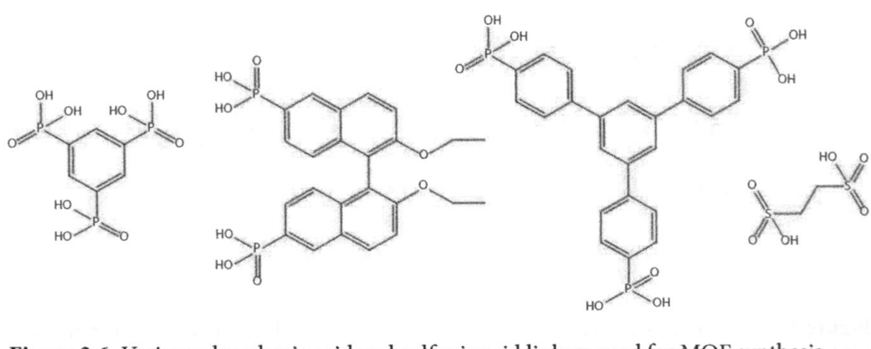

Figure 3.6 Various phosphoric acid and sulfonic acid linkers used for MOF synthesis.

1,2,4-triazole, 1,2,3-triazole, and tetrazole are some azole structures and shown in Figure 3.7. Zeolitic imidazole frameworks have high chemical and thermal stability and numerous reactive sites which permit various applications like adsorption, separation, and storage of different gases [58]. The permanent porosity also makes them attractive as an effective catalysis for different catalytic reactions [59]. 1,2,4-triazole and its derivatives are highly interested linkers due to coordinate both pyrazoles and imidazoles [60]. Another important nitrogen-containing ligand is 4,4′-bipyridyl (4,4′-bpy) and analogs that have been the most preferred ligands because of their coordination modes (ditopic bridging or monodentate). They produce numerous networks including 1D linear or zigzag-like chain, 2D square grid or interwoven honeycomb or 3D diamondoid frameworks [61]. The $[Cu(4,4'\text{-}bpy)(BF_4)_2(H_2O)_2](4,4'\text{-}bpy)$ comprised octahedral Cu^{II} sites constructed by 4,4′-bipyridine groups as a linear chain, when the $[Cu(4,4'\text{-}bpy)(MeCN)_2](BF_4)$ comprised Cu^I sites that exhibit zigzag chains. The $[Ag(4,4'\text{-}bpy)_2](CF_3SO_3)$ was the 3D diamondoid framework while the $[Cd(4,4'\text{-}bpy)_2(NO_3)_2].2C_6H_4Br_2$ had two-dimensional motif of fused square grids [62]. Amines, amides, and cyanides have been also applied as a nitrogen donor ligand [63] and some nitrogen donor ligands were given in Figure 3.8. As a result of the two types of hydrogen-bonding sites, the amide groups have been an attractive ligand. The amine groups have treated as an electron acceptor and the carbonyl groups have acted as an electron donor and have formed hydrogen bonds easily among themselves to construct MOFs. After that they have inconsequentially interacted with guest molecules and that gave an opportunity for sensor applications. For example, Biswas and coworkers produced an amide-functionalized Cd(II)-based MOF ($[Cd_5Cl_6(L)(HL)_2].7H_2O$) by solvothermal method using cadmium(II) chloride and 4-(1H-tetrazol-5-yl)-N-[4-(1H-tetrazol-5-yl)phenyl]benzamide(L) in dimethylformamide/methanol solution and it was found as a highly selective sensing tool for detection of trinitrophenol [64].

Recently, porphyrin-based frameworks have achieved considerable attention due to their extraordinary ability as high coordination number, and large size. For instance, two different porphyrin ligands reacted with

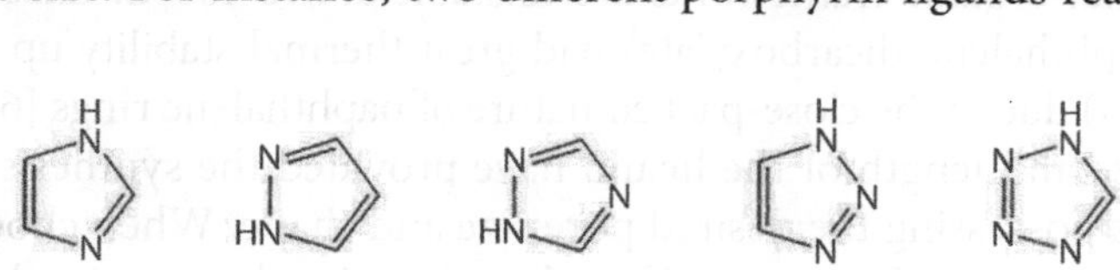

Figure 3.7 The most used azole structures.

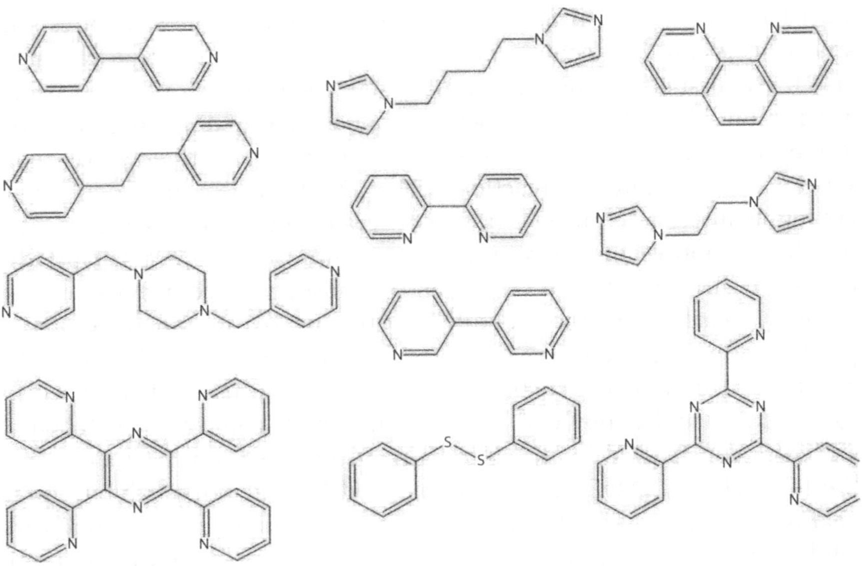

Figure 3.8 Some of the nitrogen donor ligands.

different lanthanides under hydrothermal conditions. The tetradentate porphyrin units were intercoordinated by the bridging metal ions and the open three-dimensional single-framework was produced [65].

Besides the preference of ligand type, modifying the properties of the ligand (shape, functionality, flexibility, symmetry, length, and substituent group) has been an alternative method to design new MOFs with special properties. The ligand must be strong enough to form a framework having a permanent hollow structure, and the composed coordination bond must also be strong. As a result of these, the final products have been highly stable but their thermal stability has been weaker than other porous inorganic solids. Actually, the thermal stability of MOFs has depended on the metal-ligand bond strength and the number of metal-ligand connections. High valence metal nodes have generally formed thermally stable MOFs. Most of the metal-organic frameworks are unstable above 400°C. Exceptionally, Li_2(2,6-naphthalene-dicarboxylate) had great thermal stability up to 610°C (Figure 3.9) due to the close-packed nature of naphthalene rings [66].

The size and length of the ligand have provided the synthesis of MOF structures possessing the desired pore size and shape. When choosing the ligand, the positional isomeric effect, the effect of end groups, and the effect of rings should be considered [63]. Short ligands contribute to the formation of narrow channels and small windows, which supports the separation

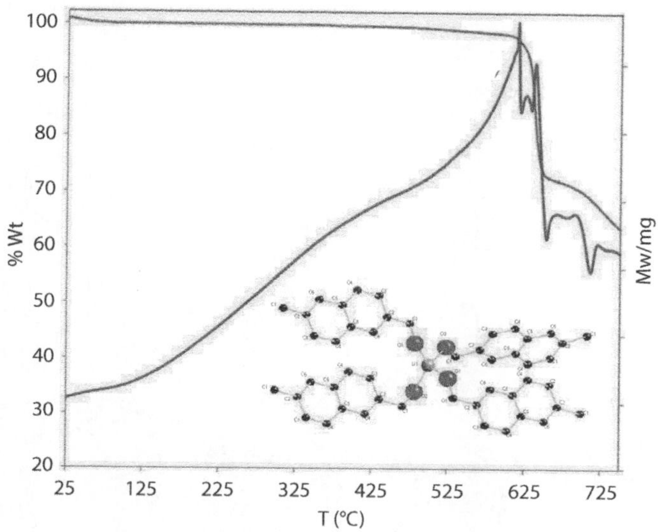

Figure 3.9 Thermogravimetric analysis of Li_2(2,6-naphthalene-dicarboxylate). Reprinted with permission from Ref. [66], copyright (2009) ACS.

of small molecules. Manganese(II) formate framework (formic acid as a short ligand) was constructed from cages that combined and formed 1D zigzag channel. The open hole was sufficient to isolate H_2 molecules from a gas mixture [67]. Bulk-shaped large ligands form large voids. Long ligands often lead to the formation of layered structures. Elongation of the length of the ligand is an impressive technique to construct MOFs with different pore magnitudes. With zinc nitrate and a series of aromatic dicarboxylic acid ligands, MOF-5 and its derivatives were prepared and as to the ligand dimension, pore volume and empty area were changed (Figure 3.10) [68]. Longer ligands have provided larger pore apertures and the largest pore aperture (98 Å) were found in IR-MOF-74-XI [69].

For design new functional MOFs, incorporated two and more ligands is an effective method [70]. $Zn_4O(BDC)_x(ABDC)_{3-x}$ was prepared using terephthalic acid and 2-aminobenzene-1,4-dicarboxylate (ABDC). In accordance with the incorporation of ABDC, the BET surface area of the complex was on the decline from 1250 to 800 m^2/g and thermal stability decreased with increasing substitution degree. However, the number of amino groups as accessible basic locations for catalytic treatments was increased the catalytic activity for propylene carbonate synthesis [71]. The selection of rigid or flexible mixed ligands is a practical method to prepare new MOFs with self-penetration, interpenetration, and helix. Liu and coworkers produced

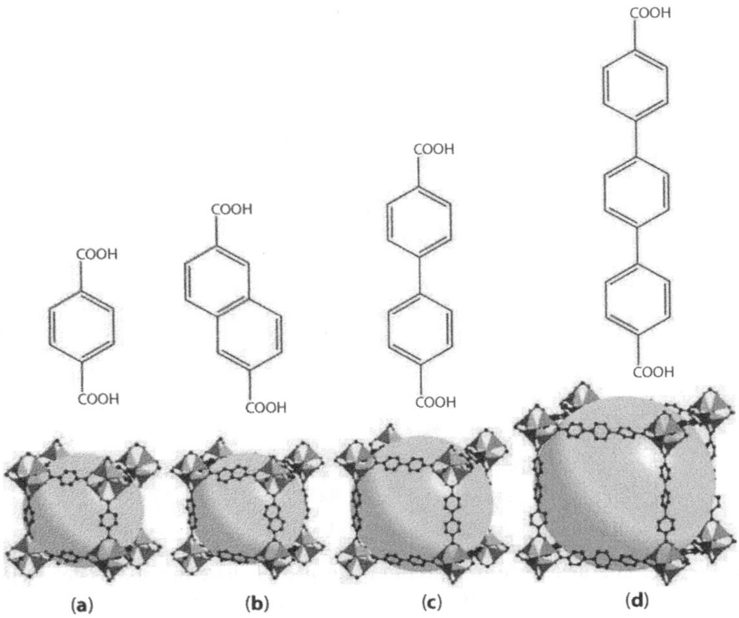

Figure 3.10 Crystal structures of some IRMOFs. Reproduced from Ref. [63], with permission from RSC.

five new MOFs using 2,4′-biphenyldicarboxylic acid as ligand and various bis(imidazoles) as coligands through hydrothermal method. The five new cobalt imidazole frameworks were obtained as 3D network, 3D diamondoid network, 2D wave-like network, a binodal (3,4)-connected network and 2D layer architecture [72]. Xue and coworkers prepared four new complexes using zinc and cobalt salt and two different ligands via a solvothermal method. The reaction of tetrakis(4-pyridyloxymethylene) methane (TPOM) with isophthalic acid, 5-hydroxyisophthalic acid, benzene-1,3,5-tricarboxylic acid, and benzene-1,2,4,5-tetracarboxylic acid and a metal salt, two different cobalt versus two different zinc MOFs were produced with unusual properties for photoelectric and magnetic applications (Figure 3.11) [73].

Attaching different functional groups to ligands is an alternative way to prepare new functional MOFs. Using terephthalic acid and aminoterephthalic acid, $[Cu_2(BDC)_2(dabco)].2DMF.2H_2O$ nanorods and $[Cu_2(BDC\text{-}NH_2)_2(dabco)].2DMF.2H_2O$ nanotubes were produced in the presence of triethylenediamine as pillar ligand and acetic acid as a modulator by the coordination modulation method. When the BDC ligand replaces with BDC-NH_2, the morphology changes from nanorods to nanotubes

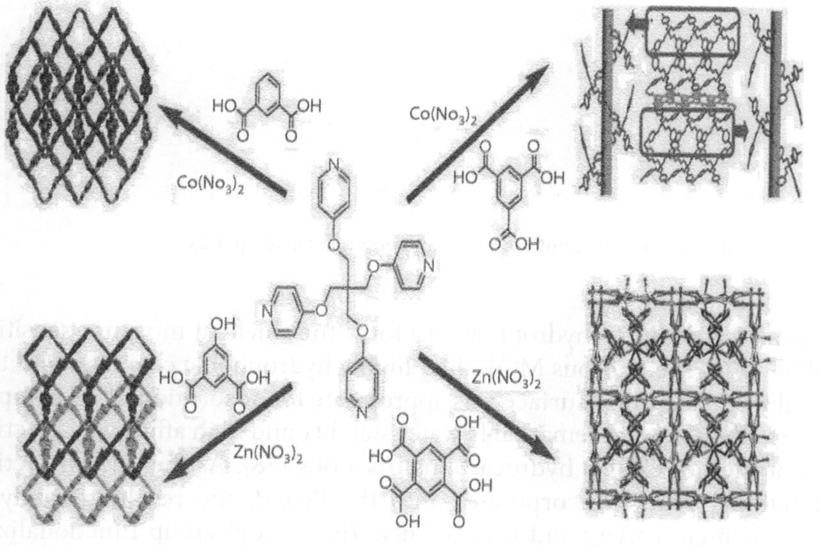

Figure 3.11 The solvothermal synthesis of four new complexes using zinc and cobalt salt and two different ligands. Reprinted with permission from Ref. [73], copyright (2011) ACS.

[74]. A new porous zinc terephthalate MOFs were synthesized using eight different terephthalic acid derivatives, and their adsorption properties were compared. According to the functional group of terephthalic acid derivatives, the selective uptake capacity of carbon dioxide over carbon monoxide was changed from 84% to 400% correlated with terephthalic acid [75]. The specific surface areas of the aluminum terephthalates $[Al_3OCl(DEF)_2(BDC\text{-}X)_3]$ changed from 1328 m^2/g to 2398 m^2/g according to functional group. The highest BET surface area was obtained from MIL101-CH_3. The BET surface areas of the complexes reduced as in turn: $-CH_3 > -NH_2 > -OCH_3 > -(OCH_3)_2 > -NO_2 \approx -(CH_3)_2 > -C_6H_4 > -F_2$ when the CO_2 uptake values reduced as in the ranking: $-CH_3 > -NH_2 > -OCH_3 > -NO_2 > -(OCH_3)_2 > -(CH_3)_2 > -F_2 > -C_6H_4$. The BET surface area has hinged on the size of functional groups and besides that the CO_2 uptake capacities of the complexes have also depended on the size and nature of functional groups that influence the interactions of CO_2 and framework [76]. The most used terephthalic acid derivatives including different functional group like halides (–Br and –Cl), $-NH_2$, $-NO_2$, $-(CH_3)_2$, $-C_4H_4$, $-(OC_3H_5)_2$, and $-(OC_7H_7)_2$ are given in Figure 3.12.

Functionalization of the ligand also improves the water stability of MOFs. With the integration of medium-long alkyl group, IRMOF-3 was changed

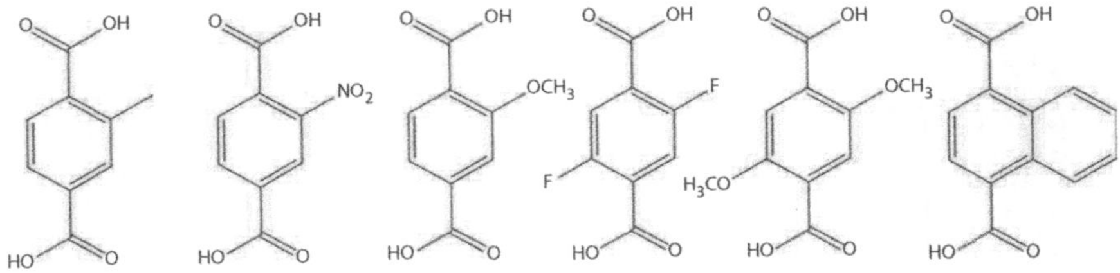

Figure 3.12 The most used functionalized terephthalic acid ligands.

from hydrophilic to hydrophobic to form the shielded moisture-sensitive MOF [77]. The fluorous MOFs had highly hydrophobic character and the perfluorinated inner surface was appropriate for adsorption of oil components. They showed remarkable water stability and high affinity to selective adsorption of C6–C8 hydrocarbons from oils [78]. As the catalytic active functional groups incorporated into the ligand, the resulting catalysts showed high activity and recyclability. The pyridyl group functionalized ligand was used to form homochiral MOF as a basic catalyst for transesterification reaction [79].

3.2.3 Secondary Building Units

The secondary building units have been proposed by Yaghi *et al.* to synthesize larger porous materials [80]. The secondary building units, which are the most common metal clusters, directly influence the geometry of the framework, the direction of polymer extension, and also support the synthesis of robust MOF with the desired properties. The most used secondary building units (SBU) are given in Figure 3.13 [81].

Secondary building units are two different kinds. One of them is the metal-containing units that range from single metal atoms to infinite groups. Second is the polyatomic organic ligands that incorporate with metal atoms [82]. The metal-containing SBUs have been regularly used and formed a well-defined shape like a square or an octahedron which is attended by polytopic linkers into periodic frameworks. MOF-808 was synthesized using $Zr_6O_4(OH)_4(CO_2)_6(HCOO)_6$ as SBU and trimesic acid as ligand and it consisted of large adamantine-shaped cages and small tetrahedral cages [83]. Bridge ligands and multidentate ligands have also been used as secondary building units. For instance, μ4-oxotetrametal basic carboxylate ($[M_4O(CO_2)_6]$) and μ3-oxotrimetal basic carboxylate ($[M_3O(CO_2)_6]$) SBUs was applied in the most used frameworks, MOF-5 and MIL-101 [84, 85].

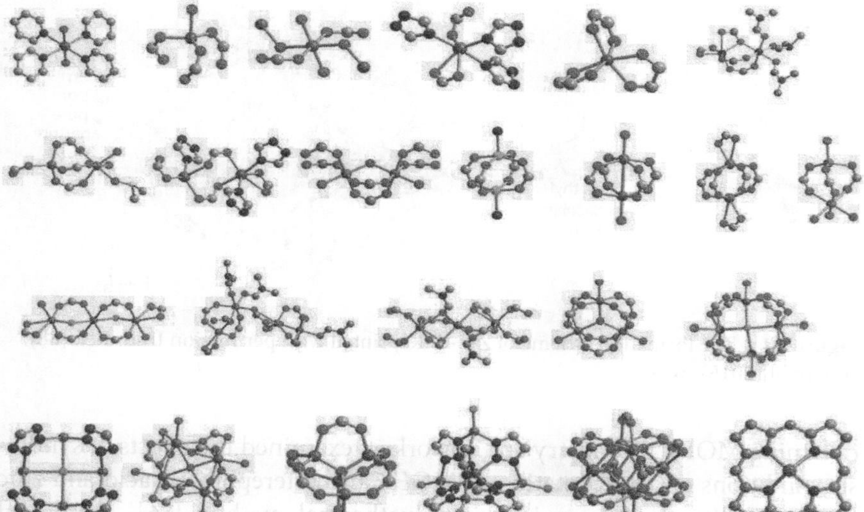

Figure 3.13 The typical secondary building units. Reprinted with permission from Ref. [81], copyright (2019) ACS.

3.2.4 Synthesis Methods

For the design of new MOFs with different crystal structures, surface properties like particle size, pore size distribution, and morphologies, various synthesis methods as the hydrothermal method, solvothermal method, microwave method, electrochemical method, mechanochemical synthesis, sonochemical (ultrasonic assisted) method, diffusion method, and template method have been developed.

3.2.4.1 Hydrothermal and Solvothermal Method

The hydrothermal (solvothermal) method is the most preferred synthesis technique to fabricate MOFs. In a closed autoclave under autogenous pressure and temperature, the self-assembly products are obtained from soluble reactants in polar solvents using conventional electric heating. If the solvent is an organic solvent such as DMF, DMSO, acetone, acetonitrile, and alcohols, the reaction is called as a solvothermal method. If the solvent is water, the reaction is termed as a hydrothermal method. The solvent polarity, reactants solubility, and interactions of reactants lead to the formation of good quality MOF crystals. Besides these, the reaction time, reaction temperature, stirring rate, and metal/ligand mole ratio determine the morphology of

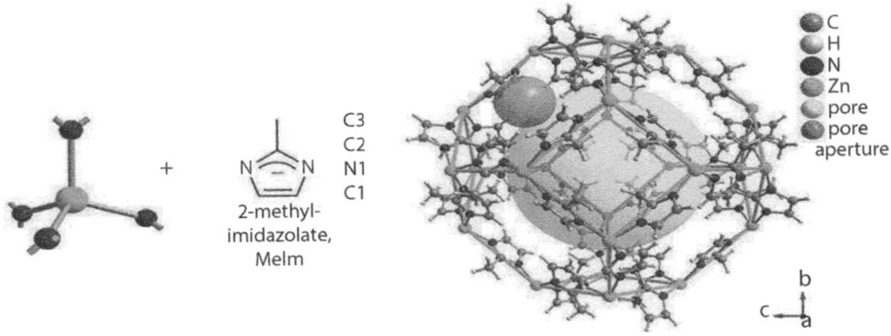

Figure 3.14 The formation scheme of ZIF-8. Reprinted with permission from Ref. [88], copyright (2015) ACS.

obtaining MOFs. McKinstry and coworkers examined the effects of synthesis conditions on the formation of MOF-5 using terephthalic acid and zinc nitrate in dimethylformamide via solvothermal method [86]. Nanosized (228 nm) ZIF-67 was synthesized using cobalt nitrate and 2-methylimidazole by a hydrothermal method [87]. ZIF-8 ($Zn(mim)_2$, mim = 2-methylimidazolate) was produced through both hydrothermal and solvothermal method. Tetrahedral nitrogen coordinated Zn(II) ions and 2-methylimidazolate was used as building block and arranged with each other to form a cage-like unit cell of ZIF-8 (Figure 3.14) [88, 89]. In DMF solvent, ZIF-8 was thermally stable up to 550 °C and had large surface area (BET: 1630 m^2/g) and high porosity and exhibited high CO_2 storage capacity and selectivity [90]. Dong and coworkers prepared ZIF-8 in water and showed that the water acted as a structure-directing agent for the preparation of open and porous ZIF-8 which has a potential for industrial applications as lubricant additives [91].

Solvothermal (hydrothermal) method has some limitations. The use of non-polar solvents causes the solubility problem for inorganic precursors and the use of water form hydrogen bonds that prevent the interactions. The ionothermal method is an alternative to hinder the limitations. A low volatility ionic liquid is used as both solvent and structure-directing agent to dissolve the precursors. For instance, Xu and coworkers prepared zinc trimesate ($[Zn_4(BTC)_2(\mu_4\text{-}O)(H_2O)_2]$) with $Zn(NO_3)_2.6H_2O$ and trimesic acid using 1-ethyl-3-methylimidazolium bromide ionic liquid as solvent by ionothermal method [92].

3.2.4.2 *Microwave Synthesis*

The microwave synthesis is a time-saving and energy-efficient heating method. The metal–organic frameworks, which was obtained long annealing time, have been produced by fast crystallization only a few minutes because of high nucleation rate [93]. The obtained crystals size change according to the reactant concentrations. In the microwave method, crystals form throughout the bulk of the solution in fast growth while crystals nucleate near the walls or on dust particles in slow growth in the solvothermal method. Choi and coworkers investigated the structural and surface properties of MOF-5 by changing the power level, irradiation time, temperature, solvent concentration, and substrate composition and compared with the solvothermal method. The uniform cubic MOF-5 was obtained in 30 minutes by microwave heating with 20–25 μm average size and 3008 m^2/g surface area, while it was synthesized at 105 °C for 24 hours by a conventional solvothermal method with a 400–500 μm size range and 3200 m^2/g surface area [94]. Ni and Masel prepared three IRMOFs, which was synthesized at 120°C for 21 h by the solvothermal method [95], in a microwave synthesizer at 150 W for 25 s [96]. Jhung and coworkers synthesized MIL-53 via solvothermal (ST), ultrasound (US) and microwave method (MW) and compared the effects of the method. They showed that the nucleation and crystal growth rates were predicted using the changes in crystallinity from crystallization curves [97]. The effects of heating on crystal morphology were shown in Figure 3.15 and the crystal size obtained following order ST>MW>US while the rate of crystallization changed following order US>MW>>ST. The US and MW irradiation ended up homogeneous nucleation and formed small crystals at a short time [98].

One of the other advantages of microwave method is used for large scale MOFs. Bag and coworkers prepared isostructural microporous lanthanide MOFs by a microwave-assisted solvothermal method only 5 minutes which prepared by solvothermal method at the same temperature for a long reaction (2 days) and evaporation time (5 days) [99].

Figure 3.15 The SEM images of MIL-53-Fe synthesized by Ultrasound for 35 min (a), microwave for 2 h (b), solvothermal for 3d (c). Reprinted from Ref. [98], copyright (2015) with Elsevier permission.

3.2.4.3 *Electrochemical Method*

The electrochemical method has been used to prepare metal-organic frameworks on an industrial scale and it offers fast synthesis and mild reaction conditions. Metal ions are used instead of metal salts, and high reactive metal species are supplied to the mixture of organic ligand and electrolyte by anodic dissolution. Faster synthesis at lower temperatures without anionic residues is the advantage of these methods [100]. The reaction time and solvent composition also have an extensive effect on the structure and morphology of the obtained MOFs. If time is short, the incomplete formation of a dense crystal layer can form and if time is long, the supporting mesh can be structurally damaged. HKUST-1 (CuBTC) was synthesized by the electrochemical method at mild conditions as low temperatures (50°C) and short reaction times. The obtained crystal layers overgrow the porous copper support [101]. Sachdeva and coworkers produced CuTATB (H_3TATB=4,4′,4″-s-triazine-2,4,6-triyl-tribenzoic acid) bulk powder and grown electrode surface and the SEM pictures were given in Figure 3.16 [102]. The bulk powder (Figure 3.16a and b) had needle-like structures and agglomerated particles with 570 m^2/g BET surface area. When CuTATB was grown on the copper surface that hinders the interpenetration, smaller particles formed due to a fast nucleation process (Figure 3.16c and d).

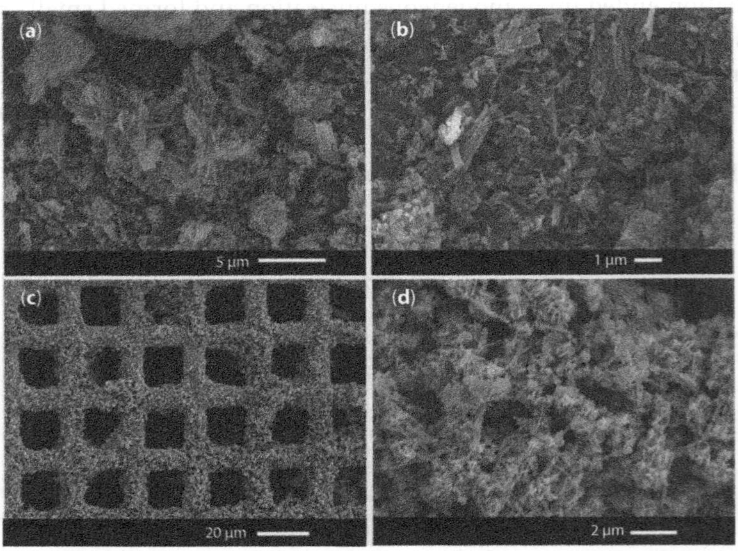

Figure 3.16 The SEM images of CuTATB as bulk powder (a, b) and grown on the copper surface (c, d). Reproduced with permission from Ref. [102], published by RSC.

3.2.4.4 *Mechanochemical Synthesis*

Mechanochemical synthesis has great potential for the clean and energy-efficient production of metal-organic frameworks [103]. While the metal salts and organic ligands were grinding in a mechanical ball milling system, the intramolecular bonds were breaking and the new bonds formed. The advantages of this method are solvent-free, harmless by-products, short reaction time, and room temperature. Klimakow and coworkers prepared HKUST-1 using copper acetate and trimesic acid for 25 min by the mechanochemical method and they showed that the remaining acetic acid molecules block the pores and mesopores were formed. After removing the gaseous by-product using one single post-synthesis activation, the high surface area (1713 m^2/g) HKUST-1 was obtained (Figure 3.17) [104].

3.2.4.5 *Sonochemical (Ultrasonic Assisted) Method*

The use of high energy ultrasound (20 kHz–15 MHz) provides a facile cost-effective and environmentally friendly synthetic tool for metal-organic frameworks. Nano-sized MOFs can be easily produced at ambient temperature and pressure by homogeneous nucleation. The homogeneous nucleation can cause a substantial reduction in crystallization time. This method depends on the acoustic cavitation that formed,

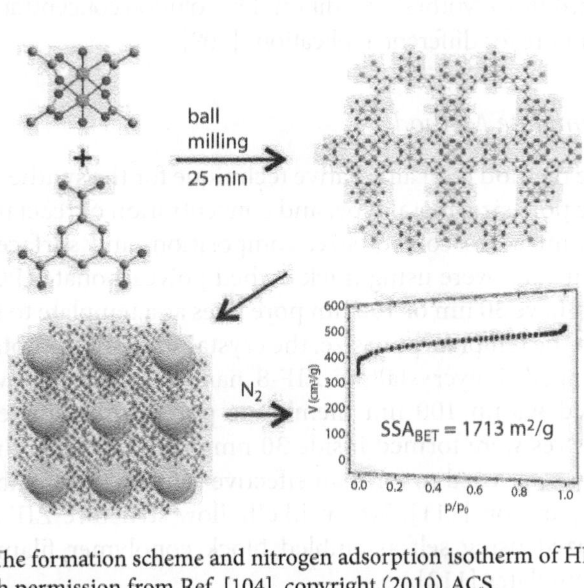

Figure 3.17 The formation scheme and nitrogen adsorption isotherm of HKUST-1. Reprinted with permission from Ref. [104], copyright (2010) ACS.

grown, and collapsed of bubbles in the liquid and generated high temperatures (5000°C) and pressures (500 atm) as well as an extraordinary heating and cooling rate with short lifetimes [105]. MOF-5 crystals were prepared by a sonochemical method that was widely decreased the reaction time (30 min) contrast to the solvothermal method (24 h) [106]. The microporous HKUST-1 was obtained in high yield using cupric acetate and trimesic acid in DMF/EtOH/H_2O solution under ultrasonic irradiation at ambient temperature, atmospheric pressure and short reaction times (5–60 min). The increase in sonication time changed the surface areas and hydrogen storage capacities of the obtained MOFs [107].

3.2.4.6 *Diffusion Method*

The diffusion method is a rapid approach and generally applies with the other synthesis methods. Qiu and coworkers used the ultrasound-vapor phase diffusion method which is time and energy-saving method to prepare the terbium based MOFs which show excellent luminescence properties for detection of picric acid in remarkably high yields [108]. Yao *et al.* developed an effective and facile contra-diffusion method and prepared ZIF-8 films all over the nylon substrate, where different synthesis solutions were separated by the porous nylon membrane and the crystallization occurred on the membrane surfaces by solution contra-diffusion. The properties of the obtained film changed according to synthesis conditions like solution concentration and synthesis temperature for different applications [109].

3.2.4.7 *Template Method*

The template method is an alternative technique for the synthesis of MOFs. The template pore size, metal type, and concentration of reactant were successfully optimized to control size, composition, and surface properties [110]. Arbulu *et al.* were using track-etched polycarbonate (PCTE) membranes which have 30 nm or 100 nm pore sizes as a template to form ZIF-8. According to the template pore size, the crystal shape of the obtained product was changed. Polycrystalline ZIF-8 nanorods and hallow nanotube were obtained within 100 nm membrane pores while single-crystalline ZIF-8 nanowires were formed inside 30 nm pores. Hence utilizing different membrane was used to form an effective sensing for membrane-based gas/liquid separation [111]. Worm-like hollow structure ZIF-8 nanotube was synthesized using self-assembled block copolymer filament-shaped micelles as templates [112].

3.2.5 Synthesis Strategies

The size and shape of the metal–organic frameworks depend on the composition and synthesis parameters. The composition parameters are solvent type, solvent compositions, metal source, the concentration of reagents, pH, and additives. The synthesis parameters are temperature, pressure, heating time, and heating source. The optimization of the composition parameters and synthesis parameters is crucial because these parameters directly affect the crystal size and morphology of the obtained product.

The metal source directly affects the crystal shape of the obtained material. In ($Zn_4O[(OOC)_2C_6H_4]_3$) complex, large crystals were obtained using zinc nitrate and zinc oxide as a metal source and small well-shaped cubic crystals were formed when zinc acetate was used as the metal source at 373 K under solvothermal conditions for 24 h (Figure 3.18) [113]. By controlling the metal source and pH values, the coordination of molecules has also changed. In another study, six binary MOF was synthesized using 4-(Pyridin-3-ylmethoxy)benzoic acid as a ligand (L) and different metal source at various pH by hydrothermal method. At pH = 5-5.5, α-[Cd(L)$_2$(H$_2$O)] complex was obtained using $CdCl_2$ and β-[Cd(L)$_2$(H$_2$O)] complex was obtained using $Cd(NO_3)_2.2H_2O$. With $Cd(OAc)_2.2H_2O$ as a metal source, the [Cd(L)$_2$] complex was prepared at pH = 1, the [Cd(L)$_2$–(H$_2$O)] complex was formed at pH=3-4 and the γ-[Cd(L)$_2$(H$_2$O)$_2$] complex was obtained at pH = 6 [114]. The pH conditions take a substantial role in the self-assembly synthesis of the MOFs. The selection of proper acid–base environment directly affects the deprotonation of the organic linker, the nucleation process and the formation of product [115]. Chen and coworkers prepared lead organic frameworks at different pH. At pH 8.5, the [Pb$_2$(tza)(μ$_3$–OH)(μ$_3$-Cl)(μ$_4$-Cl)]n complex was prepared using tetrazole-1-acetic acid (Htza) and lead(II) chloride by the hydro-solvent condition. At pH = 5 and 7, [Pb(tza)$_2$]$_n$ and [Pb(tza)(μ$_3$-Cl)]$_n$ complexes occurred [116].

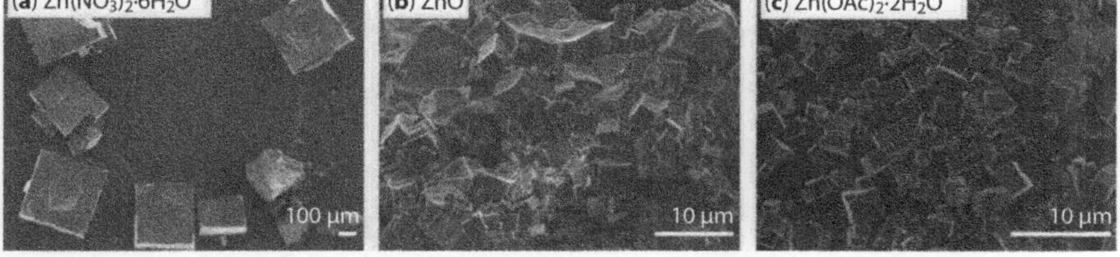

Figure 3.18 The SEM images of MOF-5 obtained using different metal salts. Reprinted from Ref. [113], copyright (2009) with Elsevier permission.

The concentration of reagents generates different coordination environments. Kim and coworkers prepared $[Co_3(BTB)_2(DMA)_4]$ (1) and $[Co_3(BTB)_2(DMA)_2]$ (2) using $CoCl_2{\cdot}6H_2O$ and 1,3,5-benzene-tribenzoic acid (H_3BTB) in DMA by a solvothermal method under different the reaction temperature and the reactant concentration (Figure 3.19). When the temperature was 100°C, 1 was synthesized for 10 days and with increasing temperature to 120°C the crystalline 1 and 2 mixture was formed. At 120°C, when the concentration was changed and increased four times, two were synthesized for 7 days. [117]. Lü and coworkers investigated the impacts of metal/ligand ratio to the MOF structure by changing the amount of the ligands. As a result of the reactions $CuSO_4.5H_2O$ and the mixed ligand (nitrilotriacetic acid/4,4′-bipyridyl), a 3D pillared-layer $Na_3[Cu_2(NTA)_2(4,4'\text{-}bpy)]ClO_4.5H_2O$ and a 2D undulated brick-wall architecture $[Cu_2(NTA)(4,4'\text{-}bpy)_2]ClO_4.4H_2O$ were produced [61]. In another study, CuI and 2,20-bisimidazole was used in different molar ratios of 1:1 and 3:4 and obtained two different luminescent MOFs [118].

For the synthesis of functional MOF, one of the important parameters is to find a convenient solvent that directly affects the coordination behavior of metal and ligand. Dimethylformamide, dimethylsulphoxide, toluene, acetone, chloroform, ethanol, methanol, and acetonitrile are generally used as solvents. With copper salt and 4,4′-tucker-box-diphenylamine ligand (L), in dimethylformamide-acetonitrile-water mixture $[Cu_2(L)_2(H_2O)_2.3/2H_2O]$ was synthesized; in dimethylformamide-water mixture, $[Cu(L)(DMF).5/2H_2O]_n$ was formed; in acetonitrile-water mixture, $[Cu(L)(H_2O).(H_2O)]_n$ was produced. More than one solvent is also utilized to examine the polarity of solvent and solvent-ligand exchange

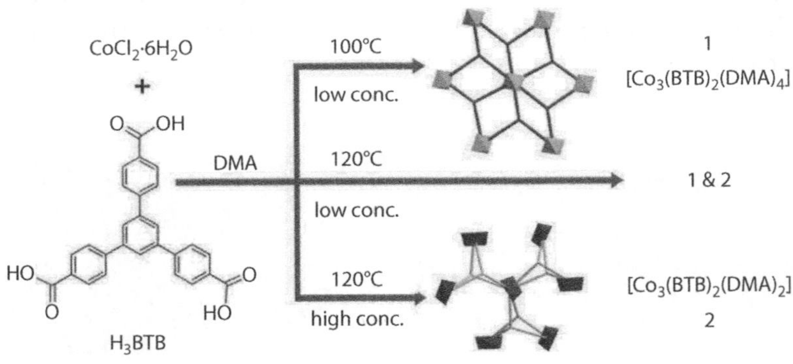

Figure 3.19 The formation scheme of cobalt-MOFs. Reprinted with permission from Ref. [117], copyright (2012) ACS.

kinetics [38]. The polarity of the solvent (acetonitrile < dimethylformamide < water) specified the structure of the obtained complex, and new complexes were formed using different types of solvent [119]. The solvents act as a coordinating agent that determined from the coordination behaviors of solvent molecules. Li and coworkers prepared Cd(II) metal-organic framework using different solvents and $[Cd_3(BPT)_2(DMF)_2].2H_2O$ [1], $[Cd_3(BPT)_2(DMA)_2]$ [2], and $[(CH_3CH_2)_2NH_2].[Cd(BPT)]\cdot 2H_2O$ [3] were synthesized using biphenyl-3,4′,5-tricarboxylic acid, and cadmium nitrate in the mixed solvents of water with N,N-dimethylformamide, N,Ndimethylacetamide, and dimethylformamide via the hydrothermal method (Figure 3.20) [120]. The solvents are also used as template reagents to generate new structures with different properties. Zuo and coworkers prepared two different Mg-based MOFs which had the same two-fold interpenetrated ReO_3 net combined by same ligand (4,4′,4″-benzene-1,3,5-triyl-tribenzoate), but distinct nodes (Mg2 and Mg6 units) in a different solvent mixture. In DMSO/water, hexanuclear clusters formed, while super-octahedral cavities constructed in DMSO/DMF/H_2O [121].

The solvent structure affects the obtained MOFs. The steric effects of the coordination solvent can change the crystal growth process. Dong and coworkers synthesized two microporous yttrium–organic frameworks using trimethyl-1,3,5-benzenetricarboxylate and yttrium ion in DMF/H_2O and DEF/H_2O mixture to show the effect of molecular size of solvents. The solvent type changes the crystal symmetry and arrangements of the ligands around the Y centers though they have similar three-dimensional structures. When $Y_2(BTC)_2(DMF)(H_2O)$ was synthesized in DMF/H_2O solvent,

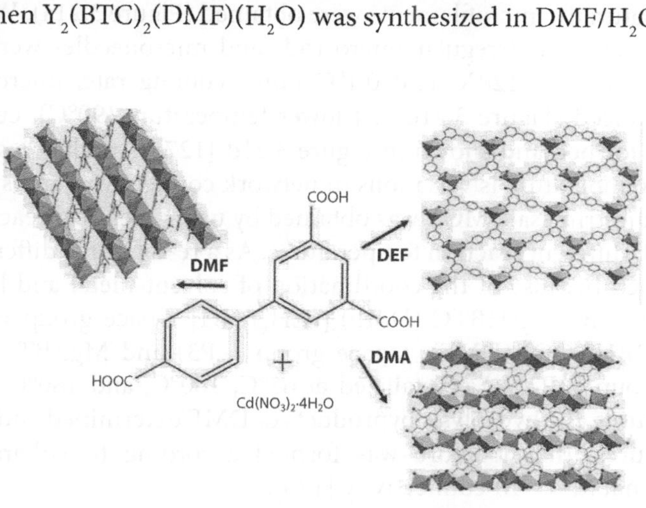

Figure 3.20 The formation scheme of three solvent-dependent Cd(II) metal–organic framework. Reprinted with permission from Ref. [120], copyright (2012) ACS.

$Y_2(BTC)_2(H_2O)_2$ was formed in DEF/H_2O solvents because of the steric effect of DEF molecules which prevent the direct coordination with metal atoms [122]. The size of channels of the MOFs is also influenced by the solvent structure. Huang and coworkers prepared three different cobalt complexes using cobalt(II) nitrate, 4,4′-((5-carboxy-1,3-phenylene)bis(oxy)) dibenzoic acid and different solvents. The channel sizes of the three complexes (76.84Å >74.37Å >72.76Å) were changed according to the sizes of the solvent molecules (DMP>DMA>DMF) [123].

The reaction time, reaction temperature, and cooling rate influence the nucleation and crystal growth rate, allowing the shape-controlled synthesis of MOF with different sizes. The reaction times probe role of thermodynamic and kinetic factors in the formation of extended MOF structures [124] and the structure dimension increases with time. Burrows and coworkers produced four cadmium trimesate from $Cd(NO_3)_2.4H_2O$ and 1,3,5-benzenetricarboxylic acid (H_3btc) reaction in DMF at 95°C at different reaction times. After heating for 10 min, the two-dimensional networks $[Cd(Hbtc)(H_2O)_2]$ and $[Cd(Hbtc)(DMF)_2]$ were formed. After heating for 1 h, the bilayer network $[Cd_3(btc)_2(H_2O)_9]4H_2O$ were isolated and heating for 2 days gave the three-dimensional network $[Cd_{12}(btc)_8$-$(DMF)_{14}(OH_2)_2]1.5DMF$ [125].

The reaction temperatures directly affect the reaction rate and energy barrier, and hence, the conformation, coordination mode, the topology, dimensional and structural properties of MOFs change [126]. When MOF-177 (Zn_4O(1,3,5-benzentribenzoate)$_2$) was synthesized at 100°C and cooling naturally, microfilaments were obtained (Figure 3.21a). With 0.5°C min^{-1} cooling rate, irregular micro rods and microneedles were formed (Figure 3.21b). At 120°C and 0.1°C min^{-1} cooling rate, micro cuboids were produced (Figure 3.21c). At lower temperature (90°C), cubic crystals were formed and shown in Figure 3.21d [127]. Another result of the adopted temperature is variations in network connectivity and symmetry. Magnesium trimesate MOF was obtained by using identical reaction mixtures with different reaction temperatures. As a result, three different MOF was formed because of the coordination of solvent-metal and hydrolysis of solvent. The $Mg(HBTC)(DMF)_2[(CH_3)_2NH]$ (space group = P63/m), $Mg_3(BTC)(HCOO)_3(DMF)_3$ (space group = P3) and $Mg_3(BTC)_2(DMF)_4$ (space group = P21/c) crystallized at 65°C, 100°C, and 180°C. At lower temperatures, the hydrolysis byproducts of DMF determined and at higher temperatures, the structure was formed according to enhanced high dimensional M–O–M connectivity [128].

The additives are adsorbed onto the crystal surface, which slows down the rate of crystal nucleation. The addition of polymer additive (polyvinyl

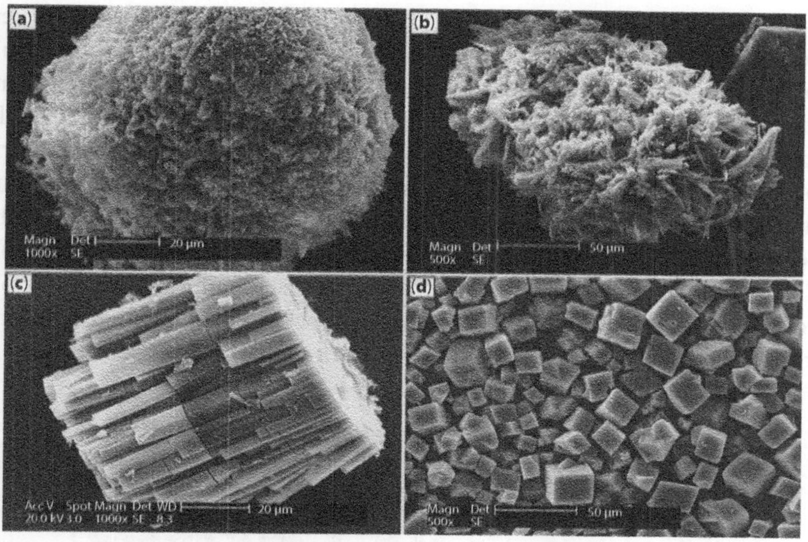

Figure 3.21 The SEM images of MOF-177 synthesized in different conditions (a) at 100°C, (b) cooling rate of 0.5°C min^{-1}, (c) at 120°C and 0.1°C min^{-1} cooling rate, (d) at 90°C. Reprinted from Ref. [127], copyright (2006) with Elsevier permission.

sulfonic acid, which is complexing copper by electrostatic interactions), the nucleation rate of the crystal $[Cu_2(Pzdc)_2(Pyz)]$ was slowed down because of the electrostatic and steric stabilization polymer and larger crystals were obtained [129]. Decorating with permeable polydimethylsiloxane with thermal vapor deposition technique, the protective hydrophobic layer form on the MOF surface and the moisture/water stability of MOF was greatly improved for potential applications [130]. The addition of modulators competed with ligands to combine with the metal clusters and also improved water stability. The concentration of modulator affects the nucleation rate hence change the crystal size. The high concentration of additive provides slow nucleation and big crystals [131]. The usage of monocarboxylic acid as a modulator, competitive coordination interactions occur and the crystallization process influence. The palmitic acid formed a hydrophobic layer of alkyl chains around the MOF particles, allowing it to disperse in the emulsified organic droplets during encapsulation and the MOFs encapsulated in photocleavable capsules were used as UV-light triggered catalysis for catalytic tetramethylbenzidine oxidation [132]. Amorphous carbon layer was used to shield the framework from decomposition under humid conditions and prevent hydrolysis and carbon-coating IRMOF yields a remarkable enhancement in moisture resistance [133]. The addition of counter ions tunes to form different coordination networks. Four new MOFs with three different coordination topologies were obtained using various counter ions. Chloride and nitrate anions caused the formation of four- and six-coordinated Zn(II) ions. $[Zn(L)(Cl)_2]$ has been the first 1D helical complex in triazole compounds and $[Zn(L)_2(H_2O)_2].(NO_3)_2$ was formed as 2D grids. Using cadmium(II) ions, nitrate and chlorate anions caused the formation of 2D to 1D coordination polymers from $[Cd(L)_2(H_2O)_2].(NO_3)_2$ to $[Cd(L)_2(H_2O)_2].(ClO_4)_2.(H_2O)_2$ [60]. Awaleh and coworkers used different anions (non-coordinating spherical anion, moderately coordinating sulfonate anions, strongly coordinating carboxylate anions, the small planar NO_3^- anion) to

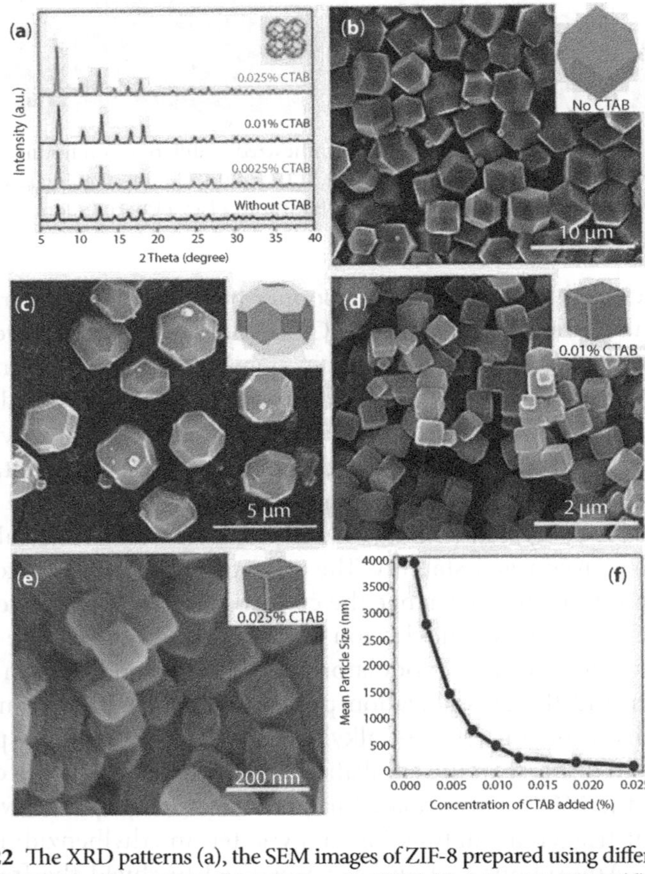

Figure 3.22 The XRD patterns (a), the SEM images of ZIF-8 prepared using different amounts of CTAB (without CTAB (b), 0.0025 wt % CTAB (c), 0.01 wt % CTAB (d), 0.025 wt % CTAB (e)), and (f) the mean particle size plot of ZIF-8. Reproduced from Ref. [135] with RSC permission.

rationalize the effects of anion size and they indicated that the anion size has a remarkable effect on the MOFs [134]. The addition of surfactants influences the reaction solution at the initial stage. The CTAB surfactant was used as an effective capping agent due to its longer hydrophobic hydrocarbon chain to control the size and morphology of ZIF-8 and the particle sizes changed from 100 nm to 4 μm (Figure 3.22f) and the SEM images were also shown in Figure 3.22 [135].

3.3 MOF Derived Materials

The derivation of MOF is an impressive subject for the preparation of various porous materials. Direct pyrolysis of MOF precursors creates different types of material with different topologies, various ordered porous structures, and pore size, high surface area, and controllable morphology [136]. The most advantages of this method are simple and convenient without additional templates. MOF-derived materials are widely used in different important applications like electrochemical storage, sensing applications, and heterogeneous catalytic reactions. Metal–organic frameworks are a sacrificial template for producing different materials like carbons, metal oxides, metal carbides, metal phosphides, and others. At different calcination temperatures, MOF precursors turn into a different type of material with different topologies for several applications.

The carbon nanomaterials have extraordinary electrical, mechanical, and thermal properties and the carbon materials derived from MOFs are widely used in gas separation and adsorption, energy storage, and conversion applications. The first example of MOF-derived carbon was reported by Liu *et al.* and nanoporous carbon was obtained by thermal transformation with high surface area and high hydrogen adsorption capacity. The obtained carbon was used as an electrode material and showed excellent electrochemical performance for double-layered capacitor [137]. Pachfule *et al.* prepared one-dimensional carbon nanorods by self-templated, catalyst-free strategy, and they used sonochemical treatment followed by chemical activation to transform nanorods into two- to six-layered graphene nanoribbons which showed excellent supercapacitor performance [138]. In another example, N-decorated nanoporous carbons were formed from the metal-organic framework (ZIF-8) with high surface areas and the CO_2 adsorption capacities were measured and found as an excellent adsorbent for CO_2 over other gases. Besides this, it showed good electrochemical performance as a platin free catalyst in oxygen reduction reaction [139]. The $Ru_3(CO)_{12}$ doped ZIF-8 metal-organic framework was used as a template and after calcination at 800°C,

Ru_3 clusters were formed and used as a catalyst for the oxidation of 2-amino benzyl alcohol. The obtained Ru_3 catalyst exhibited 100% conversion and 100% selectivity and much higher TON compared Ru catalyst [140].

As a MOF-derived metal oxide, Co_3O_4 nanoparticles were produced from cobalt MOF ($Co_3(NDC)_3(DMF)_4$) through pyrolysis and used as an electrode material. The Co_3O_4 has an agglomerated structure which enhanced the rate capability and prolonged cycle life [141]. Spherical shape CuO nanoparticles were synthesized from Cu-based MOF at 550°C and they exhibited high electrochemical performances as an anode material in the way of its cyclability (40 galvanostatic cycles) and reversible capacity (538 mAh g^{-1}) as compared to several reports [142]. Spindle-like hollow CuO/C composites were prepared by the calcination of the Cu-MOF at 700°C and tried as an anode material and they showed a high reversible specific capacity of 789 mAh g^{-1} with cyclic stability [143]. The pyrolysis of iron-containing metal–organic frameworks yielded nanoparticles with a unique iron oxide@iron carbide (Fe_3O_4@Fe_5C_2) core-shell structure dispersed on carbon supports and highly active catalysts for Fischer–Tropsch synthesis were formed [144].

3.4 Conclusion

In summary, a new metal-organic framework having desired structural, thermal and surface properties can be designed for preferred usage by the arrangement of construction agents (metal nodes, secondary building units, organic linkers), the synthesis techniques (solvothermal and hydrothermal method, microwave synthesis, electrochemical method, mechanochemical synthesis, sonochemical synthesis, diffusion method, and template method), the synthesis conditions (reaction time and temperature, cooling rate, and pressure) and constitutions (metal source, concentration of reagents, solvents, additives and pH) Besides this, the obtained MOFs can be derived by pyrolysis at different temperatures and some new MOF-derived composites with specific structures can form new candidates for several applications.

References

1. Jiao, L., Seow, J.Y.R., Skinner, W.S., Wang, Z.U., Jiang, H.-L., Metal–organic frameworks: Structures and functional applications. *Mater. Today*, 2018.
2. Rosi, N.L., Eckert, J., Eddaoudi, M., Vodak, D.T., Kim, J. *et al.*, Hydrogen storage in microporous metal-organic frameworks. *Science*, 300, 1127, 2003.

3. Saha, D., Bao, Z., Jia, F., Deng, S., Adsorption of CO2, CH4, N2O, and N2 on MOF-5, MOF-177, and zeolite 5A. *Environ. Sci. Technol.*, 44, 1820, 2010.
4. Marco-Lozar, J., Juan-Juan, J., Suárez-García, F., Cazorla-Amorós, D., Linares-Solano, A., MOF-5 and activated carbons as adsorbents for gas storage. *Int. J. Hydrogen Energ.*, 37, 2370, 2012.
5. Wang, Y., Wang, X., Wang, X., Zhang, X., Fan, W. *et al.*, Effect of Functional Groups on the Adsorption of Light Hydrocarbons in fmj-type Metal–Organic Frameworks. *Cryst. Growth Des.*, 19, 832, 2018.
6. Chang, G., Huang, M., Su, Y., Xing, H., Su, B. *et al.*, Immobilization of Ag (i) into a metal–organic framework with–SO 3 H sites for highly selective olefin–paraffin separation at room temperature. *Chem. Commun.*, 51, 2859, 2015.
7. Proch, S., Herrmannsdörfer, J., Kempe, R., Kern, C., Jess, A. *et al.*, Pt@ MOF-177: Synthesis, Room-Temperature Hydrogen Storage and Oxidation Catalysis. *Chemistry–A European Journal*, 14, 8204, 2008.
8. Chen, D., Xing, H., Wang, C., Su, Z., Highly efficient visible-light-driven CO 2 reduction to formate by a new anthracene-based zirconium MOF via dual catalytic routes. *J. Mater. Chem. A*, 4, 2657, 2016.
9. Opelt, S., Türk, S., Dietzsch, E., Henschel, A., Kaskel, S., Klemm, E., Preparation of palladium supported on MOF-5 and its use as hydrogenation catalyst. *Catal. Commun.*, 9, 1286, 2008.
10. Herbst, A., Khutia, A., Janiak, C., Brønsted instead of Lewis acidity in functionalized MIL-101Cr MOFs for efficient heterogeneous (nano-MOF) catalysis in the condensation reaction of aldehydes with alcohols. *Inorg. Chem.*, 53, 7319, 2014.
11. Gao, S., Zhao, N., Shu, M., Che, S., Palladium nanoparticles supported on MOF-5: A highly active catalyst for a ligand-and copper-free Sonogashira coupling reaction. *Appl. Catal. A: Gen.*, 388, 196, 2010.
12. Dhakshinamoorthy, A., Asiri, A.M., Garcia, H., Metal–organic framework (MOF) compounds: Photocatalysts for redox reactions and solar fuel production. *Angew. Chem. Int. Edit.*, 55, 5414, 2016.
13. Zhang, C., Wang, M., Liu, L., Yang, X., Xu, X., Electrochemical investigation of a new Cu-MOF and its electrocatalytic activity towards H2O2 oxidation in alkaline solution. *Electrochem. Commun.*, 33, 131, 2013.
14. Huo, J., Aguilera-Sigalat, J., El-Hankari, S., Bradshaw, D., Magnetic MOF microreactors for recyclable size-selective biocatalysis. *Chem. Sci.*, 6, 2015, 1938.
15. Wen, Y., Zhang, J., Xu, Q., Wu, X.-T., Zhu, Q.-L., Pore surface engineering of metal–organic frameworks for heterogeneous catalysis. *Coord. Chem. Rev.*, 376, 248, 2018.
16. Pan, S., Chen, X., Li, X., Jin, M., Nonderivatization method for determination of glyphosate, glufosinate, bialaphos, and their main metabolites in environmental waters based on magnetic metal-organic framework pretreatment. *J. Sep. Sci.*, 42, 1045, 2019.

17. Li, N., Zhou, L., Jin, X., Owens, G., Chen, Z., Simultaneous removal of tetracycline and oxytetracycline antibiotics from wastewater using a ZIF-8 metal organic-framework. *J. Hazard. Mater.*, 366, 563, 2019.
18. Chowdhury, T., Zhang, L., Zhang, J., Aggarwal, S., Removal of arsenic (III) from aqueous solution using metal organic framework-graphene oxide nanocomposite. *Nanomaterials*, 8, 1062, 2018.
19. Haque, E., Jun, J.W., Jhung, S.H., Adsorptive removal of methyl orange and methylene blue from aqueous solution with a metal-organic framework material, iron terephthalate (MOF-235). *J. Hazard. Mater.*, 185, 507, 2011.
20. Xiao, Y., Cui, Y., Zheng, Q., Xiang, S., Qian, G., Chen, B., A microporous luminescent metal–organic framework for highly selective and sensitive sensing of Cu2+ in aqueous solution. *Chem. Commun.*, 46, 5503, 2010.
21. Zhu, X., Zheng, H., Wei, X., Lin, Z., Guo, L. *et al.*, Metal–organic framework (MOF): A novel sensing platform for biomolecules. *Chem. Commun.*, 49, 1276, 2013.
22. Jia, X.-X., Yao, R.-X., Zhang, F.-Q., Zhang, X.-M., A fluorescent anionic MOF with Zn4 (trz) 2 chain for highly selective visual sensing of contaminants: Cr (III) ion and TNP. *Inorg. Chem.*, 56, 2690, 2017.
23. Gole, B., Bar, A.K., Mukherjee, P.S., Fluorescent metal–organic framework for selective sensing of nitroaromatic explosives. *Chem. Commun.*, 47, 12137, 2011
24. Miller, S.R., Heurtaux, D., Baati, T., Horcajada, P., Grenèche, J.-M., Serre, C., Biodegradable therapeutic MOFs for the delivery of bioactive molecules. *Chem. Commun.*, 46, 4526, 2010.
25. Jiang, K., Zhang, L., Hu, Q., Zhao, D., Xia, T. *et al.*, Pressure controlled drug release in a Zr-cluster-based MOF. *J. Mater. Chem. B*, 4, 6398, 2016.
26. Yao, M.S., Tang, W.X., Wang, G.E., Nath, B., Xu, G., MOF Thin Film-Coated Metal Oxide Nanowire Array: Significantly Improved Chemiresistor Sensor Performance. *Adv. Mater.*, 28, 5229, 2016.
27. Xu, X., Cao, R., Jeong, S., Cho, J., Spindle-like mesoporous α-Fe2O3 anode material prepared from MOF template for high-rate lithium batteries. *Nano Lett.*, 12, 4988, 2012.
28. Sheberla, D., Bachman, J.C., Elias, J.S., Sun, C.-J., Shao-Horn, Y., Dincă, M., Conductive MOF electrodes for stable supercapacitors with high areal capacitance. *Nat. Mater.*, 16, 220, 2017.
29. Ren, Y., Chia, G.H., Gao, Z., Metal–organic frameworks in fuel cell technologies. *Nano Today*, 8, 577, 2013.
30. Calleja, G., Botas, J., Sánchez-Sánchez, M., Orcajo, M., Hydrogen adsorption over Zeolite-like MOF materials modified by ion exchange. *Int. J. Hydrogen Energ.*, 35, 9916, 2010.
31. Gascon, J. and Kapteijn, F., Metal-Organic Framework Membranes—High Potential, Bright Future? *Angew. Chem. Int. Edit.*, 49, 1530, 2010.
32. Taylor, K.M., Rieter, W.J., Lin, W., Manganese-based nanoscale metal–organic frameworks for magnetic resonance imaging. *J. Am. Chem. Soc.*, 130, 14358, 2008.

33. Patel, H.A., Mansor, N., Gadipelli, S., Brett, D.J., Guo, Z., Superacidity in Nafion/MOF hybrid membranes retains water at low humidity to enhance proton conduction for fuel cells. *ACS Appl. Mater. Interfaces*, 8, 30687, 2016.
34. Xu, H., Liu, F., Cui, Y., Chen, B., Qian, G., A luminescent nanoscale metal–organic framework for sensing of nitroaromatic explosives. *Chem. Commun.*, 47, 3153, 2011.
35. Ayoub, G., Karadeniz, B., Howarth, A.J., Farha, O.K., Đilović, I. *et al.*, Rational Synthesis of Mixed-Metal Microporous Metal–Organic Frameworks with Controlled Composition Using Mechanochemistry. *Chem. Mater.*, 31, 5494, 2019.
36. Cheong, V.F. and Moh, P.Y., Recent advancement in metal–organic framework: Synthesis, activation, functionalisation, and bulk production. *Mater. Sci. Technol.*, 34, 1025, 2018.
37. Rodrigues, M.O., de Paula, M.V., Wanderley, K.A., Vasconcelos, I.B., Alves, Jr., S., Soares, T.A., Metal organic frameworks for drug delivery and environmental remediation: A molecular docking approach. *Int. J. Quantum Chem.*, 112, 3346, 2012.
38. Suh, M.P., Park, H.J., Prasad, T.K., Lim, D.-W., Hydrogen storage in metal–organic frameworks. *Chem. Rev.*, 112, 782, 2011.
39. Lu, W., Wei, Z., Gu, Z.-Y., Liu, T.-F., Park, J. *et al.*, Tuning the structure and function of metal–organic frameworks via linker design. *Chem. Soc. Rev.*, 43, 5561, 2014.
40. Kitagawa, S., Kitaura, R., Noro, S.I., Functional porous coordination polymers. *Angew. Chem. Int. Edit.*, 43, 2334, 2004.
41. He, H., Zhu, Q.-Q., Sun, F., Zhu, G., Two 3D metal–organic frameworks based on CoII and ZnII clusters for Knoevenagel condensation reaction and highly selective luminescence sensing. *Cryst. Growth Des.*, 18, 5573, 2018.
42. Wang, X., Zhang, X., Li, P., Otake, K.-I., Cui, Y. *et al.*, Vanadium Catalyst on Isostructural Transition Metal, Lanthanide, and Actinide Based Metal–Organic Frameworks for Alcohol Oxidation. *J. Am. Chem. Soc.*, 2019.
43. Zhang, F., Shi, J., Jin, Y., Fu, Y., Zhong, Y., Zhu, W., Facile synthesis of MIL-100 (Fe) under HF-free conditions and its application in the acetalization of aldehydes with diols. *Chem. Eng. J.*, 259, 183, 2015.
44. Ozer, D., Oztas, N.A., Köse, D.A., Şahin, O., Fabrication and characterization of magnesium and calcium trimesate complexes via ion-exchange and one-pot self-assembly reaction. *J. Mol. Struct.*, 1156, 353, 2018.
45. Bezverkhyy, I., Weber, G., Bellat, J.-P., Degradation of fluoride-free MIL-100 (Fe) and MIL-53 (Fe) in water: Effect of temperature and pH. *Microporous Mesoporous Mater.*, 219, 117, 2016.
46. Liu, Y., Eubank, J.F., Cairns, A.J., Eckert, J., Kravtsov, V.C. *et al.*, Assembly of metal–organic frameworks (MOFs) based on indium-trimer building blocks: A porous MOF with soc topology and high hydrogen storage. *Angew. Chem. Int. Edit.*, 46, 3278, 2007.

47. Botas, J.A., Calleja, G., Sánchez-Sánchez, M., Orcajo, M.G., Cobalt doping of the MOF-5 framework and its effect on gas-adsorption properties. *Langmuir*, 26, 5300, 2010.
48. Iqbal, B., Saleem, M., Arshad, S.N., Rashid, J., Hussain, N., Zaheer, M., One-pot Synthesis of Heterobimetallic Metal-Organic Frameworks (MOF) for Multifunctional Catalysis. *Chemistry–A European Journal*, 2019.
49. Cohen, S.M., Postsynthetic methods for the functionalization of metal–organic frameworks. *Chem. Rev.*, 112, 970, 2011.
50. Mason, J.A., Veenstra, M., Long, J.R., Evaluating metal–organic frameworks for natural gas storage. *Chem. Sci.*, 5, 32, 2014.
51. Gagnon, K.J., Perry, H.P., Clearfield, A., Conventional and unconventional metal–organic frameworks based on phosphonate ligands: MOFs and UMOFs. *Chem. Rev.*, 112, 1034, 2011.
52. Konar, S., Zoń, J., Prosvirin, A.V., Dunbar, K.R., Clearfield, A., Synthesis and Characterization of Four Metal- Organophosphonates with One-, Two-, and Three-Dimensional Structures. *Inorg. Chem.*, 46, 5229, 2007.
53. Taylor, J.M., Mahmoudkhani, A.H., Shimizu, G.K., A tetrahedral organophosphonate as a linker for a microporous copper framework. *Angew. Chem. Int. Edit.*, 46, 795, 2007.
54. Thuéry, P., Atoini, Y., Harrowfield, J., The sulfonate group as a ligand: A fine balance between hydrogen bonding and metal ion coordination in uranyl ion complexes. *Dalton Trans.*, 2019.
55. Horike, S., Matsuda, R., Tanaka, D., Mizuno, M., Endo, K., Kitagawa, S., Immobilization of sodium ions on the pore surface of a porous coordination polymer. *J. Am. Chem. Soc.*, 128, 4222, 2006.
56. Zhang, G., Wei, G., Liu, Z., Oliver, S.R., Fei, H., A robust sulfonate-based metal–organic framework with permanent porosity for efficient CO2 capture and conversion. *Chem. Mater.*, 28, 6276, 2016.
57. Zhang, J.-P., Zhang, Y.-B., Lin, J.-B., Chen, X.-M., Metal azolate frameworks: From crystal engineering to functional materials. *Chem. Rev.*, 112, 1001, 2011.
58. Phan, A., Doonan, C.J., Uribe-Romo, F.J., Knobler, C.B., O'keeffe, M., Yaghi, O.M., Synthesis, structure, and carbon dioxide capture properties of zeolitic imidazolate frameworks, 2009.
59. Zhou, K., Mousavi, B., Luo, Z., Phatanasri, S., Chaemchuen, S., Verpoort, F., Characterization and properties of Zn/Co zeolitic imidazolate frameworks vs. ZIF-8 and ZIF-67. *J. Mater. Chem. A*, 5, 952, 2017.
60. Yi, L., Yang, X., Lu, T., Cheng, P., Self-assembly of right-handed helical infinite chain, one-and two-dimensional coordination polymers tuned via anions. *Cryst. Growth Des.*, 5, 1215, 2005.
61. Lü, X.-Q., Jiang, J.-J., Chen, C.-L., Kang, B.-S., Su, C.-Y., 3D Coordination Polymers with Nitrilotriacetic and 4, 4 ‘-Bipyridyl Mixed Ligands: Structural Variation Based on Dinuclear or Tetranuclear Subunits Assisted by Na–O and/or O–H⊙⊙⊙ O Interactions. *Inorg. Chem.*, 44, 4515, 2005.

62. Hagrman, P.J., Hagrman, D., Zubieta, J., Organic–Inorganic Hybrid Materials: From "Simple" Coordination Polymers to Organodiamine-Templated Molybdenum Oxides. *Angew. Chem. Int. Edit.*, 38, 2638, 1999.
63. Shen, K., Zhang, M., Zheng, H., Critical factors influencing the structures and properties of metal–organic frameworks. *CrystEngComm*, 17, 981, 2015.
64. Buragohain, A., Yousufuddin, M., Sarma, M., Biswas, S., 3D luminescent amide-functionalized cadmium tetrazolate framework for selective detection of 2, 4, 6-trinitrophenol. *Cryst. Growth Des.*, 16, 842, 2016.
65. Muniappan, S., Lipstman, S., George, S., Goldberg, I., Porphyrin framework solids. Synthesis and structure of hybrid coordination polymers of tetra (carboxyphenyl) porphyrins and lanthanide-bridging ions. *Inorg. Chem.*, 46, 5544, 2007.
66. Banerjee, D., Kim, S.J., Parise, J.B., Lithium based metal–organic framework with exceptional stability. *Cryst. Growth Des.*, 9, 2500, 2009.
67. Dybtsev, D.N., Chun, H., Yoon, S.H., Kim, D., Kim, K., Microporous manganese formate: A simple metal–organic porous material with high framework stability and highly selective gas sorption properties. *J. Am. Chem. Soc.*, 126, 32, 2004.
68. Chae, H.K., Siberio-Perez, D.Y., Kim, J., Go, Y., Eddaoudi, M. *et al.*, A route to high surface area, porosity and inclusion of large molecules in crystals. *Nature*, 427, 523, 2004.
69. Deng, H., Grunder, S., Cordova, K.E., Valente, C., Furukawa, H. *et al.*, Large-pore apertures in a series of metal-organic frameworks. *Science*, 336, 1018, 2012.
70. Du, M., Li, C.-P., Liu, C.-S., Fang, S.-M., Design and construction of coordination polymers with mixed-ligand synthetic strategy. *Coord. Chem. Rev.*, 257, 1282, 2013.
71. Kleist, W., Jutz, F., Maciejewski, M., Baiker, A., Mixed-Linker Metal–Organic Frameworks as Catalysts for the Synthesis of Propylene Carbonate from Propylene Oxide and CO2. *Eur. J. Inorg. Chem.*, 2009, 3552, 2009.
72. Liu, Y., Qi, Y., Su, Y.-H., Zhao, F.-H., Che, Y.-X., Zheng, J.-M., Five novel cobalt coordination polymers: Effect of metal–ligand ratio and structure characteristics of flexible bis (imidazole) ligands. *CrystEngComm*, 12, 3283, 2010.
73. Hu, J.-S., Yao, X.-Q., Zhang, M.-D., Qin, L., Li, Y.-Z. *et al.*, Syntheses, structures, and characteristics of four new metal–organic frameworks based on flexible tetrapyridines and aromatic polycarboxylate acids. *Cryst. Growth Des.*, 12, 3426, 2012.
74. Alavi, M.A. and Morsali, A., Synthesis and characterization of different nanostructured copper (II) metal–organic frameworks by a ligand functionalization and modulation method. *CrystEngComm*, 16, 2246, 2014.
75. Deng, H., Doonan, C.J., Furukawa, H., Ferreira, R.B., Towne, J. *et al.*, Multiple functional groups of varying ratios in metal-organic frameworks. *Science*, 327, 846, 2010.

76. Buragohain, A., Van Der Voort, P., Biswas, S., Facile synthesis and gas adsorption behavior of new functionalized Al-MIL-101-X (X = –CH3,–NO2,–OCH3,–C6H4,–F2,–(CH3) 2,–(OCH3) 2) materials. *Microporous Mesoporous Mater.*, 215, 91, 2015.
77. Nguyen, J.G. and Cohen, S.M., Moisture-resistant and superhydrophobic metal–organic frameworks obtained via postsynthetic modification. *J. Am. Chem. Soc.*, 132, 4560, 2010.
78. Yang, C., Kaipa, U., Mather, Q.Z., Wang, X., Nesterov, V. *et al.*, Fluorous metal–organic frameworks with superior adsorption and hydrophobic properties toward oil spill cleanup and hydrocarbon storage. *J. Am. Chem. Soc.*, 133, 18094, 2011.
79. Seo, J.S., Whang, D., Lee, H., Im Jun, S., Oh, J. *et al.*, A homochiral metal–organic porous material for enantioselective separation and catalysis. *Nature*, 404, 982, 2000.
80. Yaghi, O.M., O'Keeffe, M., Ockwig, N.W., Chae, H.K., Eddaoudi, M., Kim, J., Reticular synthesis and the design of new materials. *Nature*, 423, 705, 2003.
81. Cui, P., Wang, P., Zhao, Y., Sun, W.-Y., Fabrication of Desired Metal–Organic Frameworks via Postsynthetic Exchange and Sequential Linker Installation. *Cryst. Growth Des.*, 19, 1454, 2019.
82. Schoedel, A., Li, M., Li, D., O'Keeffe, M., Yaghi, O.M., Structures of metal–organic frameworks with rod secondary building units. *Chem. Rev.*, 116, 12466, 2016.
83. Jiang, J., Gándara, F., Zhang, Y.-B., Na, K., Yaghi, O.M., Klemperer, W.G., Superacidity in sulfated metal–organic framework-808. *J. Am. Chem. Soc.*, 136, 12844, 2014.
84. Li, H., Eddaoudi, M., O'Keeffe, M., Yaghi, O.M., Design and synthesis of an exceptionally stable and highly porous metal-organic framework. *Nature*, 402, 276, 1999.
85. Férey, G., Mellot-Draznieks, C., Serre, C., Millange, F., Dutour, J. *et al.*, A chromium terephthalate-based solid with unusually large pore volumes and surface area. *Science*, 309, 2040, 2005.
86. McKinstry, C., Cussen, E.J., Fletcher, A.J., Patwardhan, S.V., Sefcik, J., Effect of synthesis conditions on formation pathways of metal organic framework (MOF-5) crystals. *Cryst. Growth Des.*, 13, 5481, 2013.
87. Qian, J., Sun, F., Qin, L., Hydrothermal synthesis of zeolitic imidazolate framework-67 (ZIF-67) nanocrystals. *Mater. Lett.*, 82, 220, 2012.
88. Novaković, S.B., Bogdanović, G.A., Heering, C., Makhloufi, G., Francuski, D., Janiak, C., Charge-density distribution and electrostatic flexibility of ZiF-8 based on high-resolution X-ray diffraction data and periodic calculations. *Inorg. Chem.*, 54, 2660, 2015.
89. Railey, P., Song, Y., Liu, T., Li, Y., Metal organic frameworks with immobilized nanoparticles: Synthesis and applications in photocatalytic hydrogen generation and energy storage. *Mater. Res. Bull.*, 96, 385, 2017.

90. Banerjee, R., Phan, A., Wang, B., Knobler, C., Furukawa, H. *et al.*, High-throughput synthesis of zeolitic imidazolate frameworks and application to CO2 capture. *Science*, 319, 939, 2008.
91. Shi, Q., Chen, Z., Song, Z., Li, J., Dong, J., Synthesis of ZIF-8 and ZIF-67 by steam-assisted conversion and an investigation of their tribological behaviors. *Angew. Chem. Int. Edit.*, 50, 672, 2011.
92. Xu, L., Choi, E.-Y., Kwon, Y.-U., Ionothermal synthesis of a 3D Zn–BTC metal-organic framework with distorted tetranuclear [Zn4 (μ4-O)] subunits. *Inorg. Chem. Commun.*, 11, 1190, 2008.
93. Klinowski, J., Paz, F.A.A., Silva, P., Rocha, J., Microwave-assisted synthesis of metal–organic frameworks. *Dalton Trans.*, 40, 321, 2011.
94. Choi, J.-S., Son, W.-J., Kim, J., Ahn, W.-S., Metal–organic framework MOF-5 prepared by microwave heating: Factors to be considered. *Microporous Mesoporous Mater.*, 116, 727, 2008.
95. Hafizovic, J., Bjørgen, M., Olsbye, U., Dietzel, P.D., Bordiga, S. *et al.*, The inconsistency in adsorption properties and powder XRD data of MOF-5 is rationalized by framework interpenetration and the presence of organic and inorganic species in the nanocavities. *J. Am. Chem. Soc.*, 129, 3612, 2007.
96. Ni, Z. and Masel, R.I., Rapid production of metal–organic frameworks via microwave-assisted solvothermal synthesis. *J. Am. Chem. Soc.*, 128, 12394, 2006.
97. Haque, E., Khan, N.A., Park, J.H., Jhung, S.H., Synthesis of a metal–organic framework material, iron terephthalate, by ultrasound, microwave, and conventional electric heating: A kinetic study. *Chemistry–A European Journal*, 16, 1046, 2010.
98. Khan, N.A. and Jhung, S.H., Synthesis of metal-organic frameworks (MOFs) with microwave or ultrasound: Rapid reaction, phase-selectivity, and size reduction. *Coord. Chem. Rev.*, 285, 11, 2015.
99. Bag, P.P., Wang, X.-S., Cao, R., Microwave-assisted large scale synthesis of lanthanide metal–organic frameworks (Ln-MOFs), having a preferred conformation and photoluminescence properties. *Dalton Trans.*, 44, 11954, 2015.
100. Martinez Joaristi, A., Juan-Alcañiz, J., Serra-Crespo, P., Kapteijn, F., Gascon, J., Electrochemical synthesis of some archetypical Zn2+, Cu2+, and Al3+ metal organic frameworks. *Cryst. Growth Des.*, 12, 3489, 2012.
101. Van Assche, T.R., Desmet, G., Ameloot, R., De Vos, D.E., Terryn, H., Denayer, J.F., Electrochemical synthesis of thin HKUST-1 layers on copper mesh. *Microporous Mesoporous Mater.*, 158, 209, 2012.
102. Sachdeva, S., Pustovarenko, A., Sudhölter, E.J., Kapteijn, F., de Smet, L.C., Gascon, J., Control of interpenetration of copper-based MOFs on supported surfaces by electrochemical synthesis. *CrystEngComm*, 18, 4018, 2016.
103. Beldon, P.J., Fábián, L., Stein, R.S., Thirumurugan, A., Cheetham, A.K., Friščić, T., Rapid room-temperature synthesis of zeolitic imidazolate frameworks by using mechanochemistry. *Angew. Chem. Int. Edit.*, 49, 9640, 2010.

104. Klimakow, M., Klobes, P., Thünemann, A.F., Rademann, K., Emmerling, F., Mechanochemical synthesis of metal- organic frameworks: A fast and facile approach toward quantitative yields and high specific surface areas. *Chem. Mater.*, 22, 5216, 2010.
105. Masoomi, M.Y. and Morsali, A., Sonochemical synthesis of nanoplates of two Cd (II) based metal–organic frameworks and their applications as precursors for preparation of nano-materials. *Ultrason. Sonochem.*, 28, 240, 2016.
106. Son, W.-J., Kim, J., Kim, J., Ahn, W.-S., Sonochemical synthesis of MOF-5. *Chem. Commun.*, 6336, 2008.
107. Li, Z.-Q., Qiu, L.-G., Xu, T., Wu, Y., Wang, W. *et al.*, Ultrasonic synthesis of the microporous metal–organic framework Cu3 (BTC) 2 at ambient temperature and pressure: An efficient and environmentally friendly method. *Mater. Lett.*, 63, 78, 2009.
108. Xiao, J.-D., Qiu, L.-G., Ke, F., Yuan, Y.-P., Xu, G.-S. *et al.*, Rapid synthesis of nanoscale terbium-based metal–organic frameworks by a combined ultrasound-vapour phase diffusion method for highly selective sensing of picric acid. *J. Mater. Chem. A*, 1, 8745, 2013.
109. Yao, J., Dong, D., Li, D., He, L., Xu, G., Wang, H., Contra-diffusion synthesis of ZIF-8 films on a polymer substrate. *Chem. Commun.*, 47, 2559, 2011.
110. Liu, W., Yin, R., Xu, X., Zhang, L., Shi, W., Cao, X., Structural Engineering of Low-Dimensional Metal–Organic Frameworks: Synthesis, Properties, and Applications. *Adv. Sci.*, 1802373, 2019.
111. Arbulu, R.C., Jiang, Y.B., Peterson, E.J., Qin, Y., Metal–Organic Framework (MOF) Nanorods, Nanotubes, and Nanowires. *Angew. Chem. Int. Edit.*, 57, 5813, 2018.
112. Yu, H., Qiu, X., Neelakanda, P., Deng, L., Khashab, N.M. *et al.*, Hollow ZIF-8 nanoworms from block copolymer templates. *Sci. Rep.*, 5, 15275, 2015.
113. Biemmi, E., Christian, S., Stock, N., Bein, T., High-throughput screening of synthesis parameters in the formation of the metal-organic frameworks MOF-5 and HKUST-1. *Microporous Mesoporous Mater.*, 117, 111, 2009.
114. Li, S.-L., Tan, K., Lan, Y.-Q., Qin, J.-S., Li, M.-N. *et al.*, pH-Dependent Binary Metal- Organic Compounds Assembled from Different Helical Units: Structural Variation and Supramolecular Isomers. *Cryst. Growth Des.*, 10, 1699, 2010.
115. Guo, H., Zhu, Y., Wang, S., Su, S., Zhou, L., Zhang, H., Combining coordination modulation with acid–base adjustment for the control over size of metal–organic frameworks. *Chem. Mater.*, 24, 444, 2012.
116. Chen, L., Jia, H.-Y., Hong, X.-J., Chen, D.-H., Zheng, Z.-P. *et al.*, Construction of one pH-independent 3-D pillar-layer lead-organic framework containing tetrazole-1-acetic acid. *Inorg. Chem. Commun.*, 27, 22, 2013.
117. Kim, D., Song, X., Yoon, J.H., Lah, M.S., 3, 6-Connected metal–organic frameworks based on triscarboxylate as a 3-connected organic node and a linear trinuclear Co3 (COO) 6 secondary building unit as a 6-connected node. *Cryst. Growth Des.*, 12, 4186, 2012.

118. Cheng, Y., Xu, P., Ding, Y.-B., Yin, Y.-G., Stoichiometry-dominated *in situ* formation of iodocuprate clusters and dimethyl-2, 2′-biimidazoles as building units of coordination architectures. *CrystEngComm*, 13, 2644, 2011.
119. Yang, Q., Chen, X., Chen, Z., Hao, Y., Li, Y. *et al.*, Metal–organic frameworks constructed from flexible V-shaped ligands: Adjustment of the topology, interpenetration and porosity via a solvent system. *Chem. Commun.*, 48, 10016, 2012.
120. Li, L., Wang, S., Chen, T., Sun, Z., Luo, J., Hong, M., Solvent-dependent formation of Cd (II) coordination polymers based on a C 2-symmetric tricarboxylate linker. *Cryst. Growth Des.*, 12, 4109, 2012.
121. Zuo, C., Lu, Z., Zhang, M., Solvent-induced generation of two magnesium-based metal–organic frameworks with doubly inter-penetrated ReO3 nets constructed from the same linkers but distinct inorganic nodes. *Inorg. Chem. Commun.*, 52, 41, 2015.
122. Dong, B.-X., Gu, X.-J., Xu, Q., Solvent effect on the construction of two microporous yttrium–organic frameworks with high thermostability via *in situ* ligand hydrolysis. *Dalton Trans.*, 39, 5683, 2010.
123. Huang, W.-H., Yang, G.-P., Chen, J., Chen, X., Zhang, C.-P. *et al.*, Solvent influence on sizes of channels in three new Co (II) complexes, exhibiting an active replaceable coordinated site. *Cryst. Growth Des.*, 13, 66, 2012.
124. Mahata, P., Prabu, M., Natarajan, S., Role of temperature and time in the formation of infinite- M- O- M- linkages and isolated clusters in MOFs: A few illustrative examples. *Inorg. Chem.*, 47, 8451, 2008.
125. Burrows, A.D., Frost, C.G., Kandiah, M., Keenan, L.L., Mahon, M.F. *et al.*, The effect of reaction conditions on the nature of cadmium 1, 3, 5-benzenetricarboxylate metal–organic frameworks. *Inorganica Chim. Acta*, 366, 303, 2011.
126. Sun, Y.-X. and Sun, W.-Y., Influence of temperature on metal-organic frameworks. *Chin. Chem. Lett.*, 25, 823, 2014.
127. Li, X., Cheng, F., Zhang, S., Chen, J., Shape-controlled synthesis and lithium-storage study of metal-organic frameworks Zn4O (1, 3, 5-benzenetribenzoate) 2. *J. Power Sources*, 160, 542, 2006.
128. Calderone, P.J., Banerjee, D., Plonka, A.M., Kim, S.J., Parise, J.B., Temperature dependent structure formation and photoluminescence studies of a series of magnesium-based coordination networks. *Inorganica Chim. Acta*, 394, 452, 2013.
129. Uemura, T., Hoshino, Y., Kitagawa, S., Yoshida, K., Isoda, S., Effect of organic polymer additive on crystallization of porous coordination polymer. *Chem. Mater.*, 18, 992, 2006.
130. Zhang, W., Hu, Y., Ge, J., Jiang, H.-L., Yu, S.-H., A facile and general coating approach to moisture/water-resistant metal–organic frameworks with intact porosity. *J. Am. Chem. Soc.*, 136, 16978, 2014.
131. Diring, S., Furukawa, S., Takashima, Y., Tsuruoka, T., Kitagawa, S., Controlled multiscale synthesis of porous coordination polymer in nano/micro regimes. *Chem. Mater.*, 22, 4531, 2010.

132. Deleu, W.P., Rivero, G., Teixeira, R.F., Du Prez, F.E., De Vos, D.E., Metal–organic frameworks encapsulated in photocleavable capsules for UV-light triggered catalysis. *Chem. Mater.*, 27, 5495, 2015.
133. Yang, S.J. and Park, C.R., Preparation of highly moisture-resistant black-colored metal organic frameworks. *Adv. Mater.*, 24, 4010, 2012.
134. Awaleh, M.O., Badia, A., Brisse, F., Influence of the Anion on the Structure of Bis (methylthio) methane Supramolecular Coordination Complexes. *Cryst. Growth Des.*, 6, 2674, 2006.
135. Pan, Y., Heryadi, D., Zhou, F., Zhao, L., Lestari, G. *et al.*, Tuning the crystal morphology and size of zeolitic imidazolate framework-8 in aqueous solution by surfactants. *CrystEngComm*, 13, 6937, 2011.
136. Chen, Y.-Z., Zhang, R., Jiao, L., Jiang, H.-L., Metal–organic framework-derived porous materials for catalysis. *Coord. Chem. Rev.*, 362, 1, 2018.
137. Liu, B., Shioyama, H., Akita, T., Xu, Q., Metal-organic framework as a template for porous carbon synthesis. *J. Am. Chem. Soc.*, 130, 5390, 2008.
138. Pachfule, P., Shinde, D., Majumder, M., Xu, Q., Fabrication of carbon nanorods and graphene nanoribbons from a metal–organic framework. *Nat. Chem.*, 8, 718, 2016.
139. Aijaz, A., Fujiwara, N., Xu, Q., From metal–organic framework to nitrogen-decorated nanoporous carbons: High CO2 uptake and efficient catalytic oxygen reduction. *J. Am. Chem. Soc.*, 136, 6790, 2014.
140. Ji, S., Chen, Y., Fu, Q., Chen, Y., Dong, J. *et al.*, Confined pyrolysis within metal–organic frameworks to form uniform Ru3 clusters for efficient oxidation of alcohols. *J. Am. Chem. Soc.*, 139, 9795, 2017.
141. Liu, B., Zhang, X., Shioyama, H., Mukai, T., Sakai, T., Xu, Q., Converting cobalt oxide subunits in cobalt metal-organic framework into agglomerated Co3O4 nanoparticles as an electrode material for lithium ion battery. *J. Power Sources*, 195, 857, 2010.
142. Banerjee, A., Singh, U., Aravindan, V., Srinivasan, M., Ogale, S., Synthesis of CuO nanostructures from Cu-based metal organic framework (MOF-199) for application as anode for Li-ion batteries. *Nano Energy*, 2, 1158, 2013.
143. Peng, H.-J., Hao, G.-X., Chu, Z.-H., He, C.-L., Lin, X.-M., Cai, Y.-P., Mesoporous spindle-like hollow CuO/C fabricated from a Cu-based metal-organic framework as anodes for high-performance lithium storage. *J. Alloys Compd.*, 727, 1020, 2017.
144. An, B., Cheng, K., Wang, C., Wang, Y., Lin, W., Pyrolysis of metal–organic frameworks to Fe3O4@ Fe5C2 core–shell nanoparticles for Fischer–Tropsch synthesis. *ACS Catal.*, 6, 3610, 2016.

4

Application of MOFs and Their Derived Materials in Molecular Transport

Arka Bagchi[1], Partha Saha[1], Arunima Biswas[1*] and SK Manirul Islam[2†]

[1]Molecular Cell Biology Laboratory, Department of Zoology, University of Kalyani, Kalyani, Nadia, India

[2]Department of Chemistry, University of Kalyani, Kalyani, Nadia, India

Abstract

Metal–organic frameworks (MOFs) are crystalline porous materials having unique properties that can be exploited for drug encapsulation and membrane transport. Their surface can be modified for different drugs and better membrane transport, controlled release of drugs can be ensured by their usage and they can be detected and studied by conventional imaging techniques. Reverse-phase microemulsions, solve-hydrothermal synthesis, mechanochemical synthesis, ultrasonic irradiation are few methods by which MOFs are prepared. Their compositional diversity is of great interest to scientists as they can be explored to have various different shape, size, and chemically distinct MOFs. In this chapter we have focused on use of MOFs in membrane transport and discussed the recent developments on the same. Last, it has been pointed out in the concluding remarks that MOFs have immense potential to be used in clinical and theronaustic applications, though researches are required to practically achieve the same.

Keywords: Metal organic framework, drug delivery, membrane transport, nanocarrier, organic linker, bioimaging, nanomaterial, biocompatibility

**Corresponding author*: arunima10@gmail.com; arunima10@klyuniv.ac.in

†*Corresponding author*: manir65@rediffmail.com

Inamuddin, Rajender Boddula, Mohd Imran Ahamed and Abdullah M. Asiri (eds.) Applications of Metal–Organic Frameworks and Their Derived Materials, (101–108) © 2020 Scrivener Publishing LLC

4.1 Introduction

Different classes of porous materials are gradually becoming more popular among researchers dealing with nanomaterials. Their unique structure and functional capabilities are posing them as a good candidate for the use in different fields. One of the recent candidates of crystalline porous materials is the Metal–organic frameworks (MOFs) or commonly known as MOFs. Initial works with MOFs were confined to the region of separation techniques, selective absorption, catalysis, gas storage, etc. [1]. Due to their regular porosity and other unique properties, recent researches are focusing on their use in drug encapsulation and controlled delivery [2]. There are several important factors that the researchers need to keep in mind while using a particular nanomaterial for targeted delivery of drugs within a cell. The nanomaterial must have a surface that can be easily modified for different drugs and better membrane transport, it should also ensure controlled release of drugs within the cell and it should also be detectable by conventional imaging techniques so that the course of it can be monitored [3]. Presence of all these properties within the MOFs makes it a more useful tool for membrane transport and controlled drug delivery than conventional approaches. There are several studies made by scientists, exploring their role in molecular transport and bioimaging and the number is continuously increasing. The compositional diversity of MOFs allows scientists to construct MOFs of different shapes, sizes, and chemical properties. Moreover, the presence of easily alterable metal ligand bonds in these structures makes them biodegradable, which helps in rapid degradation of the composite material within the cell and subsequent release of loaded drug or probe.

There are several methods to prepare such MOFs that can determine the particle size, shape, and structure, thus modulating their unique physiochemical properties. These methods include Reverse-phase microemulsions, solve-hydrothermal synthesis, mechanochemical synthesis, ultrasonic irradiation, etc. [4]. Scientists are also focusing on production such MOFs via an environment friendly method, so that the toxic effects of the organic solvents can be eliminated and the nanomaterials can be made more biocompatible.

4.2 MOFs as Nanocarriers for Membrane Transport

Scientists have worked through decades to develop carriers for different types of drugs that can increase their efficiency *in vivo*. There are several challenges faced by scientists while developing these kinds of carriers. There are three major criteria that scientists need to look after. First, the carriers must be biocompatible so that it does not cause any unwanted toxicity within the cellular environment. Second, the carrier must have high loading capacity. Commonly used drugs are highly complex and large organic materials and it is important that the carrier is able to accommodate the large sized drugs within them. Last, the carrier must be able to release the drug within the cellular environment in a controlled manner. There are numerous examples of carriers that can be biocompatible and possess but high drug loading capacity but exhibit an uncontrolled release of drug within the cellular environment. Such abrupt release of drugs or "burst effect" can cause severe damage to the cell itself and can also trigger the immune system against it. That is why scientists are constantly working on developing a drug carrier that can satisfy all the three criteria for drug delivery.

MOFs possess some important and unique properties that are very useful for utilizing them as a membrane transport agent. Various compositions of MOFs provide acceptable level of toxicity for biomedical applications. Moreover, the large surface area and large pore size of MOFs is responsible for a higher loading capacity and also higher release time in the cellular environment. The procedure of loading of drugs within the MOFs also depends on the size of the drug, if the size of the drug molecule is smaller than the pore size of the MOF, they can be encapsulated within the MOF by forming hydrogen bonds between the drug and MOF as seen in case of 5-fluorouracil (5-FU) and ibuprofen, whereas, if the size of the drug is higher than the pore size of the MOF, the drug molecules can be encapsulated by electrostatic interaction, utilizing the fact that often the MOFs are positively charged [4]. There are some examples of MOFs such as MIL-89, MIL-88A, MIL-100, MIL-101_NH_2, MIL-53, etc., which are currently used by scientists for different biomedical applications.

4.2.1 MIL-89

This MOF contains muconic acid as organic linker and exhibits structural flexibility. The pore size of this material is 11 Å and particle size ranging between 50 and 100 nm [3]. The drug loading percentage (wt%) of this MOF for Bu (Busulfun) is 9.8 and entrapment efficiency is 4.2% and for CDV (Cidofovir), the drug loading percentage is 14 and entrapment efficiency is 81% [3].

4.2.2 MIL-88A

This flexible MOF has fumaric acid as organic linker. Pore size is 6 Å with particle size 150 nm. The loading percentage (wt%) and entrapment efficiency is 8.0 and 3.3%, respectively, for Bu, 0.60 and 6.4% for AZT-TP (Azidothymidine triphosphate), 2.6 and 12% for CDV [3].

4.2.3 MIL-100

The organic linker in this framework is trimesic acid. The loading percentage (wt%) for IBU (Ibuprofen) in MIL-100(Cr) is 25.8 with a release time of 3 days (SBF) [5]. In MIL-100(Fe), the loading percentage of AZT-TP is 21% with a release time of 3 days (PBS), loading percentage (wt%) of DOX (Doxorubicin) is 9% with a release time of 14 days (PBS), and loading percentage of CDV and Bu are 29% and 25% respectively [3].

4.2.4 MIL-101

The organic linker for MIL-101_NH_2 is Aminoterephthalic acid. This nano framework has a particle size of 200 nm. The loading percentage of CDV and AZT-TP in MIL-101_NH_2 is 42% for both. MIL-101(Cr) has a loading percentage for IBU of 58% with a release time of 6 days (SBF) [4]. MIL-101(Fe) shows loading percentage of 12.8% for cisplatin prodrug with a release time of 3 days (PBS) [2]. Researchers have also demonstrated its capability as a membrane transporter in cancer cells. *In vitro* studies using HT-29 cells, which are a human colon adenocarcinoma cell line, have showed its efficacy in delivering cisplatin prodrug within the cellular environment in a controlled manner.

4.2.5 MIL-53

This flexible nano framework has Terephthalic acid as organic linker and has a particle size of 350 nm and pore size of 8.6 Å [3]. MIL-53(Fe) has IBu loading percentage of 17.4% with a release time of 21 days (SBF) [6]. Another MOF construct, Fe_2O_3@MIL-53(Al) can encapsulate IBU with a loading percentage of about 9.91% with a release time of about 5 days in PBS [7].

4.2.6 ZIF-8

Recently, researchers have shown keen interest on this type of zinc based MOFs, which are popular for their considerably low cytotoxicity. This material is very stable in water and sodium hydroxide aqueous solution but decomposes in acid solution. Many scientists have demonstrated the drug loading capacity of ZIF-8, and the alteration of release time of drugs at different pH. Sun *et al.* demonstrated that the ZIF-8 framework can incorporate anticancer drug 5-Fu and can release the drug in a pH dependent manner. The loading percentage of 5-Fu is 39.8 with a release time of 7 days in PBS [8]. The Carbon nanodots@ZIF-8 shows 5-Fu loading percentage of 23.1% with a release time of 50 h in PBS [9]. DOX also shows 4.67% loading percentage in ZIF-8 and also exhibits a sustained release which is around 30 days. Researchers have shown that the cytotoxicity of DOX is reduced in DOX-encapsulated ZIF-8 in NCI-H292, HT-29 and HL-60 cell lines, which may be due to the controlled slower release of drugs within the cells [10]. The loading percentage of DOX in Polyacrylic acid@ZIF-8 (PAA@ZIF-8) is about 65.5% with a release time of 60 hours in PBS [11]. It has also been demonstrated by scientists that ZIF-8 can accommodate CPT and Cyt c with loading percentage of 2% and 8% respectively. It has also been observed that the ZIF-8 with encapsulated CPT can deliver and release the CPT within cancer cells that can cause cytotoxicity within the cancer cell [12, 13]. Moreover, the Cyt c-encapsulated showed 10 times more peroxidase activity than that of free Cyt c.

4.2.7 Zn-TATAT

This is another nontoxic zinc based MOF formed with an achiral hexadenate ligand. This chiral nanoporous MOF shows considerably high drug loading capacity and controlled release of that loaded drug. Sun *et al.* showed that this MOF has a drug loading percentage of about 33.3% for anticancer drug 5Fu, with a release time of about 7 days in PBS [14].

4.2.8 BioMOF-1 (Zn)

This porous anionic MOF was constructed by An *et al.*, where adenine was used as a building block. They have also demonstrated that this MOF can encapsulate procainamide drugs into their pores with a loading percentage of 18%. This framework takes about 3 days to completely release drugs within PBS [15].

4.2.9 UiO (Zr)

He *et al.* have constructed a UiO MOF that can not only deliver drugs but can also carry different siRNAs that can silence some key drug resistance genes. They have demonstrated that they can efficiently carry siRNAs for silencing MDR gene that accounts for multiple drug resistance in ovarian cancer. They have also shown that this MOF can incorporate cisplatin with a loading percentage of about 12.3% which can be completely released within 24 h within the cells. Thus this MOF can co-deliver different therapeutic agents that can successfully combat ovarian cancer or may be many other types of cancers [16]. Kuangda Lu *et al.* synthesized a UiO(Hf) MOF, encapsulating DBP (5,15-di(*p*-benzoato)porphyrin). This MOF exhibits a loading percentage of about 77%. They have also demonstrated its effectiveness in photo dermal therapy of head and neck cancer [17]. Moreover, Morris and group have synthesized the first MOF-nucleic acid conjugate using a UiO-66-N3 MOF, the surface of which was covalently functionalized with oligonucleotides [18].

4.3 Conclusion

The versatility of MOFs has attracted researchers to use them in the biomedical field as drug carrier and theranostic system. Their large surface area, highly alterable pore size, shape, and composition and their relatively easy methods of synthesis are the main reasons behind their popularity in recent science, specifically in the field of biomedical applications. Although there is a significant amount of research works done with these MOFs, the toxic effect and biocompatibility of these materials are still a great concern for scientists before using it clinically. Some of the researchers have raised the fact that use of organic molecules already present in our body as the organic linker can significantly lower the toxicity of these MOFs. Some of the researchers have also suggested the use of the therapeutic agent itself as the organic linker to reduce the amount of adverse effect caused by the linker in addition [19]. Moreover, the stability of MOFs and its solubility or dispersibility in water is very essential for its biomedical applications because these properties accounts for the controlled release of drugs in the cellular system and also its biocompatibility. Extensive researches are going on in this field so that the half-life of such MOFs in the blood can be increased and also the drug delivery can be more controlled and efficient. There is still an extensive amount of research is needed before using these MOFs for clinical and theranostic applications. More *in vivo* studies are needed to be carried out so that all the adverse effects that are still present in these materials can be encountered. Researchers should also look for providing more cost-effective synthesis methods of these MOFs and also to improve the pharmacokinetics and targeting efficiency.

References

1. Brandt, P., Brimah, A.K., Fischer, R.D., Partially Reversible Intercalation of Ferrocene in a Zeolite-like Host Lattice Consisting of Fe(CN)6 and Me3Sn

Units. *Angew. Chem. Int. Ed. Engl.*, 27, 1521–1522, 1988, doi: 10.1002/anie.198815211.

2. Taylor-Pashow, K.M.L., Della Rocca, J., Xie, Z., Tran, S., Lin, W., Postsynthetic modifications of iron-carboxylate nanoscale metal–organic frameworks for imaging and drug delivery. *J. Am. Chem. Soc.*, 131, 14261–14263, 2009, doi: 10.1021/ja906198y.
3. Horcajada, P., Chalati, T., Serre, C., Gillet, B., Sebrie, C., Baati, T., Eubank, J.F., Heurtaux, D., Clayette, P., Kreuz, C., Chang, J.-S., Hwang, Y.K., Marsaud, V., Bories, P.-N., Cynober, L., Gil, S., Férey, G., Couvreur, P., Gref, R., Porous metal–organic-framework nanoscale carriers as a potential platform for drug delivery and imaging. *Nat. Mater.*, 9, 172–178, 2010, doi: 10.1038/nmat2608.
4. Cai, W.T., Chu, C.-C., Liu, G., Wang, Y.-X.J., Metal–Organic Framework-Based Nanomedicine Platforms for Drug Delivery and Molecular Imaging. *Small*, 11, 4806–4822, 2015, doi: 10.1002/smll.201500802.
5. Horcajada, P., Serre, C., Vallet-Regí, M., Sebban, M., Taulelle, F., Férey, G., Metal–Organic Frameworks as Efficient Materials for Drug Delivery. *Angew. Chem. Int. Ed.*, 45, 5974–5978, 2006, doi: 10.1002/anie.200601878.
6. Horcajada, P., Serre, C., Maurin, G., Ramsahye, N.A., Balas, F., Vallet-Regí, M., Sebban, M., Taulelle, F., Férey, G., Flexible porous metal–organic frameworks for a controlled drug delivery. *J. Am. Chem. Soc.*, 130, 6774–6780, 2008, doi: 10.1021/ja710973k.
7. Wu, Y., Zhou, M., Li, S., Li, Z., Li, J., Wu, B., Li, G., Li, F., Guan, X., Magnetic metal–organic frameworks: γ-Fe2O3@MOFs via confined *in situ* pyrolysis method for drug delivery. *Small*, 10, 2927–2936, 2014, doi: 10.1002/smll.201400362.
8. Sun, C.-Y., Qin, C., Wang, X.-L., Yang, G.-S., Shao, K.-Z., Lan, Y.-Q., Su, Z.-M., Huang, P., Wang, C.-G., Wang, E.-B., Zeolitic Imidazolate framework-8 as efficient pH-sensitive drug delivery vehicle. *Dalton Trans.*, 41, 6906–6909, 2012, doi: 10.1039/c2dt30357d.
9. He, L., Wang, T., An, J., Li, X., Zhang, L., Li, L., Li, G., Wu, X., Su, Z., Wang, C., Carbon nanodots@zeolitic imidazolate framework-8 nanoparticles for simultaneous pH-responsive drug delivery and fluorescence imaging. *CrystEngComm.*, 16, 3259–3263, 2014, doi: 10.1039/C3CE42506A.
10. Vasconcelos, I.B., da Silva, T.G., Militão, G.C.G., Soares, T.A., Rodrigues, N.M., Rodrigues, M.O., da Costa, N.B., Freire, R.O., Junior, S.A., Cytotoxicity and slow release of the anti-cancer drug doxorubicin from ZIF-8. *RSC Adv.*, 2, 9437–9442, 2012, doi: 10.1039/C2RA21087H.
11. Ren, H., Zhang, L., An, J., Wang, T., Li, L., Si, X., He, L., Wu, X., Wang, C., Su, Z., Polyacrylic acid@zeolitic imidazolate framework-8 nanoparticles with ultrahigh drug loading capability for pH-sensitive drug release. *Chem. Commun.*, 50, 1000–1002, 2013, doi: 10.1039/C3CC47666A.
12. Zhuang, J., Kuo, C.-H., Chou, L.-Y., Liu, D.-Y., Weerapana, E., Tsung, C.-K., Optimized metal–organic-framework nanospheres for drug delivery:

Evaluation of small-molecule encapsulation. *ACS Nano*, 8, 2812–2819, 2014, doi: 10.1021/nn406590q.

13. Lyu, F., Zhang, Y., Zare, R.N., Ge, J., Liu, Z., One-pot synthesis of protein-embedded metal–organic frameworks with enhanced biological activities. *Nano Lett.*, 14, 5761–5765, 2014, doi: 10.1021/nl5026419.
14. Sun, C.-Y., Qin, C., Wang, C.-G., Su, Z.-M., Wang, S., Wang, X.-L., Yang, G.-S., Shao, K.-Z., Lan, Y.-Q., Wang, E.-B., Chiral nanoporous metal–organic frameworks with high porosity as materials for drug delivery. *Adv. Mater. Weinheim*, 23, 5629–5632, 2011, doi: 10.1002/adma.201102538.
15. An, J., Geib, S.J., Rosi, N.L., Cation-triggered drug release from a porous zinc-adeninate metal–organic framework. *J. Am. Chem. Soc.*, 131, 8376–8377, 2009, doi: 10.1021/ja902972w.
16. He, C., Lu, K., Liu, D., Lin, W., Nanoscale Metal–Organic Frameworks for the Co-Delivery of Cisplatin and Pooled siRNAs to Enhance Therapeutic Efficacy in Drug-Resistant Ovarian Cancer Cells. *J. Am. Chem. Soc.*, 136, 5181–5184, 2014, doi: 10.1021/ja4098862.
17. Lu, K., He, C., Lin, W., Nanoscale Metal–Organic Framework for Highly Effective Photodynamic Therapy of Resistant Head and Neck Cancer. *J. Am. Chem. Soc.*, 136, 16712–16715, 2014, doi: 10.1021/ja508679h.
18. Morris, W., Briley, W.E., Auyeung, E., Cabezas, M.D., Mirkin, C.A., Nucleic acid–metal organic framework (MOF) nanoparticle conjugates. *J. Am. Chem. Soc.*, 136, 7261–7264, 2014, doi: 10.1021/ja503215w.
19. Sun, C.-Y., Qin, C., Wang, X.-L., Su, Z.-M., Metal–organic frameworks as potential drug delivery systems. *Expert Opin. Drug Deliv.*, 10, 89–101, 2013, doi: 10.1517/17425247.2013.741583.

5

Role of MOFs as Electro/-Organic Catalysts

Manorama Singh[1]*, Ankita Rai[2], Vijai K. Rai[1], Smita R. Bhardiya[1] and Ambika Asati[1]

[1]*Department of Chemistry, Guru Ghasidas Vishwavidyalaya, Bilaspur (C G), India*
[2]*Ankita Rai, School of Physical Sciences, Jawaharlal Nehru University, New Delhi, India*

Abstract
Recently, MOFs are fascinating materials due to exclusive qualities (i.e., metallic sites and large surface area, etc.), which makes them to be potentially suitable for applying in catalysis, gas storage, drug delivery, sensors, etc. Also, functionalized MOFs are "electroactive frameworks" with unique electrochemical activity and electrocatalytic behavior and acts as an excellent electrocatalyst. The salient features of MOF's materials are controllable porosity, good stability, good conductivity, diversity, and tunability in their structures, highly robustness, etc. Several types of interaction such as H-bond, π–π interactions, van der Waals interaction, etc. are possible in framework of MOFs materials. This chapter efforts on basically using MOFs materials in electrochemical sensing as well as electro-organic synthesis.

Keywords: Metal–organic framework, electrocatalysis, organic, sensor

5.1 What Is MOFs

MOFs refers to metal–organic frameworks, which are hybrid constituents of new generation, which consists of inorganic transition metal ions (may or may not be in cluster form) and organic molecules (as in linker form or as in bridging ligands), in which the organic ligands (or linkers) are groups that can donate multiple lone pairs of electrons (act as Lewis base) to the metal ion (act as Lewis acid) [1]. They are like porous coordination polymeric materials comprised of metal ion lumps and organic ligands. The structure of MOFs

**Corresponding author*: manoramabhu@gmail.com

Inamuddin, Rajender Boddula, Mohd Imran Ahamed and Abdullah M. Asiri (eds.) Applications of Metal–Organic Frameworks and Their Derived Materials, (109–120) © 2020 Scrivener Publishing LLC

permits it for synthetic tenability and provides good structural and chemical control. Generally, transitions metal ions offer various coordination numbers with different geometries, like Y- or T-shaped, linear, square pyramidal/planar, tetrahedral, trigonal–bipyramidal, trigonal–prismatic, pentagonal–bipyramidal, and octahedral. There are a wide variety of adoptions for the organic linker and firmed-skeleton containing ligands [2].

The classification of MOFs was generally done on the basis of synthesis, crystal structures, frameworks, stimuli, etc. Conventional synthesis of MOFs can be done by solvothermal synthesis while unconventional synthesis by microwave assisted synthesis, spray drying, microfluidic synthesis, microemulsion synthesis, direct coupling, electrospinning, hydrothermal, and solvothermal synthesis [3, 4].

The significant properties of MOF's are high porosity, good stability, particle morphology, good conductivity, diversity, tunability in their structures, high robustness, etc., which can be used for applications in specific area at various conditions [5, 6]. In recent years, MOFs have been extensively explored applying in the variety of field: catalysis [6], chemical sensing and gas storage [7], electro-chemical sensing [8], biomedicine [9]. The stability aspects of MOFs can be defined in the terms of chemical, thermal and mechanical stability. MOFs show inertness in terms of ability to maintain their structure, behavior, when exposed to chemicals, high temperature, and mechanical force, etc. In recent trends, porous structure of MOFs and their stability, larger active surface area and capability to get functionalized easily make them a good candidate for all above mentioned applications, they are being used as selective catalysts in different reactions [10]. The first themed issue dedicated to MOFs was published in 2009 and got a tremendous success with record number of citations received [11]. Some more additional themed issues and monographs focusing on MOFs have also been published in various research journals [12, 13].

Extensive studies of MOFs have been done in the organic catalysis as heterogeneous catalysts and in electrocatalysis for electrochemical sensing [6]. Certain conduits of MOFs, being of different sizes and shapes, make it a promising material as catalyst [14]. There are several advantages of MOFs to be used in electro-/catalysis by using its extra ordinary properties like large surface areas, ordered structures, adjustable chemical functionality, and abundance of active sites [7] make them as a promising modifier in electrochemical sensors. Sensing devices are required for sensitive detection of analyte and organic pollutants in various applications including industrial processes as well as research field, such as chemically hazardous materials detection, bio molecule detection, diagnostics, and environmental hazardous elements detection.

In this book chapter, it is our aim to offer highlights of new contributions of MOFs and their hybrid materials in electrochemical applications in electro/-organo catalysis. There are coordination bonds or/and other weak supportive interactions present in MOFs-framework i.e. pi-pi interactions, hydrogen bond, and van der Waals forces. Poor water stability of MOFs inhibits its use for electrochemical applications. Looking to these challenges, MOFs based hybrid materials are in interest [15].

5.2 MOFs as Electrocatalyst in Sensing Applications

Several MOFs exhibit electrochemical activity if appropriate metal ions and ligands are associated [16]. Nowadays, electroconductive MOFs have been designed and prepared but few works are explored as "electrocatalyst" in electrochemical sensing. Several additional components such as conductive carbon materials, conducting polymers, etc., can further improve the electrochemical nature of MOFs [17].

In this regard, Fernandes *et al.* 2014 reported the new composite material prepared by the combination of the negatively charged Fe-doped silicon tungstate and MIL-101(Cr) framework. The modified electrode exhibited admirable electro-catalytic behavior for the sensing of both nitrite, iodate, and ascorbic acid, which confirming its high flexibility in several electrochemical reactions [18]. Hosseini *et al.* 2013a reported Gold-SH-Silica NPs@ Cu-MOF based electrochemical sensor for the sensing of L-cysteine [19]. Hosseini *et al.* 2013b has also reported a new electrocatalyst, Gold-SH-silica NPs/MOFs modified glassy carbon electrode to fabricate a sensor toward the efficient electrocatalysis of hydrazine to nitrogen in water samples. The salient properties of this sensor were low limit of detection 0.01 μM in a linear range of 0.04–500 μM with good stability, good repeatability, and high sensitivity [20]. Campbell *et al.* 2015 prepared a new electroconductive 2D-MOFs Cu_3(Hexaiminotriphenylene)$_2$ for chemi-resistive sensing [21]. Wang *et al.*, 2013 has synthesized a sensitive ZnO_4(1,4-benzenedicarboxylate)$_3$(Metal Organic framework-V) modified CPE applying for sensing of "lead" a heavy metal ion [22].

Zhang *et al.* 2014 reported a nanocomposite using Co-MOF and microporous-C for the electrocatalysis of N_2H_4 as well as $C_6H_5NO_2$. Incorporation of microporous-C significantly facilitated the electrochemical properties of the material. More importantly, this material displayed an exceptional electrocatalysis for electrochemical sensing N_2H_4 as well as $C_6H_5NO_2$. DPV for reduction of $C_6H_5NO_2$ was recorded in the linear range of 0.5×10^{-6}–15×10^{-6} M and 15×10^{-6} to 235×10^{-6}. The limit of detection was obtained

0.21×10^{-6} [23]. Wen *et al.* 2010 reported a new conjugated MOFs using 2,20,4,40-biphenyltetracarboxylic acid and a uni-nodal boron nitride net (BNN) and this was the first time when MOFs was employed as a pesticide sensor *via* stripping voltammetry analysis. This composite has detected methyl parathion from 0.01 to 0.5 μg/mL and limit of detection 0.0132 μg/mL at CPE [24]. Another highly selective and sensitive Cu/(2,2′-bipyridine)/1,3,5-tricarboxylate was applied on MWCNTs by Zhou *et al.* 2014 and successfully used for the hydrogen peroxide (H_2O_2) detection. The sensing of H_2O_2 was recorded with a linear range of 3×10^{-6}–70×10^{-6} M and 70×10^{-6}–$30{,}000 \times 10^{-6}$ M. The limit of detection was obtained to be 0.46×10^{-6} M with good stability and reproducibility [25]. The electrochemically stable Cu-MOFs in macroporous carbon (MPC) was reported by Zhang and his team, 2014, and further the detection of hemoglobin and ascorbic acid was done as electrochemical sensor [26]. Nasir *et al.* 2017 has published a research article in which they established the utilization of *t*-butyllithium exfoliated TMDs ($MoSe_2$, WSe_2, MoS_2, and WS_2) as a display place for the enzymatic electrocatalytic determination of "fenitrothion" (organophosphate) [27]. Wei *et al.* 2018 has proved that large amount of enzymes could be loaded in MOFs based porous carbon, which showed the outstanding conductivity improving the electron transfer and analytical performance of modified electrode. They prepared Fe_2O_3@C from strong heating of Fe-1,3,5-benzenentricarboxylate. Further, sensing of paraoxon and H_2O_2 was recorded by immobilization of acetylcholinesterase (AChE) and mayoglobin (Mb) on Fe_2O_3@C/ionic liquid (IL)/nafion modified CPE, respectively [28].

A novel core–shell nanocomposite was reported by Zhang *et al.*, 2017 using mesoporous Fe_3O_4@C nanocapsules (NCps) and Fe (III)-based MOF fabricating an electrochemical probe for sensing of trace amounts of heavy metal ions in water. Synthesis of mesoporous Fe_3O_4@mC NCps was done by calcination of hollow Fe_3O_4@C comprising with Fe-MOF. The Fe-MOF@ mesoporous Fe_3O_4@mC nanocomposite showed large active surface area and outstanding electrochemical activity with good water stability. Furthermore, the fabricated apta-sensor revealed a linear relation between response current as a function of log C (of As^{3+} & Pb^{2+}) with the limit of detection 2.27 (Pb^{2+}) and 6.73 pM (As^{3+}) in the linear range 0.01×10^{-9} to 10.0×10^{-9} M. The reported apta-sensor exhibited negligible influence of foreign substances with satisfactory repeatability, signifying capable potential in monitoring environmental issues [29].

Microperoxidase-XI was encapsulated in porous coordination network-333 based MOFs to employ it in the determination of H_2O_2. In order to characterize the electrochemical activity of material, differential pulse voltammetry and cyclic voltammetry were recorded and performance of encapsulated microperoxidase-XI was found to be better as compared to enzyme. The limit of detection was obtained 0.127×10^{-6} M in a calibration range from 0.387×10^{-3} to 1.725×10^{-3} M with high selectivity and good storage stability achieved [30].

An efficient immobilization of horseradish peroxidase (HRP) was reported in a durable, simple and sensitive boronic acid-MOFs (MIL-100(Cr)-B) nanocomposite to develop a H_2O_2 biosensor. Salient features of MIL-100(Cr)-B are awfully large surface area, hierarchical porous structure, and adequate recognition centers, which increased the loading of HRP with negligible leakage. In the meantime, the immobilized HRP exhibited outstanding electro-reduction of H_2O_2 and enhanced stability. A calibration range of 0.5×10^{-6}–3000×10^{-6} M was observed and limit of detection was 0.1×10^{-6} M with good selectivity as well as stability. Real sample analysis was also performed for H_2O_2 detection in living cells [31]. A unique multi-functional MOFs based composite "Ag@Zn-thiosalicylate (TSA)," was reported to immobilize glucose oxidase and myo-globin for electrochemical sensing application. Ag@Zn-thiosalicylate/ionic liquid (IL)/CPE were prepared, which was utilized as an excellent platform for the sensing of nitrite (1.3×10^{-6} M–1660×10^{-6} M and 2262×10^{-6} M–$133{,}000 \times 10^{-6}$ M), hydrogen peroxide (0.3–$20{,}000 \times 10^{-6}$ M), glucose (2.0–1022×10^{-6} M) with a limit of detection 0.5×10^{-6} M (nitrite), 0.08×10^{-6} M (H_2O_2), 0.8×10^{-6} M (glucose). Therefore, Ag@Zn-thiosalicylate was found to be a perfect material for sensitive electrochemical application [32].

In other work, Copper-hemin MOFs was utilized on chitosan-reduced graphene oxide to fabricate an electrochemical H_2O_2 sensor, which showed exceptional electrical conductivity and significant peroxidase-like biological activity. The crystalline structure of the Copper-hemin MOFs boosted the electrical conductivity and therefore, Copper-hemin MOFs/Chitosan-reduced graphene oxide nanocomposite was employed in electrocatalytic reduction of H_2O_2, which was better than other mimic enzymes. The Copper-hemin MOF/Chitosan-reduced graphene oxide modified electrode exhibited low limit of detection 0.019×10^{-6} M in linear range of 0.065×10^{-6}–410×10^{-6} M [33].

Sherino *et al.* 2018 reported a novel Nickel-MOFs (using ligand piperazine and linker adipic acid) based nanocomposite to fabricate a H_2O_2 probe Under optimized conditions, the fabricated carbon paste electrode determined the concentrations of hydrogen peroxide in a calibration range of 0.004×10^{-3}–60×10^{-3} M with detection limit of 0.0009×10^{-3} M. The performance of modified electrode was excellent in terms of good repeatability, high selectivity, and storage stability. Real sample analysis was also performed on modified CPE for the H_2O_2 detection with satisfactory recoveries [34].

5.3 MOFs as Organic Catalysts in Organic Transformations

MOFs are the most fascinating heterogeneous catalysts due to their degree of tunability which makes them better in comparison to other solid catalysts. An exponential growth in MOF complexity has been witnessed in the last decade. Their performances are outstanding in the ranges of temperature due to different topology and pore size. To alter the properties of MOF in a predictable manner, controlled insertion of defects is required, as organic catalysis is sluggish in the presence of MOFs. Advanced synthesis processes *viz.*, enantioselective catalysis, olefin metathesis, or C–H activation are still not attempted up to satisfactory level [35, 36]. First, Fujita *et al.* 1994 reported and discussed preparation of 2D- MOFs with Cd (II) and 4,4′-bipyridine for cyanosilylation of aldehydes [37]. MOFs not only provide larger surface areas with boosted activity in comparison to other metal oxides materials (basic), but they also offer selectivity of size and shape, which are the significant parameter in organic catalysis and separation both. Literature also reports that approximately 90% of chemical reactions utilize heterogeneous catalysts [38].

Wang *et al.* 2009 has reported that MOF is assembled by a post-synthetic modification (PSM) scheme followed by functionalization of chemical reagents along with its retained lattice structure *via* isocyanate condensations, amide couplings, "click" chemistry and other reactions. In addition, PSM is successful with MOFs ranging from IRMOF-3 to ZIF-90, which indicates the wide applicability of this new catalyst. A huge literature reports on PSM opens-up a new synthetic methodology, which would offer a significant part in the progress of efficient catalysis based on MOFs [39].

Jia *et al.* 2013 has studied electrocatalytic activity of copper-MOF for oxidative carbonylation of CH_3OH [40]. Dakhashinamoorthy *et al.* 2014 discussed MOFs as heterogeneous catalysts to develop synthetic processes for various *N*-heterocycles [41]. Xu *et al.* 2015 utilized host-guest supramolecular interactions between varied metallic ions (positively charged) and polyoxometalate (negatively charged). In addition, composite of MOF with noble metal can be constructed using this reaction scheme to improve the performance of the noble metal for catalysis [42]. Rao *et al.* 2017 discussed beautifully the transformation of indoles with nitroalkenes *via* Friedel–Crafts alkylation by donation of H-bond (in MOFs) [43]. Larasati *et al.*

2017 have proposed manufacturing biodiesels using biomass *via* catalytic-esterification and transesterification. Herein, MOFs was synthesized Zr (IV) and benzene-1,3,5-tricarboxylic acid as a linker by refluxing followed by solvothermal method [44].

Recently, Iqbal *et al.* 2019 reported one-pot synthesis of Cobalt/Iron-MOFs by taking 2-aminoterephthalic acid and stoichiometric ratio of Fe and Co salts, which was utilized to perform the dehydrogenation reactions and in transformation of nitroarenes [45]. Furthermore, Arnanz *et al.* 2012 have also synthesized a bifunctional MOF catalyst containing Pd with Copper/benzene-1,3,5-tricarboxylate, which offers sequential click/Sonogashira reaction starting from NaN_3, 2-iodobenzylbromide and alkynes to get satisfactory yields of 8H-[1,2,3] triazolo[5,1-a] isoindoles [46]. Fu *et al.* 2012 has disclosed photocatalysis of CO using simple light irradiation and biodegradable amine functionalized Titanium based-MOFs [47]. Huang *et al.* 2012 reported preparation of PdNPs encapsulated in MIL-101(Cr)–NH_2 catalyst, which proved itself as an excellent catalyst for dehalogenation of aryl chlorides in aqueous medium [48]. Noh *et al.* 2016 reported a very stable Mo (VI) oxide-MOFs for epoxidation of cyclohexene [49]. 1-Butene was prepared by selecting dimerizing ethylene using an efficient catalyst [50].

Yang *et al.* 2015 reported MOFs for supporting novel molecular catalysts of Iridium complexes with NU-1000 and UiO-66, respectively [51]. Comito *et al* 2016 discussed the preparation of heterogeneous catalyst by cation exchange in a MOFs for olefin polymerization [52]. One of the interesting applications of MOFs in heterogeneous catalysis is metalloporphyrin based PIZA-3[$Mn_2(TpCPP)_2Mn_3$] [53]. PIZA-3 is proficiently used for hydroxylating alkanes and in olefins epoxidation. Binaphthyl metal complexes and Schiff-base are reported, which have also been united into MOFs utilizing in addition of diethyl zinc ($ZnEt_2$) to aromatic aldehydes and epoxidation of olefin, respectively [54]. Recently, an excellent review was reported by Remya and Kurian, 2019 on synthesis and catalytic applications of MOFs in various transformations [55].

5.4 Conclusion and Future Prospects

MOFs find multifaceted applications catalysts as well as in the fabrication of sensors. Biomolecules (as organic linkers) containing MOFs are interesting as compared to traditional organic linkers. However, biomolecules are biocompatible and easily recyclable. They also possess exclusive characteristics such as self-assembly characteristic, chirality, and catalytic properties. In forthcoming, these properties can be used to prepare green route catalyst and sensor composite. Designing MOFs offers platform to use their constructions and properties to apply in the area of electro-/catalytic purpose. However, several challenges are still there that need to be reduced so as to synthesize MOFs materials. Future works should figure out the finest way to transfer other components onto the MOFs. We are expecting to discover new structures of MOFs-derived materials and we have faith in that the ensuing novel materials will be extended to other unexplored research areas to assorted applications.

References

1. Li, B., Li, H., Wen, M., Cui, Y., Zhou, W., Qian, G., Chen, B., Emerging Multifunctional Metal–Organic Framework Materials. *Adv. Mater.*, 28, 8819–8860, 2016.
2. Li, J., Wang, X., Zhao, G., Chen, C., Chai, Z., Alsaedi, A., Hayat, T., Wang, X., Metal–organic framework-based materials: Superior adsorbents for the capture of toxic and radioactive metal ions. *Chem. Soc. Rev.*, 2017. 10.1039/c7cs00543a.
3. Tella, A.C. and Aaron, I.S., Synthesis and application of metal–organic frameworks materials. *Acta Chim. Pharm. Indica*, 2, 75–81, 2012.
4. Baumann, A.E., Burns, D.A., Liu, B., Thoi, V.S., Metal–organic framework functionalization and design strategies for advanced electrochemical energy storage devices. *Commun. Chem.*, 2019.
5. Lee, Y., Kim, J., Ahn, W., Synthesis of metal organic frameworks: A mini review. *Korean J. Chem. Eng.*, 30, 1667–1680, 2013.
6. Furukawa, H., Cordova, K.E., Keeffe, M.O., Yaghi, O.M., The Chemistry and Applications of Metal–Organic Frameworks. *Science*, 341, 1230444–1230456, 2013.
7. Kreno, L.E., Leong, K., Farha, O.K., Allendorf, M., Duyne, R.P.V., Hupp, J.T., Metal_Organic Framework Materials as Chemical Sensors. *Chem. Rev.*, 112, 1105–1125, 2012.
8. Fotouhi, L. and Naseri, M., Recent Electroanalytical Studies of Metal–Organic Frameworks: A Mini-Review. *Crit. Rev. Anal. Chem.*, 46, 323–331, 2016.
9. Singh, R. and Geetanjali, Applications of Nanocomposite Materials in Drug Delivery, doi: http://doi.org/10.1016.B978-0-12-813741-3.00026-1.
10. Butova, V.V., Soldatov, M.A., Guda, A.A., Lomachenko, K.A., Lamberti, C., Metal-organic frameworks: Structure, properties, methods of synthesis and characterization. *Russ. Chem. Rev.*, 85, 280–307, 2016.
11. Long, J.R. and Yaghi, O.M., The pervasive chemistry of Metal Oraganic Frameworks. *Chem. Soc. Rev.*, 38, 1213–1214, 2009.

12. Falcaro, P., Ricco, R., Doherty, C.M., Liang, K., Hill, A.J., Styles, M.J., MOF positioning technology and devices. *Chem. Soc. Rev.*, 43, 5513, 2014.
13. Zhou, H.C.J. and Kitagawa, S., Metal–Organic Frameworks (MOFs). *Chem. Soc. Rev.*, 43, 5415–5418, 2014.
14. Dey, C., Kundu, T., Biswal, B.P., Mallick, A., Banerjee, R., Crystalline metal–organic frameworks (MOFs): Synthesis, structure and functions. *Acta Crystallogr.*, 70, 3–10, 2014.
15. Cai, H., Zhou, J., Kitagawa, S., Metal Organic Frameworks. *Chem. Soc. Rev.*, 43, 5415–5418, 2014.
16. Mac-Gillivray, L.R., *Metal Organic Frameworks: Design and application*, John Wiley & Sons, 2010.
17. Marpuang, F., Konstantinov, K., Yamauchi, Y., Hossain, M.S.A., Na, J., Kim, J., Recent advances of MOF derived nanoporous carbon materials. *Chemistry—An Asian Journal*, 2019, doi: 10.1002/asia.201900026.
18. Fernandes, D.M., Granadeiro, C.M., Sousa, P.M.P., Grazina, R., Moura, J.J., Silva, G.P., Paz, F.A.A., Silva, L.C., Balula, S.S., Freire, C., SiW11Fe@MIL-101(Cr) Composite: A Novel and Versatile Electrocatalyst. *ChemElectroChem*, 21, 1293–1300, 2014.
19. Hosseini, H., Ahmar, H., Dehghani, A., Bagheri, A., Tadjarodi, A., Fakhari, A.R.A., Novel Electrochemical Sensor Based on Metal–Organic Framework for Electro-Catalytic Oxidation of L-Cysteine. *Biosens. Bioelectron.*, 42, 426–429, 2013b.
20. Hosseini, H., Ahmar, H., Dehghani, A., Bagheri, A., Fakhari, A.R., Amini, M.M., Au-SH-SiO_2 Nanoparticles Supported on Metal–Organic Framework (Au-SH-SiO2@ Cu-MOF) as a Sensor for Electrocatalytic Oxidation and Determination of Hydrazine. *Electrochim. Acta*, 88, 301–309, 2013a.
21. Campbell, M.G., Sheberla, D., Liu, S.F., Swager, T.M., Dincă, M., Cu3(Hexaiminotriphenylene)2: An electrically conductive 2d metal–organic framework for chemiresistive sensing. *Angew. Chem. Int. Ed.*, 54, 4349–4352, 2015.
22. Wang, Y., Wua, Y., Xie, J., Hua, X., Metal–organic framework modified carbon paste electrode for lead sensor. *Sens. Actuators B*, 177, 1161–1166, 2013.
23. Zhang, Y., Bo, X., Nsabimana, A., Han, C., Li, M., Guo, L., Electrocatalytically active cobalt based metal organic framework with incorporated macroporous carbon composite for electrochemical applications. *J. Mater. Chem. A*, 3, 732–738, 2015.
24. Wen, L.L., Wang, F., Leng, X., Wang, L., Wang, Y., Gong, J.M., Li, D.M., Efficient Detection of Organophosphate Pesticide Based on a Metal–Organic Framework Derived from Biphenyl tetracarboxylic Acid. *Cryst. Growth Des.*, 10, 2835–2818, 2010.
25. Zhou, E., Zhang, Y., Li, Y., He, X., Cu(II)-Based MOF Immobilized on Multiwalled Carbon Nanotubes: Synthesis and Application for Nonenzymatic Detection of Hydrogen Peroxide with High Sensitivity. *Electroanalysis*, 26, 2526–2533, 2014.

26. Zhang, Y., Nsabiman, A., Zhu, L., Bo, X., Han, C., Li, M., Guo, L., Metal organic frameworks/macroporous carbon composites with enhanced stability properties and good electrocatalytic ability for ascorbic acid and hemoglobin. *Talanta*, 129, 55–62, 2014.
27. Nasir, M.Z.M., Martinez, C.C.M., Sofer, J.K., Pumera, M., Two-dimensional 1T-Phase transition metal dichalcogenides as nanocarriers to enhance and stabilize enzyme activity for electrochemical pesticide detection. *ACS Nano*, 11, 5774–5784, 2017.
28. Wei, W., Dong, S., Huang, G., Xie, Q., Huang, T., MOF-derived Fe_2O_3 nanoparticle embedded in porous carbon as electrode materials for two enzyme-based biosensors. *Sens. and Actauat B: Chem.*, 260, 189–197, 2018.
29. Zhan, Z., Ji, H., Song, Y., Zhang, S., Liu, S.C., Fe(III)-based metal–organic framework-derived core-shell nanostructure: Sensitive electrochemical platform for high trace determination of heavy metal ions. *Biosens. Bioelectron.*, 94, 358–364.
30. Gong, C., Shen, Y., Chen, J., Song, Y., Chen, S., Song, Y., Wang, L., Microperoxidase-11@PCN-333 (Al)/three-dimensional microporous carbon electrode for sensing hydrogen peroxide. *Sens. and Actauat B: Chem.*, 239, 890–897, 2017.
31. Wang, Y., Qing, X., Zhou, Q., Zhang, Y., Liu, Q., Liu, K., Wang, W., Li, M., Lu, Z., Chen, Y., Wang, D., A novel biosensor based on boronic acid functionalized metal organic frameworks for the determination of hydrogen peroxide released from living cells. *Biosens. Bioelectron.*, 95, 131, 2017.
32. Dong, S., Zhang, D., Suo, G., Wei, D., Huang, T., Exploiting multi-function Metal–Organic Framework nanocomposite Ag@Zn-TSA as highly efficient immobilization matrixes for sensitive electrochemical biosensing. *Anal. Chim. Acta*, 934, 203–211, 2016.
33. Wang, L., Yang, H., He, J., Zhang, Y., Yu, J., Song, Y., Cu-Hemin metal organic frameworks/chitosan-reduced graphene oxide nanocomposites with peroxidase like activity for electrochemical sensing. *ECA*, 213, 691–697, 2016.
34. Sherino, B., Mohamad, S., Abdul Halim, S.N., Abdul Manan, N.S., Electrochemical detection of hydrogen peroxide on a new microporous Ni–metal organic framework material–carbon paste electrode. *Sens. and Actauat B: Chem.*, 254, 1148–1156, 2018.
35. Fotouhi, L. and Naseri, M., Recent electroanalytical studies of Metal Organic Frameworks: Mini review. *Crit. Rev. Anal. Chem.*, 46, 323–331, 2016.
36. Liu, J., Chen, L., Cui, H., Zhang, J., Zhang, L., Su, C.Y., Applications of Metal Organic Frameworks in heterogeneous supramolecular catalysis. *Chem. Soc. Rev.*, 43, 6011–6061, 2014.
37. Fujita, M., Kwon, Y.J., Washizu, S., Ogura, K., Preparation, clathration ability, and catalysis of a two-dimensional square network material composed of cadmium(II) and 4,4′-bipyridine. *J. Am. Chem. Soc.*, 116, 1151–1152, 1994.
38. Czaja, A.U., Trukhan, N., Muller, U., Industrial applications of metal–organic frameworks. *Chem. Soc. Rev.*, 38, 1284–1293, 2009.

39. Wang, Z. and Cohen, S.M., Posts synthetic modification of metal–organic framework. *Chem. Soc. Rev.*, 38, 1315–1329, 2009.
40. Jia, G., Gao, Y., Zhang, W., Wang, H., Cao, Z., Li, C., Liu, J., Metal Organic Frameworks as Heterogeneous Catalysts for Electrocatalytic Oxidative Carbonylation of Methanol to Dimethyl Carbonate. *Electrochem. Commun.*, 34, 211–214, 2013.
41. Dhakshinamoorthy, A. and Garcia, H., Metal Organic Frameworks as solid catalysts for the synthesis of nitrogen containing heterocycles. 43, 5750–5765, 2014.
42. Xu, X., Lu, Y., Yang, Y., Nosheen, F., Wang, X., Tuning the growth of metal–organic framework nano crystals by using polyoxometalates as coordination modulators. *Sci. China Mater.*, 58, 370–377, 2015.
43. Rao, P.C. and Mandal, S., Friedel–Crafts alkylation of indoles with nitroalkenes through hydrogen-bond-donating metal–organic framework. *ChemCatChem*, 9, 1172–1176, 2017.
44. Larasati, I.D., Winarni, F.R., Putri, Q.A., Hanif, W.W.L., Synthesis of Metal–organic Frameworks based on Zr4+ and Benzene 1,3,5-Tricarboxylate linker as heterogeneous catalyst in the esterification reaction of palmitic acid. *Mater. Sci. Eng.*, 214, 01200–01207, 2017.
45. Iqbal, B., Saleem, M., Arshad, S.N., Rashid, J., Hussain, N., Zaheer, M., One-Pot Synthesis of Heterobimetallic Metal–Organic Frameworks (MOFs) for Multifunctional Catalysis. *Chem. Eur. J.*, 25, 1–10, 2019.
46. Arnanz, A., Sierra, M.P., Corma, A., Iglesias, M., Sánchez, F., Bifunctional Metal Organic Framework Catalysts for multistep reactions: MOF-Cu(BTC)-[Pd] catalyst for one-pot heteroannulation of acetylenic compounds. *Adv. Synth. Catal.*, 354, 1347–1355, 2012.
47. Fu, Y., Sun, D., Chen, Y., Huang, R., Ding, Z., Fu, X., Li, Z., An Amine-Functionalized Titanium Metal–Organic Framework Photocatalyst with Visible Light Induced Activity for CO Reduction. *Angew. Chem.*, 124, 3420–3423, 2012.
48. Huang, Y., Liu, S., Lin, Z., Li, W., Li, X., Cao, R., Facile Synthesis of Palladium Nanoparticles Encapsulated in Amine-Functionalized Mesoporous Metal–Organic Frameworks and Catalytic for Dehalogenation of Aryl Chlorides. *J. Catal.*, 292, 111–117, 2012.
49. Noh, H., Cui, Y., Peters, A.W., Pahls, D.R., Ortuño, M.A., Vermeulen, N.A., Cramer, C.J., Gagliardi, L., Hupp, J.T., Farha, O.K., An exceptionally stable metal–organic framework supported molybdenum (vi) oxide catalyst for cyclohexene epoxidation. *J. Am. Chem. Soc.*, 138, 14720–14726, 2016.
50. Metzger, E.D., Brozek, C.K., Comito, R.J., Dincă, M., Selective dimerization of ethylene to 1-butene with a porous catalyst. *ACS Cent. Sci.*, 2, 148–153, 2016.
51. Yang, D., Odoh, S.O., Wang, T.C., Farha, O.K., Hupp, J.T., Cramer, C.J., Gagliardi, L., Gates, B.C., Metal–organic framework nodes as nearly ideal supports for molecular catalysts: nu-1000 and uio-66-supported iridium complexes. *J. Am. Chem. Soc.*, 137, 7391–7396, 2015.

52. Comito, R.J., Fritzsching, K.J., Sundell, B.J., Schmidt-Rohr, K., Dincă, M., Single-site heterogeneous catalysts for olefin polymerization enabled by cation exchange in a metal–organic framework. *J. Am. Chem. Soc.*, 138, 10232–10237, 2016.
53. Suslick, K. *et al.*, Microporous porphyrin solids. *Acc. Chem. Res.*, 38, 283–291, 2005.
54. Wu, C.D., Hu, A., Zhang, L., Lin, W., A homochiral porous metal–organic framework for highly enantioselective heterogeneous asymmetric catalysis. *J. Am. Chem. Soc.*, 127, 8940–8941, 2005.
55. Remya, V.R. and Kurian, M., Synthesis and catalytic applications of Metal Organic Frameworks: A review on recent literature. *Int. Nano Lett.*, 9, 17–29, 2019.

6

Application of MOFs and Their Derived Materials in Batteries

Rituraj Dutta and Ashok Kumar*

Department of Physics, Tezpur University, Napaam, Tezpur, Assam, India

Abstract

This chapter emphasizes the emerging field of metal–organic frameworks (MOFs) and their derived materials as ion conducting composite dielectric membranes in rechargeable batteries. The chapter deals with the structure, classification and properties of MOF and Ionic Liquid (IL) and their potential applications as composite ionic conductors in rechargeable batteries. MOFs incorporated with ILs can be treated as suitable ionic conductors as the interaction of IL ions with metal nodes and linker moieties of MOFs can control the ion dynamics of the nanocomposites. The chapter highlights the classification of polymer electrolytes and their historical perspectives with different strategies to develop nanocomposite polymer electrolytes. The theoretical aspects of ion transport mechanism in disordered composite polymer electrolytes have also been discussed. In last part of the chapter, authors emphasize the aspect of IL incorporated MOF based composite polymer electrolytes to be used as ion conducting membranes in rechargeable batteries. The dynamics of dielectric relaxation, ionic conductivity, electrochemical stability, and scaling behavior of ac conductivity of the IL incorporated different MOFs based composite polymer electrolyte membranes have been studied to understand the ion transport dynamics of the composite polymer electrolyte systems.

Keywords: Metal–organic framework, ionic liquid, polymer electrolyte, dielectric relaxation, ionic conductivity, electrochemical stability

**Corresponding author*: ask@tezu.ernet.in

Inamuddin, Rajender Boddula, Mohd Imran Ahamed and Abdullah M. Asiri (eds.) Applications of Metal–Organic Frameworks and Their Derived Materials, (121–176) © 2020 Scrivener Publishing LLC

6.1 Introduction

In today's rapidly growing society, energy storage and conversion devices are playing significant role for human and social development. Rechargeable batteries are inherent components of electronic devices to provide stable power supply. An electrolyte is the key component of an electrochemical device that provides a medium for ionic transport between anode and cathode. Also the electrolyte acts as dielectric or electrical insulator that enables electronic transport only through the outer circuit of the cell. Electrolytes are essentially ion conducting substances consisting of salts dissolved in any suitable medium. Electrolytes are formed by dissolving salts into a solvent where the individual components get dissociated by the thermodynamic interactions of solute molecules and the solvent. Liquid electrolytes have several disadvantages like bulky size and leakage of electrolyte and to overcome these disabilities solid electrolytes have been developed to be used in rechargeable batteries [1]. Polymer electrolytes are emergent solid electrolytes with physicochemical stability, shape versatility, and sustainability [2]. In recent times different types of ion conducting polymer electrolytes are being developed by researchers with enhanced ionic conductivity as well as electrochemical stability to be used in batteries.

Solid polymer electrolytes are complexes of lithium based salts such as lithium tetrafluoroborate ($LiBF_4$), lithium perchlorate ($LiClO_4$), lithium hexafluorophosphate ($LiPF_6$), etc., incorporated in polymer matrix such as polyvinylidene fluoride (PVdF), polyethylene oxide (PEO), poly(methyl methacrylate) (PMMA), etc. Solid polymer electrolytes possess several advantages such as mechanical and electrochemical stability, flexibility, and easy processing [3]. In Li salt based solid polymer electrolytes ionic motion takes place through the amorphous region of the segmental polymer network that in turn reduce the crystallinity of the system. The ionic transport phenomenon in solid electrolyte system is governed by various factors including degree of crystallinity, glass transition temperature, ion pair formation, interfacial polarization, etc. [4]. Incorporation of nanosized filler particles can increase the electrochemical performance of the composite polymer electrolytes. Filler particles may suppress the growth of resistive layers at the interface of electrodes and the electrolyte. The electrochemical properties of the composite polymer electrolytes are dependent of the structural topology of the filler particles. Weston and Steele [5] first reported the idea to develop the mechanical and electrochemical stability of conventional polymer electrolytes by incorporation of nanosized filler materials. They used α-$A1_2O_3$ as fillers to $LiC1O_4$–PEO complexes and achieved mechanical and electrochemical improvement in the composite polymer electrolytes. B. Scrosati *et al.* [6] reported the addition of $\gamma LiAlO_2$ in $Li[(CF_3SO_2)_2N]$ and PEO based polymer electrolytes that reveals enhancement in interfacial stability and reduction in crystallization rate. Liquan [7] reported the enhancement in ionic conductivity and mechanical stability in NaSCN–PEO complex on incorporation of powder α-Al_2O_3. W. Wieczorek *et al.* [8] investigated the effect of incorporation of Al_2O_3 filler particles on the electrochemical properties of NaI–PEO based polymer electrolytes. They proposed that the increase in ionic conductivity is governed by Lewis acid–base interaction of filler material and the host polymer. F. Croce *et al.* [9] investigated the effects of ceramic nanofillers of Al_2O_3 (5.8 nm particle size) and TiO_2 (13 nm particle size) on $LiClO_4$–PEO polymer electrolytes and achieved ionic conductivity of ~10^{-5} S cm^{-1} at room temperature and that was stable up to 15 days. J. Przyluski *et al.* [10] observed the enhancement in ionic conductivity in Al_2O_3 ceramic filler incorporated NaI–PEO polymer electrolytes upon reduction of size of filler below 4 μm.

Metal–organic frameworks (MOFs) are three dimensional crystalline microporous materials consisting of unsaturated metal clusters linked with organic molecules [11]. The organic linker molecules may have ditopic, tritopic, or polytopic functionalities depending on number of coordination sites. The organic linker moieties may contain different functional groups such as carboxylate, phosphate, sulfonate, nitrile, amine, etc., to form coordination bonds with the metal nodes [12]. Due to their large pore volume and high surface area MOFs are promising materials for thin film electrochromic devices, light harvesting, gas storage and sequestration, sensing, biomedical imaging and optical luminescence [13]. MOFs are considered as efficient filler materials to incorporate in polymer electrolytes. N. Angulakshmi *et al.* [14] reported the synthesis of PEO based composite polymer electrolytes by incorporating Magnesium based MOF (Mg-BTC) with lithium bistrifluoromethanesulfonylimide (LiTFSI). Incorporation of nanosized Mg-BTC MOF in polymer electrolyte matrix can improve ionic conductivity of the nanocomposite membranes up to two orders at 0°C as well as compatibility and thermal stability are also enhanced. The electrochemical cell based on the composite electrolyte membrane gives 110 mAh g^{-1} discharge capacity at 70°C with a current rate of 1 C.

Due to their tunable structural topology and excellent pore textures MOFs are preferable host materials as composite ionic conductors [15]. Ionic liquids (ILs) are weakly coordinated molten salts with a variety of anions and cations [16]. They are eco-friendly green solvents with specific properties such as negligible vapor pressure (10^{-10} Pa at 300 K), high ionic conductivity (~10^{-3} S Cm^{-1} at 300 K), large electrochemical window

(~6 V) and wide solubility range with high thermal stability (~300°C). ILs are considered as suitable guest materials to incorporate within the pores of MOFs to control the phase dynamics and ion mobility by tunable IL-MOF interaction [17].

ILs can be incorporated in polymer electrolytes be used as efficient ion conducting composite polymer electrolyte membranes in rechargeable batteries. In 1994, Carlin and co-workers [18] reported use of non-chloroaluminate IL in PVdF–HFP based gel polymer electrolytes to enhance the ionic conductivity. J. Fuller *et al.* [19] reported the effects of incorporation of two ILs $EMIMBF_4$ and $BMIMPF_6$ in PVdF–HFP based polymer electrolytes. Scott *et al.* [20] reported that imidazolium based ILs are excellent plasticizers that can improve thermal stability and reduce the glass transition temperature in poly (methyl methacrylate) based polymer electrolytes. T.E. Sutto *et al.* [21] investigated the effects of IL 1-*n*-propyl-2,3-dimethylimadizolium tetrafluoroborate and 1-*n*-propyl-2,3-dimethylimadizolium hexaflorophosphate separately into PEO based polymer electrolyte matrix with and without Li salt complex. They reported that PVdF–HFP based composite polymer electrolytes revealed better electrochemical performance than PEO based composite electrolytes. A. Noda and M. Watanabe [22] reported the *in situ* polymerization of vinyl monomers in ILs 1-butylpyridinium tetrafluoroborate and 1-ethyl-3-methylimidazolium tetrafluoroborate that can form mechanically stable polymer electrolyte membranes with ionic conductivity of $\sim 10^{-3}$ S cm^{-1} at 300 K. Lewandowski *et al.* [23] reported the electrochemical performance of polymer electrolytes based on poly(ethyleneoxide) (PEO), poly(vinylalcohol) (PVA), and poly(acrylonitrile) (PAN) upon incorporation of ILs 1-butyl-1-methyl-pyrrolidinium bis(trifluoromethanesulfonyl) imide, 1-butyl-3-methylimidazolium hexafluorophosphate and 1-ethyl-3-methylimidazolium tetrafluoroborate. Y. Kumar *et al.* [24] studied the ionic transport in PEO based Li ion conducting polymer electrolyte membranes plasticized with IL 1-ethyl 3-methyl imidazolium trifluoromethanesulfonate and obtained ionic conductivity of $\sim 3 \times 10^{-4}$ S cm^{-1} at 300 K. S. K. Chaurasia *et al.* [25] developed PEO based polymer electrolytes at different concentration of 1-ethyl-3-methylimidazolium tosylate (EMI-TY) and observed that the IL can act as a plasticizer to reduce the crystallinity by ~18%. They obtained ionic conductivity of $\sim 2.87 \times 10^{-5}$ S cm^{-1} at 300 K for the composite electrolyte at maximum 40 wt% of IL incorporation. Y. Kumar *et al.* [26] reported the ion conduction in magnesium trifluoromethanesulfonate based PEO matrix incorporated with different concentrations of IL 1-ethyl-3-methylimidazolium trifluoromethanesulfonate. They obtained optimum room temperature ionic conductivity of 5.6×10^{-4}

S cm^{-1} upon 50 wt% IL with improved thermal and mechanical stabilities. A. Hofmann *et al.* [27] reported ionic conductivity of ~10^{-3} S cm^{-1} at 300 K and electrochemical stability up to 5 V in a composite polymer electrolyte based on lithium bis(trifluoromethylsulfonyl) azanide, organic carbonates and PVdF–HFP incorporated with IL N-methyl-N-propylpyrrolidinium bis(trifluoromethylsulfonyl) azanide. A. R. Polu *et al.* [28] obtained ionic conductivity of 1.85×10^{-4} S cm^{-1} at 30°C by addition of IL EMIMTFSI into poly (ethylene oxide) (PEO) based composite polymer electrolyte.

However, in the polymer electrolyte, the mobility of both anions and cations of IL restrict the ionic conduction by forming neutral ion-pairs via electrostatic interaction. The mobile anions can also form passivation (SEI) layer at the interface of the electrodes and the electrolyte due to irreversible oxidation reaction of anions that in turn minimize the anodic stability [29]. So, specific strategies need to be followed to immobilize the anions so that only cations are free to move throughout segments of the polymer network to enrich the ionic conductivity. The cations of IL can interact with the organic linker moieties through electrostatic interaction, and the anions are closely packed or confined with the inorganic metal nodes. Considering this aspect of immobilization of anions of IL by incorporating in the pores of MOF, composite polymer electrolyte membranes can be developed by dispersing the IL incorporated MOF composites in polymer matrix. IL incorporation in MOF can create sufficient free space to enhance the mobility of cations along the MOF channels. The cations tend to stay at the open pores of the MOF and are free to hop to the adjacent polymer segments. The cations are able to migrate to the interconnected polymer segments along the free volume pathways and hence the mobility of cation increases by IL incorporation leading to intensify ionic conductivity. The interchain hopping of ions also takes place along with the intrachain hopping through the interconnected segments of polymer network leading to increase the electrochemical properties of the composite polymer electrolytes.

In this chapter, the aspects of MOFs and their derived materials have been discussed for their potential application as ion conducting composite polymer electrolyte membranes in rechargeable batteries. The structural topology, classification, and properties of MOFs have also been discussed to use them as efficient host materials for uptaking Ionic Liquids (ILs) within their pores. The historical perspectives and the classification of polymer electrolytes and different strategies to develop composite polymer electrolytes with compatible electrochemical performance have been emphasized. Different theoretical models to investigate the ionic conduction in composite polymer electrolytes have also been discussed. The authors emphasize the aspects of IL 1-butyl-3methylimidazolium tetrafluoroborate ($BMIMBF_4$)

incorporated NiBTC–MOF based PVdF–HFP composite polymer electrolytes and IL 1-butyl-3 methylimidazolium bromide (BMIMBr) incorporated CuBTC–MOF based poly (ethylene oxide) (PEO) composite polymer electrolytes as ion conducting electrolyte membranes in rechargeable batteries. The ionic conductivity, dielectric relaxation dynamics, electrochemical stability and ac conductivity scaling of the IL incorporated MOF based composite polymer membranes are also discussed to investigate the ion transport dynamics in composite polymer electrolytes.

6.2 Metal–Organic Frameworks

Metal–Organic Frameworks (MOFs) are three dimensional crystalline, microporous, and hybrid nanomaterials, known as porous coordination polymers (PCPs), consisting of inorganic metal cluster and organic ligands [30]. The organic ligands may be of different types such as benzene 1,3,5-tricarboxylic acid, benzene 1,4-dicarboxylic acid (terephthalic acid), biphenyl-4,4′-dicarboxylic acid (BPDC), benzoic acid, azoles, etc. [31]. The structure of Zn based MOF (ZnBDC–MOF or MOF-5) is presented in Figure 6.1. In the structure of MOF, secondary building units (metal clusters or ions) are linked with organic molecules to arrange the periodic porous topology. The secondary building units involve the inorganic metal atoms, finite polyatomic nodes of metal ions or a periodic infinite array of inorganic metal atoms. In general, the metal nodes in MOFs include the transition metal ions such as, Fe^{3+}, Cr^{3+}, Zn^{2+}, Co^{2+}, etc., and alkali and alkaline earth metal ions. Also, nitrate, sulfate, acetate, oxide, and chloride of different metals are used as precursors in the synthesis of MOFs. Variant combinations of metal nodes with linker moieties lead to form enormous structures of MOF. In each MOF geometry, the secondary building units can be substituted keeping the linker material same leading to produce quite different structures of MOFs. One of the most common MOFs, MOF-5 is consisted of tetrahedral Zn_4O metal clusters linked with benzene 1,4-dicarboxylic

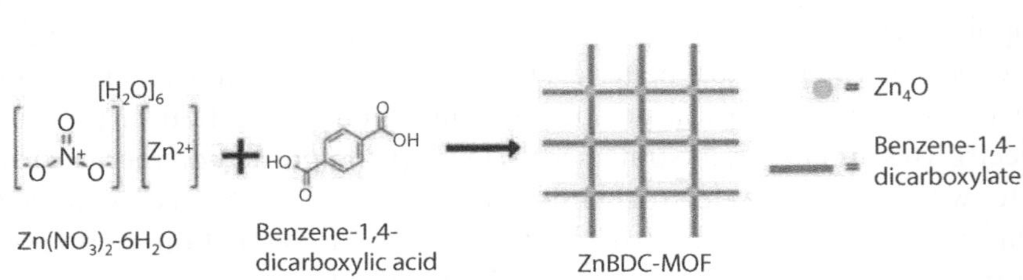

Figure 6.1 Structure of ZnBDC-MOF (MOF-5).

acid (terephthalic acid). By elongating the carbon network of the linker molecules different MOF structures can be arranged with similar topology and symmetry but with different pore parameters known as isoreticular MOFs (IRMOFs) [32]. According to the classification of MOFs reported by Tranchemontagne *et al.* [33], the secondary building blocks may possess varying coordination numbers ranging from 3 to 66. Substitution of linker molecules to the secondary building units may lead to change the unit cell parameters keeping the symmetry of MOFs unchanged due to carbon chain elongation as occurring in the series of UiO-66 to UiO-68 [34, 35]. The pore parameters of MOFs are determined by the length of the carbon chains as well as the number of benzene rings present in the linker molecules. Addition of different functional groups or substituents to the linker molecules is responsible for selectivity and chemical functionalities of the pores [36, 37]. The term MOF indicates not only the porous structure but also the rigid coordinated framework with well-defined topology in which the secondary building units and the linkers can be altered in the course of synthesis process. The term "MOF" is used as general notation of the whole class of metal–organic frameworks, whereas, individual ones are indicated followed by a particular number such as, MOF-74 (Zn_2DOT), MOF-101[Cu_2(BDC-BR)$_2$(H_2O)$_2$], MOF-253 [(Al(OH)(BPYDC)], and MOF-177 [Zn_4O(BTB)$_2$] [38, 39]. The isoreticular MOFs with same topological symmetry are placed in the series of IRMOF-1(MOF-5 or Zn_4O(BDC)$_3$.7DEF.3H_2O] to IRMOF-16 [Zn_4O(BDC)$_3$.7DEF.3H_2O] [40]. Particular MOFs are named according to the institute of their origin. Such MOFs include UiO (Universitetet i Oslo), MIL (Materials Institut Lavoisier), HKUST (Hong Kong University of Science and Technology), LIC (Leiden Institute of Chemistry). Several metal ions such as, Co, Cu, Zn, Fe, etc., are surrounded by tetrahedra of nitrogen atoms and they are linked by imidazole rings to have variant functionalities. Such series of MOFs are designated as ZIF (zeolite imidazole framework) followed by a number such as, ZIF-8 [Zn(MIM)$_2$], ZIF-90 [Zn(FIM)$_2$], etc. Some MOFs are designated according to the nomenclature proposed by the researchers those were involved in the synthesis processes. Examples of these series of MOFs are CPL-2 (Coordination Polymer with pillared Layer structure), F-MOF-1 (Fluorinated MOF), MOP-1 (Metal–Organic Polyhedra), etc.

6.2.1 Classification and Properties of Metal–Organic Frameworks

Because of the fascinating coordination geometry and porous topology of MOFs extensive research is going on to study their physicochemical properties. MOFs have been categorized into the following classes as per their structural features.

(i) Rigid Frameworks MOFs: Rigid frameworks have usually robust and stable topology with ultrahigh porosity. On adsorption or desorption of any guest molecule they can retain their porous structures. In general, rigid MOFs display type-I isotherms upon adsorption of gas molecules. The cubic structure of MOF-5 [Zn_4O (terephthalate)$_3$] exhibits surface area of 3800 m^2/g as reported by Kaye SS *et al.* [41]. Other highly porous MOFs include MOF-177) [$Zn_4O(btb)_2$] and MOF-200 [$Zn_4O(bbc)_2$] with respective BET surface area of 4746 and 6260 m^2 g^{-1} [42, 43]. However, incorporation of any small side group into a carbon network may lead to decrease porosity. By inducing mixed pillar ligands at layered unsaturated coordination sites highly porous MOFs with adequate surface area can be synthesized. Pair of different ligands linked with Zn_4O clusters can form $Zn_4O(t_2dc)(btb)_{4/3}$ (UMCM-2) framework with surface area of 5200 m^2 g^{-1}. The framework material exhibits micropores of 1.4–1.8 nm with mesopores of 2.4–3.0 nm [44]. Replacement of the Zn_4O cluster by chromium species connected with terephthalate ligand may give rise to an alternative structure of MIL-101 [$Cr_3F(H_2O)O(bdc)_3$]. The framework contains pair of pore cages of 1.2 and 1.45 nm diameters with BET surface area of 4100 m^2 g^{-1} [45]. A Cu based MOF PCN-66 [$Cu(H_2O)]_3$(ntei) reveals surface area of 4000 m^2 g^{-1} [46]. By using hexatopic carboxylate linkers alternative isostructures of highly porous MOFs with have also been synthesized.

(ii) Flexible or dynamic frameworks MOFs: Flexible or dynamic MOFs do not retain their original topology upon incorporation of guest materials and their structural features are affected by external temperature and pressure. At high pressure, upon adsorption of gas molecule, dynamic MOFs retain their pristine pore parameters. During gas adsorption and desorption, some MOFs such as, MIL-5, MIL-8, etc., reveal breathing effect that shows severe change in unit cell parameters. After adsorption of guest molecules Al based MOF [Al(bdc)(OH)] (MIL-53) displays 1D diamond like channels due to the modification of Al–OH–Al bond angle [47]. Softness of framework is attributed to the reorientation of bond angle/distance and molecular interactions. The flexible frameworks may also generate long ranged uniformly ordered arrangements that lead to develop amorphous or quasi amorphous phases. P-xylene isomers are selectively adsorbed by flexible Al $(bdc\text{-}OH)_x$ framework for potential applications in drug delivery [48]. Flexible MOFs reveal stepwise isotherms with hysteresis behavior for adsorption of CO_2 and other gas molecules. Rearrangements of adjacent network of certain MOFs may lead to display material flexibility or softness. [$Zn_2(bdc)_2(bpy)$] (MOF-508) is a framework material that depicts twofold symmetry and depending upon the position of relative network it has both open and closed structures [49]. The type of adsorbed guest molecules can distinguish the changes from closed to open topology of the framework.

(iii) Lewis acid frameworks MOFs: Lewis acid frameworks contain unsaturated metal sites for featuring interaction sites of guest molecules for potential applications in gas storage. One of the well known MOFs with unsaturated metal sites is CuBTC–MOF [$Cu_3(btc)_2.n(H_2O)$ or HKUST-1] that contains paddle wheel units of Cu linked by benzene-1,3,5-tricarboxylic acid. The axial sites of Cu^{2+} have ability to incorporate guest molecules for applications in heterogeneous catalysis [50, 51]. The presence of H_2O molecules in the open metal sites has significant impact on CO_2 capture. The HKUST-1 MOF containing 4 wt% of water content reveals higher CO_2 storage capacity than that by the conventional zeolite material. Reversible chemisorptions with nitrogen uptake can be observed using redox-active chromium centers in the MOF. The series of MOFs containing 2,5-dihydroxyterephthalic acid (dhtp) ligand with unsaturated metal centers can be represented as M_2(dhtp) (M = Mg, Zn, Mn, Co). The MOF Mg_2(dhtp) has the capability of CO_2 uptake up to 35.2 wt% at 298 K (1 atm) that is attributed to interaction of metal cluster with oxygen molecules. Also, the compound exhibits high degree of hydrophilicity, biocompatibility with acceptable stability and can store other gases including NO [52]. Cr based MOF MIL-101 or [$Cr_3F(H_2O)O(bdc)_3$] with unsaturated Cr^{+3} centers exhibits large pores that can facilitate effective diffusion of substrate materials during catalysis processes [53].

(iv) Surface functionalized frameworks MOFs: By grafting different functional groups on surface of MOFs adsorption capability of MOFs can be increased. Functional groups with high CO_2 affinity such as, arylamine, hydroxyl, alkylamine, etc., can be considered for effective functionalization on the surface of MOFs to enhance the selectivity and capacity to CO_2 adsorption [54]. The functional groups can be grafted on the surface of MOFs by modification of linker moieties or suitable coordination with the unsaturated metal nodes. The functional groups can intensify the interaction of CO_2 with functionalized framework and the pore space of the framework decreases upon functionalization [55]. Amine functionalized Al based MOF amino-MIL-53 exhibit separation factor of 60 for uptake of CO_2 compared to methane whereas, it decreases up to 5 in case of non-functionalized MIL-53. Also, the adsorption enthalpy increases from −20.1 to −38.4 upon amine functionalization [56]. Similar improvements in adsorption characteristics have also been reported for amine functionalized different frameworks such as, USO-3-In-A and USO-2-Ni compared to the pristine frameworks [57].

6.2.2 Potential Applications of MOFs

MOFs are hybrid coordination compound with fascinating structural topology and excellent pore features. By choosing unique and specific combinations of metal nodes and linker molecules different MOFs can be synthesized tailoring for specific applications such as, catalysis, sensing, electrochemistry, gas storage and sequestration and biomedical applications [58].

Gas storage and separation: MOFs are efficient materials for the purpose of storage and separation of gases owing to their tunable pore textures and wide range of fictionalization. As reported, more than 300 MOFs with varied metal nodes and linkers have been experimentally used for hydrogen storage. Zn based MOF, MOF-177 exhibits high surface area of ~ 5000 m^2 g^{-1} and shows efficient hydrogen uptake of 7.5 wt% at the pressure of 70 bar (77 K) [59]. MOF-5 or IRMOF-1 reveals the surface area up to 3800 m^2 g^{-1} with the capability of hydrogen uptake of 7.1 wt% at 40 bar (77 K). Usually the MOFs that contain open cluster sites can provide adequate area to facilitate interaction of metal clusters and H_2 molecules. Cu based MOF (NU-100) and Zn based MOF (MOF-210) have the H_2 storage capacity of 99.5 mg g^{-1} (56 bar) and 176 mg g^{-1} (80 bar), respectively, at 77K [60, 61]. However, MOFs require small amount of energy to release the adsorbed H_2 content as compared to other H_2 storage compounds including transition metal hydrides [62]. MOFs are efficient materials to be used to reduce atmospheric CO_2 level. MOF-210 has the capacity to uptake of CO_2 up to 2400 mg g^{-1} (74.2 wt%) at 298 K (50 bar) [61]. CO_2 uptake is experimentally tested for different MOFs such as, Mg-MOF-74 (68.9 wt%) at 278 K (36 bar), NU-100 (69.8 wt% at 40 bar, 298 K), HKUST-1 (19.8 wt% at 1 bar, 298 K) and MOF-5 (58 wt% at 10 bar, 273 K). It has been reported that presence of nitrogen containing polar groups or heterocyclic compounds in the pores of MOF influences the CO_2 uptake capacity as compared to any unfunctionalized MOF. Bio-MOF-11 exhibits maximum CO_2 uptake of 15.2 wt% at 298 K (1 bar) upon the effect of N-heterocycle agents [63]. Zhou and his group reported the CH_4 uptake of 16 wt% at 35 bar by the Cu based MOF (PCN-14) with surface area of 1753 m^2 g^{-1} [64]. HKUST-1 and MIL-101 show methane uptake of 15.7 wt% (150 bar) and 14.2 wt% (125 bar), respectively [61]. Presence of open cluster sites in Ni-MOF-74 is responsible for high rate of methane uptake of 190 $cm^3 g^{-1}$ (35 bar) at 298 K [65]. As reported, MOFs can separate hazardous gases like NO and CO from mixture of gases. Research is going on to capture CO using MOFs having open metal sites. Ni-MOF-74 and Co-MOF-74 can uptake ~7.0 mmol g^{-1} of NO at room temperature [66].

Catalysis: MOFs with high surface area and open metal sites can be treated as homogeneous and selective catalysts. By removing large substrate molecules from the pores, MOFs can be used as size selective catalysts. In 1994, the catalytic behavior of a Cd based MOF was reported to be used in cyanosilylation of aldehydes by removing the axial organic ligands [67]. In 2006, it has been reported that, removal of solvent molecules from HKUST-1 can expose open metal sites that can act as suitable Lewis acid catalysts [68]. MIL-101 has also been reported as effective Lewis acid catalyst where the active metal oxides act as catalytic sites upon removal of linker molecules [69]. Vanadium based MOF (MOF-48) is a promising material as homogeneous catalyst for methane oxidation [70]. A variety of organic reactions can be catalyzed by using microporous MOFs. Cd based MOF $[Cd(4\text{-btapa})_2(NO_3)_2]$ or Cr based MOF $[Cr_3F(H_2O)_2O(bdc)_3]$ are effective catalysts for Knoevenagel condensation reaction [71, 72]. MOFs can be used in photodegradation processes for removal of organic pollutants and water splitting.

Sensing: Aromatic linkers in MOFs are responsible for luminescent characteristics as they show UV-visible excitation. MOFs have wide range of applications as luminescent materials in cathode ray tubes, fluorescent tubes, pH sensors, small-molecule sensors, photovoltaic devices, and optoelectronic devices. Lanthanide elements such as, Tb, Eu, Sm, Dy, Nd, Gd, etc., exhibit possible transitions from d- to f-shell that enable them to be used as metal ions in luminescent MOFs. Naphthalene, pyrene, anthracene, and perylene types of organic ligands are commonly treated in production of luminescent MOFs. The luminescent MOFs can be doped with Ln^{3+} to be used as multicolored emission for imaging and detection of therapeutic cells [73]. Accommodation of particular guest solvent molecules can emit specific colored signal to be used as transducer to detect the properties of the solvent molecules. Recent studies show that oxoanionic pollutants such as $Cr_2O_7^{2-}$, MnO^{4-}, etc., be absorbed by water stable cationic MOFs [74]. Urotropin based MOF (Ur-MOF) can be used as chemical sensor for fluorescence quenching of tri-nitro phenol (TNP) in water [75]. Biosensing applications of Cd based nano MOF include detection of organophosphate in crops [76]. Photochromic MOFs containing Ca, Sr, Mg, etc., metal nodes and 1,4,5,8-naphthalenediimide ligand can be used to print erasable and inkless medium for minimizing paper wastage.

Biomedical Applications: Biomedical applications of MOFs include drug storage and delivery and anti bacterial activities. MOFs have the drug loading capacity four times greater than mesoporous silica with prolonged delivery time up to 21 days [77]. Due to the adequate porous topology, MIL group of MOFs are effective materials for storage and controlled delivery of biomolecules. Cr based MOFs, MOF-100 and MOF-101 can encapsulate Ibuprofen drug molecules with 35 wt% and 140 wt% capacities respectively with controlled delivery time of 5 to 6 days [78]. Iron based MOF such as, MIL-88A, MIL-8, etc., are capable of trapping anticancer, antiretroviral and antitumor agents. A magnetic MOF complex Fe_3O_4/$Cu_3(BTC)_2$ can uptake anticancer drug Nimesulide up to 0.2 g per g of MOF and takes 11 days for complete drug delivery at 37°C [79]. Zhuang *et al.* [80] studied the incorporation of camptothecin into ZIF-8 to enhance cellular uptake. Bernini *et al.* [81] reported the theoretical simulation work on adsorption performance of MOFs in drug delivery application. Based on their simulated work, the adsorption capacity of the BioMOF-100 is estimated to be 1969 mg g^{-1}, which is six times greater than that of mesoporous silica. Electrostatic interaction of charge compensating ions of MOFs can influence the drug loading capacity at low pressure. The presence of dimethylammonium cations in the pores of BioMOF-100 can reinforce the interaction with the ibuprofen drug molecules. Monte Carlo simulation studies have also been reported to compare the capacities of drug uptake of different MOFs including MIL-101 and UMCM-1 [82].

Electrochemical applications: MOFs are potential electro-active materials for application in batteries and supercapacitors. MOFs and derived composites of MOFs can be treated as cathode, anode and electrolyte membranes for ion conducting batteries. Some MOFs of MIL series can be used directly as cathode materials in Li ion batteries. Particular MOFs can be used as highly active sulfur materials and some other MOFs can be of potential interest in synthesizing metal-oxides. Tarascon *et al.* [83] first studied the electrochemical performance of Li ion battery using MIL-53 (Fe) mixed with 15 wt% carbon as the cathode electrode. The reversible capacitance of the cell attained 70 mA h g^{-1} at 50 cycles under scan rate of C/40. Vittal and co-workers [84] reported MOF based cathode material of $Li_2[(VO)_2(HPO_4)1.5(PO_4)0.5(C_2O_4)]$, which delivered reversible capacitance of 80 mA h g^{-1} at 25 cycles. Composites of MOFs have also been reported as efficient anode materials in Li ion batteries. Chen *et al.* [85] investigated the performance of Zn based MOF (MOF-177) as anode material for Li ion battery that exhibited 425 mA h g^{-1} and of 110 mA h g^{-1} discharge and charge capacity, respectively, at first cycle. Diaz *et al.* [86] reported the use of Co doped MOF-5 as electrode material for supercapacitor applications. Lee *et al.* [87] reported the pseudocapacitor behavior of Co based MOF and observed the specific capacitance of 206.76 F g^{-1} and energy density of 7.18 Wh kg^{-1} at current density of 0.6 A g^{-1}. In 2014, different nanocrystalline MOFs (nMOFs) have been reported for their capacitive behavior. Among them, Zr based MOF (nMOF-867) exhibited the stack and areal capacities of 0.64 and 5.09 mF cm^{-2}, respectively, which are six times greater than that observed for activated carbon materials [88].

6.2.3 Synthesis of MOFs

Topological design of the MOFs is assisted by intermolecular forces as well as synthesis conditions. Solvent polarity, adjustment in pH and temperature and concentration of precursors can optimize the quality of secondary building units in MOF. Different synthesis procedures of MOFs can be adapted to gain maximum yield with limited time and energy utilization.

Solvothermal and hydrothermal synthesis: MOFs can be synthesized through solvothermal and hydrothermal procedures. These are most common, well-established and uncomplicated approaches which were being adapted from the synthesis of zeolites. In both of the processes, carboxylic acid units are combined with the metal salts in solvent media and the solution is either stirred or kept in a autoclave up to a specific duration at ambient temperature to complete the reaction process. After centrifugation and washing treatment the precipitate is dried under vacuum at certain temperature to dry off the solvent part and to obtain the crystalline yield of MOF. Some advantages of these approaches are the formation of large crystals of MOF, efficient scalability, etc. However, these processes are limited by the production of less quantity of yield, excess reaction time, need of high temperature, and use of toxic solvent media such as, N,N-Dimethylformamide (DMF) [89].

Microwave assisted synthesis: Microwave assisted synthesis process reveals the interaction of charge molecules of the solvent material with the electromagnetic wave that facilitates the reaction rate. Quicker rate of reaction can influence in reduction of particle size, phase selectivity, and maintenance of morphology of MOF crystals. Preparation of MIL-101(Cr) has been reported using microwave at 210°C [90]. MIL-101 (Cr) synthesized by microwave assisted approach revealed 100 nm particle size and benzene uptake up to 16.5 m mol g^{-1} (56 mbar) at 288 K [91]. Research on microwave synthesis approach on Zr based MOFs reveals significant chemical and thermal stability due to coordination of Zr ions with the oxygen ions of the organic linker molecules. Liang *et al.* [92] reported the microwave assisted synthesis for Zn based MOF (MIL-140) at less time than used in conventional approaches. Ren *et al.* [93] reported the synthesis of highly crystalline, octahedral crystals of UiO-66 framework through microwave approach that exhibited hydrogen storage capacity of 1.26 wt%.

Mechanochemical Synthesis: In synthesis of MOFs, one important issue is to use toxic materials such as, DMF as solvents for metal salts and organic linkers. Mechanochemical synthesis is an approach to reduce the toxicity of solvents used in reaction mechanisms. In this process, MOFs are prepared by placing ball bearings into a vessel containing stoichiometric quantities of reagents. The vessel is kept closed where the mixing and pulverization of reagents take place to complete the synthesis process. This non-toxic solventless synthesis process is relatively efficient as it can produce nanocrystalline MOF particles at less reaction time compared to other synthesis approaches. Cu based MOF (HKUST-1) has been synthesized by mechanochemical approach using benzene 1,3,5-tricarboxylic acid with copper acetate under solventless conditions.

Sonochemical synthesis: Sonochemical synthesis process of MOFs is subjected to ultrasonicate the reaction mixture at 20 kHz to 10 MHz frequency range. This approach is suitable for membrane coating purposes as it can reveal several advantages that include the preparation of homogeneous mixture at rapid reaction time and lack of need of extra heating during reaction process. For the production of nanosized crystals of MOFs sonochemical synthesis process can be considered as appropriate one as the crystals can be produced instantly within the cavity of solvent media. Under the cavitation process, the bubbles of the reactants can expand and collapse to enable quick production of nanosized crystals of MOFs. Room temperature sonochemical synthesis of different MOFs including HKUST-1, MOF-5, and MOF-177 have been reported several research groups. Fard *et al.* [94] reported sonochemically synthesized Pb based MOF $[Pb_2(N_3)(NO_3)L_2]$ in aqueous medium and the synthesized nanocrystals are further calcinated so as to prepare nanosized PbO.

Electrochemical Synthesis: In this process of MOFs, metal ions can be formed not from any corresponding salt solution or by the reaction of any metal with any acidic solution. Hence, there is no need of acidic deprotonation in electrochemical synthesis method. The linker molecules are dissolved in the reaction mixture containing an electrolyte and the metal ions are produced by the dissolution of the electrode material. Several researchers have reported the formation of well dispersed MOF membranes through the electrochemical approach. Joaristi *et al.* [95] reported the electrochemical synthesis of HKUST-1 by the direct nucleation on the surface of anode at ambient temperature. Campagnol *et al.* [96] first reported the electrochemical synthesis of MIL-100(Fe) framework by dissolving 1,3,5-benzenetricarboxylic acid in the mixture of Milli-Q water and ethanol using an electrochemical cell. Stassen *et al.* [97] reported the electrochemical deposition of UiO-66 film on zirconium foil by applying current of 80 mA at 383 K.

Vapour diffusion: Vapor diffusion technique is a relatively efficient approach to synthesize MOFs. The metal salt with organic ligand is dissolved in any solvent such as, DMF in an open container surrounded by any volatile solvent such as trimethylamine. The synthesis of MOF is caused by acid deprotonation monitored by diffusion of volatile solvent into the mixture of reagents to increase the reaction rate. By this technique, large crystals of MOFs can be produced at ambient temperatures and prolonged reaction time. Wu *et al.* [98] adopted the vapor diffusion approach to synthesize Pb based MOF [Pb(1,4-NDC)(DMF)] dissolving lead nitrate and naphthalene dicarboxylate in DMF solvent in a vessel surrounded by an another vessel of trimethylamine. The steady diffusion process of trimethylamine from the surrounding vessel to inner one resulted in the formation of MOF crystals as reported.

6.3 Polymer Electrolytes

A polymer electrolyte is an ion conducting system consisted of metal salts dissolved in polymer matrix of high molecular weight. In 1973, P. V. Wright and his group first synthesized the poly(ethylene) oxide (PEO) matrix mixed with sodium iodide and sodium and potassium thiocyanates and correlated the ionic conductivity with the amorphous phase of polymer chain [99]. The oxygen atoms of the PEO chain can form multiple coordination bonds with cations that exhibits low bond rotation barriers to facilitate the segmental ionic motion of the polymer chain. Polymer electrolytes have large variety of applications in different emerging fields including rechargeable batteries, fuel cells, sensors and actuators, supercapacitors, electrochromic displays, etc. The polymer electrolytes should be capable of monitoring fast ionic transport and supply electrical insulation between the cathode and the anode.

To be suitable for solid state electrochromic devices, a polymer electrolyte must reveal some basic requirements such as:

1. Polymer electrolytes should possess ionic conductivity of ~10^{-3} S cm^{-1} and should be electrically insulator to facilitate fast ion conduction to reduce the self discharge. Higher ionic conductivity reveals the upper limit of the output power obtained from any electrochemical device.
2. Polymer electrolytes should have wide electrochemical stability which is crucial for solid state ionic devices like rechargeable batteries. Electrochemical stability of Polymer electrolyte must be high (~6 V) and compatible with the electrodes.
3. Polymer electrolyte should be thermally stable up to high temperature (~300°C) so as to maintain the rated performance and the operating conditions of the device.
4. Polymer electrolytes must be robust against mechanical, electrical, and thermal abuses. It should maintain favorable mechanical strength so as to carry on good electrochemical performance. Polymer electrolyte membranes should be flexible enough for different miniaturized applications.

The conventional polymer electrolytes prepared by dissolving salt in ion conducting solvent exhibit less ionic conductivity at room temperature (~10^{-7} to 10^{-8} S cm^{-1}) that is not efficient for practical electrochemical applications. To overcome these drawbacks modified forms of polymer electrolyte membranes are being developed.

6.3.1 Historical Perspectives and Classification of Polymer Electrolytes

In general, polymer electrolytes are classified into three sections: (i) solid polymer electrolytes, (ii) gel polymer electrolytes and (iii) composite polymer electrolytes. Solid polymer electrolytes are formed by mixing low lattice energy metal salts with polymer matrix. Solid polymer electrolytes exhibit typically low room temperature ionic conductivity of 10^{-6} to 10^{-8} S cm^{-1}. To increase the ionic conductivity of polymer electrolytes, gel polymer electrolytes are developed by adding liquid plasticizer into salt–polymer matrix. However, the gel polymer electrolytes are capable to attain ambient ionic conductivity but less mechanical stability as comparison with solid polymer electrolytes. The composite polymer electrolytes are synthesized by incorporating inorganic materials to the salt–polymer complex and the composite electrolyte system exhibits both higher ionic conductivity and electrochemical stability as compared to gel polymer electrolytes.

Solid Polymer Electrolytes: These are solvent free system of metal salt–polymer complexes. Armand, Chabagno and Duclot [100] reported the Vogel–Tamman–Fulcher approximation for CsSCN and LiSCN complexes. D. Payne [101] and C.C. Lee [102] investigated the ionic conductivity of sodium thiocyanate, sodium iodide and lithium ion complexes with PEO and reported that the ionic conduction occurs through the amorphous free volume channels of PEO. Killis, Le Nest, and Cheredame [103] synthesized the urethane based PEO solid electrolytes and investigated the ionic motion through the amorphous PEO chain. C.A. P.M. Blonsky and his group developed the polyphosphazene complex [104] and Christian V. Nicholas *et al.* [105] developed the methoxy copolymer electrolytes. Cheradame [106] and Watanabe [107] correlated glass transition temperature, ionic conductivity, and the mechanical relaxation of solid polymer electrolytes. However, the performance of PEO based solid polymer electrolytes in rechargeable batteries is limited by its poor ionic conductivity.

Gel Polymer Electrolytes (GPEs): In gel polymer electrolytes, liquid plasticizers such as propylene carbonate (PC), ethylene carbonate (EC), diethyl carbonate (DEC), etc., are added in substantial amounts to the salt–polymer complexes. Liquid plasticizers are low-molecular weight materials that are used to increase segmental motion and flexibility of the polymer chains and thereby to enhance the ionic conductivity of the electrolyte system. Gel polymer electrolytes are ionically conducting two-phase systems that possess both the cohesive nature of solids and the diffusive behavior of liquids. Gel polymer electrolytes possess high room temperature ionic conductivity of ~10^{-3} S cm^{-1} but comparatively poor mechanical stability because of the presence of liquid plasticizers. Benedict *et al.* [108] investigated the reduction in activation energy of PEO–$LiAsF_6$ electrolyte complexes on addition of plasticizer dibutyl phthalate (DBP). Peraninage *et al.* [109] studied the addition of mixture of liquid plasticizers EC, PC and 3-methyl-2-oxazolodinon (MEOX) into PAN based electrolyte system to increase ionic conductivity of gel polymer electrolytes. Watanabe *et al.* [110] studied the use of propylene carbonate (PC) and ethylene carbonate (EC) as liquid plasticizers in PAN–$LiClO_4$ polymer electrolyte complex and obtained ionic conductivity upto ~10^{-4} S cm^{-1} at room temperature upon increasing the plasticizer concentration. Salt $LiClO_4$ (8%) and polymer PAN (12%) based Gel polymer electrolyte with addition of plasticizers PC (40%) with EC (40%) was synthesized by Sun *et al.* [111] and obtained ionic conductivity up to 2×10^{-3} S cm^{-1} at 300 K. Choe *et al.* [112] studied the ionic transport in PAN based electrolytes using $LiAsF_6$, $LiN(CF_3SO_2)_2$, $LiPF_6$ and $LiCF_3SO_3$ salts with plasticizers EC and PC having electrochemical stability more than 5 V.

Composite Polymer Electrolytes (CPEs): These electrolytes are synthesized by incorporating inorganic filler materials into any solid or gel polymer electrolyte with a view to enhance the mechanical as well as electrochemical stability along with ionic conductivity. The ceramic fillers can increase the recrystallization rate of free volume polymer chain leading to form localized amorphous zones that in turn influence the ionic transport. F. Capuano *et al.* [113] observed the increased ionic conductivity of $(PEO)_8$–$LiC1O_4$ upon dispersion of 10 wt% of ceramic filler $\gamma LiAlO_2$ of grain size of 1 μm. They reported that highly dispersed ceramic fillers in polymer electrolytes can decrease the agglomeration of polymer chains and thereby crystallization rate is affected.

To enhance the ionic conductivity as well as the electrochemical stability of $LiBF_4$, $LiClO_4$, $LiCF_3SO_3$, and $LiPF_6$ based PEO complexes can be dispersed with sub-micron sized ferroelectric filler particles such as $PbTiO_3$, $BaTiO_3$, and $LiNBO_3$. The concentration and size of filler particles affect the ionic conductivity, mechanical and electrochemical stability of the composite polymer electrolytes. Considering the nature of filler particles and the host polymers composite polymer electrolytes can be divided in two different categories.

(i) Inorganic-in-organic: In this composite polymer electrolyte system, different inorganic nanoparticles such as TiO_2, Al_2O_3, SiO_2, $BaTiO_3$, etc., are dispersed in polymer matrix. The nanosized fillers can decrease the crystallinity of the host polymer matrix that facilitates the ionic transport leading to increase the ionic conductivity. In 1994, Kumar and Scanlon [114] reported that nanosized ceramic filler particles are more compatible to Li metal than microsized filler materials. Upon incorporation of 10 wt% of SiO_2 filler in PEO–$LiN(CF_3SO_2)_2$ electrolyte system, the cationic transport number increases from 0.1 to 0.2. Kim *et al.* [115] studied the effect of SiO_2 coated trimethylsilyl on salt LiBETI based PEO electrolyte and achieved ionic conductivity of 1.5×10^{-5} S cm^{-1} at room temperature and electrochemical stability up to 4.8 V. Bloise and his group reported the effects of α-Al_2O_3 and γ-Al_2O_3 filler materials on the ionic transport mechanism in PEO based composite polymer electrolytes [116]. Liu *et al.* [117] investigated the improvement in ion transport number upto 0.56 in $LiBF_4$–PEO electrolyte system upon incorporation SiO_2 as filler material. Chung *et al.* [118] reported that the incorporation of TiO_2/Al_2O_3 nanoparticles in $LiClO_4$–PEO can dissociate the cation–polymer bonding leading to influence the ionic conductivity of the composite polymer electrolytes.

(ii) Organic-in-inorganic: In this electrolyte system, polymers are intercalated inside the interlayer nanometric galleries of silicate clay such as hectorite, montmorillonite saponite etc. Polymer intercalation into layered silicate provides mechanical and thermal stability of the composite electrolyte system [119]. In organic-in-inorganic composite polymer electrolytes, both of the layered silicate phase and host polymer phases are available. Alkylammonium surfactants can be used to alter the inorganic cations present in the interlayer galleries of the silicate clay for better compatibility of the host polymer electrolytes. Montmorillonite (MMT) is one of the commonly used layered silicates with significant features such as good cation-exchange capability (CEC ~80 mequiv./100 g), aspect ratio (~1000), high surface area (~31.82 $m^2\ g^{-1}$), sufficient interlayer charge (~0.55), and clay channel width (~16 Å).

6.3.2 MOF Based Polymer Electrolytes

MOFs can be incorporated as filler materials in polymer electrolytes to enhance the electrochemical properties of the composite electrolyte membranes. Dian-Dian Han *et al.* [120] reported the modification of PVdF based gel polymer electrolytes using Mg (II)-based MOF (MOF-74) to stabilize the Li anode in Li–S battery. The schematic representation of functionality of the MOF-74 based gel polymer electrolytes is depicted in Figure 6.2. The polysulfide anions are immobilized in the pores of MOF facilitating uniform distribution of Li ions in the polymer electrolytes as depicted in SEM image in Figure 6.3. The MOF–PVdF based gel polymer electrolyte membrane revealed ionic conductivity upto 6.72×10^{-4} S cm^{-1} at 300 K with uptake capacity of organic electrolyte up to 256.1% as presented in Figure 6.3. The MOF–PVdF based cell revealed high cyclic performance with discharge capacities of 1195.3, 1083.9, 996.7, and 860.5 mA h g^{-1} at 0.2/0.5/1.0/2.0 C rates. After 250 cycles the cell exhibits capacitance of 778.4 mA h g^{-1} at per cycle capacity fading of 0.09%. Claudio Gerbaldi *et al.* [121] reported the incorporation of Al based MOF (Al-BTC MOF) as an efficient nanofiller into PEO based solid polymer electrolytes and

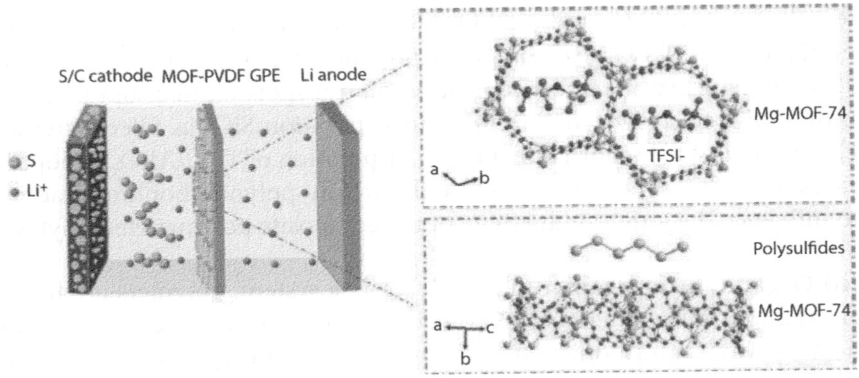

Figure 6.2 Schematic representation of MOF-74 based PVDF GPE gel polymer electrolytes for Li–S battery. Reprinted with permission from Ref. [120]. Copyright 2019 American Chemical Society.

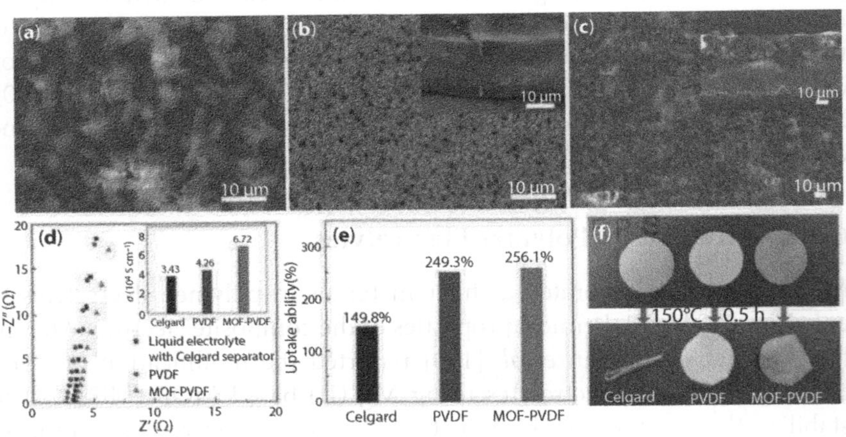

Figure 6.3 SEM micrographs of the synthesized (a) Mg–MOF-74, (b) PVdF membrane, and (c) MOF-74–PVdF membrane. Insets in (b) and (c) display the respective cross sections of the membranes. (d) The electrochemical impedance spectra and room temperature ionic conductivity of different samples. (e) The uptake ability (%) of different membranes. (f) Digital photographs of the membranes upon heating for 30 min at 150°C. Reprinted with permission from Ref. [120]. Copyright 2019 American Chemical Society.

achieved enhancement of two orders of in the ionic conductivity at ambient temperature. The prepared nanocomposite membranes are observed to be mechanically stable with excellent cyclic and interfacial efficiency. The nanocomposite electrolyte cell exhibited capacitance of 40 mA h g^{-1} at rate of 5 C that indicates fast Li ion exchange through the composite polymer electrolyte membrane. The Coulombic efficiency increased more than 98% from the second cycle and remained stable throughout all the cycles indicating excellent mechanical as well as interfacial stability of the composite polymer electrolyte membrane. After 100 cycles the cell attained specific capacity of 100 mA h g^{-1} at rate of 2 C. Hanyu Huoa *et al.* [122] reported the incorporation of cationic MOF (CMOF) in PEO based solid polymer electrolytes in order to immobilize anions and facilitates uniform distribution of Li^+ cations through the nanocomposite membranes. The schematic representation of Li deposition characteristics of PEO (LiTFSI) composite electrolyte and P@CMOF composite electrolyte upon immobilization of anion is depicted in Figure 6.4. The fabricated P@CMOF had been grafted with $-NH_2$ functional group with a view to keep the ether oxygen of PEO chains protected by hydrogen bonds, that enhances the electrochemical stability to 4.97 V. The P@CMOF composite electrolyte membrane exhibited ionic conductivity up to 6.3×10^{-4} S cm^{-1} at 60°C having increased Li^+ ion transference number up to 0.72 as presented in Figure 6.5. The cell consisted of $LiFePO_4$ cathodes reveals capacity retention up to 85.4% at 300 cycles and 1 C rate. R. Senthil Kumar *et al.* [123] reported Cu–BDC MOF incorporated in complex electrolyte of PEO and Li salt LiTFSI. At 30°C ionic conductivity increases to one order magnitude and above 50°C up to two order magnitude. The electrochemical cell made up of the nanocomposite membrane exhibited capacitance of 132 mAh g^{-1} at C/20-rate and 126 mAh g^{-1} at C/10-rate, respectively. The cell delivered columbic efficiency of 98% with discharge capacitance of 120 mAh g^{-1} at the rate of

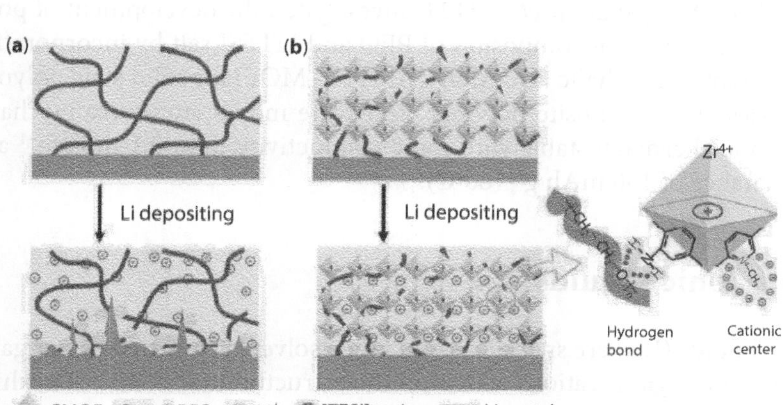

Figure 6.4 Schematic representation of Li deposition characteristics of (a) PEO (LiTFSI) composite electrolyte and (b) P@CMOF composite electrolyte upon immobilization of anion. Reprinted with permission from Ref. [122]. Copyright 2019 Elsevier.

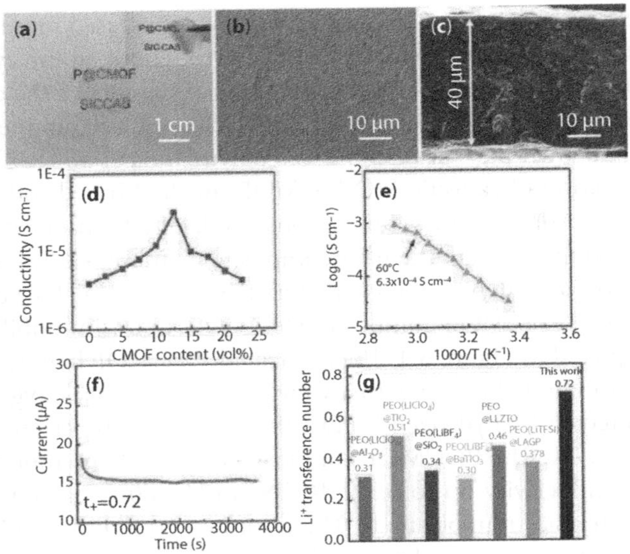

Figure 6.5 (a) Digital photographs of P@CMOF in flat and bended positions (inset). (b) SEM micrograph of cross sectional area of P@CMOF and (c) SEM micrograph of plane-view of P@CMOF. (d) Room temperature ionic conductivity of P@CMOF at different CMOF contents. (e) Ionic conductivity of P@CMOF at 12.5 vol% CMOF at different temperatures. (f) Li+ transference number of P@CMOF. (g) Comparison of Li+ transference number of P@CMOF with different reported polymer electrolytes. Reprinted with permission from Ref. [122]. Copyright 2019 Elsevier.

1C. Shruti Suriyakumar *et al.* [124] investigated the development of polymer electrolyte nanocomposites of PEO and LiTFSI salt by incorporating aluminum terephthalic acid MOF (Al–TPA–MOF) for solid state polymer batteries. The composite polymer electrolyte membranes were mechanically and thermally stable with ionic conductivity upto 0.1 mS cm^{-1} and capacitance of 130 mAh g^{-1} (60°C).

6.4 Ionic Liquids

Ionic liquids (ILs) are special class of green solvents consisted of inorganic anions and organic cations with unlimited structural variations. ILs exhibit weak electrostatic interaction in between the asymmetric organic cations and symmetric charge delocalized anions. Some of the (ILs) are already liquid at room temperature with melting temperature below 100°C known as room temperature ILs. Unlike common inorganic salts, ILs do not require

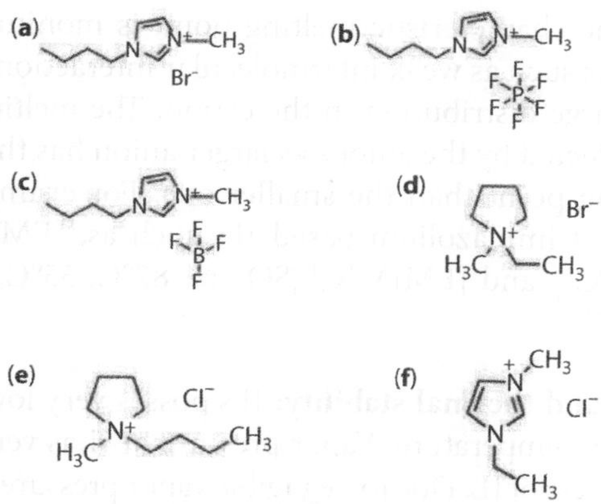

Figure 6.6 Structures of different imidazolium and pyrrolidinium based ILs where, (a) 1-butyl-3-methylimidazolium bromide, (b) 1-butyl-3-methylimidazolium hexafluorophosphate, (c) 1-butyl-3-methylimidazolium tetrafluoroborate, (d) 1-ethyl-1 methylpyrrilodinium bromide, (e) 1-butyl-1 methylpyrrilodinium chloride, and (f) 1-ethyl-3-methylimidazolium chloride.

any salvation process as they are already dissociated as pair of ions in liquid state. In 1914, P. Walden reported the IL ethylammonium nitrate [125] and later on in 1992 Wilkes *et al.* [126] discovered water and air stable ILs for commercial uses. ILs are consisted of inorganic anions and organic cations tailoring for specific applications. The commonly used cations include imidazolium, pyrrolidinium, and various kinds of quaternary ammonium salts. The structure of different imidazolium and pyrrolidinium based ILs is shown in Figure 6.6. The anions may include bromide, chloride, tetrafluoroborate, hexafluorophosphate, bis(trifluoromethanesulfonyl)imide etc.

6.4.1 Properties of Ionic Liquids

Ionic liquids possess several significant properties that make them advantageous over conventional solvents to be used in different catalytic, sensing and electrochromic applications. The physicochemical properties of ILs can be widely varied by choosing selective cations and anions. The structural features of ILs can be related to the different physical and chemical properties such as:

Melting Point: One of the key properties of an IL is its melting point, which has particular significance in the chemical composition of the IL. The ILs with chloride anions and organic cations possess melting temperatures below 150°C. The characteristic melting point is monitored by different features of cations such as weak intermolecular interaction, low symmetry, and uniform charge distribution in the cation. The melting temperatures of ILs are also affected by the anions as larger anion has the ability to minimize the melting point than the smaller one. For example, the melting points of different imidazolium based ILs such as, [EMIM]Cl, [EMIM]NO_2, [EMIM]$AlCl_4$, and [EMIM]CF_3SO_3 are 87°C, 55°C, 7°C, and −9°C, respectively [127].

Vapor Pressure and thermal stability: ILs posses very low vapor pressure $\sim 10^{-10}$ Pa at room temperature. Vapor pressure of IL is very crucial during the synthesis process of IL. Due to negligible vapor pressure of IL the distillation of reaction mixture can be effectively done during the product isolation process. Low vapor pressure of IL can also hinder the formation of azeotrope between the solute and solvent molecules. The thermal stability of ILs is controlled by the strength of the intermolecular carbon and hydrogen bonds. ILs synthesized by protonation of phosphane or amine exhibit restricted thermal stability. The ILs prepared by phosphane or amine alkylation reactions possess the affinity to undergo thermally activated dealkylation or transalkylation processes that are strongly dependent on their anionic properties. Two imidazolium based ILs [EMIM]BF_4 and [EMIM][$(CF_3SO_2)_2N$] have the thermal stability up to 300°C and 400°C, respectively [127].

Density: The density of an IL is dependent on the nature of its cations and anions. Studies on different imidazolium based chloroaluminate ILs reveals linear variation of density with the N-alkyl cationic chain of the ILs. Comparable studies on cations of ILs depict that the density decreases with increasing bulkiness of the cations of different ILs. Also, adjustment in density can be performed by controlled structural changes of the cations of ILs. Variation of anions may also affect the density characteristics of ILs. For example, ILs with anions such as trifluoroacetate or triflate may exhibit different density ranges that can be adjusted by choosing specific cations [127].

Viscosity: The viscosity of ILs is limited by their affinity to form hydrogen bonds through van der Waals interaction. For example, in chloroaluminate based ILs, viscosity is determined by the generation of hydrogen bonds in between hydrogen atoms of imidazolium ring and the chloride ion. Comparative studies of viscosity of ILs with a variety of cations and anions have been studied by several researchers. The transition from $CF_3SO_3^-$ to n-$C_4F_9SO_3^-$ ion and CF_3COO^- to n-$C_3F_7COO^-$ ion depicts enhancement in viscosity due to dominant van der Waals interaction. However, the comparison of [BMIM]CF_3SO_3 with [BMIM]$(CF_3SO_2)_2N$ reveals reduction in viscosity as the hydrogen bonds are completely suppressed despite the affect of van der Waals interactions [127]. The viscosity of ILs is also influenced by the nature of the cations. Fluorinated alkyl species can reveal higher viscosity than the imidazolium ones due to dominant affect of van der Waals interaction. By adding small quantity of organic solvents or increasing temperature, the viscosity of ILs can be minimized.

Environmental Aspects: ILs are eco-friendly solvents and can be use as non-volatile medium for reaction purposes to replace toxic volatile solvents. ILs hold special solubility characteristics to enable the biphasic reactions in catalysis processes. Also, the non-volatile ILs can act as effective solvents to isolate the reaction mixtures by distillation. Use of ILs in superacid catalysis processes has significant environmental aspects as the ILs can be treated as suitable alternatives to substitute HF in this regard. On the basis of their environmental impact ILs can be categorized as green solvents for clean processes.

6.4.2 Ionic Liquid Incorporated MOF

Several researchers have reported the incorporation of IL into the MOF to study the ion dynamics and to use the composite system as compatible ionic conductor in electrochemical solid state devices including batteries. Yukihiro Yoshida *et al.* [128] reported the processes of incorporation of an IL in a MOF including wet impregnation process and capillary action process as schematically represented in Figure 6.7. Yifei Chen *et al.* [129] reported the theoretical and experimental investigations on IL $BMIMPF_6$ incorporated IRMOF-1 that revealed the interaction of $BMIM^+$ cations with linker moieties and that of PF_6^- anions with Zn metal nodes. K. Fujie *et al.* [130] investigated the IL incorporation in pores of MOF to enhance the ionic conductivity of the composite system at low temperature. The phase dynamics of ions of IL can be maintained by the nanosizing of IL by the tunable interaction of IL ions with the MOF. They used the IL 1-ethyl-3-methylimidazolium bis(trifluoromethylsulfonyl) amide (EMI-TFSA) to be incorporated in the Zn based MOF ZIF-8. The bulk EMI-TFSA showed sharp decrease in ionic conductivity at low temperature (below 250 K) while the EMI-TFSA incorporated ZIF-8 nanocomposite showed no decrease due to lack of phase transition of ions of IL. Li-Hong Chen *et al.* [131] reported the incorporation of 1-Ethyl-3-methylimidazolium Chloride (EMIMCl) into the pores of Zr based MOF UiO-67 as schematically represented in Figure 6.8. The IL incorporated

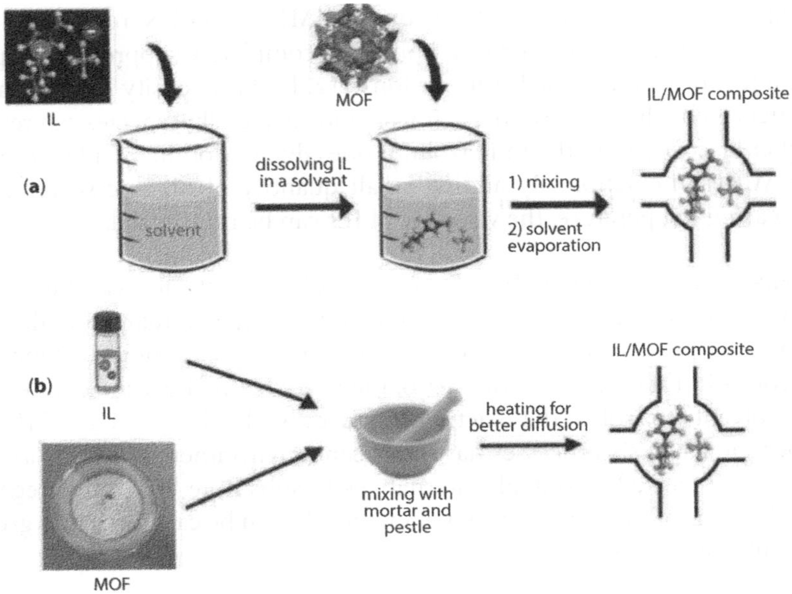

Figure 6.7 Schematic representation of different processes of incorporation of IL in MOF (a) wet impregnation process and (b) capillary action process. Reprinted with permission from Ref. [128]. Copyright 2019 American Chemical Society.

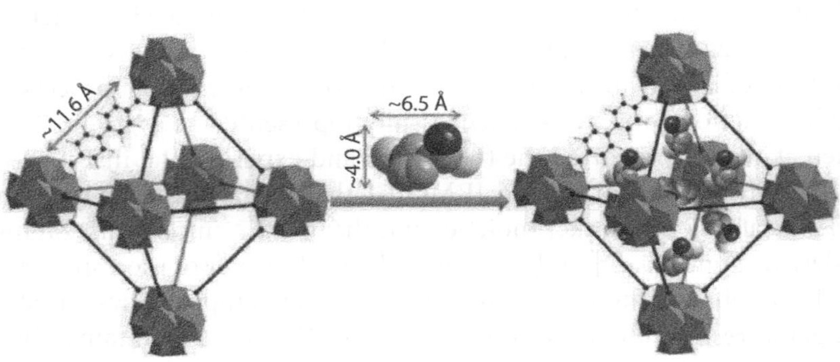

Figure 6.8 The process of incorporation of EMIMCl in the pores of UiO-67. Reprinted with permission from Ref. [131]. Copyright 2017 Elsevier.

UiO-67 nanocomposite showed ionic conductivity upto 1.67×10^{-3} S cm^{-1} at 200°C and 0.37 eV activation energy as presented in Figure 6.9. The humidity-independent and thermally stable hybrid ionic conductor can be potentially used in electrochemical devices. Qiuxia Xu *et al.* [132] investigated the incorporation of two ILs with same cations and different anions,

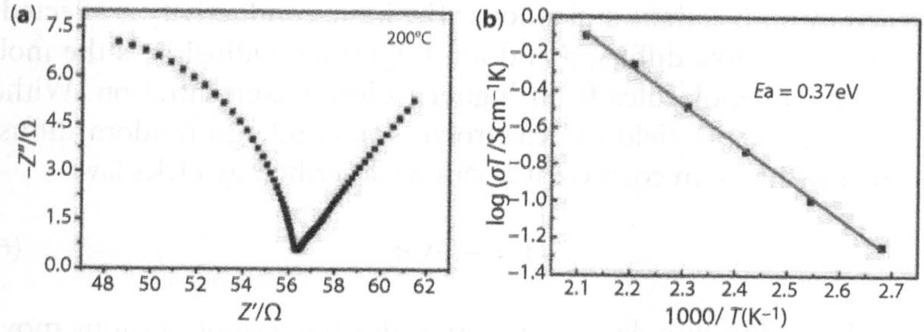

Figure 6.9 (a) Nyquist impedance plots of IL EMIMCl@UiO-67 at 200°C and (b) Arrhenius plots of IL EMIMCl@UiO-67. Reprinted with permission from Ref. [131]. Copyright 2017 Elsevier.

1-ethyl-3-methylimidazolium dicyanamide and 1-ethyl-3-methylimidazolium thiocyanate, into the pores of Cr based MOF MIL-101 at different concentration of IL. They reported that with increasing IL content in MOF pores, ionic conductivity increases. They obtained the ionic conductivity of 6.21×10^{-3} S cm^{-1} and activation energy up to 0.18 eV at 150°C.

6.5 Ion Transport in Polymer Electrolytes

The ionic transport dynamics of polymer electrolyte is a fascinating topic of research interest which depicts the key role of electrochemical performance. To develop polymer electrolytes with high ionic conductivity and electrochemical stability understanding of ion transport mechanism is utmost necessary. Ion conduction in polymer electrolytes is due to segmental motion of polymer network. The phenomenon of ion conduction is due to two main reasons: the first reason is the activated hopping of ions from any coordinated site to adjacent one within the vicinity of the polymer matrix and the second one is the dynamics of polymer segments. The flexing of the polymeric segments can facilitate the ion mobility by breaking the old linkage and forming the new attachments with the adjacent polymer segments. The mobility of ions can generate electrical responses that can be detected or verified by different experimental techniques such as electrochemical impedance spectroscopy.

6.5.1 General Description of Ionic Conductivity

Study on the mechanism of ionic conductivity is the prime aspect of polymer electrolyte research. In polymer electrolytes the ionic motion is governed by both cations and anions. The ionic conductivity is affected by different factors like diffusion and ion migration. Diffusion is the mobility of atoms or molecules from higher to lower concentration. Without any external electric field the electrolyte ions undergo random diffusive motion. The diffusion coefficient D can be described by Fick's law:

$$J = -D\nabla n \tag{6.1}$$

where J is the ion flux density that indicates the number of ions moving through unit area through unit time and n is the density of ions. At low density regions, uniform diffusion of ions takes and the diffusive coefficient D can be expressed by Einstein's relation:

$$D = \mu kT/q \tag{6.2}$$

where μ is the mobility of ions at temperature T, k is the Boltzmann's constant with ion charge q.

Under any external electric field, migration of ions occurs by hopping of ions to nearby coordination sites. Usually ions are trapped by the lattice and they must acquire enough thermal energy to be free from the lattice sites. Ionic conduction occurs when the ions are able to migrate from on lattice site to any adjacent one. The mobility of ions through the lattice sites can be explained by vacancy and interstitial mechanisms. In vacancy mechanism, hopping of any mobile ion occurs from its normal site to an equivalent empty site. In interstitial mechanism, an interstitial ion hops or jumps to any nearby equivalent site. The ionic conductivity (σ) of polymer electrolyte is approximated by,

$$\sigma = \sum nq\mu \tag{6.3}$$

where ion concentration is denoted by n with charge q and ion mobility μ. It can be concluded from the Eq. (6.3) that ionic conduction increases with increasing quantity of ions as well as the mobility of ions.

6.5.2 Models for Ionic Transport in Polymer Electrolytes

The ionic transport dynamics in polymer electrolytes is governed by correlated hopping processes in the amorphous region. Ion conduction in polymer electrolytes can be described by different ion transport models such as: (i) Arrhenius model, (ii) William–Landel–Ferry (WLF) model,

(iii) Vogel–Tamman–Fulcher (VTF) model, and (iv) Dynamic Bond Percolation (DBP) model.

(i) Arrhenius model: In Arrhenius model, the ionic motion in solid polymer electrolytes obeys Arrhenius formalism. With increasing temperature ion mobility increases following Arrhenius model as depicted by Eq. (6.4). This can be attributed to the increasing ion hopping with increasing temperature resulting in enhanced ionic conductivity. The temperature dependence of ionic conductivity (σ) can be expressed by Arrhenius model:

$$\sigma = \sigma_o \exp^{\left(\frac{-E_a}{K_B T}\right)} \tag{6.4}$$

where σ_o is a pre-exponential term with activation energy E_a, K_B is Boltzmann constant, and T is temperature. Linear variation of log σ vs. $1/T$ is called as Arrhenius plot and the slop of the curve denotes the activation energy of the thermally activated process. Activation energy (E_a) is the sum of binding energy (E_b) and strain energy (E_s) of the charge carriers. An atom requires the binding energy (E_b) to leave its original position and needs the strain energy (E_s) to generate a path to migrate to neighboring sites. Increasing concentration of charge carriers can reduce the strain energy (E_s) by inducing segmental expansion of the electrolyte membranes. Anderson–Stuart model can explain E_b and E_s for any ionic material as,

$$E_b = \frac{\beta z z_O e^2}{\gamma(r + r_O)} \tag{6.5}$$

$$E_s = 4\pi.Gr_d(r - r_d)^2 \tag{6.6}$$

where r is radius and z is cationic charge state. β is the Madelung constant, r_O is radius, z_O is charge state for any non-bridging O^{2-} ion, and e is charge of an electron. γ is covalency parameter that signifies charge neutralization rate, r_d is radius of doorway having elastic modulus term G. The activation energy can be formulated considering E_b and E_s as,

$$E_a = \frac{\beta z z_O e^2}{\gamma(r + r_O)} + 4\pi.Gr_d(r - r_d)^2 \tag{6.7}$$

(ii) William–Landel–Ferry (WLF) model: William–Landel–Ferry (WLF) approximation is valid for ionic conduction in highly amorphous system. In highly amorphous polymer electrolyte, the electrical and mechanical relaxation processes occurring above the glass transition temperature (T_g) can be dealt with an empirical function. William–Landel–Ferry (WLF) model can correlate any number of temperature dependent relaxations processes as follows:

$$log\left[\frac{\eta(T)}{\eta(T_s)}\right] = loga_T = -\frac{C_1(T-T_s)}{C_2+T-T_s} \tag{6.8}$$

where η is the viscosity, a_T is the shift factor that denotes the mechanical fluidity, relaxation rate, or inverse relaxation time), C_1 and C_2 are constants with the reference temperature T_S (in general, $T_S = T_g + 50\ K$).

The mechanical shift factor can be denoted as the ratio of viscosity at a particular temperature and that measured at a reference temperature. It can be also measured as the ratio of relaxation time at any temperature and that at a chosen reference temperature. Williams, Landel, and Ferry [133] reported that among seventeen various polymer composites, the variations of shift factor (a_T) with temperature superimposed with one another.

The diffusivity (D) and viscosity can be related by Walden approximation:

$$\eta = \frac{Const}{D} \tag{6.9}$$

The diffusivity can be related to ionic conductivity by Nernst–Einstein relation as follows:

$$\sigma = \frac{DNq^2}{kT} \tag{6.10}$$

where σ is the ionic conductivity at temperature T, k is the Boltzmann's constant, N is the number, and q is the charge state of charge carriers. As the viscosity is inversely proportional to ionic conductivity, hence the shift factor can be formulated as follows:

$$loga_T = log\left[\frac{\eta(T)}{\eta(T_s)}\right] = log\left[\frac{\sigma(T_s)}{\sigma(T)}\right] \tag{6.11}$$

(iii) Vogel–Tamman–Fulcher (VTF) model: This model is applicable to semi-solid gel polymer electrolytes, where the ionic motion takes place above glass transition temperature through the liquid phase of the polymer segments. In highly amorphous polymer materials the ion dynamics can be approximated by VTF model as follows:

$$\sigma = \sigma_o T^{\frac{1}{2}} \exp\left[\frac{-B}{k(T - T_o)}\right] \quad (6.12)$$

where B is pseudo-activation energy and σ_o the pre exponential factor. At temperature T_0, configurational transition probability attains zero. In general, T_0 holds a value in between 20 and 50 K below T_g. When T_0 approaches T, the ionic conductivity becomes vanishingly small.

By fitting the conductivity data, the VTF parameters can be obtained according to the Eq.

$$log_{10}\left(\sigma T^{1/2}\right) = log_{10}\sigma_0 - 0.43\left[\frac{-E_a}{k(T - T_o)}\right] \quad (6.13)$$

The VTF equation can be derived from configurational entropy and quasi-thermodynamic free volume models and generally VTF behavior is observed in electrolyte organic solutions, solid polymer electrolytes and in ionic liquids.

(iv) Dynamic bond percolation (DBP) model: Dynamic Bond Percolation (DBP) model is applicable to disordered dense polymer electrolyte where ion can hop from one site to another in a regular lattice only if the ion path is free or open. This is a microscopic model that explains the dynamical diffusion processes in disordered systems.

Assuming first-order hopping for an ion between adjacent sites, the probability of an ion at site j can be expressed as:

$$\frac{dP_i}{dT} = \sum_j \left(P_j W_{ji} - P_i W_{ij}\right) \quad (6.14)$$

where W_{ji} is rate of hopping from site j to i, $P(t)$ is the probability of an ion at time t at a site i.

In polymer electrolytes, above glass-transition temperature where the polymeric segmental relaxation is possible, the time-dependent hopping becomes:

$$W = W(t) \tag{6.15}$$

As the probability of rates of correlated hopping is a function of time, time dependence W can determine the time-dependence of ion transport properties. In general, a renewal time τ_{ren} can be considered as a characteristic time to reassign the jumps as permitted or forbidden and the diffusion coefficient can be expressed as:

$$D \approx \frac{\langle x^2 \rangle_o}{\tau_{ren}} \tag{6.16}$$

In Eq. (6.16) the numerator denotes the mean-square displacement of an ion in absent of reassigned hopping and the denominator depicts the average renewal time corresponding to the time of segmental relaxation in polymer electrolytes.

6.5.3 Impedance Spectroscopy and Ionic Conductivity Measurements

Electrochemical impedance spectroscopy is an analytical technique to investigate the electrochemical phenomena in the electrode/electrolyte interfaces in solid state devices. It is a perturbative tool to elucidate the response of the non linear electrochemical processes as ac signals of small amplitudes. The physical parameters measured by electrochemical impedance spectroscopy can be categorized as:

1. The parameters pertinent to the material itself such as dielectric constant, conductivity, mobility or concentration of charges.
2. The parameters pertinent to the interface of electrodes and electrolyte such as capacitance of interface, oxidation of electrode material, etc.

The contributions from each of the regions are represented by a combination of R–C components of the cell. At any particular relaxation frequency certain relaxation peaks may appear as the R–C circuit responses to the applied electric signal. Using impedance spectroscopy the ionic conductivity σ of polymer electrolytes can be expressed as follows:

$$\sigma = G.\frac{l}{A} = \frac{1}{R}.\left(\frac{l}{A}\right) \tag{6.17}$$

where G (1/Resistance) is the sample's conductance, R is the sample's resistance, l is the width or thickness of the sample and A is the area of the electrode/electrolyte interface. As like electronic conductors, dc potential cannot be applied directly in ionic conductors as it can rapidly increase resistance as a function of time due to interfacial polarization effect. Therefore, ac potential is applied to determine the ionic conductivity in polymer electrolyte based ionic conductors. In ionic conductivity measurement by ac potential, the resistance term (R) is replaced by impedance (Z) that can is the sum of reactance and resistance. The reactance term has capacitive and inductive components. The polymer electrolyte is sandwiched between two non-blocking or blocking electrodes and one sinusoidal ac potential is applied to the cell. The resulting current occurs corresponding to the applied potential that can be related by the parameters as follows:

1. The ratio of maximum voltage to maximum current (V_{max}/I_{max}).
2. The phase difference between voltage and current (θ).

The impedance (Z) is represented by the combination of the two parameters. The impedance magnitude $\left(|Z| = V_{max}/I_{max}\right)$ with phase angle (θ) can be denoted as functions of angular frequency in the frequency range of 1 mHz–1 MHz.

The complex impedance (Z) can be expressed in terms of angular frequency (ω) as follows:

$$Z = R - \frac{j}{\omega C} = R + \frac{1}{j\omega C} = Z' - jZ'' \tag{6.18}$$

where $j = \sqrt{-1}$, C is the capacitance, Z' and Z'' are the real and imaginary parts of impedance.

Considering capacitance (C) and resistance (R) are in parallel, the complex dielectric impedance Z can be written as:

$$Z = \frac{R\left(\frac{1}{j\omega C}\right)}{R + \frac{1}{j\omega C}} \tag{6.19}$$

or,

$$Z = \frac{R}{1+\omega^2 C^2 R^2} + j\frac{-R^2\,\omega C}{1+\omega^2 C^2 R^2} \tag{6.20}$$

Hence, Z′ and Z″ terms can be represented as:

$$Z' = \frac{R}{1+\omega^2 C^2 R^2} \tag{6.21}$$

and,

$$Z'' = \frac{-R^2\,\omega C}{1+\omega^2 C^2 R^2} \tag{6.22}$$

Removing ω from both of the Eqs:

$$(Z' - R)^2 + Z''^2 = \frac{R^2}{4} \tag{6.23}$$

This Eq. (6.23) represents the equation of any circle at centre ($R/2$, 0) with radius $R/2$. The total impedance of the circuit with flattening of the semicircle and the tilting of the spike can be modeled by the impedance of a constant phase element (Z_{CPE}) as follows:

$$Z_{Total} = \frac{1}{R_b + i\omega C_g} - Z_{CPE} \tag{6.24}$$

The CPE acts as a leaky capacitor responsible to the capacitive dispersion at the interface of electrode and electrolyte and can be denoted as:

$$Z_{CPE}(\omega)=\frac{1}{(i\omega C_{dl})^n} \tag{6.25}$$

Hence, the total dielectric impedance of the circuit can be represented as:

$$Z_{Total}=\frac{1}{R_b+i\omega C_g}-\frac{1}{(i\omega C_{dl})^n} \tag{6.26}$$

where n holds the value in between ½ and 1.

6.5.4 Concept of Mismatch and Relaxation

The mismatch and ion hopping mechanism caused by correlated time dependent factor W (t) and normalized function g (t). W (t) can be expressed as:

$$W(t)=\left(\Gamma_0 x_0^2\right)^{-1}\frac{d}{dt}\langle r^2(t)\rangle \tag{6.27}$$

where Γ_0 is the elementary hopping rate, x_0 is the hopping distance, and $r^2(t)$ the mean square displacement of ions involved in hopping processes. If at time $t = 0$ mismatch is generated then the normalized distance in between the actual and relaxed time of the hopping ion is represented by normalized function g (t). The mismatch is reduced by the rearrangement of neighboring ions and the process of decaying mismatch is monitored by the term g (t). The rate of relaxation of mismatch is denoted by the negative time derivative term $\left[-\frac{d}{dt}g(t)\right]$ in case of any successful hopping made by the rearrangement of neighboring ions. When the effective potential V(r) remains the same and the ions make backward hopping to previous sites the rate of relaxation of mismatch is expressed by the term:

$$-\frac{d}{dt}W(t).\frac{1}{W(t)} \tag{6.28}$$

Thus the rate of forward hopping is denoted as:

$$-\frac{d}{dt}g(t) = Ag(t)W(t) \tag{6.29}$$

and the rate of backward hopping is denoted as:

$$-\frac{d}{dt}W(t) = -BW(t)\frac{d}{dt}g(t) \tag{6.30}$$

where the parameter A is the characteristic frequency, and B represents the conductivities at high and low frequencies.

6.5.5 Scaling of ac Conductivity

The ac conductivity $\sigma(\omega)$ is determined by charge carrier hopping from one coordinated site to another. The ac conductivity shows frequency independent behavior at low frequency but after crossing certain hopping frequency it follows frequency dependent behavior. However, at different ion concentrations and different temperatures ac conductivity profiles follow time-temperature superposition by displaying identical scaling behavior. The superposition of the ac conductivity curves is approximated by scaling law as follows:

$$\sigma(\omega) = \sigma_{dc}F\left(\omega/\omega_s\right) \tag{6.31}$$

where F is the temperature independent term with scaling parameter ω_s, which is temperature dependent. The ac conductivity curves showed scaling behavior as follows:

$$\sigma/\sigma_0 = F(f/f_0) \tag{6.32}$$

Considering f_0 as the characteristic frequency and can be denoted as

$$f_0 = \sigma_0 T/x \tag{6.33}$$

where σ_0 is the dc conductivity, T is the temperature, and x is the ion concentration. They used the Summerfield scaling approach according to

Eq. (6.33) that revealed the superposition of ac conductivity curves into a master one.

In 1999, D.L. Sidebottom proposed a particular scaling approach considering the relaxation strength as $\Delta\varepsilon = \varepsilon_s - \varepsilon_\infty$ that demonstrated the relation between the changes in ionic diffusion path and dielectric permittivity [134].

The frequency of AC conduction can be considered as:

$$f_0 = \sigma_0/\varepsilon_0\Delta\varepsilon, \tag{6.34}$$

the scaling condition can be formulated as,

$$\sigma/\sigma_0 = \mathrm{F}\ (f\varepsilon_0\Delta\varepsilon/\sigma_0) \tag{6.35}$$

Using this scaling approach by D.L. Sidebottom the ac conductivity profiles in ion conducting polymer electrolyte membranes can be successfully scaled.

6.6 IL Incorporated MOF Based Composite Polymer Electrolytes

IL incorporated MOF can be dispersed in polymer electrolytes to prepare composite polymer electrolyte membrane to be used as ion conducting membranes between anode and cathode in any rechargeable battery. Incorporation of IL generates extra free space in pores of MOF for cationic transport as anions of IL are immobilized due to electrostatic interaction with metal nodes. The cations have the tendency to reside at the open pores MOF near to the linker molecules. The cations can hop through the MOF channels in vicinity of the polymer matrix. In polymer electrolyte membranes, both interchain and intrachain hopping of cations can occur to intensify the electrochemical performance of the batteries.

R. Dutta and A. Kumar [135] reported the synthesis of NiBTC–MOF by solvothermal process upon incorporation of IL 1-butyl-3methylimidazolium tetrafluoroborate. The IL incorporated NiBTC–MOF nanocomposites had been dispersed in the polar polymer matrix of poly (vinylidene fluoride-co-hexafluoropropylene) (PVdF–HFP). The schematic representation of the synthesis of IL $BMIMBF_4$ incorporated NiBTC–MOF–PVdF–HFP membranes is depicted in Figure 6.10. The dielectric permittivity and modulus relaxation dynamics, ionic conductivity, scaling behavior of AC conductivity and electrochemical stability of $BMIMBF_4$ incorporated NiBTC–MOF–PVdF–HFP membranes have also been studied. The real and imaginary plots of room temperature permittivity curves depicted

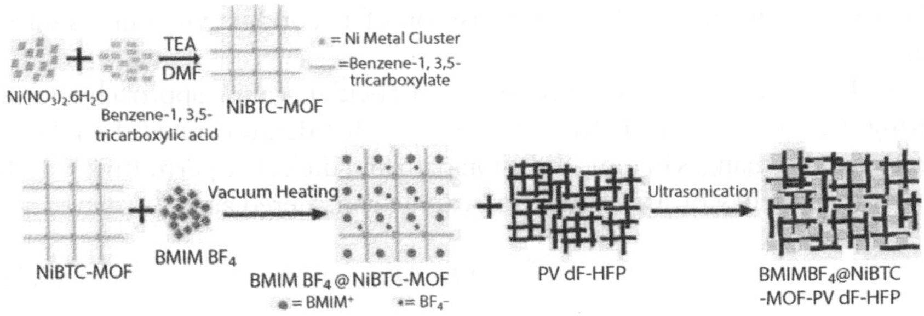

Figure 6.10 Synthesis of IL BMIMBF4 incorporated NiBTC-MOF based PVdF–HFP composite polymer electrolyte. Reprinted with permission from Ref. [135]. Copyright 2018 Springer Nature.

non-Debye relaxation dynamics as shown in Figure 6.11. The real part of permittivity has been fitted by HN approximation and the fitted parameters are depicted in Table 6.1. The real part of modulus spectra was not showing any significant relaxation peak while the imaginary one showed relaxation peaks upon increasing wt% of IL. The imaginary part of room temperature dielectric modulus spectra at different concentration of IL with the temperature dependent modulus spectra at 50 wt% of $BMIMBF_4$ is presented in Figure 6.12. The imaginary part of dielectric modulus data had been fitted according to KWW approximation and the fitted parameters are presented in Table 6.2. The 50 wt% of $BMIMBF_4$ incorporated

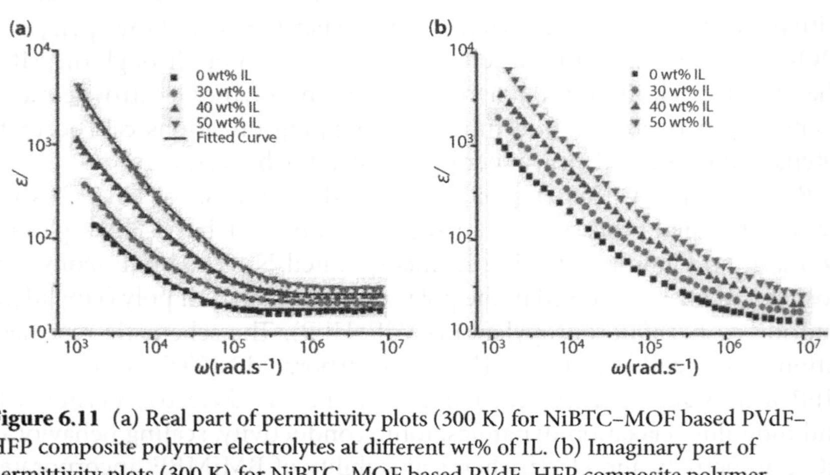

Figure 6.11 (a) Real part of permittivity plots (300 K) for NiBTC–MOF based PVdF–HFP composite polymer electrolytes at different wt% of IL. (b) Imaginary part of permittivity plots (300 K) for NiBTC–MOF based PVdF–HFP composite polymer electrolytes at different wt% of IL. Reprinted with permission from Ref. [135]. Copyright 2018 Springer Nature.

Table 6.1 HN parameters of variation of ε' with frequency (300 K) for NiBTC-MOF based composite electrolytes at different wt% of $BMIMBF_4$.

Wt% of IL	ε_∞	$\Delta\varepsilon$	β_{HN}	α_{HN}
0%	25	136	0.62	0.92
30%	33	576	0.59	0.94
40%	42	945	0.55	0.96
50%	51	1789	0.51	0.98

Reprinted with permission from Ref. [135]. Copyright 2018 Springer Nature.

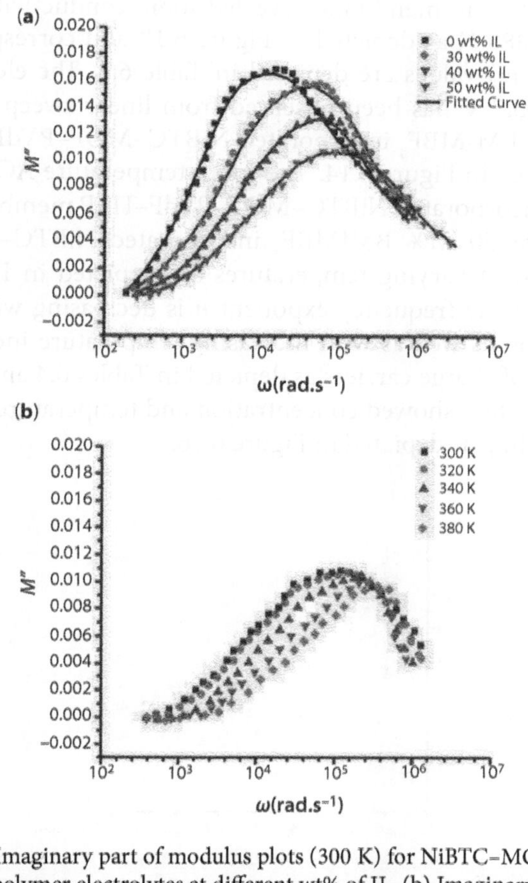

Figure 6.12 (a) Imaginary part of modulus plots (300 K) for NiBTC–MOF based PVdF–HFP composite polymer electrolytes at different wt% of IL. (b) Imaginary part of modulus plots for 50 wt% of $BMIMBF_4$ incorporated NiBTC–MOF based PVdF–HFP composite polymer electrolytes at different temperatures. Reprinted with permission from Ref. [135]. Copyright 2018 Springer Nature.

Table 6.2 M''_{max}, $\log\omega_{max}$, relaxation time (τ), and β_{KWW} for NiBTC–MOF based composite electrolytes at different wt% of $BMIMBF_4$ (300 K).

Wt% of IL	M''_{max}	$\log\omega_{max}$	Relaxation time (τ)	β_{KWW}
0%	0.0169	7.659	2.19×10^{-8}	0.78
30%	0.0159	7.745	1.79×10^{-8}	0.76
40%	0.0148	7.772	1.68×10^{-8}	0.73
50%	0.0109	7.793	1.60×10^{-8}	0.70

composite electrolyte membrane revealed ionic conductivity upto 6.5×10^{-3} S cm^{-1} at 380 K as depicted in Figure 6.13 and corresponding activation energy (E_a) values are depicted in Table 6.3. The electrochemical stability up to 5.7 V has been observed from linear sweep voltammetry for the 50 wt% $BMIMBF_4$ incorporated NiBTC–MOF–PVdF–HFP membrane as depicted in Figure 6.14. The room temperature AC conductivity of $BMIMBF_4$ incorporated NiBTC–MOF–PVdF–HFP membranes and AC conductivity for 50 wt% $BMIMBF_4$ incorporated NiBTC–MOF–PVdF–HFP membrane at varying temperatures are depicted in Figure 6.15. It is observed that the frequency exponent n is decreasing with increasing IL concentration as well as with increasing temperature indicating short range hopping of charge carriers as depicted in Tables 6.4 and 6.5. The AC conductivity profiles showed concentration and temperature independent behavior of scaling as depicted in Figure 6.16.

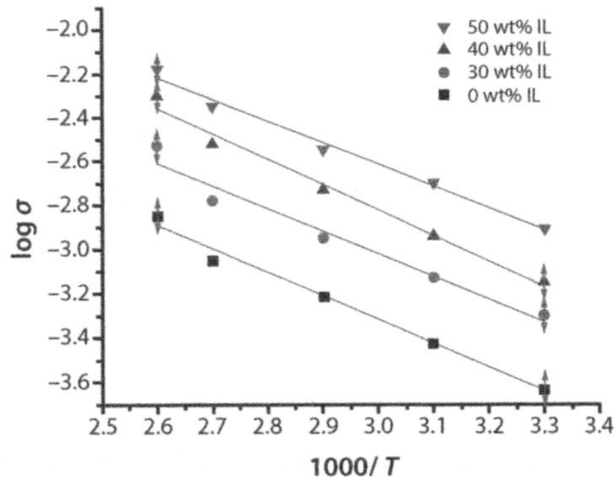

Figure 6.13 Temperature dependent ionic conductivity plots for NiBTC–MOF based PVdF–HFP composite polymer electrolytes at different wt% of IL. Reprinted with permission from Ref. [135]. Copyright 2018 Springer Nature.

Table 6.3 Activation energy (E_a) with regression line (intercept) of NiBTC–MOF based PVdF–HFP composite electrolytes at different wt% of $BMIMBF_4$.

Wt% of IL	Activation energy (E_a) (eV)	Regression line (intercept) value
0%	0.31	−0.135
30%	0.29	0.063
40%	0.26	0.186
50%	0.23	0.345

Reprinted with permission from Ref. [135]. Copyright 2018 Springer Nature.

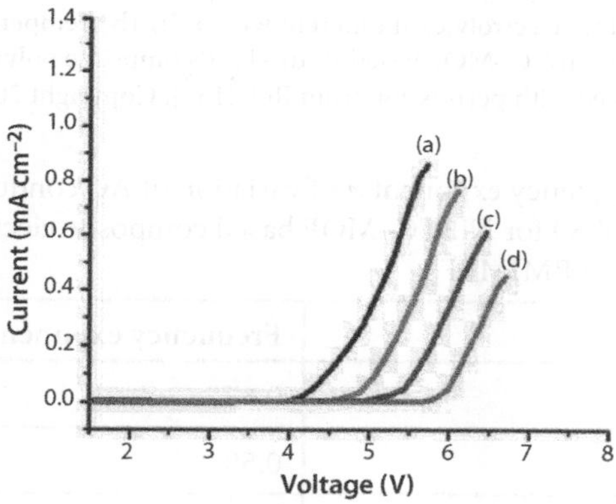

Figure 6.14 Electrochemical stability profiles for NiBTC–MOF based PVdF–HFP composite polymer electrolytes at different wt% of IL where (a) 0 wt% IL, (b) 30 wt% IL, (c) 40 wt% IL, and (d) 50 wt% IL. Reprinted with permission from Ref. [135]. Copyright 2018 Springer Nature.

R. Dutta and A. Kumar [136] reported the synthesis of 1-butyl-3 methylimidazolium bromide incorporated CuBTC–MOF based poly (ethylene oxide) composite polymer electrolytes and investigated their structural and electrochemical properties. Significant changes in binding energies of the spin–orbit peaks of Cu species were obtained from XPS data as depicted in Figure 6.17 indicating interaction between Br^- anions and Cu metal nodes of the CuBTC–MOF. The periodic oscillation characteristics and coordination geometry of Cu K-edge were observed from the *k*-space as well as R-space data of scanning XANES and EXAFS as depicted in Figures 6.18 and 6.19. Asymmetric periodic oscillation and changes in coordination number had been observed from *k*-space and R-space data owing to the dominant interaction of Br^- with unsaturated Cu metal cluster of MOF. The composite polymer electrolyte exhibited ionic conductivity of 5×10^{-3} S cm^{-1} with

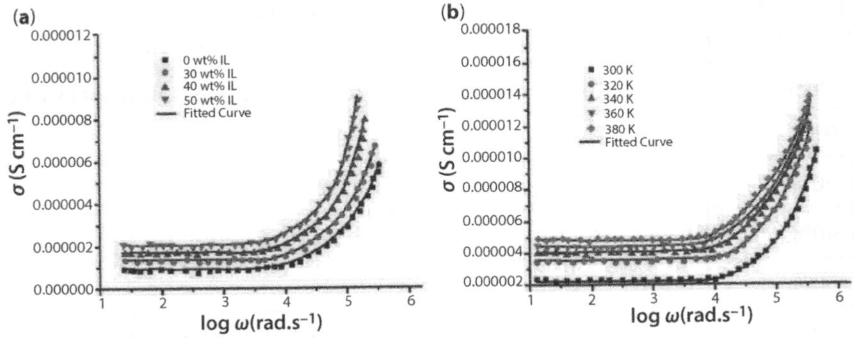

Figure 6.15 (a) Room temperature AC conductivity plots for NiBTC–MOF based PVdF–HFP composite polymer electrolytes at different wt% of IL. (b) Temperature dependent AC conductivity plots for NiBTC–MOF based PVdF–HFP composite polymer electrolytes at 50 wt% of IL. Reprinted with permission from Ref. [135]. Copyright 2018 Springer Nature.

Table 6.4 Frequency exponent *n* of variation of AC conductivity with frequency (300 K) for NiBTC–MOF based composite electrolytes at different wt% of $BMIMBF_4$.

Wt% of IL	**Frequency exponent (*n*)**
0%	0.67
30%	0.58
40%	0.49
50%	0.38

Reprinted with permission from Ref. [135]. Copyright 2018 Springer Nature.

Table 6.5 Frequency exponent *n* of variation of AC conductivity with frequency for 50 wt% $BMIMBF_4$ incorporated NiBTC–MOF based composite electrolytes at different temperatures.

Temperature (K)	**Frequency exponent (*n*)**
300	0.38
320	0.34
340	0.31
360	0.28
380	0.25

Reprinted with permission from Ref. [135]. Copyright 2018 Springer Nature.

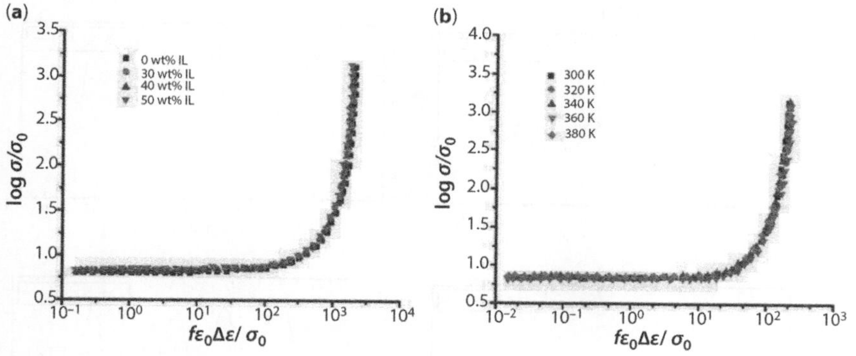

Figure 6.16 (a) Room temperature AC conductivity scaling plots for NiBTC–MOF based PVdF–HFP composite polymer electrolytes at different wt% of IL. (b) Temperature dependent AC conductivity scaling plots for NiBTC–MOF based PVdF–HFP composite polymer electrolytes at 50 wt% of IL. Reprinted with permission from Ref. [135]. Copyright 2018 Springer Nature

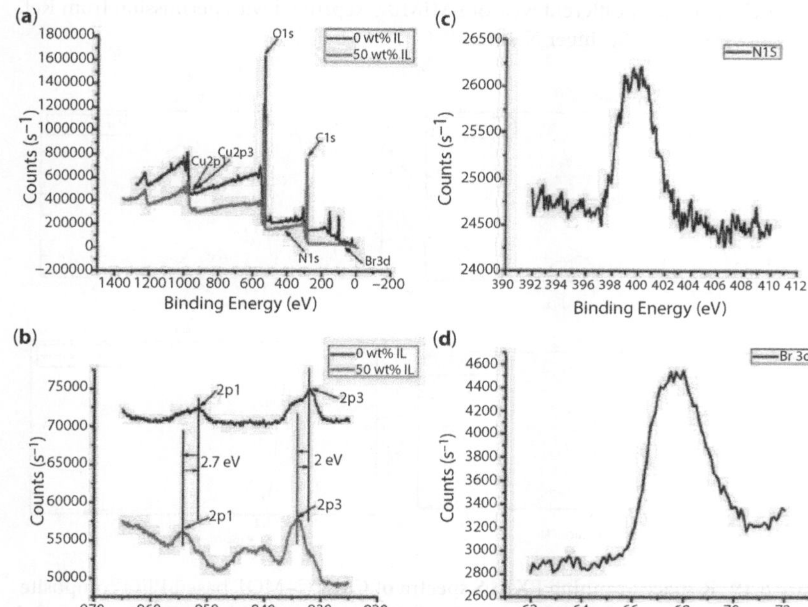

Figure 6.17 (a) XPS spectra for CuBTC–MOF based PEO composite polymer electrolytes at 0 wt% and 50 wt% of BMIMBr. (b) XPS spectra of Cu 2p peaks of CuBTC–MOF based PEO composite polymer electrolytes at 0 wt% and 50 wt% of BMIMBr. (c) XPS spectrum of N 1s peak of CuBTC–MOF based PEO composite polymer electrolytes at 50 wt% of BMIMBr. (d) XPS spectrum of Br 3d peak of CuBTC–MOF based PEO composite polymer electrolytes at 50 wt% of BMIMBr. Reprinted with permission from Ref. [136]. Copyright 2018 Springer Nature.

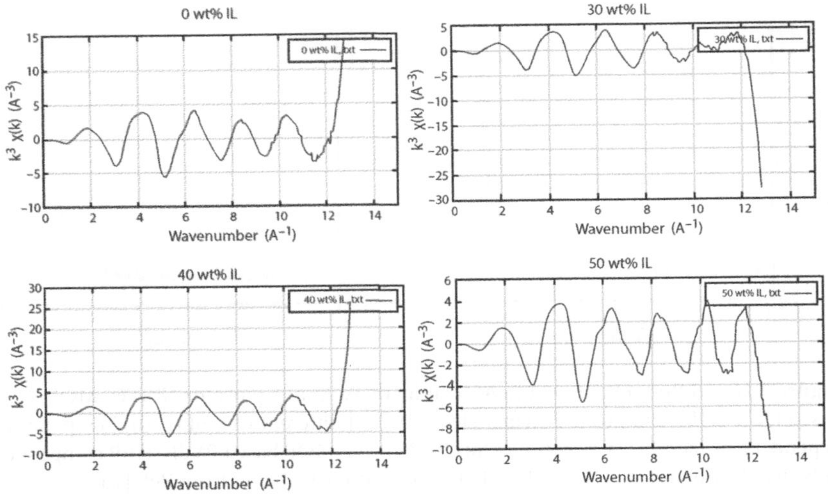

Figure 6.18 *k*-space Scanning XANES spectra of CuBTC–MOF based PEO composite polymer electrolytes at different wt% of BMIMBr. Reprinted with permission from Ref. [136]. Copyright 2018 Springer Nature.

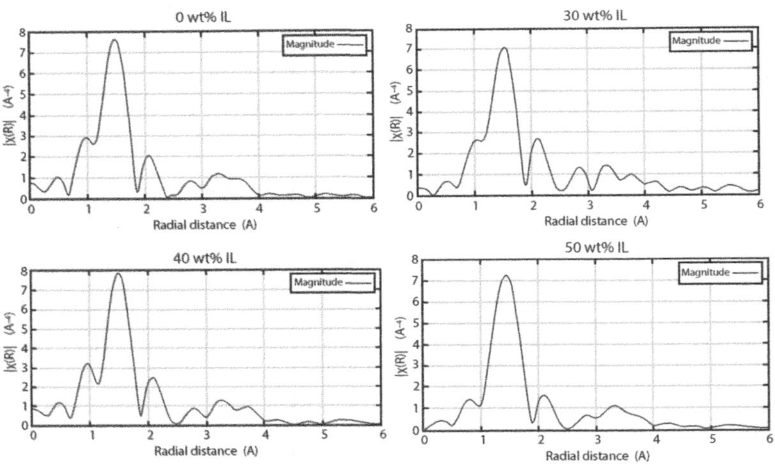

Figure 6.19 R-space Scanning EXAFS spectra of CuBTC–MOF based PEO composite polymer electrolytes at different wt% of BMIMBr. Reprinted with permission from Ref. [136]. Copyright 2018 Springer Nature.

activation energy of 0.24 eV at 50 wt% of BMIMBr incorporation at 380 K as depicted in Figure 6.20 and Table 6.6. The electrochemical stability profiles as depicted in Figure 6.21 had been observed to increase with increasing IL incorporation and attained 6.1 V upon incorporation of 50 wt% of BMIMBr.

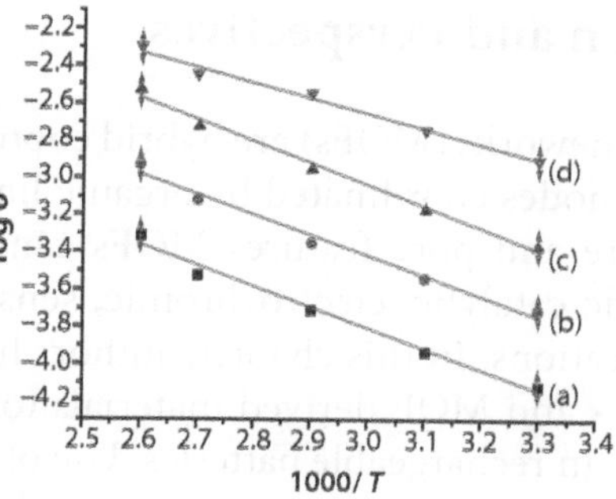

Figure 6.20 Temperature dependent ionic conductivity profiles of CuBTC–MOF based PEO nanocomposites at different wt% of BMIMBr. Reprinted with permission from Ref. [136]. Copyright 2018 Springer Nature.

Table 6.6 Activation energy (E_a) with regression line (intercept) for CuBTC–MOF based PEO composite electrolytes at different wt% of BMIMBr.

Wt% of IL	Activation energy (E_a)(eV)	Regression line (intercept) value
0%	0.32	−0.466
30%	0.30	−0.161
40%	0.27	−0.441
50%	0.24	−0.191

Reprinted with permission from Ref. [136]. Copyright 2018 Springer Nature.

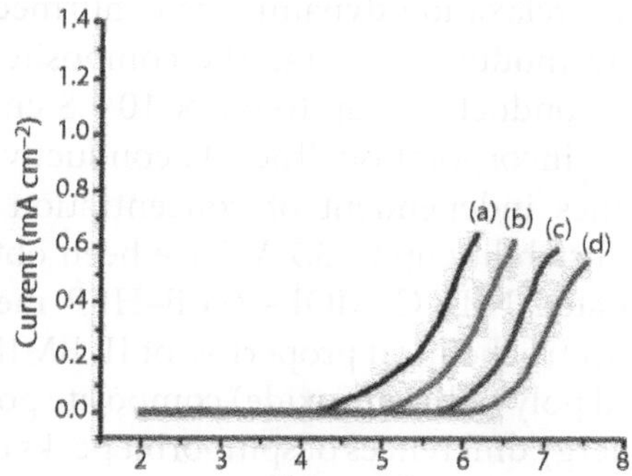

Figure 6.21 Electrochemical stability of CuBTC–MOF based PEO nanocomposites at different wt% of BMIMBr where, (a) 0 wt% BMIMBr, (b) 30 wt% BMIMBr, (c) 40 wt% BMIMBr and (d) 50 wt% BMIMBr. Reprinted with permission from Ref. [136]. Copyright 2018 Springer Nature.

6.7 Conclusion and Perspectives

Metal–Organic Frameworks (MOFs) are hybrid coordination compounds consisting of metal nodes coordinated by organic linkers. Based on their topological structure and pore features MOFs can be used as efficient materials for specific catalytic, electrochromic, sensing, gas storage and sequestration applications. In this chapter, authors have discussed different aspects of MOFs and MOF derived materials to be used as ion conducting membranes in rechargeable batteries. Use of polymer electrolytes as ion conducting membranes between anodes and cathodes in batteries is a fascinating field of research in the field of electrochemical solid state devices including rechargeable batteries. Authors have reviewed the historical perspectives of ion conducting polymer electrolytes with different strategies to increase the electrochemical performance of composite polymer electrolytes. MOFs can be embedded as filler materials in polar polymer electrolyte matrix to synthesize composite electrolyte membranes to use as ion conducting dielectric membranes in rechargeable batteries. MOFs can be incorporated with Ionic Liquids (ILs) to be used as composite ionic conductors as their ion dynamics can be controlled by dominant MOF–IL interaction. IL incorporation can facilitate cationic transport through MOF channels as the IL anions are immobilized by the unsaturated metal centers through dominant electrostatic interaction. MOF incorporated with IL can be dispersed in polar polymer electrolyte to develop composite electrolyte membranes with compatible electrochemical performance. Authors have reviewed the development of IL $BMIMBF_4$ incorporated NiBTC–MOF based PVdF–HFP membranes that exhibited non-Debye dielectric relaxation dynamics as confirmed from the dielectric permittivity and modulus spectra. The composite polymer electrolytes revealed ionic conductivity up to 6.5×10^{-3} S cm^{-1} at 380 K upon 50 wt% of $BMIMBF_4$ incorporation. The AC conductivity profiles exhibited scaling dynamics independent of concentration and temperature and electrochemical stability up to 5.7 V have been obtained for 50 wt% $BMIMBF_4$ incorporated NiBTC–MOF–PVdF–HFP membranes. Studies on structural and electrochemical properties of IL BMIMBr incorporated CuBTC–MOF based poly (ethylene oxide) composite polymer electrolytes revealed binding energy differences of spin–orbit peaks of Cu species upon IL incorporation as observed from XPS spectra. Heterogeneity in oscillation periodicity and changes in coordination number had been observed from the *k*-space and R-space data indicating dominant interaction of

Br^- with metal clusters of CuBTC–MOF. At 50 wt% of BMIMBr incorporation in CuBTC–MOF, the composite poly (ethylene oxide) membrane exhibited ionic conductivity of 5×10^{-3} S cm^{-1} with activation energy of 0.24 eV at 380 K. The electrochemical stability had been observed to increase with increasing IL incorporation and attained 6.1 V at 50 wt% of BMIMBr incorporation in CuBTC–MOF–PEO composite electrolytes. Synthesis of MOFs with variation of metal clusters and organic ligands and incorporation of IL with different cations and anions are of fascinating research interest in view of their promising electrochemical performance. Theoretical modeling of the dynamics of IL ions within the pores of MOF can be carried out by Molecular Dynamics (MD) and Monte Carlo (MC) simulation techniques to investigate their interactions at different binding sites of MOFs. Inelastic Neutron Scattering (INS) spectroscopy can elucidate the phonon assisted diffusion processes in MOFs. INS spectroscopy can investigate the local coordination structure of the metal clusters of MOFs and probable interaction of IL ions with metal nodes or linker moieties of MOFs. The shifting or modulation of any defect site present in the metal nodes or linker molecules of MOFs can also be investigated to increase the microporosity for entrapping the ions of ILs. INS experiment on the IL incorporated MOF nanocomposites can provide information on the density of states of different phonon modes associated with the linker molecules of MOFs and corresponding changes in the scattering intensity profiles with IL incorporation. Change in scattering intensity of a particular phonon mode indicates the probable displacement of IL ions within the pores of MOF. Swift heavy ion (SHI) irradiation is a technique to modify structural, morphological, and electrochemical parameters of polymer electrolytes in a selective and controlled manner. Interaction of energetic ions with polymer electrolytes may cause cross-linking, chain-scission with shifting of molecular fragments in polymer chains. In the process of irradiation, high energy ions can transmit energy to the atomic lattice through electron–phonon coupling to create amorphous columnar phases. This irradiation can induce structural defects in MOF based composite electrolyte membranes by creating vacancies at the node or linker position modifying their electrochemical properties. SHI irradiation in MOF can cause certain defects including vacancies and dislocations in organic linker moieties resulting in increased pore volume. Increased concentration of IL can be accommodated in the enlarged MOF pores affected by the SHI irradiation induced linker defects to increase ion mobility in composite polymer electrolyte membranes.

References

1. Chinnam, P.R. and Wunder, S.L., Polyoctahedralsilsesquioxane-nanoparticle electrolytes for lithium batteries: POSS–Lithium salts and POSS–PEGs. *Chem. Mater.*, 23, 5111, 2011.
2. Quartarone, E. and Mustarelli, P., Electrolytes for solid-state lithium rechargeable batteries: Recent advances and perspectives. *Chem. Soc. Rev.*, 40, 2525, 2011.
3. Tarascon, J.M. and Armand, M., Issues and challenges facing rechargeable lithium batteries. *Nature*, 414, 359, 2001.
4. Zhang, H., Kulkarni, S., Wunder, S.L., Blends of POSS PEO $(n = 4)_8$ and high molecular weight poly(ethylene oxide) as solid polymer electrolytes for lithium batteries. *J. Phys. Chem. B*, 111, 3583, 2007.
5. Weston, J.E. and Steele, B.C.H., Effects of inert fillers on the mechanical and electrochemical properties of lithium salt-poly(ethylene oxide) polymer electrolytes. *Solid State Ionics*, 7, 75, 1982.
6. Borghini, M.C., Mastragostino, M., Passerini, S., Scrosati, B., Electrochemical properties of Polyethylene Oxide-Li$(CF_3SO_2)_2N$]-gamma $LiAlO_2$ composite polymer electrolytes. *J. Electrochem. Soc.*, 142, 2118, 1995.
7. Liquan, C. (Ed.), *Materials for Solid State Batteries*, World Scientific, Singapore, 1988.
8. Wieczorek, W., Florjanczykt, Z., Stevens, J.R., Composite polyether based solid electrolytes. *Electrochim. Acta*, 40, 2251, 1995.
9. Croce, F., Appetecchi, G.B., Persi, L. *et al.*, Nanocomposite Polymer Electrolytes for Lithium Batteries. *Nature*, 394, 456, 1998.
10. Przyluski, J., Siekierski, M., Wieczorek, W., Effective medium theory in studies of conductivity of composite polymeric electrolytes. *Electrochim. Acta*, 40, 2101, 1995.
11. Shekhah, O., Liu, J., Fischer, R. *et al.*, MOF thin films: Existing and future applications. *Chem. Soc. Rev.*, 40, 1081, 2011.
12. Li, J., Kuppler, R.J., Zhou, H., Selective gas adsorption and separation in metal–organic frameworks. *Chem. Soc. Rev.*, 38, 1477, 2009.
13. Tranchemontagne, D.J., Hunt, J.R., Yaghi, O.M., Room temperature synthesis of metal–organic frameworks: MOF-5, MOF-74, MOF-177, MOF-199, and IRMOF-0. *Tetrahedron*, 64, 8553, 2008.
14. Angulakshmi, N., Kumar, R.S., Kulandainathan, M.A., Stephan, A.M., Composite Polymer Electrolytes Encompassing Metal Organic Frameworks: A New Strategy for All-Solid-State Lithium Batteries. *J. Phys. Chem. C*, 118, 24240, 2014.
15. Chui, S.S., Lo, S.M., Charmant, J.P. *et al.*, A chemically functionalizable nanoporous material. *Science*, 283, 1148, 1999.
16. Welton, T., Room-Temperature Ionic Liquids. Solvents for Synthesis and Catalysis. *Chem. Rev.*, 99, 2071, 1997.

17. Sang, X., Zhang, J. *et al.*, Ionic liquid accelerates the crystallization of Zr-based metal–organic frameworks. *Nat. Commun.*, 8, 1, 2017.
18. Carlin, R.T., Delong, H.C., Fuller, J., Trulove, P.C.J., Dual Intercalating Molten Electrolyte Batteries. *Electrochem. Soc.*, 141, 73, 1994.
19. Fuller, J., Breda, A.C., Carlin, R.T., Ionic liquid–polymer gel electrolytes from hydrophilic and hydrophobic ionic liquids. *J. Electroanal. Chem.*, 459, 29, 1998.
20. Scott, M.P., Brazel, C.S., Benton, M.G., Mays, J.W., Holbrey, J.D., Rogers, R.D., Application of ionic liquids as plasticizers for poly(methylmethacrylate). *Chem. Commun.*, 13, 1370, 2002.
21. Sutto, T.E., Hydrophobic and hydrophilic interactions of ionic liquids and polymers in solid polymer gel electrolytes. *J. Electrochem. Soc.*, 154, 101, 2007.
22. Noda, A. and Watanabe, M., Highly conductive polymer electrolytes prepared by *in-situ* polymerization of vinyl monomers in room temperature molten salts. *Electrochim. Acta*, 45, 1265, 2000.
23. Lewandowski, A. and Swiderska, A., New composite solid electrolytes based on a polymer and ionic liquids. *Solid State Ionics*, 169, 21, 2004.
24. Kumar, Y., Hashmi, S.A., Pandey, G.P., Lithium ion transport and ion–polymer interaction in PEO based polymer electrolyte plasticized with ionic liquid. *Solid State Ionics*, 201, 73, 2011.
25. Chaurasia, S.K., Singh, R.K., Chandra, S., Structural and transport studies on polymeric membranes of PEO containing ionic liquid, EMIM-TY: Evidence of complexation. *Solid State Ionics*, 183, 32, 2011.
26. Kumar, Y., Hashmi, S.A., Pandey, G.P., Ionic liquid mediated magnesium ion conduction in poly(ethylene oxide) based polymer electrolyte. *Electrochim. Acta*, 56, 3864, 2011.
27. Hofmann, A., Schulz, M., Hanemann, T., Gel electrolytes based on ionic liquids for advanced lithium polymer batteries. *Electrochim. Acta*, 89, 823, 2013.
28. Polu, A.R. and Rhee, H.W., Ionic Liquid Doped PEO-Based Solid Polymer Electrolytes for Lithium-Ion Polymer Batteries. *Int. J. Hydrog. Energy*, 42, 7212, 2017.
29. Park, C.H., Kim, D.W., Prakash, J. *et al.*, Electrochemical stability and conductivity enhancement of composite polymer electrolytes. *Solid State Ionics*, 159, 111, 2003.
30. Bohrmanand, J.A. and Carreon, M.A., Synthesis and CO_2/CH_4 separation performance of Bio-MOF-1 membranes. *Chem. Commun.*, 48, 5130, 2012.
31. Prestipino, C., Regli, L., Vitillo, J.G. *et al.*, Local Structure of Framework Cu(II) in HKUST-1 Metallorganic Framework: Spectroscopic Characterization upon Activation and Interaction with Adsorbates. *Chem. Mater.*, 18, 1337, 2006.
32. Chavan, S., Vitillo, J.G., Gianolio, D. *et al.*, H_2 storage in isostructural UiO-67 and UiO-66 MOFs. *Phys. Chem. Chem. Phys.*, 14, 1614, 2012.
33. Tranchemontagne, D.J., Hunt, J.R., Yaghi, O.M., Room temperature synthesis of metal-organic frameworks: MOF-5, MOF-74, MOF-177, MOF-199, and IRMOF-0. *Tetrahedron*, 64, 8553, 2008.

34. Cavka, J.H., Jakobsen, S., Olsbye, U. *et al.*, A new zirconium inorganic building brick forming metal organic frameworks with exceptional stability. *J. Am. Chem. Soc.*, 130, 13850, 2008.
35. Valenzano, L., Civalleri, B., Chavan, S. *et al.*, Disclosing the Complex Structure of UiO-66 Metal Organic Framework: A Synergic Combination of Experiment and Theory. *Chem. Mater.*, 23, 1700, 2011.
36. Chavan, S., Vitillo, J.G., Uddin., M.J. *et al.*, Functionalization of UiO-66 metal–organic framework and highly crosslinked polystyrene with $Cr(CO)_3$: *In situ* formation, stability, and photoreactivity. *Chem. Mater.*, 22, 4602, 2010.
37. Kandiah, M., Usseglio, S., Svelle, S. *et al.*, Post-synthetic modification of the metal–organic framework compound UiO-66. *J. Mater. Chem.*, 20, 9848, 2010.
38. Rowsell, J.L.C. and Yaghi, O.M., Metal–organic frameworks: A new class of porous materials. *Microporous Mesoporous Mater.*, 73, 3, 2004.
39. Yaghi, O.M., O'Keeffe, M., Ockwig, N.W. *et al.*, Reticular synthesis and the design of new materials. *Nature*, 423, 705, 2003.
40. Eddaoudi, M., Kim, J., Rosi, N. *et al.*, Systematic design of pore size and functionality in isoreticular MOFs and their application in methane storage. *Science*, 295, 469, 2002.
41. Kaye, S.S., Dailly, A., Yaghi, O.M., Long, J.R., Impact of preparation and handling on the hydrogen storage properties of Zn4O (1,4-benzenedicarboxylate)3 (MOF-5). *J. Am. Chem. Soc.*, 129, 14176, 2007.
42. Chae, H.K., Siberio-Perez, D.Y., Kim, J. *et al.*, A route to high surface area, porosity and inclusion of large molecules in crystals. *Nature*, 427, 523, 2004.
43. Furukawa, H., Ko, N., Go, Y.B. *et al.*, Ultrahigh porosity in metal-organic frameworks. *Science*, 329, 424, 2010.
44. Ferey, G., Mellot-Draznieks, C., Serre, C. *et al.*, A chromium terephthalate-based solid with unusually large pore volumes and surfacearea. *Science*, 309, 2040, 2005.
45. Hong, D.Y., Hwang, Y.K., Serre, C. *et al.*, Porous chromium terephthalate MIL-101 with coordinatively unsaturated sites: Surface functionalization, encapsulation, sorption and catalysis. *Adv. Funct. Mater.*, 19, 1537, 2009.
46. Chui, S.S.Y., Lo, S.M.F., Charmant, J.P.H. *et al.*, A chemically functionalizable nanoporous material. *Science*, 283, 1148, 1999.
47. Horcajada, P., Serre, C., Maurin, G. *et al.*, Flexible porous metal–organic frameworks for a controlled drug delivery. *J. Am. Chem. Soc.*, 130, 6774, 2008.
48. Finsy, V., Kirschhock, C.E.A., Vedts, G. *et al.*, Framework breathing in the vapour-phase adsorption and separation of xylene isomers with the metal–organic framework MIL-53. *Chem. Eur. J.*, 15, 7724, 2009.
49. Kitagawa, S., Kitaura, R., Noro, S., Functional porous coordination polymers. *Angew. Chem. Int. Ed.*, 43, 2334, 2004.
50. Alaerts, L., Seguin, E., Poelman, H. *et al.*, Probing the Lewis acidity and catalytic activity of the metal-organic framework Cu-3(btc)(2). (BTC = benzene-1,3,5-tricarboxylate). *Chem. Eur. J.*, 12, 7353, 2006.

51. Kramer, M., Ulrich, S.B., Kaskel, S., Synthesis and properties of the metal-organic framework Mo_3 $(BTC)_2$(TUDMOF-1). *J. Mater. Chem.*, 16, 2245, 2006.
52. Henschel, A., Gedrich, K., Kraehnert, R. *et al.*, Catalytic properties of MIL-101. *Chem. Commun.*, 4192, 2008.
53. Kitaura, R., Onoyama, G., Sakamoto, H. *et al.*, Immobilization of a metallo-schiff base into a microporous coordination polymer. *Angew. Chem. Int. Ed.*, 43, 2684, 2004.
54. Millward, A. and Yaghi, O., Metal–organic frameworks with exceptionally high capacity for storage of carbon dioxide at room temperature. *J. Am. Chem. Soc.*, 127, 17998, 2005.
55. Serre, C., Bourrelly, S., Vimont, A. *et al.*, An explanation for the very large breathing effect of a metal-organic framework during CO_2 adsorption. *Adv. Mater.*, 19, 2246, 2007.
56. Couck, S., Denayer, J., Baron, G. *et al.*, An amine functionalized MIL-53 metal-organic framework with large separation power for CO_2 and CH_4. *J. Am. Chem. Soc.*, 131, 6326, 2009.
57. Arstad, B., Fjellvåg, H., Kongshaug, K. *et al.*, Amine functionalised metal organic frameworks (MOFs) as adsorbents for carbon dioxide. *Adsorption*, 14, 755, 2008.
58. Horcajada, P., Serre, C., Maurin, G. *et al.*, Flexible porous metal–organic frameworks for a controlled drug delivery. *J. Am. Chem. Soc.*, 130, 6774, 2008.
59. Wong-Foy, A.G., Matzger, A.J., Yaghi, O.M., Exceptional H_2 saturation in microporous metal-organic frameworks. *J. Am. Chem. Soc.*, 128, 3494, 2006.
60. Farha, O.K., Yazaydin, A.O., Eryazici, I. *et al.*, *De novo* synthesis of a metal–organic framework material featuring ultrahigh surface area and gas storage capacities. *Nat. Chem.*, 2, 944, 2010.
61. Furukawa, H., Ko, N., Go, Y.B. *et al.*, Ultrahigh porosity in metal-organic frameworks. *Science*, 329, 424, 2010.
62. Suh, M.P., Park, H.J., Prasad, T.K., Lim, D.W., Hydrogen storage in metal-organic frameworks. *Chem. Rev.*, 112, 782, 2012.
63. An, J., Geib, S.J., Rosi, N.L., High and selective CO_2 uptake in a cobalt adeninate metal-organic framework exhibiting pyrimidine and amino decorated pores. *J. Am. Chem. Soc.*, 132, 38, 2009.
64. Ma, S., Sun, D., Simmons, J.M. *et al.*, Metal organic framework from an anthracene derivative containing nanoscopic cage exhibiting high methane uptake. *J. Am. Chem. Soc.*, 130, 1012, 2008.
65. Wu., H., Zhou., W., Yildirim, T., High-capacity methane storage in metal-organic frameworks M_2 (dhtp): The important role of open metal site. *J. Am. Chem. Soc.*, 131, 4995, 2009.
66. McKinlay, A.C., Xiao, B., Wragg, D.S. *et al.*, Exceptional behaviour over the whole adsorption-storage-delivery cycle for NO in porous metal–organic frameworks. *J. Am. Chem. Soc.*, 130, 10440, 2008.

67. Fujita, M., Kwon, Y.J., Washizu, S., Ogura, K., Preparation, clathration ability, and catalysis of a two-dimensional square network material composed of cadmium (II) and 4,4′-bipyridine. *J. Am. Chem. Soc.*, 116, 1151, 1994.
68. Alaerts, L. *et al.*, Probing the Lewis acidity and catalytic activity of the metal-organic framework $Cu_3(BTC)_2$. (BTC = benzene-1,3,5-tricarboxylate). *Chemistry*, 12, 7353, 2006.
69. Horike, S., Dincă, M., Tamaki, K., Long, J.R., Size-selective Lewis acid catalysis in a microporous metal–organic framework with exposed Mn^{2+} coordination sites. *J. Am. Chem. Soc.*, 130, 5854, 2008.
70. Phan, A., Czaja, A.U., Gándara, F., Knobler, C.B., Yaghi, O.M., Metal-organic frameworks of vanadium as catalysts for conversion of methane to acetic acid. *Inorg. Chem.*, 50, 7388, 2011.
71. Hasegawa, S., Horike, S., Matsuda, R., Furukawa, S., Mochizuki, K.Y., Kinoshita, Y., Kitagawa, S., Three dimensional porous coordination polymer functionalized with amide group based on tridentate ligand: Selective sorption and catalysis. *J. Am. Chem. Soc.*, 129, 2607.
72. Hwang, Y.K., Hong, D.Y., Chang, J.S., Jhung, H.S., Seo, Y.K., Kim, J., Vimont, A., Daturi, M., Serre, C., Férey, G., Amine grafting on coordinatively unsaturated metal centers of MOFs: Consequences for catalysis and metal encapsulation. *Angew. Chem. Int. Ed.*, 47, 4144, 2008.
73. Lim, Y.T., Noh, Y.W., Cho, J.H., Han, J.H., Choi, B.S., Kwon, J., Hong, K.S., Gokarna, A., Cho, Y.H., Chung, B.H., Multiplexed imaging of a therapeutic cells with multispectrally encoded magnetofluorescent nanocomposite emulsions. *J. Am. Chem. Soc.*, 131, 17145, 2009.
74. Desai, A.V., Manna, B., Karmakar, A., Sahu, A., Ghosh, S.K., A Water-Stable Cationic Metal–Organic Framework as a Dual Adsorbent of Oxoanion Pollutants. *Angew. Chem. Int. Ltd.*, 55, 7811, 2016.
75. Mukherjee, S., Aamod, V., Desai, A.V. *et al.*, Exploitation of Guest Accessible Aliphatic Amine Functionality of a Metal–Organic Framework for Selective Detection of 2,4,6-Trinitrophenol (TNP) in Water. *Cryst. Growth Des.*, 15, 4627, 2015.
76. Deep, A., Bhardwaj, S.K., Paul, A.K., Kim, K.H., Kumar, P., Surface assembly of nano metal organic framework on amine functionalized indium tin oxide substrate for impedimetric sensing of parathion. *Biosens. Bioelectron.*, 65, 226, 2015.
77. Horcajada, P., Chalati, T., Serre, C. *et al.*, Porous metal-organic framework nanoscale carriers as a potential plateform for drug delivery and imaging. *Nat. Mater.*, 9, 172, 2010.
78. Horcajada, P., Serre, C., Vallet-regi, M., Sebban, M., Taulelle, F., Ferey, G., Metal–organic frameworks as efficient materials for drug delivery. *Angew. Chem. Int. Ed.*, 45, 5974, 2006.
79. Ke, F., Yuan, Y.P., Qiu, L.G., Shen, Y.H., Xie, A.J., Zhu, J.F., Tian, X.Y., Zhang, L.D., Fecile fabrication of magnetic metal-organic framework nanocomposite for potential targeted drug delivery. *J. Mater. Chem.*, 21, 3843, 2011.

80. Zhuang, J., Kuo, C.H., Chou, L.Y., Weerapana, E., Tsung, C.K., Optimized metal–organic frameworks nanospheres for drug delivery: Evaluation of small-molecule incapsulation. *ACS Nano*, 8, 3, 2812, 2014.
81. Bernini, M.C., Jimenez, D.F., Pasinetti, M., Pastor, A.J.R., Snurr, R.Q., Screening of bio-compatible metal–organic frameworks as potential drug carriers using Monte Carlo simulations. *J. Mater. Chem. B*, 2, 766, 2014.
82. Babarao, R. and Jiang, J., Unraveling the energatics and dynamics of Ibuprofen in mesoporous metal-organic frameworks. *J. Phys. Chem.*, 113, 18287, 2009.
83. Fateeva, A., Horcajada, P., Devic, T., Serre, C., Marrot, J., Greneche, J.M., Morcrette, M., Tarascon, J.M., Maurin, G., Ferey, G., Synthesis, Structure, Characterization and Redox Properties of the Porous MIL-68 (Fe) Solid. *Eur. J. Inorg. Chem.*, 3789, 2010.
84. Nagarathinam, M., Saravanan, K., Phua, E.J.H., Reddy, M.V., Chowdari, B.V.R., Vittal, J.J., Redox-active metal-centered oxalato phosphate open framework cathode materials for lithium ion batteries. *Angew. Chem. Int. Ed.*, 51, 5866, 2012.
85. Li, X.X., Cheng, F.Y., Zhang, S., Chen, J., Shape-controlled synthesis and lithium-storage study of metal-organic frameworks Zn_4O(1,3,5-benzenetribenzoate)$_2$. *J. Power Sources*, 160, 542, 2006.
86. Diaz, R., Orcajo, M.G., Botas, J.A., Calleja, G., Palma., J., Co8-MOF-5 as electrode for supercapacitors. *Mater. Lett.*, 68, 16, 2012.
87. Lee, D.Y., Yoon, S.J., Shrestha, N.K., Lee, S.H., Ahn, H., Han, S.H., Unusual energy storage and charge retention in Co-based meal–organic frameworks. *Microporous Mesoporous Mater.*, 153, 163, 2012.
88. Choi, K.M., Jeong, H.M., Park, J.H., Zhang, Y.B., Kang, J.K., Yaghi, O.M., Supercapacititors for nanocrystalline metal–organic frameworks. *ACS Nano*, 8, 7451, 2014.
89. Bauer, S., Serre, C., Devic, T., Horcajada, P., Marrot, J., Férey, G., Stock, N., High-throughput assisted rationalization of the formation of metal organic frameworks in the iron(III) aminoterephthalate solvothermal system. *Inorg. Chem.*, 47, 7568, 2008.
90. Zhao, Z., Li, X., Huang, S., Xia, Q., Li, Z., Adsorption and diffusion of benzene on chromium-based metal organic framework MIL-101 synthesized by microwave irradiation. *Ind. Eng. Chem. Res.*, 50, 2254, 2011.
91. Choi, J.Y., Kim, J., Jhung, S.H., Kim, H.K., Chang, J.-S., Chae, H.K., Microwave synthesis of a porous metal-organic framework, zinc terephthalate MOF-5. *Bull. Korean Chem. Soc.*, 27, 1523, 2006.
92. Liang, W. and D'Alessandro, D.M., Microwave-assisted solvothermal synthesis of zirconium oxide based metal-organic frameworks. *Chem. Commun.*, 49, 3706, 2013.
93. Ren, J., Segakweng, T., Langmi, H., Musyoka, N., North, B., Mathe, M., Bessarabov, D., Microwave-assisted modulated synthesis of zirconium-based metal–organic framework (Zr-MOF) for hydrogen storage applications. *Int. J. Mater. Res.*, 105, 516, 2014.

94. Fard, M.J.S., Rastaghi, F., Ghanbari, N., Sonochemical synthesis of new nano-two-dimensional lead (II) coordination polymer: As precursor for preparation of PbO nanostructure. *J. Mol. Struct.*, 1032, 133, 2013.
95. Joaristi, A.M., Juan-Alcaniz, J., Serra-Crespo, P., Kapteijn, F., Gascon, J., Electrochemical synthesis of some archetypical Zn2+, Cu2+, and Al3+ metal organic framework. *Cryst. Growth Des.*, 12, 3489, 2012.
96. Campagnol, N., Van Assche, T., Boudewijns, T., Denayer, J., Binnemans, K., De Vos, D., Fransaer, J., High pressure, high temperature electrochemical synthesis of metal–organic frameworks: films of MIL-100 (Fe) and HKUST-1 in different morphologies. *J. Mater. Chem. A*, 1, 5827, 2013.
97. Stassen, I., Styles, M., Van Assche, T., Campagnol, N., Fransaer, J., Denayer, J., Tan, J.-C., Falcaro, P., De Vos, D., Ameloot, R., Electrochemical film deposition of the zirconium metal–organic framework UiO-66 and application in a miniaturized sorbent trap. *Chem. Mater.*, 27, 1801, 2015.
98. Wu, R.F., Zhang, T.L., Qiao, X.J., Zhang, J.G., Liu, Y.H., Synthesis, crystal structure and thermal stability of a novel 3D coordination polymer Pb(1, 4-napdc)(DMF). *Chin. J. Inorg. Chem.*, 22, 1340, 2006.
99. Wright, P.V., Electrical conductivity in ionic complexes of poly (ethylene oxide). *Br. Polym. J.*, 7, 5, 319, 1975.
100. Armand, M.B., Chabagno, J.M., Duclot, M., *Fast Ion Transport in Solids*, North Holland, New York, 1979.
101. Payne, D.R. and Wright, P.V., Morphology and ionic conductivity of some lithium ion complexes with poly (ethylene oxide). *Polymer*, 23, 690, 1982.
102. Lee, C.C. and Wright, P.V., Morphology and ionic conductivity of complexes of sodium iodide and sodium thiocyanate with poly(ethylene oxide). *Polymer*, 23, 681, 1982.
103. Killis, A. *et al.*, Miseenévidenceexpérimentale de l'importance de la température de transition vitreuse sur la conductivitéionique des réseauxmacromoléculaires. *Macromol. Chem. Rapid Commun.*, 1, 595, 1980.
104. Blonsky, P.M. *et al.*, Polyphosphazene solid electrolytes. *J. Am. Chem. Soc.*, 106, 6854, 1984.
105. Nicholas, C.V. *et al.*, Improved synthesis of oxymethylene-linked poly(oxyethylene). *Br. Polym. J.*, 20, 289, 1988.
106. Cheradame, H. (Ed.), *IUPAC Macromolecules*, Pergamon, New York, 1982.
107. Watanabe, M. and Ogata, N., *Polymer Electrolytes Reviews I*, Elsevier, London, 1987.
108. Benedict, T.J. *et al.*, Characterization of plasticized solid polymer electrolyte by XRD and AC impedance methods. *J. Power Sources*, 75, 171, 1998.
109. Peraninage, D. *et al.*, PAN based electrolyte with ternary solvent mixtures as plascticizers. *J. Electrochem. Soc.*, 142, 1789, 1995.
110. Watanabe, M. *et al.*, Ionic conductivity of hybrid films composed of polyacrylonitrile, ethylene carbonate, and $LiClO_4$. *J. Polym. Sci. Part B: Polym. Phys.*, 21, 939, 1983.

111. Sun, Y.K. and Jin, S.H., Synthesis and electrochemical characteristics of spinel phase $LiMn_2O_4$-based cathode materials for lithium polymer batteries. *J. Mater. Chem.*, 8, 2399, 1998.
112. Choe, H.S. *et al.*, Characterization of some polyacrylonitrile based Electrolytes. *Chem. Mater.*, 9, 369, 1997.
113. Capuano, F. *et al.*, Composite Polymer Electrolyte. *J. Electrochem. Soc.*, 138, 1918, 1991.
114. Kumar, B. and Scanlon, L.G., Polymer–ceramic composite electrolytes. *J. Power Sources*, 52, 261, 1994.
115. Kim, J.W. *et al.*, Electrochemical characteristics of two types of PEO-based composite electrolyte with functional SiO_2. *J. Power Sources*, 119, 415, 2003.
116. Bloise, A.C. *et al.*, Nuclear magnetic resonance study of PEO-based composite polymer electrolytes. *Electrochim. Acta*, 46, 1571, 2001.
117. Liu, Y. *et al.*, *In situ* preparation of poly(ethylene oxide)-SiO_2 composite polymer electrolytes. *J. Power Sources*, 129, 303, 2004.
118. Chung, S.H. *et al.*, Enhancement of ion transport in polymer electrolytes by addition of nanoscale inorganic oxide. *J. Power Sources*, 97, 644, 2001.
119. Jeevanandam, P. and Vasudevan, S., Conductivity of a confined polymer electrolyte: Lithium-polypropylene glycol intercalated in layered $CdPS_3$. *J. Phys. Chem. B*, 102, 4753, 1998.
120. Dian-Dian, H., Zhen-Yu, W., Gui-Ling, P., Xue-Ping, G., Metal–Organic-Framework-Based Gel Polymer Electrolyte with Immobilized Anions to Stabilize a Lithium Anode for a Quasi-Solid-State Lithium–Sulfur Battery. *ACS Appl. Mater. Interfaces*, 11, 18427, 2019.
121. Gerbaldi, C., Nair, R.J., Kulandainathan, M.A. *et al.*, Innovative high performing metal organic framework (MOF)-laden nanocomposite polymer electrolytes for all-solid-state lithium batteries. *J. Mater. Chem. A*, 2, 9948, 2014.
122. Huo, H., Wu, B., Zhang, T. *et al.*, Anion-immobilized polymer electrolyte achieved by cationic metal–organic framework filler for dendrite-free solid-state batteries. *Energy Storage Mater.*, 18, 59, 2019.
123. Kumar, R.S., Rajab, M., Kulandainathana, M.A. *et al.*, Metal organic framework-laden composite polymer electrolytes for efficient and durable all solid-state-lithium batteries. *RSC Adv.*, 4, 50, 26171, 2014.
124. Suriyakumar, S., Gopi, S., Kathiresan, M. *et al.*, Metal organic framework laden poly(ethylene oxide) based composite electrolytes for all-solid-state Li–S and Li–metal polymer batteries. *Electrochim. Acta*, 285, 355, 2018.
125. Walden, P., Molecular weights and electrical conductivity of several fused salts. *Bull. Acad. Sci.*, 405, 1914.
126. Wilkes, J.S. and Zaworotko, M.J., Air and water stable 1-ethyl-3 methylimidazolium based ionic liquids. *J. Chem. Soc. Chem. Commun.*, 965, 1992.
127. Wasserscheid, P. and Wilhelm, K., Ionic Liquids-New "Solutions" for Transition Metal Catalysis. *Angew. Chem. Int. Ed.*, 39, 3772, 2000.

128. Yoshida, Y. and Kitagawa, H., Ionic Conduction in Metal–Organic Frameworks with Incorporated Ionic Liquids. *ACS Sustain. Chem. Eng.*, 7, 70, 2019.
129. Chen, Y., Hu, Z., Krishna, M.G. *et al.*, Ionic Liquid/Metal–Organic Framework Composite for CO_2 Capture: A Computational Investigation. *J. Phys. Chem. C*, 115, 21736, 2011.
130. Fujie, K., Otsubo, K., Ikeda, R. *et al.*, Low temperature ionic conductor: Ionic liquid incorporated within a metal–organic framework. *Chem. Sci.*, 6, 4306, 2015.
131. Chen, L., Wu, B., Zhao, H. *et al.*, High temperature ionic conduction mediated by ionic liquid incorporated within the metal–organic framework UiO-67(Zr). *Inorg. Chem. Commun.*, 81, 1, 2017.
132. Xu, Q., Zhang, X., Zeng, S. *et al.*, Ionic Liquid Incorporated Metal Organic Framework for High Ionic Conductivity over Extended Temperature Range. *ACS Sustain. Chem. Eng.*, 7, 7892, 2019.
133. Williams, M.L. *et al.*, The temperature dependence of relaxation mechanisms in amorphous polymers and other glass-forming liquids. *J. Am. Chem. Soc.*, 77, 3701, 1955.
134. Sidebottom, D.L., Universal Approach for Scaling the ac Conductivity in Ionic Glasses. *Phys. Rev. Lett.*, 82, 3653, 1999.
135. Dutta, R. and Kumar, A., Effect of IL incorporation on ionic transport in PVdF-HFP-based polymer electrolyte nanocomposite doped with NiBTC–metal–organic framework. *J. Solid State Electrochem.*, 22, 2945, 2018.
136. Dutta, R. and Kumar, A., Ion transport dynamics in ionic liquid incorporated CuBTC–metal-organic framework based composite polymer electrolyte. *J. Mater. Sci.: Mater. Electron.*, 30, 1132, 2019.

7

Fine Chemical Synthesis Using Metal–Organic Frameworks as Catalysts

Aasif Helal

Center of Research Excellence in Nanotechnology, King Fahd University of Petroleum and Minerals, Dhahran, Saudi Arabia

Abstract

Fine chemicals form an important part of different chemical industries. In this chapter we have discussed the synthesis of fine chemicals with the help of Metal–Organic Frameworks (MOFs) as catalysts. We have selected some of the most common reactions, such as oxidation, cycloaddition, esterification, and C–C bond formation, used in the synthesis of fine chemicals and reviewed how their yield, selectivity, and catalyst reusability are affected by MOFs and MOF composites as heterogeneous catalysts.

Keywords: Fine chemicals, metal–organic framework, catalysts, oxidation, cycloaddition, esterification

7.1 Introduction

Fine chemicals are single or complex, pure, high quality chemical substances that are scientifically derived by multistep chemical synthesis or biotechnological processes to function as ingredients in more complex specialty chemicals. They comprise of building blocks, advanced intermediates or active component used in pharmaceutical industries as an active pharmaceutical ingredients (API), food flavors, agricultural insecticides, resins, liquid crystals for TV, and more. These fine chemicals are classified according to the way they are sold. (a) Exclusive Category: These fine chemicals are the most expensive and limited in production. They are

Email: aasifh@kfupm.edu.sa

Inamuddin, Rajender Boddula, Mohd Imran Ahamed and Abdullah M. Asiri (eds.) Applications of Metal–Organic Frameworks and Their Derived Materials, (177–192) © 2020 Scrivener Publishing LLC

custom manufactured for a proprietary product. Companies who exclusively manufacture them are the sole producer and are the only ones who know exactly how to use it. (b) Standard Category: These fine chemicals are considered standard products that can be derived and sold to multiple customers. They are less expensive and produce in comparatively larger quantity. It can include chemicals which once were exclusive but no longer are patented, e.g., medications whose patent has expired. Based on the molecular structures, fine chemicals are also categorized as low molecular weight (LMW) and high molecular weight (HMW) products. (i) Low Molecular Weight (LMW) fine chemicals are small molecules that have a molecular weight less than or equal to 700. These are produced mainly by chemical synthesis, or by extraction from animals and plants. Among the different low-molecular-weight fine chemicals the aromatic N-heterocyclic compounds, that form the building block of many pharmaceuticals and agrochemicals, forms a very important part. The most common LMW fine chemicals used as building blocks in the pharmaceutical industries are imidazoles, miconazole, ketoconazole, isoconazole, tetrazoles, tetrazolidines, pyrimidines, β-lactams, sulfonamide, benzodiazepine derivatives. Several pyridines and imazapyr derivatives are used as herbicides and insecticides. (ii) High Molecular Weight (HMW) fine chemicals mainly involve large molecules that are produced by biological process, or multistep chemical reactions and mainly include proteins and peptides.

Metal–Organic Frameworks (MOFs), are pervasive class of crystalline or nanocrystalline materials with colossal internal surface areas and ultrahigh porosity. These extended crystalline structures are composed of metal cations or clusters of cations ("nodes") that are connected by multitopic organic "strut" or "linker" ions or molecules. The extraordinary degree of variability in both the organic, inorganic components and structural motifs of MOFs (allowing an essentially infinite number of possible combinations) makes them an important material with potential applications in gas storage [1, 2], separation [3], and heterogeneous catalysis [4]. MOFs have been extensively used as a heterogeneous catalyst for the following reasons. (a) MOFs are very robust and therefore well adapted for catalysis under extreme conditions. (b) The porosity of MOFs produces large internal surface areas that facilitate the catalytic activity. (c) The uniform pores and channels account for the much needed catalytic selectivity. (d) The organic components of the MOFs can be synthesized with much greater chemical varieties that may tune the catalytic reactivity and selectivity. These properties of MOFs can help in the production of high-value-added products (such as fine chemicals, delicate molecules, etc.) that can be carried out under milder conditions with increase yield and selectivity.

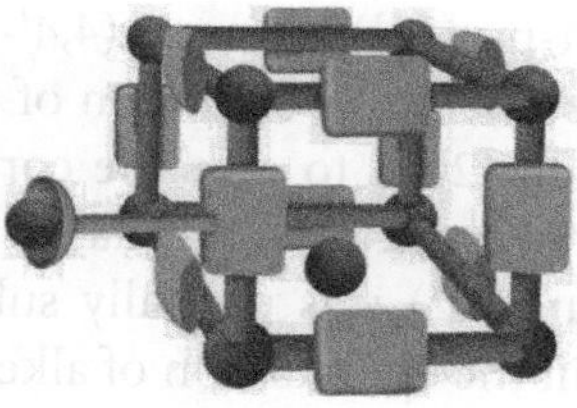

Figure 7.1 Possible location of the catalytic sites in MOF for the synthesis of fine chemicals.

In this chapter we will confine our discussion within those MOFs or MOF composites that catalyze reactions commonly used for the synthesis of fine chemicals. MOF based heterogeneous catalysts for the synthesis of fine chemicals have several locations for catalytically active sites such as metal nodes with exchangeable coordination positions, functionalized linkers with attached active sites, and inclusion of active species within the pores of the MOFs (Figure 7.1). Some of the common reactions frequently used for the synthesis of fine chemicals are (a) oxidation reactions such as epoxidation, sulfoxidation, aerobic oxidation, (b) 1,3-cycloaddition reactions, transesterification reactions, and (c) C–C bond formation reactions such as Heck reaction, Sonogashira coupling, and Suzuki coupling.

7.2 Oxidation Reaction

7.2.1 Epoxidation

Epoxidation is the chemical reaction which converts the carbon-carbon double bond into oxiranes (epoxides), using a variety of reagents. There are several MOFs reported for the epoxidation of olefins leading to the formation of epoxide based fine chemicals. The MOFs are used in two major ways for epoxidation. (i) The metal ions or the metal oxide cluster of the main framework are used as a Lewis acid in the reaction. (ii) Some metal, (molybdenum, manganese) through an organic ligand is attached to the main framework of the MOF, act as a Lewis acid. This Lewis acid binds with the oxygen from the oxidant (oxygen from the air, H_2O_2, or tert-butyl hydroperoxide (*t*-BuOOH)) and transfers it to the unsaturated site of the alkene.

In the following examples metal nodes of the framework were used as the Lewis acid. Jintana Othong *et al.* reported three new type of MOFs, based on the linkers 1,4-phenylenediacetic acid (1,4-H_2phda), 1,2-bis(4-pyridyl) ethane (4,4′-bpa), and 1,3-bis(4-pyridyl) propane (4,4′-bpp), which were used as a heterogeneous catalyst for the epoxidation of trans stilbene and styrene at 70°C in acetonitrile for 20 h to give the corresponding epoxides in good yield [5]. In another work by Amanda W. Stubbs *et al.* Zinc oxide clusters of MOF-5 (Figure 7.2) was partially substituted with the Manganese and where applied in the epoxidation of alkene. It is noted that the metal oxide clusters in MOF-5 substituted with Manganese ion interact with *t*BuSO$_2$PhIO to produce Mn(IV)-oxo intermediate that catalyzes the oxygen transfer to form epoxides from cyclic alkenes in 99% selectivity [6]. Bimetallic–organic frameworks such as the MnFe-MOF-74 also helps in the epoxidation of the alkene with H_2O_2 as an oxidant giving 100% conversion for styrene with 95.0% selectivity [7].

Composites of metal complexes immobilized on MOF can also act as Lewis acid in epoxidation reaction as elaborated by the following examples. Khalil Tabatabaeian *et al.* reported the epoxidation of chalcones or bis-chalcones with the help of a post synthetically modified (Cr)NH$_2$-MIL-101 (Figure 7.2). The MOF was postsynthetically modified with pyridyl diimino Nickel complex to act as a heterogeneous chemoselective catalyst for epoxidation under mild conditions. The catalysts can be used for consecutively four cycles without decrease in its activity [8]. The introduction of certain moieties such as Polyoxomolybdic Cobalt in Zirconium based

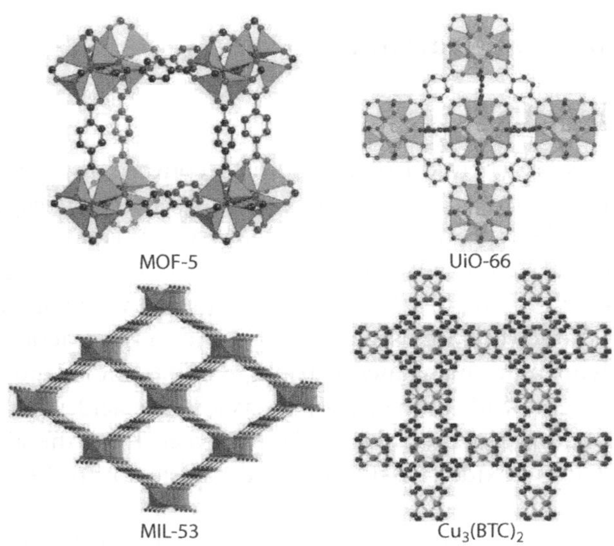

Figure 7.2 Schematic diagram of different metal–organic framework used as catalysts in the synthesis of fine chemicals.

MOF also helps in the epoxidation of alkenes with hydrogen peroxide as an oxidant or with atmospheric oxygen as oxidant and *t*-BuOOH as initiator [9]. In another work UiO-66 (Figure 7.2) and UiO-67 with amino groups when post synthetically modified with salicyaldehyde Molybdenum complex helped in the epoxidation of alkenes with *t*-BuOOH as an oxidant. This also shows that the diffusive properties of the reactants and the products are largely influenced by the effective pore size and number of amino groups [10]. There are also report of the use of chiral MOF for the enantioselective epoxidation of alkenes with *t*-BuOOH as an oxidant and selectivity up to 100% [11]. Jian Zhao *et al.* introduced copper inside a COOH functionalized UiO-66, which acted as a heterogeneous catalyst for epoxidation [12]. Size selective epoxidation was achieved on using MOF having six-coordinated tetranuclear Cu(II) secondary building units (SBUs) and 1-(3,5-dicarboxyphenyl)-2,5-dimethyl-1H-pyrrole-3,4-dicarboxylic acid ligands. This MOF uses *t*-BuOOH or molecular oxygen as oxidants [13]. Similarly, Sara Abednatanzi *et al.* reported the immobilization of amino pyridine derived Molybdenum complex in $Cu_3(BTC)_2$ (Figure 7.2) (BTC = benzene-1,3,5-tricarboxylate) metal–organic framework (MOF) that uses TBHP (*t*-butylhydrogen peroxide) as the oxidant to get size-selective epoxidation with 99% selectivity [14]. Thus in all the above examples of epoxidation MOF itself acts as a catalyst or act as a support to host a metal complex that act as a catalyst for the production of epoxides as catalysts.

7.2.2 Sulfoxidation

Sulfoxidation is the oxidation of the organic sulfur containing heterocycles to sulfoxides or sulfones. Usually in these cases TBHP (tert-butyl hydroperoxide) or H_2O_2 is used as an oxidant. MOF in these cases can act as a support for the moieties which will carry the oxygen, or the central metal ion, of the MOF framework can also act as a Lewis acid that carry the oxygen, to the oxidation site. In some cases, also the linker generates the oxygen for the oxidation. Bing-Bing Lu *et al.* reported the synthesis of a new Cobalt and polyoxovanadate-resorcin[4]arene-based metal–organic framework as a catalyst for sulfoxidation. In the presence of TBHP a peroxyvanadic acid complex is formed which transfer the oxygen to the sulfur-containing aromatic substrate that is then oxidized to sulfoxide and sulfone [15]. Subhadip Goswami *et al.* prepared two Zr-based benzothiadiazole and benzoselenadiazole containing metal-organic frameworks (MOFs) as photocatalysts for the selective oxidation 2-chloroethyl ethyl sulfide. In these two MOFs the benzothiadiazole- and benzoselenadiazole-based linkers sensitize the photochemical formation of singlet oxygen that helps in the oxidation of the sulfide [16]. The zinc or copper clusters in some MOF has a been used as a Lewis acid for the binding of the TBHP or the H_2O_2 and transfer of oxygen to the sulfur of the organic sulfide for oxidation [17, 18].

7.2.3 Aerobic Oxidation of Alcohols

The oxidation of alcohols to aldehyde or ketones in the presence of atmospheric oxygen is known as aerobic oxidation. In case of MOF there are very few information of nodal metal ion of the framework involve in the aerobic oxidation of alcohols to aldehydes or ketones. The aerobic oxidation of alcohols usually takes place with the help of noble metals (Pd, Ru, Au, Pt, etc.) present in the void volume of the MOF or attached to the modified ligands. Some of the examples are Gong-Jun Chen *et al.* reported the use of palladium incorporated copper MOF in presence of air at elevated temperature for the oxidation of alcohol [19]. Palladium nanoparticles, are also introduced into the nanoporous metal–organic framework of $Cu_2(BDC)_2$ interlinked by (DABCO) (BDC, 1,4-benzenedicarboxylate, DABCO ¼ 1,4-diazabicyclo[2.2.2]octane) [20] and silica-MIL-101NH_2 double layered support for high activity for oxidation of alcohol, endurance, and leaching control [21]. There also have been reports of the use of noble metals such as ruthenium and gold in the aerobic oxidation. Jing-Si Wang *et al.* reported the use of gold nanoparticles in the Cu(II) MOF for the selective aerobic oxidation [22]. On the other hand, Suntao Wu *et al.* reported the "*click*" postsynthetic modification of the MIL-101(Cr) with terpyridyl moieties, which on metalation with ruthenium act as a highly active and stable single-site heterogeneous metal catalysts [23].

In the absence of noble metals there should be the presence of carbon-centered radical promoters as a cocatalyst along with molecular oxygen. The most commonly used cocatalysts in the aerobic oxidation of alcohols is 2,2,6,6-Tetramethylpiperidinyloxy (TEMPO) a stable nitroxyl radical. Martin Lammert *et al.* synthesized Cerium based MOF having UiO-66 topology (Figure 7.2) with linker molecules having different size and functionalities. In this process TEMPO results in the formation of oxoammonium ion on the surface of the MOF by one electron oxidation and simultaneous reduction of Ce^{4+} to Ce^{3+}. In case of BDC due to small pore size only the Cerium at the surface can participate in the oxidation as the TEMPO cannot enter the pores resulting in 29% conversion from alcohol to aldehyde. But in case of NDC as a linker the pore size increases so TEMPO has access to the internal pore voids that results in the formation of more oxoammonium counterpart that enhances the oxidation reaction resulting in a conversion of 80% [24]. The presence of melamine as a linker of a copper/amine based bifunctional MOF helped in the aerobic oxidation with a yield of 95% in the absence of a base [25]. Another base free aerobic oxidation was reported by Abu Taher *et al.* where they introduce copper attached to the PSM alkyl amine chain of the MOF. This copper containing zinc based MOF can oxidize alcohols containing long chain alkyl units and inactive hetero-aryl moieties with 100% selectivity [26]. Similar type of work was also reported by Hui Liu and his team where they introduce cuprous iodide in pyridinyl functionalized MIL-101-NH_2 as a heterogenous catalyst for oxidation under mild condition using air as oxidant [27].

7.3 1,3-Dipolar Cycloaddition Reaction

The 1,3-dipolar cycloaddition is the most common chemical reaction used to synthesize a five membered ring by using a 1,3-dipole (azide) and a dipolarophile (alkyne) as reagents. Copper(I)-catalyzed azide–alkyne cycloaddition is one of the most common 1,3-dipolar cycloaddition reaction which is used to prepare a triazole derivatives commonly used in the synthesis of fine chemicals. There are several MOFs which catalyzes such reactions. The copper based MOFs are directly used as a catalyst in such reactions. For example, Bing-Bing Lu *et al.* reported a copper(I)-based MOF by incorporating Keggin-type polyoxometalate (POM) anions and a functionalized wheel-like resorcin[4]arene-based ligand. This MOF catalyzed several types of alkyne and azides giving 99% yield of the respective triazole products. The MOF was recyclable for five consecutive cycles [28]. Another new highly efficient, and reusable Cu-MOF based on organic linkers 2,4,6-tri-(4-pyridyl)-1,3,5-triazine (PTZ) and sodium 2,6-naphthalene disulfonate (NSA) has been developed for the solvent-free, regioselective synthesis of 1,2,3-triazoles via the reaction of organic azides to terminal alkynes. This catalyst gave excellent product yields with low catalyst loading. Moreover, this catalyst was recovered and reused efficiently up to five cycles without major loss of reactivity [29].

7.4 Transesterification Reaction

Transesterification is the process of interchanging one organic moiety of an ester with another organic moiety of an alcohol. These reactions are often catalyzed by the addition of an acid or base catalyst. There are several examples were MOF is used in the transesterification reaction. In these cases, the metal ions or the metal cluster or the linkers of the MOF framework or some moieties introduced by PSM in the MOF act as the acid or base in transesterification reaction. The following examples elaborate the use of the metal of the framework as a catalyst in the transesterification. Anirban Karmakar *et al.* reported the synthesis of three new mixed linker Zinc MOF with bipyridine as one of the linker and 2-acetamidoterephthalic acid, or 2-propionamidoterephthalic acid or 2-benzamidoterephthalic acid as the other linker. The Zinc metal clusters of this MOF act as a Lewis acid in the transesterification reaction with a good recyclable property [30]. The Zirconium cluster center of the UiO-66 was also used as a Lewis acid in the transesterification reaction by Xiao Liu *et al.* Furthermore, the UiO-66 (Figure 7.2) with an amino functionality showed an increase in catalytic activity due to the synergy effect between the Zirconium metal cluster and the amino group during catalysis [31]. Cobalt based metal center of the MOF has also been used in the transesterification of vegetable oil and other ester with high yield and good recyclability [32]. The linkers of the MOF has also been used as a Lewis base to catalyze a transesterification reaction. K. Kim *et al.* reported a Zinc based MOF with (4*R*,5*R*)-2,2-dimethyl-5-[(4-pyridinylamino)carbonyl]-1,3-dioxolane-4-carboxylic acid as a linker. In this the accessible pyridine moieties helps in the base catalyze transesterification reaction with high yield and reusability for five cycles without the loss of appreciable catalytic activity [33]. In another example of base catalyze transesterification, uncoordinated pyridyl groups of a Zinc and *N,N′*, *N″,N‴*-tetrakis[3-*tert*-butyl-5-(4-pyridinyl)salicylidene]-1,2,4,5-benzenetetraamine (bisSalen) linker produce good yield and reusability for three consecutive cycles [34]. Sometimes a MOF is functionalized with some group that helps in the catalysis, e.g., William T. Schumacher *et al.* prepared three different linkers of imidazolium-tagged biphenyldicarboxylates, and they were incorporated (6–7%) as a mixed linker into the UiO-67 framework. On activation the N-heterocylic carbene (NHC) of the linker in the MOF was found to enhance the transesterification of vinyl acetate with benzylalcohol in good yield and the catalyst was found to be reused for five consecutive cycles [35].

7.5 C–C Bond Formation Reactions

7.5.1 Heck Reactions

Heck reaction is a palladium-catalyzed C–C bond formation between less activated aryl halides or vinyl halides and activated alkenes under basic condition. MOF having palladium or palladium nanoparticles incorporated in the pores or attached to the functionalized linker are extensively used for the Heck reaction. There are several reasons to use palladium in MOF for catalysis (i) palladium can be immobilized on the functionalized linkers that would prevent loss of catalyst (leaching) and prolong the life of the catalysts, (ii) the high surface area, porosity and uniform pore size helps in the selective catalysis, (iii) in case of palladium nanoparticles, incorporation into MOFs prevents aggregation, (iv) it is possible to assemble different types of linkers and metal nodes that help in the catalysis by palladium. Palladium is incorporated in the MOF as a nanoparticle or in the form of complex, it can also be a part of the metallic node of the MOF. The following examples elaborate the use of palladium in the form of metal complex. Shuping Jia and coworkers introduce bipyridyl linker by post synthetic linker installation in a zirconium based MOF with dihydroanthracene based tetratopic linkers as the building block. Pd (II) was introduced by postsynthetic metalation in the bipyridine ligand that helps in the selective catalysis of Heck reaction [36]. Yong-Liang Wei *et al.* reported the synthesis of mixed linker MOF using biphenyl-4,4′-dicarboxylic acid and presynthesized palladium N-heterocyclic bis-carbene dicarboxylic acid with $ZrCl_4$ to give isoreticular UiO-67. This palladium bis carbene embedded MOF material was used to promote Heck coupling reaction with high selectivity and yield [37]. In order to stabilize the Pd (0) intermediate during the catalytic cycle of the Heck reaction and avoid aggregation, acid functionalization of the MOF node was done by Ken-ichi Otake *et al.* In this the palladium was incorporated into phosphate and sulfated MOFs of Hafnium [38]. In some cases the palladium complex such as Bis(tri(1-piperidinyl) phosphine) palladium chloride or Bis(triphenylphosphine) palladium dichloride are incorporated in the Ni-MOF for the Heck reaction of estragole with iodobenzene. It was found that the Ni-MOF encapsulated palladium complex gave us a better yield than the palladium complex without encapsulation [39]. As explained earlier palladium nanoparticles are also loaded in MOF for the Heck coupling. Chemical tailoring of the inner channel and cavities of a MOF with alkyne functionalization helped in the stabilizing the palladium nanoparticle and prevent leaching and aggregation during the Heck reaction [40]. Palladium nanoparticles was incorporated in Zr MOF-808, the inherent Bronsted basicity of the MOF make it an excellent heterogenous catalyst for Heck reaction without any additional base [41]. There are also report of palladium containing inorganic nodes of MOF catalyze the Heck reaction. Lixin You and coworkers prepared three heterobimetallic MOF of Lanthanide and palladium with 2,2-bipyridine-4,4-dicarboxylic acid as a heteroleptic ligand [42]. Another heterobimetallic MOF of palladium and lead with 2,2′-bipyridine-5,5′-dicarboxylic acid as a ligand was prepared by Yanwei Ren *et al.* [43]. In both cases it was found to give very good yield with a variety of substrate and with high selectivity and reusability that are very effective for the synthesis of fine chemicals.

7.5.2 Sonogashira Coupling

The Sonogashira reaction is a cross-coupling reaction between a terminal alkyne and an aryl or vinyl halide to form carbon–carbon bonds by using a palladium catalyst and copper co-catalyst. Like the Heck reaction this coupling reaction also takes place with the help of palladium complex or palladium nanoparticle incorporated in a MOF. Annapurna and coworkers introduced Pd nanoparticles in MIL101 and found that it acts as an excellent catalyst for Sonogashira coupling, with variety of alkyne, aryl halides and even with sterically hindered aryl halides, with high yield and selectivity [44]. In another work Li *et al.* introduced Pd in a ZrMOF based on 2,2′-bipyridine-5,5′-dicarboxylate linker and applied it in the carbonylative Sonogashira coupling at atmospheric pressure in presence of carbon monoxide giving good yield and the catalysts was reused for five times. The catalyst was found to bind with the bipyridine moiety of the linker as confirmed by XPS and FTIR [45]. In another similar work by Hexing Li and coworkers a heterobimetallic MOF of Yttrium with the bipyridine dicarboxylate moiety was prepared in which the bipyridine functionality bind with the palladium to be used in the Sonogashira coupling [46]. Francis Verpoort *et al.* introduce azolium ligand 1,3-bis(4-carboxyphenyl) imidazolium chloride in the MOF. This modification of the linker can be used for the generation of the N-heterocyclic carbene that binds with the palladium through the post synthetic modifications. This was then used as a successful heterogeneous catalyst in the Sonogashira coupling in air under atmospheric pressure with different derivatives of iodobenzene and bromopyridine in good yield and selectivity [47].

7.5.3 Suzuki Coupling

Suzuki coupling is a metal catalyzed coupling reaction between an organic halide and a boronic acid derivative. Palladium (0) is the most commonly used metal but other metals were also used. MOF usually act as a support for this metal catalyzed reaction. Palladium is incorporated in the MOF as a complex or in the form of nanoparticles. Xinle Li *et al.* prepared a mixed-linker MOF containing bipyridyl and biphenyl moieties and isoreticular with the UiO-67. Palladium dichloride was immobilized on the bipyridine linker. The bipyridine moiety was functionalized with electron donating methyl group to study the electronic and the steric effects of the linker on Suzuki catalysis. It was found that the yield and the selectivity increased to a large extent as compared to the pristine MOF [48]. In another work Dengrong Sun *et al.* introduced palladium nanoclusters in NH_2–UiO-66 (Zr) and used them in the Suzuki catalysis in the presence of light giving 99% conversion and selectivity of biphenyl compounds [49]. A heterobimetallic MOF of lanthanides and palladium with 1,1′-di(p-carboxybenzyl)-2,2′-diimidazole as a linker was prepared as a heterogenous catalyst. It showed excellent catalytic properties and can be reused four to eight times without significant loss of catalytic activity [50]. Nickel was also used in place of palladium as the catalysts in Suzuki coupling by Palani Elumalai *et al.* They synthesized UiO-66 linker (Figure 7.2) functionalized with triazole and bipyridine moieties which helps in the binding of the Nickel–phosphine complexes. The incorporated catalyst was found to be reusable for seven cycles without appreciable decrease in the catalytic activity [51].

7.6 Conclusion

Thus from the above all examples, elaborated in the different sections of this chapter it can be concluded that MOFs and MOF composites has played and will be playing an important role, in future, as a catalyst in the synthesis of fine chemicals. The tunable nature of the pores of MOF will not only improve the yield and selectivity but also reduce the formation of other byproducts. As a heterogeneous catalyst these MOFs can be reused for several cycles without appreciable decrease in their catalytic activities for the production of fine chemicals in an industrial scale.

References

1. Collins, D.J. and Zhou, H.-C., Hydrogen storage in metal–organic frameworks. *J. Mater. Chem.*, 17, 3154, 2007.
2. Li, J.-R., Ma, Y., McCarthy, M.C., Sculley, J., Yu, J., Jeong, H.-K., Balbuena, P.B., Zhou, H.-C., Carbon dioxide capture-related gas adsorption and separation in metal-organic frameworks. *Coord. Chem. Rev.*, 255, 1791, 2011.
3. Li, J.-R., Kuppler, R.J., Zhou, H.-C., Selective gas adsorption and separation in metal–organic frameworks. *Chem. Soc. Rev.*, 38, 1477, 2009.

4. Lee, J., Farha, O.K., Roberts, J., Scheidt, K.A., Nguyen, S.T., Hupp, J.T., Metal–organic framework materials as catalysts. *Chem. Soc. Rev.*, 38, 1450, 2009.
5. Othong, J., Boonmak, J., Ha, J., Leelasubcharoen, S., Youngme, S., Thermally Induced Single-Crystal-to-Single-Crystal Transformation and Heterogeneous Catalysts for Epoxidation Reaction of Co (II) Based Metal–Organic. *Cryst. Growth Des.*, 17, 1824, 2017.
6. Stubbs, A.W., Braglia, L., Borfecchia, E., Meyer, R.J., Roman-Leshkov, Y., Lamberti, C., Dinca, M., Selective Catalytic Olefin Epoxidation with Mn^{II}-Exchanged MOF-5. *ACS Catal.*, 8, 596, 2018.
7. Yuan, K., Song, T., Wang, D., Zou, Y., Li, J., Zhang, X., Tang, Z., Hu, W., Bimetal–organic frameworks for functionality optimization: MnFe-MOF-74 as a stable and efficient catalyst for the epoxidation of alkenes with H_2O_2. *Nanoscale*, 10, 1591, 2018.
8. Tabatabaeian, K., Zanjanchi, M.A., Mahmoodi, N.O., Eftekhari, T., Shafiei, S.M., Diimino Nickel Complex Anchored into the MOF Cavity as Catalyst for Epoxidation of Chalcones and Bischalcones. *J. Clust. Sci.*, 28, 949, 2017.
9. Song, X., Hu, D., Yang, X., Zhang, H., Zhang, W., Li, J., Jia, M., Yu, J., Polyoxomolybdic Cobalt Encapsulated within Zr-Based Metal–Organic Frameworks as Efficient Heterogeneous Catalysts for Olefins Epoxidation. *ACS Sustain. Chem. Eng.*, 7, 3624, 2019.
10. Kaposi, M., Cokoja, M., Hutterer, C.H., Hauser, S.A., Kaposi, T., Klappenberger, F., Pöthig, A., Barth, J.V., Herrmanna, W.A., Kühn, F.E., Immobilisation of a molecular epoxidation catalyst on UiO-66 and-67: The effect of pore size on catalyst activity and recycling. *Dalton Trans.*, 44, 15976, 2015.
11. Berijani, K., Morsali, A., Hupp, J.T., An effective strategy for creating asymmetric MOFs for chirality induction: A chiral Zr-based MOF for enantioselective epoxidation. *Catal. Sci. Technol.*, 9, 3388, 2019.
12. Zhao, J., Wang, W., Tang, H., Ramella, D., Luan, Y., Modification of Cu^{2+} into Zr-based metal–organic framework (MOF) with carboxylic units as an efficient heterogeneous catalyst for aerobic epoxidation of olefins. *Mol. Catal.*, 456, 57, 2018.
13. Chen, C., Shen, P., Wan, M., Ding, N., Shi, X., Wang, X., Zhang, N., Size-selective epoxidation of olefins in two new metal-organic framework constructed from six-coordinated tetranuclear Cu (II) SBUs. *Microporous Mesoporous Mater.*, 232, 167, 2016.
14. Abednatanzi, S., Abbasi, A., Masteri-Farahani, M., Size-selective epoxidation of olefins in two new metal-organic framework constructed from six-coordinated tetranuclear Cu (II) SBUs. *J. Mol. A-Chem.*, 399, 10, 2015.
15. Lu, B.-B., Yang, J., Liu, Y.Y., Ma, J.-F., A Polyoxovanadate–Resorcin[4]arene-Based Porous Metal–Organic Framework as an Efficient Multifunctional Catalyst for the Cycloaddition of CO_2 with Epoxides and the Selective Oxidation of Sulfides. *Inorg. Chem.*, 56, 11710, 2017.
16. Goswami, S., Miller, C.E., Logsdon, J.L., Buru, C.T., Wu, Y.–.L., Bowman, D.N., Islamoglu, T., Asiri, A.M., Cramer, C.J., Wasielewski, M.R., Hupp, J.T.,

Farha, O.K., Atomistic approach toward selective photocatalytic oxidation of a mustard-gas simulant: A case study with heavy-chalcogen-containing PCN-57 analogues. *ACS Appl. Mater. Interfaces*, 9, 19535, 2017.
17. Patel, P., Parmar, B., Kureshy, R.I., Khan, N.H., Suresh, E., Amine-functionalized Zn (ii) MOF as an efficient multifunctional catalyst for CO_2 utilization and sulfoxidation reaction. *Dalton Trans.*, 47, 8041, 2018.
18. Rostamnia, S., Alamgholiloo, H., Jafari, M., Rookhosh, R., Abbasi, A.R., Synthesis and catalytic study of open metal site metal–organic frameworks of $Cu_3(BTC)_2$ microbelts in selective organic sulfide oxidation. *Appl. Organomet. Chem.*, 30, 954, 2016.
19. Chen, G.-J., Wang, J.-S., Jin, F.-Z., Liu, M.-Y., Zhao, C.-W., Li, Y.-A., Dong., Y.-B., Pd@ Cu (II)-MOF-catalyzed aerobic oxidation of benzylic alcohols in air with high conversion and selectivity. *Inorg. Chem.*, 55, 3058, 2016.
20. Akbari, S., Mokhtari, J., Mirjafari, Z., Solvent-free and melt aerobic oxidation of benzyl alcohols using Pd/$Cu_2(BDC)_2$DABCO–MOF prepared by one-step and through reduction by dimethylformamide. *RSC Adv.*, 7, 40881, 2017.
21. Pascanu, V., Gómez, A.B., Ayats, C., Platero-Prats, A.E., Carson, F., Su, J., Yao, Q., Pericàs, M.À., Zou, X., Martín-Matute, B., Double-supported silica-metal–organic framework palladium nanocatalyst for the aerobic oxidation of alcohols under batch and continuous flow regimes. *ACS Catal.*, 5, 472, 2015.
22. Wang, J.-S., Jin, F.-Z., Ma, H.-C., Li, X.-B., Liu, M.-Y., Kan, J.-L., Chen, G.-J., Dong, Y.-B., Au@ Cu (II)-MOF: Highly efficient bifunctional heterogeneous catalyst for successive oxidation–condensation reactions. *Inorg. Chem.*, 55, 6685, 2016.
23. Wu, S., Chen, L., Yin, B., Li, Y., "Click" post-functionalization of a metal–organic framework for engineering active single-site heterogeneous Ru (iii) catalysts. *Chem. Commun.*, 51, 9884, 2015.
24. Lammert, M., Wharmby, M.T., Smolders, S., Bueken, B., Lie, A., Lomachenko, K.A., Vos, D.D., Stock, N., Cerium-based metal organic frameworks with UiO-66 architecture: Synthesis, properties and redox catalytic activity. *Chem. Commun.*, 51, 12578, 2015.
25. Miao, Z., Luan, Y., Qi, C., Ramella, D., The synthesis of a bifunctional copper metal organic framework and its application in the aerobic oxidation/Knoevenagel condensation sequential reaction. *Dalton Trans.*, 45, 13917, 2016.
26. Taher, A., Kim, D.W., Lee, I.-M., Highly efficient metal organic framework (MOF)-based copper catalysts for the base-free aerobic oxidation of various alcohols. *RSC Adv.*, 7, 17806, 2017.
27. Liu, H., Ramella, D., Yu, P., Luan, Y., A novel modified MIL-101-NH 2 ligand for CuI-catalyzed and air promoted oxidation of secondary alcohols. *RSC Adv.*, 7, 22353, 2017.
28. Lu, B.-B., Yang, J., Che, G.-B., Pei, W.-Y., Ma, J.–F., Highly Stable Copper(I)-Based Metal–Organic Framework Assembled with Resorcin[4]arene and

Polyoxometalate for Efficient Heterogeneous Catalysis of Azide–Alkyne "Click" Reaction. *ACS Appl. Mater. Interfaces*, 10, 2628, 2018.

29. Li, P., Regati, S., Huang, H., Arman, H., John, D., Zhao, C.-G., Chen, B., A metal-organic framework as a highly efficient and reusable catalyst for the solvent-free 1, 3-dipolar cycloaddition of organic azides to alkynes. *Inorg. Chem. Front.*, 2, 42, 2015.
30. Karmakar, A., Fátima, M., Guedes da Silva, C., Pombeiro, A.J.L., Zinc metal–organic frameworks: Efficient catalysts for the diastereoselective Henry reaction and transesterification. *Dalton Trans.*, 43, 7795, 2014.
31. Liu, X., Qi, W., Wang, Y., Su, R., He, Z., Exploration of Intrinsic Lipase-Like Activity of Zirconium-Based Metal-Organic Frameworks. *Eur. J. Inorg. Chem.*, 4579, 2018.
32. Rodríguez, R.P., Márquez-López, E., Guerrero, A., Chiñas, L.E., Hernández-González, D.F., Rivera, J.M., Hydrothermal synthesis of cobalt (II) 3D metal–organic framework acid catalyst applied in the transesterification process of vegetable oil. *Mater. Lett.*, 217, 117, 2018.
33. Seo, J.S. *et al.*, A homochiral metal–organic porous material for enantioselective separation and catalysis. *Nature*, 404, 6781, 982, 2000.
34. Kim, W.S., Lee, K.Y., Ryu, E.H., Gu, J.M., Kim, Y., Lee, S.J., Huh., S. Catalytic Transesterifications by a Zn–BisSalen MOF Containing Open Pyridyl Groups Inside 1D Channels. *Eur. J. Inorg. Chem.*, 4228, 2013.
35. Schumacher, W.T., Mathews, M.J., Larson, S.A., Lemmon, C.E., Campbell, K.A., Crabb, B.T., Chicoine, B.J.-A., Beauvais, L.G., Perry, M.C., Organocatalysis by site-isolated N-heterocyclic carbenes doped into the UIO-67 framework. *Polyhedron*, 114, 422, 2016.
36. Jia, S., Xiao, X., Li, Q., Li, Y., Duan, Z., Li, Y., Li, X., Lin, Z., Zhao, Y., Huang, W., Tuning the Connectivity, Rigidity, and Functionality of Two-Dimensional Zr-Based Metal–Organic Frameworks., 58, 19, 12748–12755, 2019. doi: 10.1021/acs.inorgchem.9b01666.
37. Wei, Y.–.L., Li, Y., Chen, Y.-Q., Dong, Y., Yao, J.-J., Han, X.-Y., Dong, Y.-B., Pd (II)-NHDC-Functionalized UiO-67 Type MOF for Catalyzing Heck Cross-Coupling and Intermolecular Benzyne–Benzyne–Alkene Insertion Reactions. *Inorg. Chem.*, 57, 4379, 2018.
38. Otake, K., Ye, J., Mandal, M., Islamoglu, T., Buru, C.T., Hupp, J.T., Delferro, M., Truhlar, D.G., Cramer, C.J., Farha, O.K., Enhanced Activity of Heterogeneous Pd(II) Catalysts on Acid-Functionalized Metal–Organic Frameworks. *ACS Catal.*, 9, 5383, 2019.
39. Tarnowicz-Ligus, S., Augustyniak, A., Trzeciak, A.M., Incorporation of $PdCl_2P_2$ Complexes in Ni-MOF for Catalyzing Heck Arylation of Functionalized Olefins. *Eur. J. Inorg. Chem.*, 2019. doi: 10.1002/ejic.201900618.
40. Gole, B., Sanyal, U., Banerjee, R., Mukherjee, P.S., High loading of Pd nanoparticles by interior functionalization of MOFs for heterogeneous catalysis. *Inorg. Chem.*, 55, 2345, 2016.

41. Yan, X., Wang, K., Xu, X., Wang, S., Ning, Q., Xiao, W., Zhang, N., Chen, Z., Chen., C., Brønsted Basicity in Metal–Organic Framework-808 and Its Application in Base-Free Catalysis. *Inorg. Chem.*, 57, 8033, 2018.
42. You, L., Zong, W., Xiong, G., Ding, F., Wang, S., Ren, B., Dragutan, I., Dragutan, V., Sun., Y., Cooperative effects of lanthanides when associated with palladium in novel, 3D Pd/Ln coordination polymers. Sustainable applications as water-stable, heterogeneous catalysts in carbon–carbon cross-coupling reactions. *Appl. Catal. A*, 511, 1, 2016.
43. Ren, Y., Jiang, O., Li, J., Zeng, H., A heterometal (Pd–Pb) organic framework: Synthesis, structure and heterogeneous catalytic application. *Appl. Organomet. Chem.*, 30, 699, 2016.
44. Annapurna, M., Parsharamulu, T., Vishnuvardhan Reddy, P., Suresh, M., Likhar, P.R., Lakshmi Kantam, M., Nano palladium supported on high-surface-area metal–organic framework MIL-101: An efficient catalyst for Sonogashira coupling of aryl and heteroaryl bromides with alkynes. *Appl. Organomet. Chem.*, 29, 234, 2015.
45. Bai, C.H., Jian, S.P., Yao, X.F., Li, Y.W., Carbonylative Sonogashira coupling of terminal alkynes with aryl iodides under atmospheric pressure of CO using Pd (II)@ MOF as the catalyst. *Catal. Sci. Technol.*, 4, 3261, 2014.
46. Huang, J., Wang, W., Li, H., Water–Medium Organic Reactions Catalyzed by Active and Reusable Pd/Y Heterobimetal–Organic Framework. *ACS Catal.*, 3, 1526, 2013.
47. Ezugwu, C.I., Mousavi, B., Asrafa, M.A., Mehta, A., Vardhana, H., Verpoort., F., An N-heterocyclic carbene based MOF catalyst for Sonogashira cross-coupling reaction. *Catal. Sci. Technol.*, 6, 2050, 2016.
48. Li, X., Zeeland, R.V., Maligal-Ganesh, R.V., Pei, Y., Power, G., Stanley, L., Huang, W., Impact of linker engineering on the catalytic activity of metal–organic frameworks containing Pd (II)–bipyridine complexes. *ACS Catal.*, 6, 6324, 2016.
49. Sun, D. and Li, Z., Double-Solvent Method to Pd Nanoclusters Encapsulated inside the Cavity of NH_2–Uio-66(Zr) for Efficient Visible-Light-Promoted Suzuki Coupling Reaction. *J. Phys. Chem. C*, 120, 19744, 2016.
50. You, L.-X., Cui, L.-X., Zhao, B.-B., Xiong, G., Ding, F., Ren, B.-Y., Shi, Z.-L., Dragutan, I., Dragutan, V., Sun, Y.-G., Tailoring the structure, pH sensitivity and catalytic performance in Suzuki–Miyaura cross-couplings of Ln/Pd MOFs based on the 1,1′-di(*p*-carboxybenzyl)-2,2′-diimidazole linker. *Dalton Trans.*, 47, 8755, 2018.
51. Elumalai, P., Mamlouk, H., Yiming, W., Feng, L., Yuan, S., Zhou, H.-C., Madrahimov, S.T., Recyclable and Reusable Heteroleptic Nickel Catalyst Immobilized on Metal–Organic Framework for Suzuki–Miyaura Coupling. *ACS Appl. Mater. Interfaces*, 10, 41431, 2018.

8

Application of Metal Organic Framework and Derived Material in Hydrogenation Catalysis

Tejaswini Sahoo, Jagannath Panda, Jnana Ranjan Sahu and Rojalin Sahu*

School of Applied Sciences, Kalinga Institute of Industrial Technology, Deemed to be University, Bhubaneswar, India

Abstract

Metal organic framework (MOF) and derived materials possess unique physiochemical properties and these properties have made them suitable to be used as catalysts for various types of hydrogenation reactions. Integrating with other functional species, MOF based catalyst provides different active sites for various types of hydrogenation reaction. In order to upgrade the interlinkages between the active catalytic centers, reactants, and products, structural properties of catalyst is very vital. Similar to traditional heterogeneous catalysts, MOF based catalyst by collaboration with various active sites and structural properties can attain enhanced selectivity and activities in hydrogenation reaction. This chapter covers about the exceptional variation and intensity of MOF based catalyst structures and their selective application in hydrogenation reaction of various compounds.

Keywords: Metal organic framework, catalyst, hydrogenation reaction

8.1 Introduction

The foundation of various chemical industries like fine chemicals [1], food [2], synthetic dyes [3], polymers [4], perfumes [5], medicines, etc., depends upon hydrogenation reaction. A wide range of compounds like aldehydes, cinnamaldehyde, furfural, alkenes, alkynes, etc., can be hydrogenated with very high yield and selectivity in mild environment. Much

**Corresponding author*: rsahufch@kiit.ac.in

Inamuddin, Rajender Boddula, Mohd Imran Ahamed and Abdullah M. Asiri (eds.) Applications of Metal–Organic Frameworks and Their Derived Materials, (193–218) © 2020 Scrivener Publishing LLC

research has been done since years to explore better methods in the field of optimizing reaction system, which consists of catalysts, solvents, and reductants. Particularly, the function of catalysts in upgrading the activities and selectivity of hydrogenation reaction is very vital. Traditionally used hydrogenation catalysts like metal oxides [6], sulfides [7] are associated with serious drawbacks and hence needs to be idealized. To mention a few the particles of metal having higher activity in comparison to metal oxides and sulfides can readily be poisoned and lose their activity due to reaction with sulfur, nitrogen, arsenic, and phosphorous [8]. Even though chemically, metal oxides and sulfides are stable but they need high temperature and pressure during hydrogenation reaction and hence this create hurdle in performing the reaction mild condition with high activity and selectivity. In order to stabilize these traditional catalysts, few supporting material like carbon materials [9], zeolites [10], silica, etc., are used but due to their poor tunability, these supporting materials again become an obstacle in achieving required selectivity and activity in mild conditions. Over the last few decades, Metal organic framework has been identified as novel category of porous material. MOFs cover wide varieties of applications such as catalytic reaction [11], separation and storage of gas [12], drug delivery [13], luminescence [14], and sensing [15] because of their special properties of high surface area and porosity, distinguishable, flexible, and adaptable. The summation of MOF with other functional ligands give well designed composite material which can enhance synergistic effect which are far better than properties of single component. The MOF based catalyst having more than one active site in single system is best candidate for hydrogenation reaction. Unlike traditional catalyst, MOF based catalyst have new properties with better adjustability and can be reasonably designed for the application in hydrogenation reaction. These catalysts have large number of specified active centers [16]. The additional active centre in these catalysts can be designed by confinement of external organic and inorganic linkers with other functional groups. The collaborative impact of these MOF based catalyst linked with guest material makes a single system, which favors hydrogenation reaction. The distribution and sizes of active centre of MOF based catalyst can be altered and managed, thereby favoring the activity and selectivity of hydrogenation reaction. The encapsulation property in MOF based catalyst promotes the durability and stability of active center, thus favoring the hydrogenation process. The catalytic activities of MOF based catalyst is supported by its high specific surface area and also contribute in improvement of adsorption of reductant and reactant favoring in increased local concentration. Both of these factors help in enhancing the activity. The exact control of hydrogenation process is favored by ease in tunable structure of MOF based catalyst. On changing the sizes of pores in MOF, causes an effective adjustment in diffusion of reactants to active center [17]. Even though MOF based catalyst have so many properties favoring catalysis in industrial application, the development of these catalyst is quite slower in comparison to other traditionally used commercial catalysts. Unfortunately, MOF based catalyst are also associated with some unavoidable disadvantages which hinders in its application. These drawbacks include poor stability of the catalyst in extreme condition like very high pressure and temperature, humidity in air and solvents which can break some of the weak frameworks; moreover, they are very expensive in comparison to traditional catalysts and reusability. Hence, it is very much necessary to design cost effective, stable, and reusable catalysts for the enhancement of MOF catalysts for various industrial uses, especially hydrogenation reaction. Modern synthesis technique and various crystal developments have favored in fabricating MOF materials with better stability by using organic ligands and hard metal ions which can tolerate temperature up to 758 K without affecting its real structure. This temperature is more than the temperature range (623–673 K) at which most of the MOF decomposes [18]. The instability of the MOF may be utilized to design MOF derived carbon materials. This chapter summarizes developments in the field of MOF based catalysts in the hydrogenation of diene, alpha beta unsaturated aldehydes, aromatics, etc. All the associated strategies and synthesis methods in the improvement of various types of hydrogenation reaction are reported including collaboration of active sites, idealization of structure, and correlations among reactants and catalysts are also discussed.

8.1.1 The Active Centers in Parent MOF Materials

The coordinated unsaturated sites and multifunctional ligands of Metal organic framework made it suitable to be used as catalysts. These active sites can also link with other species like homogeneous catalyst in order to produce new catalytic sites in catalyst having MOF foundation [19].

8.1.2 The Active Centers in MOF Catalyst

MOF materials have been universally regarded as potential heterogeneous catalysts [20]. These active sites of MOF originated catalyst as shown in Figure 8.1 can be ions, ligands, or other material. MOF material can be used in original form or it can by reconstructed by hybridization using different material by confining active species within the framework. These hybrid

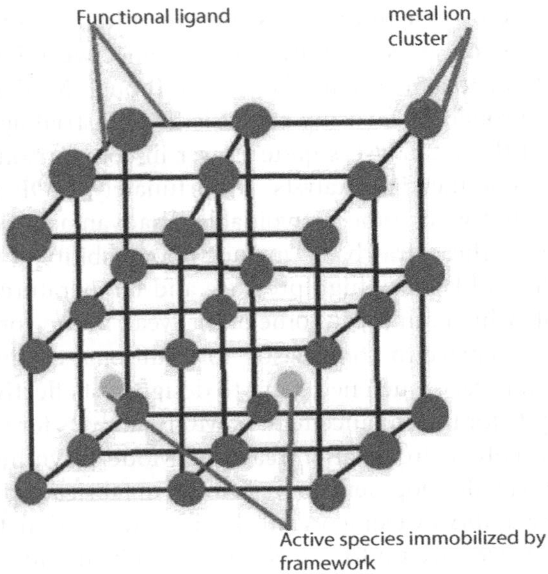

Figure 8.1 Active site in MOF based catalyst [19].

MOFs are associated with higher advantages and lesser drawbacks which were related to parent MOF materials and can be used for various types of catalytic reaction specially hydrogenation reaction.

8.1.3 Metal Nodes

In general metal ions link with organic ligands and solvents in MOF during inter collaboration and form coordinative unsaturated sites when the solvents are removed [21]. These coordinative unsaturated sites act as electron-pair receptors and hence called as Lewis acid sites, thereby increasing rate of the reaction. Table 8.1, shows various metal ions forming Coordinative unsaturated sites in metal organic framework.

HKUST-1 and MIL-101 (Cr) are used in tailoring Lewis acid catalyst due to their unique Coordinative unsaturated sites (CUSs). HKUST-1 contain four connected paddle wheel cluster where as MIL-101(Cr) contain six connected trigonal–prismatic cluster. On increasing the temperature in a vacuum, the water molecules coordinated to Cr^{3+} in can be removed easily and hence enable the CUSs to be released during activating it. Another method such as introduction of defective unstabilized organic ligands or by incomplete decomposition of MOF, CUS can be efficiently introduced into these catalysts.

Table 8.1 Metal ion for hydrogenation reaction.

MOF	IONS	References
HKUST-1	Cu^{2+}	[22]
MIL-101 (Cr)	Cr^{3+}	[23]
MIL-100 (Fe)	Fe^{3+}	[24]
MOF-74 (Mg)	Mg^{2+}	[25]
CPO-27 (Co)	Co^{2+}	[26]
UIO-66	Zr^{4+}	[27]

8.2 Hydrogenation Reactions

8.2.1 Hydrogenation of Alpha–Beta Unsaturated Aldehyde

The selective hydrogenation of carbon–oxygen group of alpha beta unsaturated aldehydes is a new trend for meeting demand of various flavoring, perfume, and medicinal sector, which is difficult to be produced from limited natural resources [28]. But, production of suitable catalysis for the hydrogenation process faces lots of hurdles because this reaction of C–C group is favored thermodynamically. Hence to overcome these problems metal organic framework is identified as effective medium for selective hydrogenation of alpha beta unsaturated aldehydes. Confining Pt nanoparticles within an inner and outer shell made of an Metal organic framework with metal ions Fe^{3+}, Cr^{3+} known as MIL-101 give rise to a stable catalyst, which converts a variety of alpha–beta unsaturated very efficiently into unsaturated alcohols [29–31]. It was reported that the encapsulation of Metal nanoparticles in MOF is identified as one of the encouraging ways to improve the sustainability of MOF supported catalysis. The Pt/MOF@MOFs nanocomposite was reported of having retained properties like pore texture, crystalline structure etc. It shows better chemoselectivity in comparison to uncoated material in the hydrogenation of cinnamaldehyde. It was found that selectivity to cinnamyl alcohol via C=O hydrogenation shown better results from 55% to 96% during conversion of cinnamaldehyde. Further this catalyst can be easily recycled while retaining its selectivity and activity [32]. Recently, for the catalytic hydrogenation of furfural copper-chromium or metal catalysts are widely used. Studies have been done on the selective hydrogenation of furfural to furfuryl alcohol over alumina–carbon matrix implanted Ni

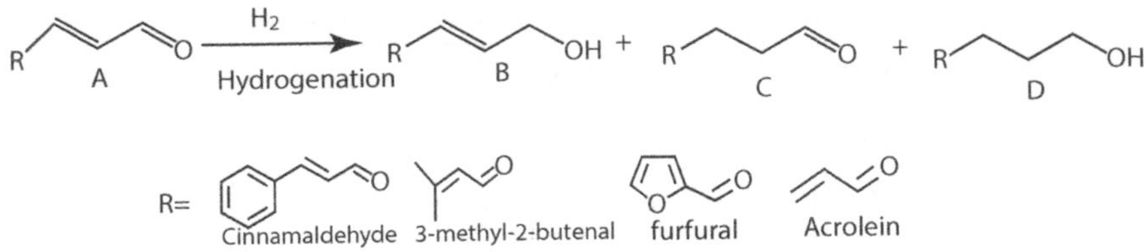

Figure 8.2 Hydrogenation reaction [28].

nanoparticles catalyst. Reverse microemulsion method was used to synthesize Ni-loaded metal organic framework and pyrolysis method under a nitrogen atmosphere was used to synthesize Ni/Al_2O_3 [33]. Zirconium based metal organic framework such as Uio-66, UiO-67, MIL-140B, MIL-140-A, MIL-140C, etc., with Ru nanoparticles encapsulated in it were reported as effective agent for selective hydrogenation of furfural to furfuryl alcohol. The characterization of Ru catalyst was done by TEM, XPS, etc., to determine particle size and oxidation state. This catalyst produced 94.9% yield of furfuryl alcohol and is reusable for five consecutive reaction cycles with high performance [34].

The Figure 8.2 shows example of hydrogenation of alpha beta unsaturated aldehydes as reported from literature survey [28].

8.2.2 Hydrogenation of Cinnamaldehyde

Another such strong and selective catalyst based on platinum nanocluster, which was encapsulated in an amino functionalized Zr-terephthalated metal organic framework (UiO-66–NH_2) was constructed. It contributed in the hydrogenation of ethylene, 1-hexene and 1,3-cyclooctadiene. Further, it was also reported that it was used in chemoselective hydrogenation of cinnamaldehyde and was found that using this MOF high conversion of cinnamaldehyde and high selectivity of cinnamyl alcohol can be done. This was done when the Pt nanocluster was confined in the amino functionalized MOF and not when Pt nanocluster was present externally on MOF. It was also found that this catalyst can be recycled number of times without affecting its activity and selectivity. These facts were confirmed by kinetic studies for 20 h on the catalyst [35].

Zeolitic imidazolate framework (ZIF) was used as a new class of porous crystalline system, which has a combination of desired properties of both MOF as well as Zeolites. In comparison to other MOFs, ZIFs have far better thermal, hydrothermal, and chemical stability [36–39]. The pore diameter of ZIF compounds are 1.16 nm which can be easily accessed by small apertures. By using different methods ZIF-8 have been attached by gold, silver, palladium, platinum, ruthenium, and gold–nickel bimetal nanoparticles and used for the hydrogenation of 1,4-butynediol selectively, amino carbonylation, alkene hydrogenation in gas phase and acetophenone asymmetric hydrogenation [40–44].

Cinnamaldehyde hydrogenation to unsaturated alcohols or saturated aldehydes as shown in Figure 8.3, is very vital in chemical industries like medicine, perfume and food additives [45]. Recently, researchers reported use of ZIF-8 supported palladium catalyst for the hydrogenation of cinnamaldehyde. These catalysts demonstrated better selectivity and activity to hydro cinnamaldehyde (HCAL) in comparison to traditional supports SiO_2 and Al_2O_3. Table 8.2, displays various MOF based catalyst for cinnamaldehyde hydrogenation as per literature survey [45].

8.2.3 Hydrogenation of Nitroarene

Nitroarenes hydrogenation results in anilines which are an important raw material in various chemical industries like polymers, medicine, etc., hence this process is gaining fame among researchers [46]. The process of hydrogenation of nitroarenes involves breaking of hydrogen, nitro group reduction, hydroxylamine condensation, and nitroso species. Along with the nitro group reduction other functional group like halide, cyano, carbonyl, etc., also gets reduced by hydrogenation. Hence, this create hurdle in selective hydrogenation of nitroarenes due to uncontrolled side reaction. Few chemoselective catalysts have been reported but they were found to be associated with low catalytic activity [47, 48]. In this context MOF based catalyst emerged as effective agents with higher activities in hydrogenation reaction. Lanthanide based Metal organic framework were reported to be utilized as

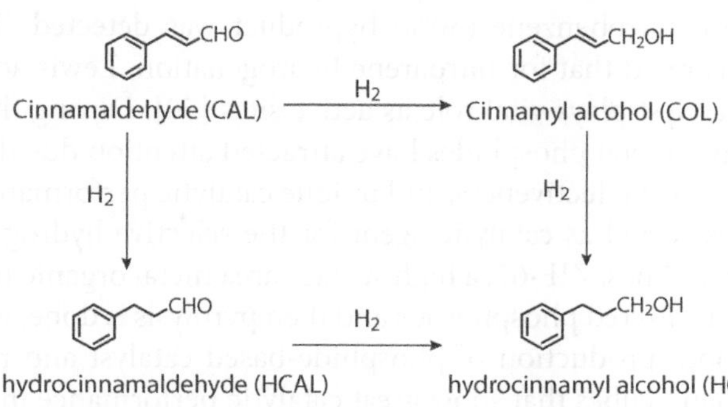

Figure 8.3 Hydrogenation reaction mechanism of cinnamaldehyde (CAL) [45].

Table 8.2 Catalytic performance of different MOF based catalyst for selective hydrogenation of cinnamaldehyde [45].

Sl. no	Catalyst	Conversion (%)	Selectivity (%)		
			Hydro cinnamaldehyde	Cinnamyl alcohol	Hydro cinnamyl alcohol
1.	Pd/ZIF-8 (0.25%)	39.5	90.2	0.9	8.9
2.	Pd/ZIF-8 (0.5%)	61.6	90.1	0.6	9.3
3.	Pd/ZIF-8 (1%)	>99.9	90.4	0	9.6
4.	Pd-MIL-101 (Cr)	99.8	78.2	0.1	21.7
5.	Pd-MIL-53 (Al)	99.9	67.8	0.7	31.5
6.	Pd/Al_2O_3	99.9	70.6	0.5	28.9

catalyst for nitrobenzene hydrogenation. La, Pr, Nd, Sm are few lanthanides which were used for synthesis of the required catalyst [49]. This catalyst produced aniline as only product within the desired time period during the hydrogenation of nitro arenes. The activity order of the corresponding lanthanide metal was found in tune with Lewis acid strength, the order is La < Sm < Nd < Pr. Nitroarene with NH_2, CH_2CN, CHO, Br, I as substituent at Para position gave rise to 100% conversions to corresponding amines with high selectivity. It was identified that only for 4-iodonitrobenzene, dehalogenation with nitrobenzene (30%) byproduct was detected. The above results demonstrated that for nitroarene hydrogenation, Lewis acid sites in MOF based catalyst plays vital role as active sites [50]. During the last few years, transition–metal phosphides have attracted attention due their abundance on earth, cost effectiveness, and unique catalytic performance. Cobalt phosphide is reported as catalytic agent for the selective hydrogenation of nitroarenes to anilines. ZIF-67, a high surface area metal organic framework is encapsulated with red phosphorous and then pyrolysis is done, which supports the smooth production of phosphide-based catalyst and resulted in (Co_2P/CNx) nano cubes that show great catalytic performance in the selective hydrogenation of nitroarenes to aniline as shown in Figure 8.4 and Co_2P/

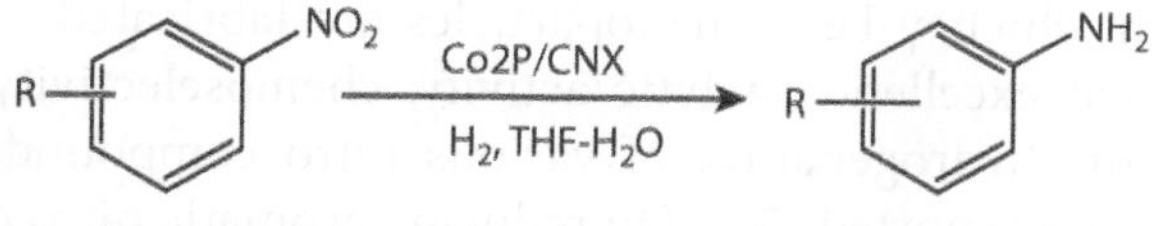

Figure 8.4 Hydrogenation reaction of substituted nitroarenes [51].

Table 8.3 Hydrogenation reaction of various substituted nitroarenes and their result data through literature survey [51].

Entry	Substrate	Product	Time (h)	Conversion (%)	Selectivity (%)
1.	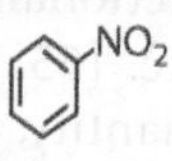	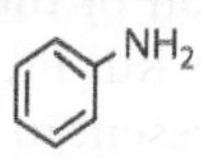	6	>99	>99
2.	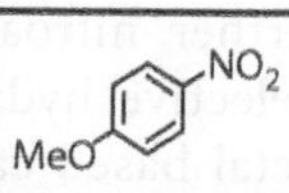	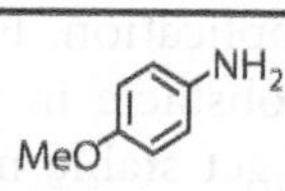	6 12	67 >99	>99 >99
3.	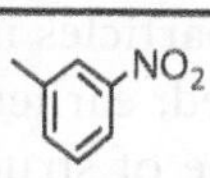	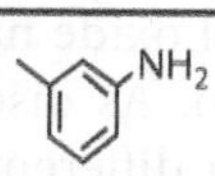	6	>99	>99
4.	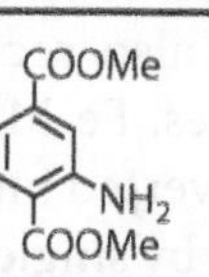	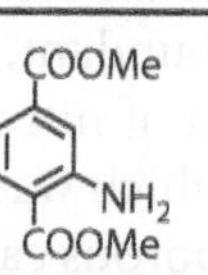	6	>99	94
5.	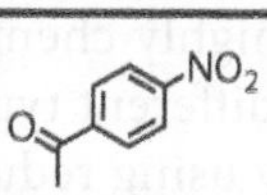	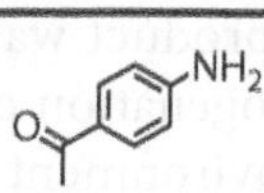	6	>99	97
6.	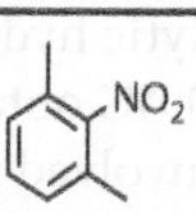	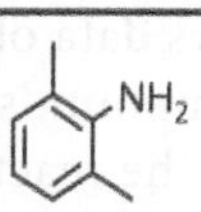	6 12	72 >99	>99 >99
7.	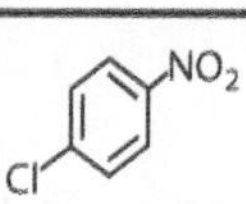	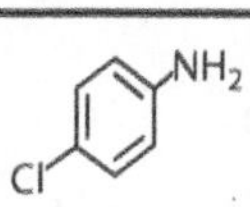	6	>99	>99

CNx role as a catalyst and related data of hydrogenation reaction of various substituted nitroarenes is shown in Table 8.3 [51].

8.2.4 Hydrogenation of Nitro Compounds

The γ-Fe_2O_3 nanoparticles distributed in porous carbon were designed through Fe-based meal organic framework. The Fe based metal organic framework in which γ-Fe_2O_3 nanoparticles are fabricated and the final product carried excellent catalytic activity, chemoselectivity, and recyclability for the hydrogenation of various nitro compounds in a mild environment was reported [52]. On reducing aromatic nitro compounds, aromatic amines are produced, which are important raw materials for the production of dyes, pigments, medicines, chemicals, agro herbicides, etc. [53, 54]. Due to increase in the rate of pollution hydrazine hydrate is the best candidate for reducing agent as it produces only nitrogen as byproduct which reduces risk on the environment and also quality of the resulting anilines. Raney Ni, has been reported to be good for the reduction of nitroarenes but the hydrogenation of their functionalized derivatives requires metal based catalyst like Ru, Pt, Pd, etc. [55, 56]. These metal based catalysts are costly as well present in less quantity, which limits their use in different industrial application. Further, nitroarenes with reducible functional groups create obstacle in selective hydrogenation of nitro group [57–60]. In order to get stable metal based catalyst, the confinement of highly distributed metal oxide nanoparticles in a porous host framework will be a great solution. As discussed, earlier MOF are described as a crystalline material with different type of structure. Iron based nanocrystals are economical, abundant, and also recyclable and are considered active for hydrogenation of nitroarenes. Fe-MIL-88A was reported [56] as a MOF based catalyst which was converted into magnetic γ-Fe_2O_3 nanoparticles encapsulated in porous carbon by single step pyrolysis at a temperature of 500°C. This product was highly chemoselective, recyclable, and has potential for hydrogenation of different types of nitro compounds into anilines in a mild environment by using reducing agent as hydrazine hydrate. The Table 8.4 shows data of catalytic hydrogenation of aromatic as well as aliphatic nitro compounds at 500°C obtained from literature survey and Figure 8.5 displays the reaction involved [52].

8.2.5 Hydrogenation of Benzene

Thermally stable MOF, MIL-120 incorporated with Ni particles forming heterogeneous catalyst Ni/MIL-120 and contributed gas-phase benzene hydrogenation showing better result over Ni/Al_2O_3 catalyst [61]. The characterization result obtained from X-ray diffraction, hydrogen temperature-programmed reduction and transmission electron microscope shows that nickel metal were properly distributed on MOF and the interaction with nickel was weaker in comparison to Al_2O_3 and metal which supported the performance of the catalyst in benzene hydrogenation. Cyclohexane,

Table 8.4 Data collected from literature survey for catalytic hydrogenation of aromatic nitro compounds to corresponding amine [52].

Entry	Substrate	Time (h)	Conversion (%)	Selectivity (%)
1.	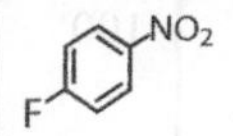	1.5	100	99
2.	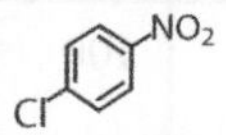	1.5	100	98
3.	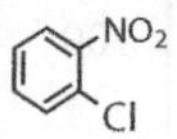	0.75	100	100
4.	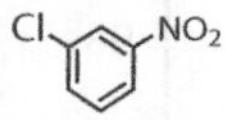	6	81	96.7%
5.	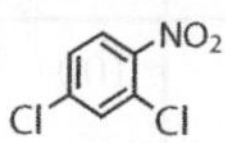	0.75	100	>98
6.	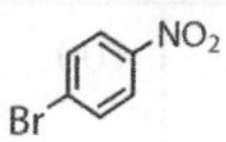	1.25	100	100
7.	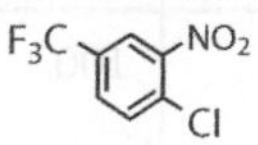	2	100	100
8.	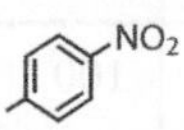	1.5	100	100
9.	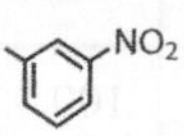	6	100	100
10.	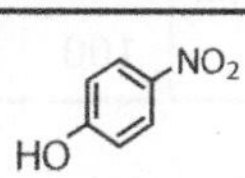	6	100	100
11.	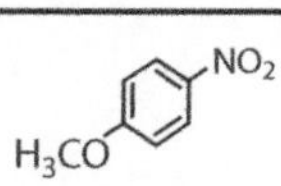	2	100	100
12.	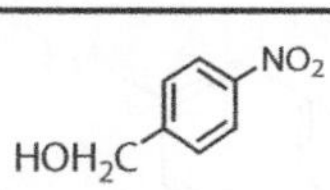	5	100	100

(*Continued*)

Table 8.4 Data collected from literature survey for catalytic hydrogenation of aromatic nitro compounds to corresponding amine [52]. (*Continued*)

Entry	Substrate	Time (h)	Conversion (%)	Selectivity (%)
13.	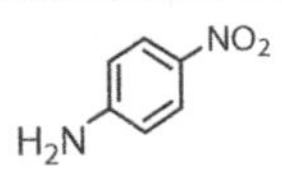	4	100	100
14.	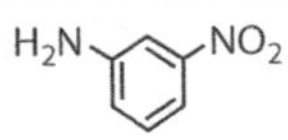	1.5	100	100
15.	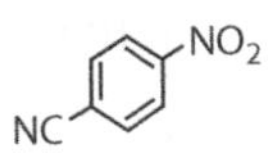	2	98	100
16.	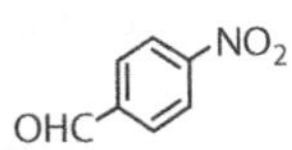	2	100	98
17.	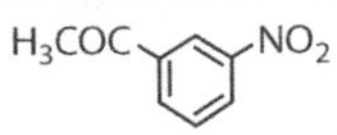	6	100	100
18.	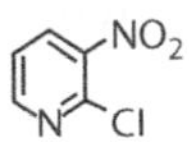	12	62	100
19.	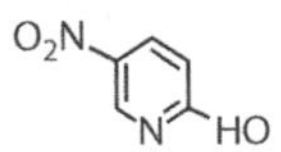	12	69	100
20.	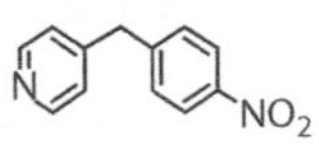	8	100	100
21.	CH_3NO_2	1	100	100
22.	CH_3CH2NO_2	2	81	100

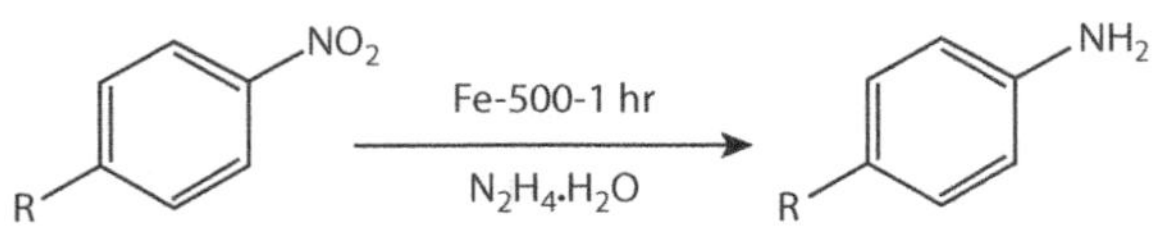

Figure 8.5 Reaction showing catalytic hydrogenation of aromatic nitro compounds to corresponding amine [52].

the hydrogenation product of benzene plays an important role as aromatic content in petroleum industry as well as in the production of Nylon-6 and Nylon-66 [62]. Other group VIII metals used as catalyst are Fe [63], Pt [64, 65], Pd [66, 67], Ru [68], and Co [69] but Ni [70–75] is preferred over others because it is cheap and abundant. Zeolites and mesoporous molecular sieves have been recently reported with high catalytic activity in the hydrogenation of benzene because of their well-defined structure, diversified cage units, and high surface area [76, 77].

8.2.6 Hydrogenation of Quinoline

The Ru^{3+} and free amine groups coordination at the skeleton of MOF is reported as a semi homogeneous catalyst for effective chemoselective hydrogenation of quinolones. The absence of amino group causes aggregation of Ru precursor during the process of pyrolysis giving rise to formation of clusters of Ru. The distribution of Ru on N-doped carbon can be checked by various characterization techniques like electron microscopy and X-ray absorption for spherical aberration correction and fine structure measurements, respectively [78]. Figure 8.6 represents the reaction and Table 8.5 displays data of catalytic performance of Ru single atoms coordinated in carbon architecture (Ru SAs/N–C) and Ru NCs/C catalysts for the regioselective hydrogenation of quinolone [78].

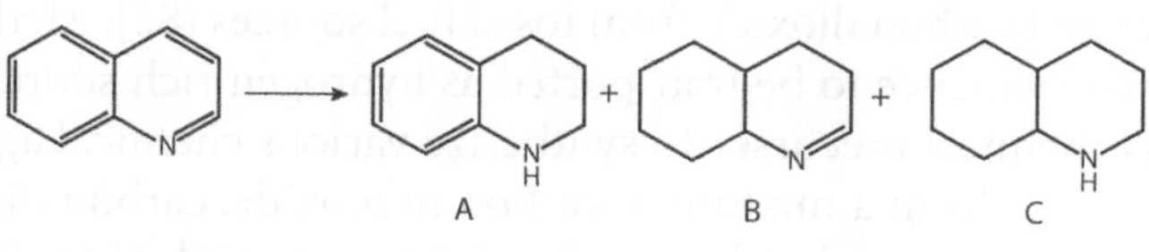

Figure 8.6 Catalytic performance of Ru single atoms coordinated in carbon architecture (Ru SAs/N–C) and Ru NCs/C catalysts for the regioselective hydrogenation of quinolone [78].

Table 8.5 Reaction data collected from literature survey for catalytic performance of Ru single atoms coordinated in carbon architecture (Ru SAs/N–C) and Ru NCs/C catalysts for the regioselective hydrogenation of quinolone [78].

			Selectivity (%)		
Entry	**Catalysts**	**Conversion (%)**	**A**	**B**	**C**
1.	Ru SAs/N-C	>99	>99	<1	0
2.	Ru NCs/C	>99	79	21	0

8.2.7 Hydrogenation of Carbon Dioxide

It is well known that carbon dioxide is one of the pollutants but it is abundant, cheap, and non-toxic [79]. The conversion of CO_2 into useful chemical substances is very helpful but not easy to achieve due to its low reactivity. Lin and Co-worker were first to report in 2011 the doped UiO-67 MOF for photocatalytic reduction of carbon dioxide to carbon monoxide [80].

The presence of amino functional group enhanced absorption of light as well as carbon dioxide [80]. Lee constructed modified Uio-66(Zr)–NH_2, which was used as photocatalyst for the reduction of carbon dioxide to formic acid [80]. Wang and his group reported imidazolate MOF, Co-ZIF-9 which catalyzed the conversion of carbon dioxide to carbon monoxide under mild condition. These results show role of metal organic framework in artificial photosynthesis [80]. Kubaik and co-worker used Mn (I)@UiO-67-bpydc catalyst for carbon dioxide reduction [80].

The selectivity and activity of Cu metal catalyst can be enhanced by a Zirconium based metal organic framework, UiO-66, for the hydrogenation of carbon dioxide to methanol [81]. The catalyst is formed by encapsulating 18 nm single copper nanocrystal within UiO-66. The strong interaction between Cu nanocrystal and zirconium oxide resulted in lowering of zirconium binding energy towards lower oxidation state which leads to the formation of active MOF based copper catalyst. The carbon dioxide hydrogenation product, methanol is an important way for the recyclization of trapped carbon dioxide from fossil fuel sources [82]. Methanol as a fuel adds convenience to be transported as hydrogen rich source and can be used as chemical precursor to synthesize various chemical agents [83]. Industrially it is from a mixture of carbon monoxide, carbon dioxide and hydrogen gas using catalyst like copper, zinc oxide and Al_2O_3 [84]. From isotope labeling experiments it was found that the main carbon source for methanol production is carbon dioxide and the active oxidation state of copper is done by carbon monoxide [85]. The main challenge involves in finding such catalyst which only uses carbon dioxide and hydrogen for the production of methanol [86]. Recently, mesoporous Co_3O_4 catalyst were reported to produce highest percentage of methanol under mild environment, i.e., 250°C and 6 bars [87]. But these catalysts are associated with lesser selectivity because along with methanol, either hydrocarbons or carbon monoxide is produced. To ease this problem a Cu nanocrystal confined within a metal organic framework was reported with 100% conversion of Carbon dioxide to methanol with high activity and selectivity. The hydrogenation of carbon dioxide to methanol is structure sensitive. In this context, MOF act as boon because it interfaces with other catalytically metal due to their nanostructured metal oxide units, which provide ease in the systematic investigation of various effects of catalytic interface [88].

8.2.8 Hydrogenation of Aromatics

An iridium zirconium-based metal organic framework (multifunctional heterogeneous catalyst) was reported to be effective as catalyst for the hydrogenation of aromatic compounds in high amount in presence of mild environment [89]. Figure 8.7, represent conventional method for hydrogenation of aromatic substances. This catalyst proved to be recyclable, active and can keep its reactivity and selectivity intact for the five consecutive applications. The zirconium-based MOF (UiO-66–NH_2) due to its large specific surface area, pore size, and resistance to organic solvent and water was preferred as promising candidate for various catalytic applications [90]. The combined properties of soluble organometallic complexes with MOF as supporting medium enable these new highly stable materials to perform as heterogeneous catalyst after post synthetic changes. The application of Ir–Zr–MOF also involved aromatic hydrogenation forming cyclohexane derivatives which has an important role in preservation of environment, petrochemicals, and clean diesel generation and also in medicine sector for safe drug designing [91, 92].

8.2.9 Hydrogenation of Levulinic Acid

A hafnium-based metal organic framework has been reported to be used as catalyst in the hydrogenation reaction of levulinic ester into γ-valerolactone in the presence of isopropanol as hydrogen donor. The catalyst Hf-MOF-808 can be reused number of times with a negligible loss in catalytic activity. By studying the NMR data, it was found that the

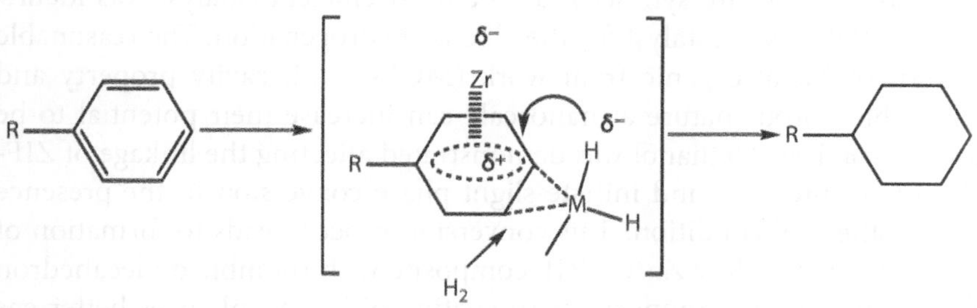

Figure 8.7 Reaction mechanism of hydrogenation of aromatic compounds [89].

hydrogenation was done through direct intermolecular hydrogen transfer method. The catalyst Hf-MOF-808 was combined with Bronsted acid Al-Beta zeolite in order to perform a four step, single pot conversion of furfural to γ-valerolactone with 75% high yield [93].

An ultra-stable Ir@ZrO_2@C single atom catalyst was synthesized during the assembly of UiO-66 followed by pyrolysis and was used in hydrogenation of Levulinic acid to c-valerolactone, which is a prime reaction in the synthesis of renewable chemicals and fuels.

The Ir@ZrO2@C has excellent stability during recycling [94]. For biorefinery industry for the synthesis of renewable fuels and chemicals, hydrogenation of levulinic acid is very important. Ru/ZrO_2@C catalyst based on UiO-66 (Zr-MOF) material was used for hydrogenation of levulinic acid to valerolactone and its performance was compared with that of commercial Ru/C [95]. Both the catalyst shows full conversion of Levulinic acid with quantitative yield. But the catalyst Ru/C exhibited less resistance towards deactivation after the first cycle. Various characterization methods such as TPR, XPS, ICP, etc., along with physisorption data revealed that early deactivation of Ru/C was due to leaching of Ru and loss of surface area through deposition of carbonaceous material in the micropores [95].

The self-prepared Ru/ZrO_2@C catalyst did not exhibit any loss in catalytic performance during continuous use either in water or in protic aqueous solution. In comparison to Ru/C catalyst, leaching of ruthenium was absent in case of Ru/ZrO_2@C in the presence harsh chemical conditions. This excellent catalytic performance was reported by TPR, is due to strong interaction between Ru and ZrO_2 (nano tetragonal in shape) [95].

8.2.10 Hydrogenation of Alkenes and Alkynes

In chemical industry hydrogenation of alkene is very important for organic synthesis [96]. For this purpose, a wide range of homogeneous and heterogeneous catalysts are synthesized. The most efficient catalyst was identified as MOF based catalyst for the alkenes hydrogenation. The reasonable pattern of Metal organic framework (MOFs) with cavity property and adjustable porous nature at nanoscale can increase their potential to be used as catalyst. Methanol was demonstrated affecting the linkage of ZIF-67 in Co^{2+} presence and initiate slight phase conversion in the presence of solvothermal condition. This conversion process leads to formation of a well-defined hollow Zn/Co ZIF composite with rhombic dodecahedron shape. This hollow composite with coating of ZIF-8 enhances better gas storage and porous encapsulation, which facilitate excellent selectivity and activity in the semi hydrogenation of acetylene [97].

The selective hydrogenation of alkynes is very necessary for industries in order to enhance the condition of alkenes streams by eliminating acetylene which is an impurity being produced when hydrocarbon undergo steam cracking [98]. During, the alkynes hydrogenation, the catalyst causes chemisorption of hydrogen which undergoes dissociation and this dissociated hydrogen gets sequentially added to unsaturated bond, thereby leading to semi hydrogenation of alkenes and over hydrogenation of alkanes [99]. Previous, reports suggested Iridium metal nanoparticles encapsulated in MOF ZIF-8 was used in the hydrogenation of phenylacetylene. This catalyst loaded with metal (Ir) by 1.5% weight resulted complete conversion [100] and exhibited more than 90% selectivity during alkyne hydrogenation, but also resulted excess hydrogenation during the reaction. This excess hydrogenation during catalysis occurred due to presence of nearby centers in large nanoparticles. This was solved by a Pd–Ag catalyst encapsulated in MOF. It showed efficient partial hydrogenation of phenylacetylene.

Introducing selective agents with metal nanoparticles inside MOF enhanced the selectivity of hydrogenation products. Compounds containing nitrogen, which are used for the modification of the surface increased the selectivity and activity of catalyst for semi hydrogenation of alkyne was reported [101]. A kind of ionic liquid, 1,1,3,3-Teremethyl guanidinium trifluoroacetate, which contained nitrogen and was coordinated into $Cu(btc)_2$ metal organic framework with palladium nanoparticles to obtain Pd-ionic liquid(IL)-MOF catalyst as shown in Figure 8.8 which when used with palladium loading % by 13.2, resulted in more than 99% selectivity in hydrogenation of phenylacetylene displayed in Figure 8.9 [98].

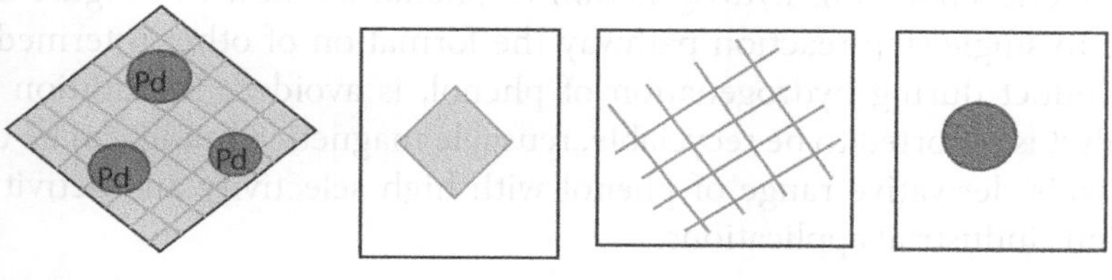

Figure 8.8 Palladium and ionic liquid based MOF based catalyst.

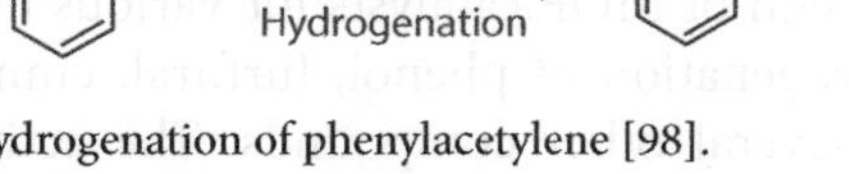

Figure 8.9 Selective hydrogenation of phenylacetylene [98].

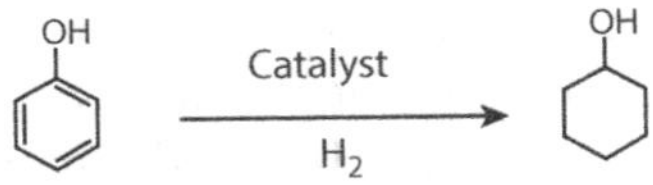

Figure 8.10 Hydrogenation reaction of phenol [102].

8.2.11 Hydrogenation of Phenol

The selective hydrogenation of phenol gives rise to cyclohexanol, which is an important raw material in many chemical industries [102]. The main challenge is to design an economical selective catalyst for this purpose. Nickel–Cobalt alloy bimetallic nanoparticles confined in N-doped carbon model MOF is reported as one such selective catalyst which gives only cyclohexanol as product during phenol hydrogenation [102]. The integration of all the characterization showed cobalt and nickel particles were properly alloyed and dispersed in each nanoparticle, which was encapsulated by N-doped carbon layer. Various catalytic experiments results revealed the catalytic performance during selective hydrogenation of phenol was controlled by temperature of pyrolysis and alloy composition of these nanoparticles. In comparison to Cu, Co–Cu alloy, and Ni–Cu alloy (N-doped carbon) catalytic system, Co–Ni@NC catalyst exhibited best catalytic performance in the hydrogenation of phenol [102]. The cobalt–nickel alloyed, N-doped carbon catalyst displaced complete hydrogenation of phenol (more than 99%) to cyclohexanol, which was two to four times higher than that of individual Co and Ni, N-doped carbon catalyst [102]. The involved mechanism revealed only single step in the formation of cyclohexanol from hydrogenation of phenol as shown in Figure 8.10. Due to single step reaction pathway the formation of other intermediate byproduct during hydrogenation of phenol, is avoided. In addition this catalyst is reported to be recyclable, reusable magnetically and can be used for wide derivative range of phenol with high selectivity and activity in various industrial applications.

8.3 Conclusion

Metal organic framework-based catalysts have been identified as potential candidate for hydrogenation due to their unique properties. This chapter summarizes design of MOF catalysts for various hydrogenation reactions like the hydrogenation of phenol, furfural, cinnamaldehyde, nitro compounds, and several other compounds. The active center in MOF and tunable structure assisted in achieving high selectivities and activities during hydrogenation reaction. Research is going on to improve the present MOF catalytic systems in order to facilitate their performance in various types of purposes. In presence of moist environment, these catalyst with MOF background exhibited low activity which limits its application in hydrogenation reaction in aqueous solution. To solve this, the core catalyst can be coated with hydrophobic material like polydimethylsiloxane to increase the efficiency of catalyst. Noble metals like gold, platinum, platinum, palladium, and ruthenium are mostly preferred for encapsulating into MOF based catalyst for hydrogenation reaction but due to their limiting source and high costs they are not widely used for industrial application. Hence it metals like iron and cobalt is preferable used in place of them due to their low cost and high abundances. For future application in hydrogenation reaction, it is necessary to find substitute of noble metals, which will be equally active like them but economical and largely available. Further, by controlling the morphology of metal nanoparticles, which are encapsulated in MOF based catalyst, will be effective for hydrogenation reaction. MOF based catalyst combined with enzymes as active species can enhance the catalytic performance in liquid solution, provided the stability of the catalysts in water is better. These enzymes increase chemo and enantioselectivity on combining with MOF based catalyst. In order to achieve this target, suitable catalyst need to be produced on wide scale. Literature survey reveals that using MOF based catalyst for hydrogenation reaction has an encouraging future, though many hurdles are associated with it.

References

1. Su, J. and Chen, J.-S., Synthetic porous materials applied in hydrogenation reactions. *Microporous Mesoporous Mater.*, 237, 246–259, 2017.
2. Edwards, H.D., Shelver, W.L., Choi, S., Nisbet, D.J., Krueger, N.A., Anderson, R.C., Smith, S.B., Immunogenic inhibition of prominent ruminal bacteria as a means to reduce lipolysis and biohydrogenation activity *in vitro*. *Food Chem.*, 218, 372–377, 2017.
3. Amir, Md, Kurtan, U., Baykal, A., Synthesis and application of magnetically recyclable nanocatalyst Fe3O4@ Nico@ Cu in the reduction of azo dyes. *Chin. J. Catal.*, 36, 8, 1280–1286, 2015.
4. Wang, H. and Rempel, G.L., Aqueous-phase catalytic hydrogenation of unsaturated polymers. *Catal. Today*, 247, 117–123, 2015.
5. Linares, N., Moreno-Marrodan, C., Barbaro, P., PdNP@ Titanate Nanotubes as Effective Catalyst for Continuous-Flow Partial Hydrogenation Reactions. *ChemCatChem*, 8, 5, 1001–1011, 2016.

6. Yokoyama, T. and Yamagata, N., Hydrogenation of carboxylic acids to the corresponding aldehydes. *Appl. Catal. A: Gen.*, 221, 1–2, 227–239, 2001.
7. Stanislaus, A. and Cooper, B. H., Aromatic hydrogenation catalysis: A review. *Cat. Rev.—Sci. Eng.*, 36, 1, 75–123, 1994.
8. Stanley, J.N.G., Worthington, K., Heinroth, F., Masters, A.F., Maschmeyer, T., Designing nanoscopic, fluxional bimetallic Pt–Ru alloy hydrogenation catalysts for improved sulfur tolerance. *Catal. Today*, 178, 1, 164–171, 2011.
9. Louie, Y.L., Tang, J., Hell, A.M.L., Bell, A.T., Kinetics of hydrogenation and hydrogenolysis of 2, 5-dimethylfuran over noble metals catalysts under mild conditions. *Appl. Catal. B: Environmental*, 202, 557–568, 2017.
10. Chen, N., Ren, Y., Qian, E.W., Elucidation of the active phase in PtSn/SAPO-11 for hydrodeoxygenation of methyl palmitate. *J. Catal.*, 334, 79–88, 2016.
11. Liu, J., Chen, L., Cui, H., Zhang, J., Zhang, L., Su, C.-Y., Applications of metal–organic frameworks in heterogeneous supramolecular catalysis. *Chem. Soc. Rev.*, 43, 16, 6011–6061, 2014.
12. Nandasiri, M.I., Jambovane, S.R., Peter McGrail, B., Schaef, H.T., Nune, S.K., Adsorption, separation, and catalytic properties of densified metal-organic frameworks. *Coord. Chem. Rev.*, 311, 38–52, 2016.
13. Gimenez-Marques, M., Hidalgo, T., Serre, C., Horcajada, P., *Coord. Chem. Rev.*, 307, 342–360, 2016.
14. Rocha, J., Brites, C.D.S., Carlos, L.D., Lanthanide organic framework luminescent thermometers. *Chem. Eur. J.*, 22, 42, 14782–14795, 2016.
15. Lin, R.-B., Liu, S.-Y., Ye, J.-W., Li, X.-Y., Zhang, J.-P., Photoluminescent metal–organic frameworks for gas sensing. *Adv. Sci.*, 3, 7, 1500434, 2016.
16. Chughtai, A.H., Ahmad, N., Younus, H.A., Laypkov, A., Verpoort, F., Metal–organic frameworks: Versatile heterogeneous catalysts for efficient catalytic organic transformations. *Chem. Soc. Rev.*, 44, 19, 6804–6849, 2015.
17. Eddaoudi, M., Kim, J., Rosi, N., Vodak, D., Wachter, J., O'Keeffe, M., Yaghi, O.M., Systematic design of pore size and functionality in isoreticular MOFs and their application in methane storage. *Science*, 295, 5554, 469–472, 2002.
18. Li, R.-J., Li, M., Zhou, X.-P., Li, D., O'Keeffe, M., A highly stable MOF with a rod SBU and a tetra carboxylate linker: Unusual topology and CO 2 adsorption behaviour under ambient conditions. *Chem. Commun.*, 50, 31, 4047–4049, 2014.
19. Evans, J.D., Sumby, C.J., Doonan, C.J., Post-synthetic metalation of metal–organic frameworks. *Chem. Soc. Rev.*, 43, 16, 5933–5951, 2014.
20. Corma, A., García, H., Llabrési Xamena, F.X., Engineering metal organic frameworks for heterogeneous catalysis. *Chem. Rev.*, 110, 8, 4606–4655, 2010.
21. Koh, H.S., Rana, M.K., Hwang, J., Siegel, D.J., Thermodynamic screening of metal-substituted MOFs for carbon capture. *Phys. Chem. Chem. Phys.*, 15, 13, 4573–4581, 2013.
22. Alaerts, L., Séguin, E., Poelman, H., Thibault-Starzyk, F., Jacobs, P.A., De Vos, D.E., Probing the Lewis Acidity and Catalytic Activity of the Metal–Organic

Framework [Cu3 (btc) 2](BTC = Benzene-1, 3, 5-tricarboxylate). *Chem. Eur. J.*, 12, 28, 7353–7363, 2006.

23. Henschel, A., Gedrich, K., Kraehnert, R., Kaskel, S., Catalytic properties of MIL-101. *Chem. Commun.*, 35, 4192–4194, 2008.
24. Anand, R., Borghi, F., Manoli, F., Manet, I., Agostoni, V., Reschiglian, P., Gref, R., Monti, S., Host–guest interactions in Fe (III)-trimesate MOF nanoparticles loaded with doxorubicin. *J. Phys. Chem. B*, 118, 29, 8532–8539, 2014.
25. Cabello, C.P., Gómez-Pozuelo, G., Opanasenko, M., Nachtigall, P., Čejka, J., Metal–Organic Frameworks M-MOF-74 and M-MIL-100: Comparison of textural, acidic, and catalytic properties. *ChemPlusChem*, 81, 8, 828–835, 2016.
26. Jhung, S.H., Khan, N.A., Hasan, Z., Analogous porous metal–organic frameworks: Synthesis, stability and application in adsorption. *CrystEngComm*, 14, 21, 7099–7109, 2012.
27. Piscopo, C.G., Polyzoidis, A., Schwarzer, M., Loebbecke, S., Stability of UiO-66 under acidic treatment: Opportunities and limitations for post-synthetic modifications. *Microporous Mesoporous Mater.*, 208, 30–35, 2015.
28. Zhao, M., Yuan, K., Wang, Y., Li, G., Guo, J., Gu, L., Hu, W., Zhao, H., Tang, Z., Metal–organic frameworks as selectivity regulators for hydrogenation reactions. *Nature*, 539, 7627, 76, 2016.
29. Férey, G., Mellot-Draznieks, C., Serre, C., Millange, F., Dutour, J., Surblé, S., Margiolaki, I., A chromium terephthalate-based solid with unusually large pore volumes and surface area. *Science*, 309, 5743, 2040–2042, 2005.
30. Taylor-Pashow, K.M.L., Della Rocca, J., Xie, Z., Tran, S., Lin, W., Postsynthetic modifications of iron-carboxylate nanoscale metal– organic frameworks for imaging and drug delivery. *J. Am. Chem. Soc.*, 131, 40, 14261–14263, 2009.
31. Bauer, S., Serre, C., Devic, T., Horcajada, P., Marrot, J., Férey, G., Stock, N., High-throughput assisted rationalization of the formation of metal organic frameworks in the iron (III) aminoterephthalate solvothermal system. *Inorg. Chem.*, 47, 17, 7568–7576, 2008.
32. Liu, H., Chang, L., Chen, L., Li, Y., Nanocomposites of Platinum/Metal–Organic Frameworks Coated with Metal–Organic Frameworks with Remarkably Enhanced Chemoselectivity for Cinnamaldehyde Hydrogenation. *ChemCatChem*, 8, 5, 946–951, 2016.
33. Hu, F., Wang, Y., Xu, S., Zhang, Z., Chen, Y., Fan, J., Yuan, H., Gao, L., Xiao, G., Efficient and Selective Ni/Al 2 O 3-C Catalyst Derived from Metal–Organic Frameworks for the Hydrogenation of Furfural to Furfuryl Alcohol. *Catal. Letters*, 1–11, 2019.
34. Yuan, Q., Zhang, D., van Haandel, L., Ye, F., Xue, T., Hensen, E.J.M., Guan, Y., Selective liquid phase hydrogenation of furfural to furfuryl alcohol by Ru/Zr-MOFs. *J. Mol. Catal. A: Chem.*, 406, 58–64, 2015.
35. Guo, Z., Xiao, C., Maligal-Ganesh, R.V., Zhou, L., Goh, T.W., Li, X., Tesfagaber, D., Thiel, A., Huang, W., Pt nanoclusters confined within metal–organic framework cavities for chemoselective cinnamaldehyde hydrogenation. *ACS Catal.*, 4, 5, 1340–1348, 2014.

36. Yaghi, O.M., O'Keeffe, M., Ockwig, N.W., Chae, H.K., Eddaoudi, M., Kim, J., Reticular synthesis and the design of new materials. *Nature*, 423, 6941, 705, 2003.
37. Férey, G., Hybrid porous solids: past, present, future. *Chem. Soc. Rev.*, 37, 1, 191–214, 2008.
38. Phan, A., Doonan, C.J., Uribe-Romo, F.J., Knobler, C.B., O'keeffe, M., Yaghi, O.M., Synthesis, structure, and carbon dioxide capture properties of zeolitic imidazolate frameworks, 2009.
39. Park, K.S., Ni, Z., Côté, A.P., Choi, J.Y., Huang, R., Uribe-Romo, F.J., Chae, H.K., O'Keeffe, M., Yaghi, O.M., Exceptional chemical and thermal stability of zeolitic imidazolate frameworks. *Proc. Natl. Acad. Sci.*, 103, 27, 10186–10191, 2006.
40. Jiang, H.-L., Akita, T., Ishida, T., Haruta, M., Xu, Q., Synergistic catalysis of Au@ Ag core– shell nanoparticles stabilized on metal– organic framework. *J. Am. Chem. Soc.*, 133, 5, 1304–1306, 2011.
41. Zhang, M., Yang, Y., Li, C., Liu, Q., Williams, C.T., Liang, C., PVP–Pd@ ZIF-8 as highly efficient and stable catalysts for selective hydrogenation of 1, 4-butynediol. *Catal. Sci. Technol.*, 4, 2, 329–332, 2014.
42. Dang, T.T., Zhu, Y., Ngiam, J.S.Y., Ghosh, S.C., Chen, A., Seayad, A.M., Palladium nanoparticles supported on ZIF-8 as an efficient heterogeneous catalyst for aminocarbonylation. *ACS Catal.*, 3, 6, 1406–1410, 2013.
43. Jiang, H.-L., Liu, B., Akita, T., Haruta, M., Sakurai, H., Xu, Q., Au@ ZIF-8: CO oxidation over gold nanoparticles deposited to metal–organic framework. *J. Am. Chem. Soc.*, 131, 32, 11302–11303, 2009.
44. Liu, M., Fan, B., Shi, X., Li, R., Ru/ZIF-8 with a chiral modifier for asymmetric hydrogenation of acetophenone. *Catal. Commun.*, 42, 20–24, 2013.
45. Zhao, Y., Liu, M., Fan, B., Chen, Y., Lv, W., Lu, N., Li, R., Pd nanoparticles supported on ZIF-8 as an efficient heterogeneous catalyst for the selective hydrogenation of cinnamaldehyde. *Catal. Commun.*, 57, 119–123, 2014.
46. Chen, L., Li, H., Zhan, W., Cao, Z., Chen, J., Jiang, Q., Jiang, Y., Xie, Z., Kuang, Q., Zheng, L., Controlled encapsulation of flower-like Rh–Ni alloys with MOFs via tunable template Dealloying for enhanced selective hydrogenation of alkyne. *ACS Appl. Mater. Interfaces*, 8, 45, 31059–31066, 2016.
47. Reis, P.M. and Royo, B., Chemoselective hydrogenation of nitroarenes and deoxygenation of pyridine N-oxides with H2 catalyzed by MoO2Cl2. *Tetrahedron Lett.*, 50, 8, 949–952, 2009.
48. Corma, A., González-Arellano, C., Iglesias, M., Sánchez, F., Gold complexes as catalysts: Chemoselective hydrogenation of nitroarenes. *Appl. Catal. A: Gen.*, 356, 1, 99–102, 2009.
49. D'Vries, R.F., Iglesias, M., Snejko, N., Alvarez-Garcia, S., Gutiérrez-Puebla, E., Angeles Monge, M., Mixed lanthanide succinate–sulfate 3D MOFs: Catalysts in nitroaromatic reduction reactions and emitting materials. *J. Mater. Chem.*, 22, 3, 1191–1198, 2012.
50. Chen, Z., Chen, J., Li, Y., Metal–organic-framework-based catalysts for hydrogenation reactions. *Chin. J. Catal.*, 38, 7, 1108–1126, 2017.

51. Yang, S., Peng, L., Oveisi, E., Bulut, S., Sun, D.T., Asgari, M., Trukhina, O., Queen, W.L., MOF-Derived Cobalt Phosphide/Carbon Nanocubes for Selective Hydrogenation of Nitroarenes to Anilines. *Chemistry–A European Journal*, 24, 17, 4234–4238, 2018.
52. Li, Y., Zhou, Y.-X., Ma, X., Jiang, H.-L., A metal–organic framework-templated synthesis of γ-Fe 2 O 3 nanoparticles encapsulated in porous carbon for efficient and chemoselective hydrogenation of nitro compounds. *Chem. Commun.*, 52, 22, 4199–4202, 2016.
53. Downing, R.S., Kunkeler, P.J., Van Bekkum, H., Catalytic syntheses of aromatic amines. *Catal. Today*, 37, 2, 121–136, 1997.
54. Blaser, H.-U., Steiner, H., Studer, M., Selective catalytic hydrogenation of functionalized nitroarenes: An update. *ChemCatChem*, 1, 2, 210–221, 2009.
55. Nishimura, S., *Handbook of heterogeneous catalytic hydrogenation for organic synthesis*, Wiley, New York, 2001.
56. Cornils, B., Herrmann, W.A., Beller, M., Paciello, R. (Eds.), *Applied Homogeneous Catalysis with Organometallic Compounds: A Comprehensive Handbook in Four Volumes*, vol. 4, John Wiley & Sons, 2017.
57. Corma, A. and Serna, P., Chemoselective hydrogenation of nitro compounds with supported gold catalysts. *Science*, 313, 5785, 332–334, 2006.
58. Grirrane, A., Corma, A., García, H., Gold-catalyzed synthesis of aromatic azo compounds from anilines and nitroaromatics. *Science*, 322, 5908, 1661–1664, 2008.
59. Li, S.-C. and Diebold, U., Reactivity of TiO2 rutile and anatase surfaces toward nitroaromatics. *J. Am. Chem. Soc.*, 132, 1, 64–66, 2009.
60. Liu, X., Li, H.-Q., Ye, S., Liu, Y.-M., He, H.-Y., Cao, Y., Gold-catalyzed direct hydrogenative coupling of nitroarenes to synthesize aromatic azo compounds. *Angew. Chem. Int. Ed.*, 53, 29, 7624–7628, 2014.
61. Wan, Y., Chen, C., Xiao, W., Jian, L., Zhang, N., Ni/MIL-120: An efficient metal–organic framework catalyst for hydrogenation of benzene to cyclohexane. *Microporous Mesoporous Mater.*, 171, 9–13, 2013.
62. Yasuda, H., Kameoka, T., Sato, T., Kijima, N., Yoshimura, Y., Sulfur-tolerant Pd–Pt/Al2O3–B2O3 catalyst for aromatic hydrogenation. *Appl. Catal. A: Gen.*, 185, 2, L199–L201, 1999.
63. Phillips, M.I. and Emmett, P.H., Iron as a catalyst for the hydrogenation of benzene. *J. Catal.*, 101, 2, 268–272, 1986.
64. Tsai, K.-Y., Wang, I., Tsai, T.-C., Zeolite supported platinum catalysts for benzene hydrogenation and naphthene isomerization. *Catal. Today*, 166, 1, 73–78, 2011.
65. Simon, L.J., Van Ommen, J.G., Jentys, A., Lercher, J.A., Sulfur-tolerant Pt-supported zeolite catalysts for benzene hydrogenation: I. Influence of the support. *J. Catal.*, 201, 1, 60–69, 2001.
66. Chou, P. and Albert Vannice, M., Benzene hydrogenation over supported and unsupported palladium: I. Kinetic behavior. *J. Catal.*, 107, 1, 129–139, 1987.
67. Chou, P. and Albert Vannice, M., Benzene hydrogenation over supported and unsupported palladium: II. Reaction model. *J. Catal.*, 107, 1, 140–153, 1987.

68. Milone, C., Neri, G., Donato, A., Musolino, M.G., Mercadante, L., Selective hydrogenation of benzene to cyclohexene on Ru/γ-Al2O3. *J. Catal.*, 159, 2, 253–258, 1996.
69. Lu, S., Lonergan, W.W., Bosco, J.P., Wang, S., Zhu, Y., Xie, Y., Chen, J.G., Low temperature hydrogenation of benzene and cyclohexene: A comparative study between γ-Al2O3 supported PtCo and PtNi bimetallic catalysts. *J. Catal.*, 259, 2, 260–268, 2008.
70. Molina, R. and Poncelet, G., Hydrogenation of benzene over alumina-supported nickel catalysts prepared from Ni (II) acetylacetonate. *J. Catal.*, 199, 2, 162–170, 2001.
71. Savva, P.G., Goundani, K., Vakros, J., Bourikas, K., Fountzoula, C., Vattis, D., Lycourghiotis, A., Kordulis, C., Benzene hydrogenation over Ni/Al2O3 catalysts prepared by conventional and sol–gel techniques. *Appl. Catal. B: Environmental*, 79, 3, 199–207, 2008.
72. Jasik, A., Wojcieszak, R., Monteverdi, S., Ziolek, M., Bettahar, M.M., Study of nickel catalysts supported on Al2O3, SiO2 or Nb2O5 oxides. *J. Mol. Catal. A: Chem.*, 242, 1–2, 81–90, 2005.
73. Yang, H., Song, S., Rao, R., Wang, X., Yu, Q., Zhang, A., Enhanced catalytic activity of benzene hydrogenation over nickel confined in carbon nanotubes. *J. Mol. Catal. A: Chem.*, 323, 1–2, 33–39, 2010.
74. Louloudi, A., Michalopoulos, J., Gangas, N.-H., Papayannakos, N., Hydrogenation of benzene on Ni/Al-pillared saponite catalysts. *Appl. Catal. A: Gen.*, 242, 1, 41–49, 2003.
75. Chettibi, M., Boudjahem, A.-G., Bettahar, M., Synthesis of Ni/SiO 2 nanoparticles for catalytic benzene hydrogenation. *Transit. Met. Chem.*, 36, 2, 163–169, 2011.
76. Ismagilov, Z.R., Yashnik, S.A., Startsev, A.N., Boronin, A.I., Stadnichenko, A.I., Kriventsov, V.V., Kasztelan, S., Guillaume, D., Makkee, M., Moulijn, J.A., Deep desulphurization of diesel fuels on bifunctional monolithic nanostructured Pt-zeolite catalysts. *Catal. Today*, 144, 3–4, 235–250, 2009.
77. Abu Bakar, N.H.H., Bettahar, M.M., Abu Bakar, M., Monteverdi, S., Ismail, J., Low temperature activation of Pt/Ni supported MCM-41 catalysts for hydrogenation of benzene. *J. Mol. Catal. A: Chem.*, 333, 1–2, 11–19, 2010.
78. Wang, X., Chen, W., Zhang, L., Yao, T., Liu, W., Lin, Y., Ju, H. *et al.*, Uncoordinated amine groups of metal–organic frameworks to anchor single Ru sites as chemoselective catalysts toward the hydrogenation of quinoline. *J. Am. Chem. Soc.*, 139, 28, 9419–9422, 2017.
79. Kreuzer, L.B., Kenyon, N.D., Patel, C.K.N., Air pollution: Sensitive detection of ten pollutant gases by carbon monoxide and carbon dioxide lasers. *Science*, 177, 4046, 347–349, 1972.
80. Kang, Y.-S., Lu, Y., Chen, K., Zhao, Y., Wang, P., Sun, W.-Y., Metal–organic frameworks with catalytic centers: From synthesis to catalytic application. *Coord. Chem. Rev.*, 378, 262–280, 2019.

81. Rungtaweevoranit, B., Baek, J., Araujo, J.R., Archanjo, B.S., Choi, K.M., Yaghi, O.M., Somorjai, G.A., Copper nanocrystals encapsulated in Zr-based metal–organic frameworks for highly selective CO2 hydrogenation to methanol. *Nano Lett.*, 16, 12, 7645–7649, 2016.
82. Schoedel, A., Ji, Z., Yaghi, O.M., The role of metal–organic frameworks in a carbon-neutral energy cycle. *Nat. Energy*, 1, 4, 16034, 2016.
83. Stöcker, M., Methanol-to-hydrocarbons: Catalytic materials and their behavior. *Microporous Mesoporous Mater.*, 29, 1–2, 3–48, 1999.
84. Behrens, M., Studt, F., Kasatkin, I., Kühl, S., Hävecker, M., Abild-Pedersen, F., Zander, S. *et al.*, The active site of methanol synthesis over Cu/ZnO/Al2O3 industrial catalysts. *Science*, 336, 6083, 893–897, 2012.
85. Manrique, R., Jiménez, R., Rodríguez-Pereira, J., Baldovino-Medrano, V.G., Karelovic, A., Insights into the role of Zn and Ga in the hydrogenation of CO2 to methanol over Pd. *Int. J. Hydrog. Energy*, 44, 31, 16526–16536, 2019.
86. Chinchen, G.C., Denny, P.J., Parker, D.G., Spencer, M.S., Whan, D.A., Mechanism of methanol synthesis from CO2/CO/H2 mixtures over copper/zinc oxide/alumina catalysts: use of14C-labelled reactants. *Appl. Catal.*, 30, 2, 333–338, 1987.
87. Li, C.-S., Melaet, G., Ralston, W.T., An, K., Brooks, C., Ye, Y., Liu, Y.-S. *et al.*, High-performance hybrid oxide catalyst of manganese and cobalt for low-pressure methanol synthesis. *Nat. Commun.*, 6, 6538, 2015.
88. Rungtaweevoranit, B., Zhao, Y., Choi, K.M., Yaghi, O.M., Cooperative effects at the interface of nanocrystalline metal–organic frameworks. *Nano Res.*, 9, 1, 47–58, 2016.
89. Rasero-Almansa, A.M., Corma, A., Iglesias, M., Sánchez, F., Post-functionalized iridium–Zr-MOF as a promising recyclable catalyst for the hydrogenation of aromatics. *Green Chem.*, 16, 7, 3522–3527, 2014.
90. Morris, W., Doonan, C.J., Yaghi, O.M., Postsynthetic modification of a metal–organic framework for stabilization of a hemiaminal and ammonia uptake. *Inorg. Chem.*, 50, 15, 6853–6855, 2011.
91. James, B.R., Wang, Y., Hu, T.Q., *Hydrogenation of substituted aromatics*, pp. 423–428, Chemical Industries, Marcel Dekker, New York, 1996.
92. Enya, T., Suzuki, H., Watanabe, T., Hirayama, T., Hisamatsu, Y., 3-Nitrobenzanthrone, a powerful bacterial mutagen and suspected human carcinogen found in diesel exhaust and airborne particulates. *Environ. Sci. Technol.*, 31, 10, 2772–2776, 1997.
93. Rojas-Buzo, S., García-García, P., Corma, A., Catalytic Transfer Hydrogenation of Biomass-Derived Carbonyls over Hafnium-Based Metal–Organic Frameworks. *ChemSusChem*, 11, 2, 432–438, 2018.
94. Chen, H., Wang, L., Yang, J., Yang, R.T., Investigation on hydrogenation of metal–organic frameworks HKUST-1, MIL-53, and ZIF-8 by hydrogen spillover. *J. Phys. Chem. C*, 117, 15, 7565–7576, 2013.
95. Cao, W., Luo, W., Ge, H., Su, Y., Wang, A., Zhang, T., UiO-66 derived Ru/ZrO 2@ C as a highly stable catalyst for hydrogenation of levulinic acid to γ-valerolactone. *Green Chem.*, 19, 9, 2201–2211, 2017.

96. Chirik, P.J., Iron- and cobalt-catalyzed alkene hydrogenation: Catalysis with both redox-active and strong field ligands. *Acc. Chem. Res.*, 48, 6, 1687–1695, 2015.
97. Yang, J., Zhang, F., Lu, H., Hong, X., Jiang, H., Wu, Y., Li, Y., Hollow Zn/Co ZIF Particles Derived from Core–Shell ZIF-67@ ZIF-8 as Selective Catalyst for the Semi-Hydrogenation of Acetylene. *Angew. Chem. Int. Ed.*, 54, 37, 10889–10893, 2015.
98. Chen, L., Chen, X., Liu, H., Li, Y., Encapsulation of Mono-or Bimetal Nanoparticles Inside Metal–Organic Frameworks via *In situ* Incorporation of Metal Precursors. *Small*, 11, 22, 2642–2648, 2015.
99. Wang, X., Li, M., Cao, C., Liu, C., Liu, J., Zhu, Y., Zhang, S., Song, W., Surfactant-Free Palladium Nanoparticles Encapsulated in ZIF-8 Hollow Nanospheres for Size-Selective Catalysis in Liquid-Phase Solution. *ChemCatChem*, 8, 20, 3224–3228, 2016.
100. Zahmakiran, M., Iridium nanoparticles stabilized by metal organic frameworks (IrNPs@ ZIF-8): Synthesis, structural properties and catalytic performance. *Dalton Trans.*, 41, 41, 12690–12696, 2012.
101. Peng, L., Zhang, J., Yang, S., Han, B., Sang, X., Liu, C., Yang, G., The ionic liquid microphase enhances the catalytic activity of Pd nanoparticles supported by a metal–organic framework. *Green Chem.*, 17, 8, 4178–4182, 2015.
102. Huang, G., Yang, Q., Xu, Q., Yu, S.-H., Jiang, H.-L., Polydimethylsiloxane coating for a palladium/MOF composite: Highly improved catalytic performance by surface hydrophobization. *Angew. Chem. Int. Ed.*, 55, 26, 7379–7383, 2016.

9

Application of MOFs and Their Derived Materials in Solid-Phase Extraction

Adrián Gutiérrez-Serpa[1], Iván Taima-Mancera[1], Jorge Pasán[2], Juan H. Ayala[1] and Verónica Pino[1,3*]

[1]Departamento de Química (Unidad Departamental de Química Analítica), Universidad de La Laguna, La Laguna (Tenerife), Spain
[2]Departamento de Física (Laboratorio de Rayos X y Materiales Moleculares), Universidad de La Laguna, La Laguna (Tenerife), Spain
[3]University Institute of Tropical Diseases and Public Health, Universidad de La Laguna (ULL), La Laguna, Tenerife, Spain

Abstract

Solid-phase extraction (SPE) is a sample preparation technique and clean-up procedure widely used in analytical laboratories worldwide. The procedure requires relatively large amounts of samples, which pass through a small amount of sorbent immobilized into a device (mainly cartridges or disks) previously activated. The analytes and in some cases other interfering species experience retention in the stationary phase of the device. After proper washing to remove interfering species, analytes experience desorption using low amounts of a proper elution solvent (usually amounts lower than 3 mL). Although there is a wide number of commercial materials available as sorbents, the current trends on miniaturization, the increasing need of more selective materials, and the requirement of higher sensitivities for target analytes despite analyzing complex samples, undoubtedly shift to the development of novel and high efficient sorbents for SPE, greener if possible.

Therefore, recent trends focus their efforts on the development of metal–organic frameworks (MOFs) and their derived materials as novel extractant materials, thus emerging a powerful alternative material into analytical applications. The interest relates to their outstanding properties such as high porosity, astonishing surface areas, tunability, and the possibility of designee highly specific materials by reticular chemistry, among others.

**Corresponding author*: veropino@ull.edu.es

Inamuddin, Rajender Boddula, Mohd Imran Ahamed and Abdullah M. Asiri (eds.) Applications of Metal–Organic Frameworks and Their Derived Materials, (219–262) © 2020 Scrivener Publishing LLC

This chapter describes main recent analytical applications of SPE reported for MOFs and their derived materials, including covalent organic frameworks (COFs), while covering the dispersive and the magnetic-assisted operational modes of SPE.

Keywords: Metal–organic frameworks, solid-phase extraction, miniaturization, environmental friendly, novel materials, state of the art

9.1 Solid-Phase Extraction

Solid-phase extraction (SPE) is a sample preparation and clean-up technique that emerged in the middle of the 1970s as an alternative to the classical liquid–liquid extraction (LLE) and Soxhlet procedures. The main advantages offered by this technique include its easier operationally performance, lower solvents volume consumption requirements, and the potential for automation [1, 2]. These characteristics make it an effective method for the extraction and enrichment of analytes from complex matrices in different fields such as environmental, food, and biological analysis, among others [3–5].

Conventional SPE is an exhaustive extraction technique based on the transference of almost all the analytes from the sample matrix to a sorbent. In its most known configuration, this transference takes place in a cartridge (or disk) constituted of a polypropylene or glass syringe barrel-like body and a packaged sorbent material between two frits [6]. Generally, the SPE operational procedure for the extraction of target analytes takes places in four steps: i) conditioning, ii) extraction, iii) washing, and iv) desorption. Figure 9.1(a) shows a general scheme of a SPE procedure. In any case, it is important to optimize these four steps in any sample analysis to ensure the highest effectiveness [7].

Sorbent conditioning is crucial to get a proper analytical performance. In this step, a solvent activates the surface of the sorbent improving the possible interactions between the extractant material and the analytes contained on the sample. The election of a non-proper solvent can became a problem in terms of ineffectively performances due to the reduction of the extraction capability and the low reproducibility. Generally, the sorbent is first wet using a solvent with similar polarity and then exchanged for other similar to the sample matrix. It is important to maintain the sorbent wet before the sample loading to ensure the existence of active points in the sorbent.

During the extraction, the sample slowly passes through the cartridge under a constant flow rate using vacuum or positive pressure. The analytes then involve into a partition equilibrium between the sample and

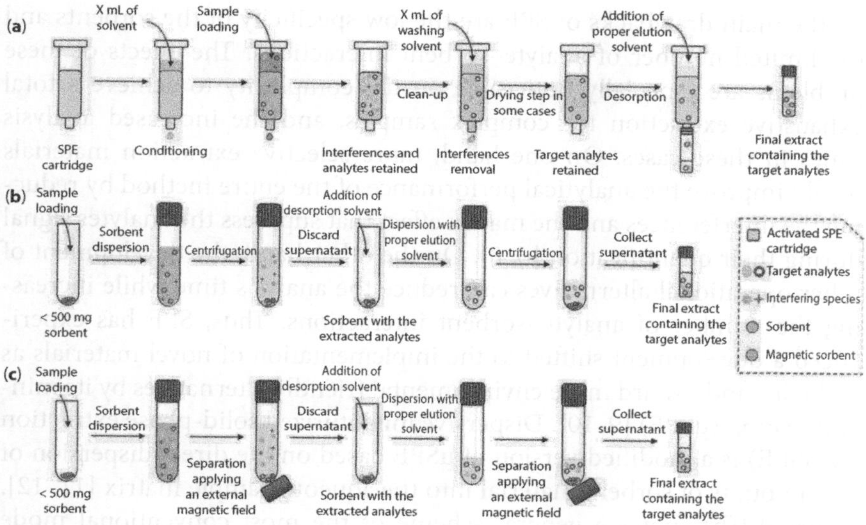

Figure 9.1 General schematic representation for: (a) solid-phase extraction (SPE), (b) miniaturized dispersive solid-phase extraction (D-μSPE), and (c) magnetic-assisted miniaturized dispersive solid-phase extraction (m-D-μSPE).

the sorbent. Thus, the analytes must have a greater affinity for the sorbent chosen to ensure their retention. Nowadays, different kinds of phases are commercially available: normal phases, reverse phases, and ion exchange phases, classified in this case depending on the nature of the interactions that take place (analyte–sorbent). In most of the cases, the mechanism of extraction of the analytes is due to intermolecular forces such as hydrogen bonding, ionic bonding, Van de Waals forces, or a combination of them. Other options of SPE imply retention of only the possible sample interfering compounds while avoiding the retention of the analytes. This way, there is a sample cleanup of interferences in the sorbent.

In common SPE procedures, the addition of a solvent ensures removal of the interferences right after the extraction step. The selected solvent must be strong enough to remove the impurities but also weak enough to leave the target analytes retained in the sorbent. In most of the cases, solvents (or solvents mixtures) with a different polarity from the analyte are useful to ensure the washing of the interferences. Finally, the elution/desorption process takes places adding a solvent which has high affinity with the analytes. At this point, the breakthrough volume is one of the most relevant parameters to control. The breakthrough volume is the maximum volume of aqueous sample that can be loaded into the sorbent avoiding losses of the analytes.

The main drawbacks of SPE are the low specificity of the sorbents and the limited number of analyte–sorbent interactions. The effects of these problems are especially noticeable on the complexity to achieve a total exhaustive extraction for complex samples, and the increased analysis times in these cases. On one hand, more selective extraction materials would improve the analytical performance of the entire method by reducing the interferences and the matrix effect that suppress the analytes signal during their quantification [6, 8]. On the other hand, the development of other operational alternatives can reduce the analysis time while increasing the number of analyte–sorbent interactions. Thus, SPE has experienced a development shifted to the implementation of novel materials as sorbents, and toward more environmental friendly alternatives by its miniaturization (μSPE) [9, 10]. Dispersive miniaturized solid-phase extraction (D-μSPE) is a modified version of μSPE based on the direct dispersion of low amounts of sorbent material into the aqueous sample matrix [11, 12]. Figure 9.1(b) shows a general scheme of the most conventional mode of the D-μSPE procedure. First, a low amount of sorbent (mg or μg) is weighted and added to the sample. Generally, the sorbent and the sample are directly placed in a glass centrifuge tube to avoid transfer steps. Then, there is a strong stirring to ensure complete dispersion of the sorbent into the sample, giving as result a high number of interactions between the analytes and the sorbent material. There are different ways to ensure this dispersion: conventional stirring, vortex, and ultrasounds, among others [13]. Afterward, a centrifugation step allows the separation of the sample—main matrix and non-extracted components—(located in the upper zone) from the added sorbent containing the retained analytes (located in the bottom of the tube). Afterward, desorption of the analytes is achieved by a re-dispersion of the sorbent in a proper low amount of elution solvent followed be a centrifugation step. Depending on the compatibility of the elution solvents with the detection system, further steps will be (or not) required. Most common steps if such compatibility is not possible include a drying step followed by solvent-exchange (with a compatible solvent) [11, 13].

Although the advantages offered by D-μSPE over μSPE have entailed an improvement, the method is still time-consuming and requires a number of steps involving relatively long analysis times. Magnetic-assisted miniaturized dispersive solid-phase extraction (m-D-μSPE) provides an easier manipulation of the sorbents by the use of external magnetic fields, avoiding centrifugation or filtration steps [14, 15]. Figure 9.1(c) shows a general schematic representation of the m-D-μSPE procedure performed in its more conventional mode. Briefly, the method starts by ensuring proper dispersion of the magnetic composite in the aqueous sample. Once the analytes partition properly to the sorbent, an external magnet ensures quickly separation of the phases. Then, an adequate desorption solvent ensures the release of the retained analytes by ensuring mass transfer from the sorbent to the solvent. Finally, an external magnet ensures separation of the sorbent and the enriched final solution containing the analytes [11, 15, 16]. Magnetic-based methods show a unique performance in terms of fastness and assurance of more simple procedures. However, the preparation of magnetic materials is tedious, requires several synthetic steps, and once formed, they tend to form agglomerates [17].

In solid-phase microextraction (SPME), the sorbent material locates over the surface of a support. In its classical and more widespread configuration, the sorbent covers one centimeter of a wire forming a fiber. For the extraction of the analytes, the fiber can be direct immersed in the sample matrix (direct immersion mode (DI)) or exposed to the upper sample gas phase (headspace mode (HS)). Once the analytes achieve the partitioning equilibrium between the donor and acceptor phase, the fiber is commonly exposed to high temperatures to ensure thermal desorption of analytes when coupled to gas chromatography, or exposed to a solvent for desorption when coupled to liquid chromatography [18, 19]. SPME is organic–solvent free when coupled to GC, which is an advantage from an environmental point of view. However, it takes long extraction times and it is a non-exhaustive extraction technique.

9.1.1 Materials in SPE

Among the recent materials proposed as alternatives to commercially available sorbents in solid-phase extraction techniques, it is important to highlight the potential of porous materials, particularly metal–organic frameworks (MOFs) and covalent organic frameworks (COFs) due to their outstanding properties such as high adsorption capacity, impressive surface area, adequate thermal stability, and striking tunability [20–23].

MOFs are highly porous crystalline materials formed by metallic ions (or metallic clusters) and organic ligands termed secondary unit buildings (SBUs). The metallic ions acts like nodes or centers and the organic linkers play the role of bridges interconnecting the nodes by coordination bonds, thus forming an ordered network full of pores, channels and cavities. In fact, MOFs are the materials with the highest surface areas known [24, 25]. The connectivity between the nodes and the linkers defines the crystal morphology and structure; the nodes determine the number of connection points based on their coordination sphere and the linker determines the number of interconnected nodes [26].

Regarding COFs, they are also crystalline structures but formed by organic monomers called linkers and linkages, non-metallic light elements such as C, H, N, B, O and Si replaces the metallic nodes used on MOFs [27]. The organic SBUs form a two-dimensional (2-D) or three-dimensional (3-D) networks by the formation of strong covalent bonds [27]. As the monomer units possess rigid structures the topology and geometry of the resulting COFs skeleton depends on the geometry matching of the SBUs [27–29]. Figure 9.2 points out the differences and similarities between MOFs and COFs.

There are several synthetic strategies reported for the preparation of MOFs and COFs including solvothermal methods, microwave-assisted approaches, mechano-chemical strategies, and room temperature synthesis. Among them, solvothermal methods are those most commonly used due to their simplicity and effectivity. The solvothermal synthesis implies the heating of a solution containing the MOF or COF precursors in a closed system under specific temperature and pressure for a certain time, followed by a cooling step, and crystals clean-up to remove the remaining non-reacted precursors. Finally, there is an activation step of the resulting crystals (normally by a simple heating). The resulting physical–chemical properties of the COFs and MOFs are influenced not only by the precursor's nature but also by the set of synthetic conditions used [25, 28].

The aim of this chapter is to describe main analytical application of MOFs and COFs as sorbent materials for miniaturized solid-phase

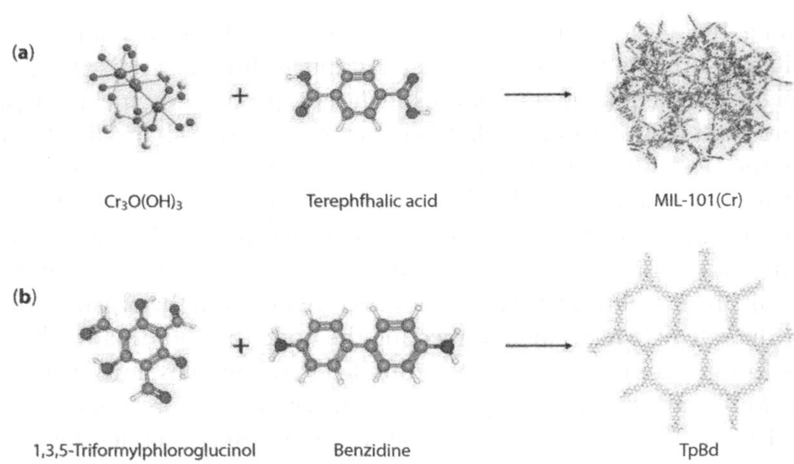

Figure 9.2 Representative scheme of SBUs structures and their corresponding (a) MOF or (b) COF.

extraction under different configurations [22, 30] in the last 3 years (2017–2019). Main modes include their use as package sorbent in SPE [31], and as dispersed sorbent in D-μSPE [32] or in m-D-μSPE [33]. This book chapter does not cover the use of MOFs and COFs as coatings in SPME [21, 34–36].

9.2 MOFs and COFs in Miniaturized Solid-Phase Extraction (μSPE)

MOFs and COFs of quite different nature have been applied successfully as sorbents in μSPE. Table 9.1 summarizes representative analytical applications of MOFs and COFs as sorbents in μSPE procedures [37–57]. In most of the cases, these materials have been used as package sorbents in cartridges [38, 40, 43, 44, 48–50, 53, 54, 57]. The amount of sorbent packed into a 3 mL empty cartridge ranges from 15 mg [38] to 100 mg [54]. Besides the classical cartridges, pipette tip [37, 39, 41, 55, 56], in-syringe [42, 47, 52], and on column [46], have been explored as alternative configurations. The reduction of the size of the packaged sorbent can significantly decrease the extraction time from 23 min [38] to 40 s [47], and the amount of solvent required for the analytes desorption is also decreased [47].

Taking account the common performance of μSPE, it is not surprising that practically all the studies have been performed mainly with high-performance liquid chromatography (HPLC) [37–40, 44, 46–50, 52–57]. Only two studies reported gas chromatography (GC) applications with MOFs μSPE devices [42, 43], for example the study of Zhang *et al.* that used *n*-hexane as desorption solvent, which was compatible with the GC column and did not affect the MOF stability [42]. The coupling between μSPE and the LC system can be off-line, by injecting the final eluent in the detection system, or performing an on-line extraction. The main advantage of on-line methods over off-line is the automatization of the entire procedure. Pang *et al.* employed a monolith column filled with HKUST-1 (Cu) and *N*-methylolacrylamide for the on-line determination of ursolic acid in Chinese herbal medicine, achieving excellent accuracies and recoveries [46].

Regarding MOFs as sorbents, their applications in μSPE include the successful analytical determination of several kinds of analytes, such as drugs [37], herbicides [38, 40], organophosphorus pesticides [44], heavy metals [41], polycyclic aromatic hydrocarbons [42], terpenoids [43], phytohormones [39], and other compounds [46, 47], in biological [37, 47], food [38–41, 44, 45], and water [37, 40, 42] matrixes. In most of these

Table 9.1 MOFs and COFs as sorbents in analytical µSPE applications.

Sorbent (mg)	Format (extraction time)/desorption solvent (mL)	Analytes (number)/ sample (amount)	LOD*	RSD* (%)	RR* (%)	Analytical technique	Ref.
MOFs							
UiO-66–NH_2(Zr) (5)	Pipette tip (≤12 min)/ CH_3OH (10 µL)	Drug (1)/urine & water (both 100 µL)	40 ng·L^{-1}	2.5	98.2–99.4	HPLC-UV*	[37]
UiO-67(Zr) (15)	Cartridge (23 min)/ CH_3OH (5% HCOOH) (1)	Herbicides (4)/food (30 mL aqueous extract)	0.1–0.5 µg·L^{-1}	≤7.7	86.1–103	HPLC-DAD*	[38]
UiO-66(Zr)@ PAN* (5)	Pipette tip (-)/ 90% ACN*-ammonia (1)	Phytohormones (4)/food (1 mL methanolic extract)	10–20 ng·L^{-1}	≤5.6	84.4–111	HPLC-FD*	[39]
MOF-808(Zr) (30)	Cartridge (20 min)/ CH_3OH (5% HCOOH) (3)	Herbicides (4)/drink & water (both 5 mL)	0.1–0.5 µg·L^{-1}	≤6.9	62.7–109	HPLC-UV*	[40]
PCN-222/MOF-545(Zr) (2)	Pipette tip (≤7 min)/ 10% HCl (50 µL)	Hg^{2+} (1)/food (1.8 mL aqueous extract)	20 ng·L^{-1}	3.1	74.3–98.7	CVAAS*	[41]

(*Continued*)

Table 9.1 MOFs and COFs as sorbents in analytical µSPE applications. (*Continued*)

Sorbent (mg)	Format (extraction time)/desorption solvent (mL)	Analytes (number)/ sample (amount)	LOD*	RSD* (%)	RR* (%)	Analytical technique	Ref.
MIL-101(Cr) (2)	In-syringe (-)/ *n*-hexane (0.5)	PAHs* (8)/ water (10 mL)	0.20–1.9 ng·L⁻¹	≤9.7	84.4–105	GC-MS*	[42]
MIL-101(Cr)@ PANI* (40)	Cartridge (10 min)/ ACN* (100 µL)	Terpenoids (2)/plants (5 mL methanolic extract) & food (8 mL)	0.1–1 µg·L⁻¹	≤9.8	90.3–122	GC-FID*	[43]
MIL-101(Cr)@ MIP* (100)	Cartridge (60 min)/ CH_3OH/water/HAc* (95:5:2, v/v/v) (2)	Pesticides (2)/food (100 mL aqueous extract)	11–15 µg·L⁻¹	≤3.7	87.2–91.7	HPLC-UV*	[44]
Chitosan/MIL-68(Al) (5.6)	Circular filter (-)/ HNO_3 (4)	Heavy metals (2)/ mollusks (80 mL aqueous extract)	0.16 µg·L⁻¹	≤4.3	95.0–97.5	ICP*-OES*	[45]

(*Continued*)

Table 9.1 MOFs and COFs as sorbents in analytical µSPE applications. (*Continued*)

Sorbent (mg)	Format (extraction time)/desorption solvent (mL)	Analytes (number)/ sample (amount)	LOD*	RSD* (%)	RR* (%)	Analytical technique	Ref.
HKUST-1(Cu) (1)	Monolithic column (-)/ 95% CH_3OH (flow rate = 1 $mL{\cdot}min^{-1}$)	UA* (1)/Chinese Herbal medicine (20 mL ethanolic extract)	0.17 $\mu g{\cdot}mL^{-1}$	≤6.4	86.3–105	HPLC-UV*	[46]
MOF-5(Zn)@ Fe_3O_4–NH_2 (2)	In-syringe (40 s)/ CH_3OH–HNO_3 (8:2, v/v) (150 µL)	Mandelic acid (1)/ urine (100 µL)	0.10 $\mu g{\cdot}mL^{-1}$	≤3.5	90.3–94.5	HPLC-UV*	[47]
MOF-5(Zn)@ SBA-15 (2)	In-syringe (40 s)/ CH_3OH–HNO_3 (8:2, v/v) (150 µL)	Mandelic acid (1)/ urine (100 µL)	0.13 $\mu g{\cdot}mL^{-1}$	≤3.5	90.3–94.5	HPLC-UV*	[47]
COFs							
TpAzo (25)	Cartridge (25 min)/ ACN* (300 µL)	BUs* (4)/drink & food (100 mL sample or aqueous extract)	0.05–0.2 $\mu g{\cdot}L^{-1}$	≤6.5	84.1–109	HPLC-VWD*	[48]

(*Continued*)

Table 9.1 MOFs and COFs as sorbents in analytical µSPE applications. (*Continued*)

Sorbent (mg)	Format (extraction time)/desorption solvent (mL)	Analytes (number)/ sample (amount)	LOD*	RSD* (%)	RR* (%)	Analytical technique	Ref.
TpPa-NO_2 (25)	Cartridge (6.7 min)/ ACN* (4)	BAs* (8)/food (25 mL TCA* extract)	0.92–2.6 $\mu g{\cdot}L^{-1}$	≤11	80.3–115	HPLC-FD*	[49]
TpBD (30)	Cartridge (6.7 min)/ ACN* (10% HAc*) (4)	PEDs* (4)/beverage (10 mL) & food (50 mL aqueous extract)	56–123 $\mu g{\cdot}L^{-1}$	≤4.9	82.0–96.3	HPLC-UV*	[50]
TpBD (20)	Cartridge (13.3 min)/ HNO_3 (2)	Heavy metals (10)/ drink & water (20 mL)	2.10–21.6 $ng{\cdot}L^{-1}$	≤4.3	81–96	ICP*-MS*	[51]
COF@PS-GMA* (20)	In-syringe (3 min)/ ethanol (0.5)	NSAIDs* (7)/ water (10 mL)	0.13–0.82 $\mu g{\cdot}L^{-1}$	≤8.9	84.3–99.6	HPLC-UV*	[52]
PCONF (100)	Cartridge (25 min)/ CH_3OH (5% NH_4OH) (6)	SAs* (8)/food & drink (100 mL sample or aqueous extract)	0.14–2.0 $ng{\cdot}L^{-1}$	≤13	83.5–109	UHPLC-MS/ MS	[53]
NH_2@COF (100)	Cartridge (4 min)/ ammonia–CH_3OH (8:92, v/v) (2)	Pesticides (6)/ water (20 mL)	0.01–0.06 $\mu g{\cdot}L^{-1}$	≤8.7	89.6–102	HPLC-DAD*	[54]

(*Continued*)

Table 9.1 MOFs and COFs as sorbents in analytical μSPE applications. (*Continued*)

Sorbent (mg)	**Format (extraction time)/desorption solvent (mL)**	**Analytes (number)/ sample (amount)**	**LOD***	**RSD* (%)**	**RR* (%)**	**Analytical technique**	**Ref.**
NH_2–MIL-68@ COF (8)	Pipette tip (10 min)/ ACN* (200 μL)	SAs* (6)/water, drink & food (2 mL sample or aqueous extract)	1 $\mu g{\cdot}L^{-1}$	≤5.2	68.9–104	HPLC-VWD*	[55]
SNW-1@PAN* (12.5)	Pipette tip (-)/ 7.5% ammonia–CH_3OH (1)	SAs* (5)/food (4 mL aqueous extract)	1.7–2.7 $\mu g{\cdot}L^{-1}$	≤9.2	86.0–111	HPLC-PDA*	[56]
DAAQ-Tfp (3)	cartridge (50 min)/ ACN*-water (1:10, v/v) (0.2)	BUs* (4)/beverage & food (100 mL sample or aqueous extract)	0.02–0.08 $\mu g{\cdot}L^{-1}$	≤6.8	85.5–112	HPLC-UV*	[57]

*Abbreviations: LOD for limit of detection, RSD for inter-day relative standard deviation, RR for relative recovery (in real samples), UV for ultraviolet detector, DAD for diode array detector, PAN for polyacrylonitrile, ACN for acetonitrile, FD for fluorescence detector, CVAAS for cold vapor atomic absorption spectroscopy, PAHs for polycyclic aromatic hydrocarbons, MS for mass spectrometry, PANI for polyaniline, FID for flame ionization detector, MIP for molecularly imprinted polymer, HAc for acetic acid, ICP for inductively coupled plasma, OES for optical emission spectrometry, UA for ursolic acid, BUs for benzoylurea insecticides, VWD for variable wavelength detector, BAs for biogenic amines, TCA for trichloroacetic acid, PEDs for phenolic endocrine disruptors, PS-GMA for poly (styrene-divinyl benzene-glycidylmethacrylate), NSAIDS for non-steroidal anti-inflammatory drugs, SAs for sulfonamides, and PDA for photodiode array detector.

applications, MOFs containing trivalent metallic nodes such as zirconium [37–41], chromium [42–44], and aluminum [45], have been used as packaged sorbent materials. This may be due to their higher chemical stability in aqueous media compared with those MOFs formed by low valence metallic nodes.

These sorbents are packaged as neat MOFs crystals [37, 38, 40, 41, 46] or incorporated into other material forming a composites [39, 43, 44, 47]. Composites are especially useful in this μSPE methodologies to reduce the back pressure generated due to the sub-micron or micron size of MOFs particles [39, 43]. Furthermore, the properties of the combined material can introduce some improvements to MOFs, such as a better chemical stability under harsh operational conditions or a more selective extraction [44, 47]. Rahimpoor *et al.* prepared a core–shell mesoporous silica nanoparticle (SBA-15)@MOF-5(Zn) to improve the hydro-stability of the MOF and the sorbent structure uniformity for better extraction reproducibility compared to neat MOF-5(Zn) [47].

Regarding COFs as sorbents, 1,3,5-triformylbenzene (Tp)-based COFs are those mainly used in analytical applications. This COFs family exhibits excellent surface area and a superior chemical resistance to a wide range of conditions including strong acidic and basic aqueous solutions [48–54, 57]. These materials have been used for the analysis of water [51–53, 55] and food [48–50, 53–57], and particularly for the determination of insecticides [48, 57], non-steroidal drugs [52], endocrine disruptors [50], and sulfonamides [53, 55, 56].

Neat packaged COFs in μSPE cartridges has demonstrated excellent analytical performances in their applications with limits of detection in the $\mu g \cdot L^{-1}$ and recoveries ranging from 80% to 115% [48–51, 53, 54, 57]. Composites of COFs mainly include their combination with polymers [52, 56]. On these applications, polymers can act as a substrate with the COF being the main extractive phase [52], or they can play a role of being an extractive material in synergy with the COF [52]. Li *et al.* proposed the use of styrene-divinyl benzene-glycidylmethacrylate microparticles in combination with COFs for the determination of non-steroidal drugs in complex matrices. The use of this composite allows a direct analysis without further previous matrix clean up procedures achieving recoveries between 84.3% and 99.6%. The reported composite demonstrated not only higher extraction capability than other sorbents, but also faster adsorption of the target analytes [52].

Undoubtedly, MOFs and COFs possess many advantages over other materials in terms of analytical applicability. Their combination can entail a new era of clever-materials for microextraction purposes. In this way,

Chen *et al.* reported a MOF decorated with a COF to improve the extraction of sulfonamides [55]. The proposed material demonstrated possess higher sensitivity than single MOF or the single COF, and even better than other sorbents reported for this group of analytes. It is also remarkable the low amount of sorbent required when combining MOF and COF (8 mg). In addition, this sorbent presented a reusability of more than 100 extraction steps without losing sensitivity.

9.3 MOFs and COFs in Miniaturized Dispersive Solid-Phase Extraction (D-μSPE)

Table 9.2 compiles representative examples of MOFs and COFs in D-μSPE applications [58–82]. This section mainly focuses on the use of MOFs as sorbent materials in D-μSPE, as there are more reported studies with MOFs than with COFs. To the best of our knowledge, there is only one publication on the use of COFs as sorbent material in D-μSPE. Li *et al.* used a molecularly imprinted COF (MICOF) for the selective extraction of nonsteroidal anti-inflammatory drugs in different water samples, including pharmaceutical wastewater and hospital wastewater. The MICOF exhibited high extraction capability with satisfactory limits of detection in the range of 0.4–2.9 $\mu g \cdot L^{-1}$. The MICOFs material combined the high surface area of the COFs and the specificity offered by molecularly imprinted materials, resulting in a powerful tool in complex samples analysis applications [82].

MOFs used in D-μSPE include (mostly) net MOFs crystals [58–61, 63–65, 70, 71, 73–75, 77, 78, 80, 81, 83–86], composites [62, 66, 68, 69, 76, 79, 87], and embedded MOFs in devices [67, 72]. Although there is not dispersion of the sorbent when using embedded MOFs, these materials are included in D-μSPE modality as the extraction takes places by strong agitation of the device into the sample. The main advantage over powder dispersion is the avoiding of centrifugation steps as the sorbent forms a single unit. As disadvantage, the extraction times using the devices are considerably longer as the interactions sorbent–analyte are more hindered.

The use of neat MOFs is the most extended mode in D-μSPE, compiling more than 20 different applications in the last 3 years. Generally, the dispersion of the MOF during the extraction is performed by vortex, requiring low extraction times (< than 10 min) [59, 63, 66, 70, 71, 73, 74, 80, 81]. More energetic agitation procedures like sonication reduce considerably the extraction times. However, the MOF have to possess good mechanical and thermal stability to ensure the integrity of its crystallinity during the extraction procedure [64, 77]. Recently, Amiri *et al.* utilized ZIF-8(Zn)

Table 9.2 MOFs and COFs as sorbents in analytical d-μSPE applications.

Sorbent (mg)	Microextraction format (agitation & time)	Analytes (number)/ sample (amount)	LOD*	RSD* (%)	RR* (%)	Analytical technique	Ref.
MOFs							
UiO-66(Zr) (40)	Bare powder (shaking & 5 min)	Insecticides (5)/ water (5 mL)	0.02–0.4 μg·L^{-1}	≤13	73.7–119	GC-MS*	[58]
UiO-66–NO_2(Zr) (20)	Bare powder (vortex & 3 min)	EDCs* (9)/water (20 mL)	1.5–90 ng·L^{-1}	≤14	79.2–126	HPLC-DAD*	[59]
UiO-66–NH_2(Zr) (15)	Bare powder (- & 10 min)	NBs* (8)/water (2 mL)	0.72–1.9 μg·L^{-1}	≤4.5	85.2–105	HPLC-DAD*	[60]
UiO-66–OH(Zr) (10)	Bare powder (shaking & 30 min)	Thorium ions (1)/ water (250 mL)	0.35 μg·L^{-1}	0.84	90.0–95.5	UV-Vis* spectroscopy	[61]
UiO-66–NH_2(Zr)/ urea-POP* (10)	Composite powder (sonication & 20 min)	U^{6+} (1)/water (25 mL)	0.6 μg·L^{-1}	1.9	96.9–98.1	UV-Vis* spectroscopy	[62]

(Continued)

Table 9.2 MOFs and COFs as sorbents in analytical d-µSPE applications. (*Continued*)

Sorbent (mg)	Microextraction format (agitation & time)	Analytes (number)/ sample (amount)	LOD*	RSD* (%)	RR* (%)	Analytical technique	Ref.
MOF-545(Zr) (10)	Bare powder (vortex & 15 min)	Pb^{2+} (1)/food, beverage, & water (250 mL)	1.78 $\mu g \cdot L^{-1}$	3.5	90.0–107	FAAS*	[63]
MIL-101(Cr) (10)	Bare powder (sonication & 8 min)	PCAs* (12)/ water (10 mL)	0.18–0.88 ng·L-1	≤3.9	95.3–106	UHPLC-MS/MS	[64]
MIL-101(Fe) (120)	Bare powder (shaking & 5 min)	PAEs* (8)/water (50 mL)	6.6–21 $ng \cdot L^{-1}$	≤19	70.0–118	HPLC-MS*	[65]
MIL-101(Cr)@ GO* (5)	Composite powder (vortex & 20 min)	SAs* (12)/ drink (1 mL)	12.0–145 $ng \cdot L^{-1}$	≤9.2	80.0–104	UHPLC-MS/MS	[66]
MIL-101(Cr)/ PVA* cryogel (60 $mg \cdot mL^{-1}$)	Device (vortex & 1 h)	NSAIDs* (4)/ water (10 mL)	7.0–37 $ng \cdot L^{-1}$	≤9.9	78.4–106	UHPLC-MS/MS	[67]
MIL-125(Ti)@ sulfide (1)	Composite powder (sonication & 35 min)	NPAHs* (16)/ water & food (10 mL)	2.9–83 $ng \cdot L^{-1}$	≤10	71.3–112	GC-NCI*-MS*	[68]

(*Continued*)

Table 9.2 MOFs and COFs as sorbents in analytical d-µSPE applications. (*Continued*)

Sorbent (mg)	Microextraction format (agitation & time)	Analytes (number)/ sample (amount)	LOD*	RSD* (%)	RR* (%)	Analytical technique	Ref.
MIL-100(Fe)@ IL* (2)	Composite powder (vortex & 1 min)	PAHs* (12)/water, food, & drink (40 mL)	2.0–5.5 $ng \cdot L^{-1}$	≤4.9	97.0–104	GC-FID*	[69]
MIL-53(Al) (8)	Bare powder (vortex & 30 min)	Hormones (8)/ water & urine (8 mL)	0.0015–1.0 $\mu g \cdot L^{-1}$	≤10	80.6–98.4	UHPLC-MS/ MS	[70]
MIL-53(Al) (5)	Bare powder (vortex & 5 min)	EDCs* (6)/ water (10 mL)	0.07–0.2 $\mu g \cdot L^{-1}$ & 13–21 $ng \cdot L^{-1}$	≤11	29.9–170 & 26.0–130	HPLC-DAD* & LC-TOF*	[71]
MIL-53(Al)/ PVDF* (150)	Device (vortex & 30 min)	Estrogens (4)/ urine (20 mL)	0.005–1 $\mu g \cdot L^{-1}$	≤11	80.4–103	HPLC-FD*	[72]
CIM-80(Al) (20)	Bare powder (vortex & 3 min)	PAHs* (15) & EDCs* (7)/ water (10 mL)	0.75–9.3 $ng \cdot L^{-1}$ (PAHs*) & 0.11–21 $\mu g \cdot L^{-1}$ (EDCs*)	≤17	–	UHPLC-UV* & UHPLC-FD*	[73]

(*Continued*)

Table 9.2 MOFs and COFs as sorbents in analytical d-μSPE applications. (*Continued*)

Sorbent (mg)	Microextraction format (agitation & time)	Analytes (number)/ sample (amount)	LOD*	RSD* (%)	RR* (%)	Analytical technique	Ref.
CIM-81(Zn) (10)	Bare powder (vortex & 1 min)	PCPs* (9)/ water (10 mL)	0.70–1.5 $\mu g{\cdot}L^{-1}$	≤13	82.2–128	UHPLC-UV*	[74]
Zn-BTC(Zn) (10)	Bare powder (shaking & 30 min)	Pesticides (6)/ water (15 mL)	0.20–1.6 $\mu g{\cdot}L^{-1}$	≤9.5	78.6–116	HPLC-UV*	[75]
D101@ZIF-8(Zn) (20)	Composite powder (shaking & 20 min)	Phenols (2)/ urine (5 mL)	1.37–1.43 $\mu g{\cdot}L^{-1}$	≤4.6	96.2–107	CE*-UV*	[76]
ZIF-8(Zn) (8)	Bare powder (sonication & 30 s)	Pesticides (5)/ drink & water (20 mL)	0.03–0.2 $\mu g{\cdot}L^{-1}$	≤8.8	91.9–99.5	GC-FID*	[77]
HKUST-1(Cu) (0.61)	Bare powder (stirring & 2 min)	MCPA* (1)/ water, soil, & food (40 mL)	10 $ng{\cdot}L^{-1}$	≤5.0	57.0–100	IMS*	[78]
HKUST-1(Cu)@GO* (20)	Composite powder (stirring & 4 min)	PAHs* (7)/ water (50 mL)	3.0–10 $ng{\cdot}L^{-1}$	≤8.9	91.8–99.5	GC-FID*	[79]

(*Continued*)

Table 9.2 MOFs and COFs as sorbents in analytical d-μSPE applications. (*Continued*)

Sorbent (mg)	Microextraction format (agitation & time)	Analytes (number)/ sample (amount)	LOD*	RSD* (%)	RR* (%)	Analytical technique	Ref.
JUC-48(Cd) (25)	Bare powder (vortex & 10 min)	PAHs* (7)/ water (12)	0.021–0.13 $\mu g \cdot L^{-1}$	≤4.3	85.8–110	HPLC-DAD*	[80]
Tb-MOF(Tb) (1.8)	Bare powder (stirring & 2 h)	MG* (1)/ food (30 mL)	1.66 $\mu g \cdot L^{-1}$	2.5	95.6–104	UV-Vis* spectroscopy	[81]
COFs							
MICOF@SiO_2 (15)	Composite powder (sonication & 10 min)	NSAIDs* (6)/ water (10 mL)	0.20–1.4 $\mu g \cdot L^{-1}$	≤9.4	77.3–112	HPLC-UV*	[82]

*Abbreviations: LOD for limit of detection, RSD for inter-day relative standard deviation, RSD for inter-day relative standard deviation, RR for relative recovery (in real samples), MS for mass spectrometry, EDCs for endocrine disrupting chemicals, DAD for diode array detector, NBs for nitrobenzene compounds, UV-Vis for ultraviolet-visible detector, FAAS for flame atomic absorption spectroscopy, PCAs for phenoxy carboxylic acids, PAEs for phthalic acid esters, GO for graphene oxide, SAs for sulfonamides, PVA for polyvinyl alcohol, NSAIDs for non-steroidal anti-inflammatory drugs, NPAHs for nitro-polycyclic aromatic hydrocarbons, NCI for negative chemical ionization, IL for ionic liquid, PAHs for polycyclic aromatic hydrocarbons, FID for flame ionization detector, TOF for time of flight detector, PVDF for polyvinylidene fluoride, FD for fluorescence detector, UV for ultraviolet detector, PCPs for personal care products, CE for capillary electrophoresis, MCPA for 2-methyl-4-chlorophenoxyacetic acid, IMS for corona discharge ion mobility spectrometry, and MG for malachite green.

bare powder for the extraction of organophosphorus pesticides in water and fruit juice in just 30 s by sonication reporting limits of detection lower than 0.21 ng·L^{-1} and enrichments factors higher than 800, thus proving the potential of this procedure [77].

The highest amount of MOF used in D-μSPE is 120 mg [65], but normally the amounts utilized in this miniaturized extraction procedure range between 10 mg [63, 74, 75] and 40 mg [58]. Mohammadnejad *et al.* have reported the lowest amount of MOF used in D-μSPE for the determination of herbicides in water, soil and agricultural products using 0.6 mg of HKUST-1(Cu) [78]. In spite of this low amount of sorbent, the limits of detection were of 10 ng·L^{-1} and the enrichment factor was 20. Recoveries assays showed satisfactory extraction recovery, between 98.0% and 104.0% in aqueous standards solutions, soils, and tomatoes samples.

MIL-101(Cr), UiO-66(Zr), and its amino (UiO-66–NH_2), nitro (UiO-66–NO_2), and hydroxyl (UiO-66–OH) derived MOFs, have been the most widely used for the determination of a broad variety of emerging pollutants in complex matrixes [58–61, 64, 83–85]. The extended use of this UiO-66(Zr) MOF family may be due to their good chemical stability in water and organic solvents, while presenting satisfactory adsorption capacity. Recently, Taima-Mancera *et al.* demonstrated the relevance of the nature of the MOF used and its degree of functionalization in the determination of different kinds of analytes using D-μSPE. These studies serve as an example to reconsider the way that MOFs have been applied in analytical applications, highlighting the need of proper design [59, 88].

All the reported applications using neat MOFs present adequate analytical performance, with limits of detection in the range of ng·L^{-1}–μg·L^{-1}, and good reproducibility with relative standard deviations lower than 17% [73]. Gao *et al.* reported the lowest limit of detection for endocrine disruptors at a concentration level of 1.5 ng·L^{-1} using MIL-53(Al) and liquid chromatography with mass spectrometry (LC-MS/MS) [70].

Regarding the use of MOF composites in D-μSPE, graphene oxide (GO) combined is the combination mostly described. The incorporation of GO provides an increasing effective surface area in the resulting composite, thus favoring the efficiency in D-μSPE [66, 79, 87]. Wang *et al.* compared the extraction efficiency of MOFs when using the same amount of the hybrid material, the neat MOF and the neat GO. The hybrid material showed higher adsorption capacity than the neat MOF or neat GO [87]. MOF@GO composites have been applied as sorbents in D-μSPE for the determination of organic pollutants in complex matrixes, including drugs residuals in food [87], sulfonamides in milk [66], and PAHs in environmental waters

[66]. Other materials used in combination with MOFs are ionic liquids (ILs) and macroporous resins [69, 76]. Nasrollahpour *et al.* reported the use of an IL modifying the MOF MIL-100(Fe) for the extraction of polycyclic aromatic hydrocarbons from environmental waters. The use of the IL@MOF material implies a significantly improvement of the extraction kinetics and the chemical stability in aqueous samples compared to the non-functionalized MOF [69].

9.4 MOFs and COFs in Magnetic-Assisted Miniaturized Dispersive Solid-Phase Extraction (m-D-μSPE)

It is important to note that most of the solid-phase extraction applications reported in the literature utilizing MOFs and/or COFs as sorbent material belongs to the magnetic-assisted version of dispersive solid-phase microextraction (m-D-μSPE). This extraction technique ensures adequate interaction between the extractant material and the analytes (for its dispersive mode) without tedious and time-wasting centrifugation or separation steps (for its magnetic response). Table 9.3 compiles several representative examples of applications of MOFs and COFs in m-D-μSPE [89–130].

In order to perform extractions by m-D-μSPE it is necessary to employ magnetic composites. In general, Fe_3O_4 nano- or micro-particles are those responsible of the magnetic properties of MOF composites [89–109, 111, 114–136]. Nevertheless, other materials such as magnetic graphene [110, 112] and nickel nanoparticles are also useful for this purpose [113]. It is necessary to perform several functionalization steps to achieve a successful incorporation of the MOFs or the COFs to the magnetic particles. Some authors do not perform any functionalization, and MOFs become magnetic simply by mixing the MOF powder with the magnetic nanoparticles [99, 103]. However, this strategy is not recommendable as not all the crystals experience magnetization, and losses of certain amount of sorbent material occur during the extraction. Undoubtedly, the preparation of the material is the main drawback of the technique, and not the m-D-μSPE analytical application itself. In the majority of the synthesis, the resulting composite is heterogeneous. Heterogeneous magnetic composites based on MOFs or COFs present magnetic nanoparticles decorating the crystals or *vice versa* [89–95, 97–109, 111, 122, 125, 131–134]. Nowadays, there is an increasing interest in the utilization of homogeneous core–shell materials. These composites present clearly two different parts. The first

Table 9.3 MOFs and COFs as sorbents in analytical m-d-μSPE applications.

Sorbent (mg)	Format (extraction time)	Analytes (number)/ sample (amount)	LOD*	RSD* (%)	RR* (%)	Analytical technique	Ref.
MOFs (heterogeneous)							
JUC-48(Cd) (25)	Fe_3O_4 NPs* + MAA* (10 min)	SAs* (5)/meat (8 mL ACN* extract)	1.73–5.23 $ng \cdot g^{-1}$	≤4.5	76.1–103	HPLC-DAD*	[89]
HKUST-1(Cu) (10)	Fe_3O_4@SiO_2@APTES*(50 min)	Insecticides (6)/ water (15 mL)	0.30–1.6 $\mu g \cdot L^{-1}$	≤9.5	81.2–113	HPLC-UV*	[90]
IRMOF(Zn) (10)	Fe_3O_4@GO* (30 min)	Fungicides (5)/ food (10 mL)	0.21–1.0 $\mu g \cdot L^{-1}$	–	74.8–99.5	HPLC-MS/ MS	[91]
MIL-100(Fe) (5)	Fe_3O_4 (5 min)	Blood lipid regulators (5)/ water (10 mL)	4.0–99 $\mu g \cdot L^{-1}$	≤3.2	70.0–112	HPLC-UV*	[92]
MIL-101(Cr) (10)	Fe_3O_4@g-C_3N_4 (30 min)	Glucocorticoids (5)/ cosmetics (40 mL)	2–5 $ng \cdot L^{-1}$	≤5.5	77.3–113	UHPLC-MS/ MS	[93]
MIL-101(Cr) (30)	Fe_3O_4 (18 min)	NSAIDs* (8)/water (50 mL)	0.01–0.2 $\mu g \cdot L^{-1}$	≤11	81.2–117	UHPLC-MS/ MS	[94]

(*Continued*)

Table 9.3 MOFs and COFs as sorbents in analytical m-d-μSPE applications. (*Continued*)

Sorbent (mg)	Format (extraction time)	Analytes (number)/ sample (amount)	LOD*	RSD* (%)	RR* (%)	Analytical technique	Ref.
MIL-101(Cr) (-)	aptamer@Fe_3O_4-NH_2 (40 min)	Ochratoxin A (1)/ food (10 mL)	0.067 ng·L^{-1}	≤6.5	82.8–108	UHPLC-MS/ MS	[95]
MIL-101(Cr) (15)	Fe_3O_4 (20 min)	PAEs* (5)/ water & plasma (both 5 mL)	0.08–0.2 μg·L^{-1}	≤10	85.1–107	GC-MS*	[96]
MIL-101(Cr) (11.5)	Fe_3O_4@MWCNTs* (13.5 min)	Parabens & PAEs*(5)/water & creams (18 mL)	0.03–0.2 μg·L^{-1}	≤9.3	38.0–70.6	HPLC-DAD*	[97]
MIL-101(Cr) (10)	Fe_3O_4 (10 min)	Pesticides (5)/water & drink (50 mL)	0.008–0.02 ng·mL^{-1}	≤5.6	78.3–104	GC-ECD*	[98]
MIL-101(Cr) (5)	Fe_3O_4@SiO_2@GO* (25 min)	Herbicides (7)/ food (7 mL hexane extract)	0.01–0.08 μg·kg^{-1}	≤7.8	83.9–104	HPLC-UV*	[99]
MIL-101(Cr) (11)	Fe_3O_4 (5 min)	Herbicides (7)/ food (1 mL)	1.08–18.1 pg·g^{-1}	≤13	79.3–117	HPLC-MS/ MS	[100]

(*Continued*)

Table 9.3 MOFs and COFs as sorbents in analytical m-d-μSPE applications. (*Continued*)

Sorbent (mg)	Format (extraction time)	Analytes (number)/ sample (amount)	LOD*	RSD* (%)	RR* (%)	Analytical technique	Ref.
MIL-101(Fe) (14.8)	Fe_3O_4@PAEDTC* (20 min)	Heavy metals (2)/ water & food (15 mL)	1.0–1.5 $ng{\cdot}L^{-1}$	≤7.5	92.5–98.2	ETAAS*	[101]
MIL-101–NH_2(Fe) (20)	Fe_3O_4@COOH (20 min)	Fungicides (4)/ water (30 mL)	0.04–0.4 $\mu g{\cdot}L^{-1}$	≤10	71.1–99.1	HPLC-UV*	[102]
MIL-101–NMe_3(Cr) (8)	Fe_3O_4@SiO_2 (25 min)	Azide (1)/drugs (20 mL)	0.24 $\mu g{\cdot}L^{-1}$	0.36	96.5–101	IC*-CCD*	[103]
MOF-1210(Zr/Cu) (5)	Fe_3O_4 (60 min)	Benzophenones (3)/ water & soil (20 mL)	0.01–0.02 $ng{\cdot}mL^{-1}$	≤2.6	87.6–114	HPLC-UV*	[104]
MOF-235(Fe) (5)	Fe_3O_4-NH_2 (5 min)	Insecticides (5)/food, drink & water (10 mL)	0.25–0.50 $\mu g{\cdot}L^{-1}$	≤4.9	76.5–99.6	HPLC-UV*	[105]
MOF-5(Zn) (7)	Fe_3O_4@APTES* (12 min)	Pesticides (4)/ water (60 mL)	0.04–0.1 $\mu g{\cdot}L^{-1}$	≤7.1	80.2–108	HPLC-DAD* HPLC-FD*	[106]

(*Continued*)

Table 9.3 MOFs and COFs as sorbents in analytical m-d-µSPE applications. (*Continued*)

Sorbent (mg)	Format (extraction time)	Analytes (number)/ sample (amount)	LOD*	RSD* (%)	RR* (%)	Analytical technique	Ref.
MOF-5(Zn) (50)	Fe_3O_4@SiO_2 (50 min)	N- & S-PAHs* (5)/ water (100 mL)	25–33 $ng \cdot L^{-1}$	≤1.9	92.6–97.3	HPLC-DAD*	[107]
MOF-74(Ni) (40)	Fe_3O_4@SiO_2 (2.5 min)	PAEs* (4)/Chinese liquor (25 mL)	0.46–2.1 $\mu g \cdot L^{-1}$	≤3.7	74.4–105	HPLC-PDA*	[108]
TMU-8(Cd) (10)	Fe_3O_4 NPs* + MAA* (11 min)	Heavy metals (7)/ water & food (185 mL)	0.3–1 $\mu g \cdot L^{-1}$	≤6.4	90.0–108	ICP-AES*	[109]
UiO-66(Zr) (40)	magG*@PD* (1 min)	Pesticides (9)/ tobacco (3.50 mL)	10.8–45.5 $ng \cdot g^{-1}$	≤13	57.9–126	GC-MS*	[110]
ZIF-67(Co) (9)	Fe_3O_4 (20 min)	Aromatic amino acids (7)/lily (3 mL)	0.04–0.9 $ng \cdot g^{-1}$	≤5.1	88.0–96.8	HPLC-DAD*	[111]
ZIF-8(Zn) (10)	magG* (10 min)	PAEs* (9)/plasma (10 mL)	0.003–0.01 $ng \cdot mL^{-1}$	≤6.5	92.1–105	GC-MS*	[112]

(*Continued*)

Table 9.3 MOFs and COFs as sorbents in analytical m-d-μSPE applications. (*Continued*)

Sorbent (mg)	Format (extraction time)	Analytes (number)/ sample (amount)	LOD*	RSD* (%)	RR* (%)	Analytical technique	Ref.
MOFs (homogeneous)							
MIL-100(Fe) (23)	Ni@MIP* (1 min)	Anti-malaria agent (1)/blood serum (10 mL)	0.2 μg·mL^{-1}	0.67	96.0–103	HPLC-UV*	[113]
MIL-100(Fe) (25)	Fe_3O_4 (45 min)	NSAIDs* (7)/ water (50 mL)	0.02–0.09 μg·L^{-1}	≤9.6	75.2–105	UHPLC-MS/ MS	[114]
MIL-101(Cr) (1.5)	Fe_3O_4 (60 min)	PAHs* (5)/ water (1 L)	0.08–0.8 ng·L^{-1}	≤4.8	–	HPLC-UV*	[115]
MIL-101(Fe) (60)	Fe_3O_4@PD* (3 min)	SUHs* (4)/water & food (25 mL)	0.12–0.34 μg·L^{-1}	≤4.8	87.1–109	HPLC-PDA*	[116]
MIL-125–NH_2 (25)	Fe_3O_4-Cys (20 min)	Fluoroquinolones (5)/water (50 mL)	0.05–0.2 μg·L^{-1}	≤8.9	83.8–109	UHPLC-UV*	[117]
MIL-53(Al) (20)	Fe_3O_4 (8 min)	Phenols & anilines (5)/water (20 mL)	0.03–0.2 μg·L^{-1}	≤13	39.5–93.3	HPLC-PDA*	[118]

(*Continued*)

Table 9.3 MOFs and COFs as sorbents in analytical m-d-μSPE applications. (*Continued*)

Sorbent (mg)	Format (extraction time)	Analytes (number)/ sample (amount)	LOD*	RSD* (%)	RR* (%)	Analytical technique	Ref.
TMU-10 (5)	Fe_3O_4 (4 min)	TCAs* (2)/plasma (6 mL)	2–5 ng·mL^{-1}	≤5.2	90.5–99.0	HPLC-UV*	[119]
TMU-21 (4)	Fe_3O_4 (6 min)	Insecticides (2)/ drink (100 mL)	0.05–0.1 μg·L^{-1}	≤4.4	93.0–105	HPLC-UV*	[120]
TMU-23/TMU-24 (6)	Fe_3O_4 (6 min)	PAEs* (3)/ water (100 mL)	0.20–0.40 μg·L^{-1}	≤5.9	90.5–99.0	GC-FID*	[121]
COFs							
CTF (50)	Fe_2O_3 (15 min)	PFCs* (6)/ water (25 mL)	0.62–1.4 ng·L^{-1}	≤9.7	81.8–114	HPLC-MS*	[122]
TpBD (20)	Fe_3O_4 10 min	PAES* (5)/human serum (10 mL)	1.00–78.1 ng·L^{-1}	≤3.4	93.0–108	HPLC-MS*	[123]
TpBD (40)	Fe_3O_4 30 min	EDCs* (4)/ drink (25 mL)	0.08–0.2 ng·mL^{-1}	≤5.2	81.3–118	HPLC-FD*	[124]
CTF/Ni (10)	Fe_3O_4 20 min	PAEs* (6)/plastic materials (20 mL)	0.02–0.09 mg·kg^{-1}	≤1.0	70.6–119	GC-FID*	[125]

(*Continued*)

Table 9.3 MOFs and COFs as sorbents in analytical m-d-μSPE applications. (*Continued*)

Sorbent (mg)	Format (extraction time)	Analytes (number)/ sample (amount)	LOD*	RSD* (%)	RR* (%)	Analytical technique	Ref.
LZU1 (5)	Fe_3O_4@PEI* 30 min	PAHs* (6)/water & soil (20 mL)	0.20–20 $pg \cdot mL^{-1}$	≤4.4	90.9–108	HPLC-FD*	[126]
Mag-COF (100)	Fe_3O_4 15 min	Allergenic disperse dyes (19)/ textile (50 mL)	0.021–0.58 $\mu g \cdot kg^{-1}$	≤7.1	72.7–107	UFLC*-MS/ MS	[127]
TbBd (20)	Fe_3O_4 30 min	Estrogens (7)/urine (20 mL)	0.20–7.7 $ng \cdot L^{-1}$	≤6.7	80.6–112	HPLC-MS*	[128]
TpBD (20)	Fe_3O_4 10 min	SAs* (10)/food (20 mL)	0.28–1.5 $\mu g \cdot L^{-1}$	≤3.3	82–94	HPLC-UV*	[129]
TpDA (10)	Fe_3O_4 10 min	PAHs* (15)/oil & meat (10 mL)	0.03–0.7 $\mu g \cdot L^{-1}$	≤4.3	85.5–104	HPLC-DAD*	[130]

*Abbreviations: LOD for limit of detection, RSD for inter-day relative standard deviation, RR for relative recovery (in real samples), NPs for nanoparticles, MAA for mercaptoacetic acid, SAs for sulfonamides, DAD for diode array detector, APTES for (3-aminopropyl)triethoxysilane, UV for ultraviolet detector, GO for graphene oxide, NSAIDs for non-steroidal anti-inflammatory drugs, PAEs for phthalic acid esters, MS for mass spectrometry, MWCNTs for multiwalled carbon nanotubes, DAD for diode array detector, ECD for electron capture detector, PAEDTC for 2-(propylamino-ethyl) dithiocarbamate, ETAAS for electrothermal atomic absorption spectrometry, IC for ion-exchange chromatography, CCD for charge-coupled device, FD for fluorescence detector, N- and S-PAHs for N- and S-containing polycyclic aromatic hydrocarbons, PDA for photodiode array detector, ICP-AES for inductively coupled plasma-mass spectrometry, magG for magnetic graphene, PD for polydopamine, MIP for molecularly imprinted polymer, PAHs for polycyclic aromatic hydrocarbons, SUHs for sulfonylurea herbicides, TCAs for tricyclic antidepressants, FID for flame ionization detector, UFLC for ultra fast liquid chromatography, EDCs for endocrine disrupting chemicals, PFCs for perfluorinated compounds, and PEI for polyethyleneimine.

(inner) part is the core, formed by the solid magnetic particle. The second (outer) part is the shell, where the MOF or the COF completely covers the core. The main advantages over heterogeneous composites are the high homogeneity of the material, thus ensuring a high inter-batch reproducibility and further robustness of the m-D-μSPE method incorporating them, and higher stability by reducing the influence of the environment over the magnetic core during the extraction. However, their preparation is still tricky and challenging, especially for MOFs [113–121, 123–130, 135, 136].

In the reported applications, the amount required for the heterogeneous composite ranges between 5 mg [105] and 50 mg [107], being the lowest amount used 2.5 mg, in an application for the determination of colchicine in biological samples. In spite of the low amount of sorbent used, the method showed excellent reproducibility with relative standard deviation lower than 3.5% at trace levels [134]. For core–shell magnetic MOFs/COFs, the amount of sorbent required is usually lower, ranging from 4 mg [120] to 25 mg [114], as the material possess a higher active surface than the heterogeneous materials.

As shown in Table 9.3, the MOF MIL-101(Cr) has been the more used in different applications among the different magnetic-MOFs composites, due to the numerous accessible coordinative unsaturated metal sites that this MOF present. The unsaturated metal sites have great potential for enhancing extraction of polar organic molecules. In addition, this MOF is stable in water, permanently porous, and easy to modify [94–103, 115, 132, 133]. Song *et al.* have used MIL-101(Cr) in combination with Fe_3O_4 and Fe_3O_4@GO for the determination of triazines in food samples, showing satisfactorily adsorption capacity at low concentration levels [99, 100]. The preparation of the material required mixing the MOF powder and Fe_3O_4@GO nanoparticles. In both applications reported, the amount of MOF was 5 mg (plus 5 mg of Fe_3O_4@GO in the GO composite). The introduction of GO improved the adsorption capability of the composite achieving LODs at $ng{\cdot}L^{-1}$ level in spite of using UV as HPLC detector. However, longer extraction times are required when using the GO composite [99]. Multi-walled carbon nanotubes have been also combined with magnetic MOF composites [97]. Recently, in order to improve the biochemical applicability, Zhang *et al.* reported an aptamer functionalized Fe_3O_4@MIL-101(Cr) composite to perform an extraction of mycotoxins via bimolecular recognition mechanisms. This approach guaranteed a highly selective method in complex matrixes [95].

Other MOFs used in a less extended way are MIL-100(Fe) [92, 113, 114, 131], MOF-5(Zn) [106, 107, 134], UiO-66(Zr) [110], the TMU family [109,

119–121], HKUST-1(Cu) [90], and ZIF-67(Co) [111]. All the reported applications using magnetic MOFs composites present adequate analytical performance, with limits of detection in the range of $ng{\cdot}L^{-1}$–$\mu g{\cdot}L^{-1}$, and proper reproducibility with relative standard deviations lower than 13% [110]. In most cases, it is interesting to note that there are non-noticeable matrix effects when using magnetic MOF composites as sorbent materials. Indeed, reported recoveries values from 75% to 120% at low concentration levels and dealing with a wide variety of matrices, such as food [89, 91, 95, 99, 100, 105, 109, 116, 131, 133], drinks [101, 108, 120], water [90, 92, 94, 96–98, 101–103, 105–107, 109, 132], tobacco [110], and biological samples [96, 112, 113, 119, 134].

Regarding COFs, their low crystal density, high porosity, and good chemical selectivity make them excellent candidates for the preparation of core–shell composites over magnetic functionalized surfaces. In addition to this, the organic structure allows an easy tuneability of the pores, while ensuring good chemical selectivity for the extraction of target analytes via hydrogen bonding, Van der Waals forces, size-exclusion effect, or π–π stacking. The use of COFs as sorbent material in this configuration is still being under development. However, they already appear in a number of applications, such as the determination of endocrine disruptors [123, 124, 136]; drugs [129, 135], polycyclic aromatic hydrocarbons [126, 130], phthalate esters [125], and estrogens [128], in water, food, and biological samples. Wang *et al.* reported the lowest limits of detection, ranging from 0.2 to 20 $pg{\cdot}mL^{-1}$, using the LZU-1 COF for the determination of polycyclic aromatic hydrocarbons in water and soil [126]. This COF is highly stable in water and offer strong π stacking interaction and hydrophobic effect with the analytes due to the benzene rings and imine groups present in the its structure, therefore being an excellent material for the extraction and enrichment of compounds with abundant π electrons.

The amount of COF@Fe_3O_4 used in the reported applications ranges between 5 mg [126, 135] and 10 mg [125, 130, 136]. The highest amount of sorbent used in m-D-μSPE is 100 mg. This amount is much higher than those reported for MOF core–shell materials. However, it is due to the sample volume and not to the sorbent adsorption capacity [127]. As a rule, a ratio of 1 $mg{\cdot}mL^{-1}$ for the COF amount to the sample volume is employed. This ratio is a clear indicator of the high adsorption capacity offered by these novel materials.

COF magnetic composites have been mainly used in combination with liquid chromatography and diverse detection systems including UV-Vis [129, 135], DAD [130], fluorescence detectors [124, 126, 136], and MS [122, 123, 127, 128]. There is only one application on the use of a magnetic

COF in combination with GC-FID for the determination of phthalates in plastic packing materials [125].

9.5 Concluding Remarks

Metal–organic frameworks (MOFs) and covalent–organic frameworks (COFs) are firm candidates as promising sorbents in solid-phase extraction applications, and indeed their use is a hotspot nowadays in analytical sample preparation. This is due to their striking properties such as high chemical stability, easy pores tuneability, synthetic versatility, and astonishing surface area. Although the applicability of COFs have not been widely explored compared to MOFs, the results obtained so far gives a clear idea of the future repercussion. In any case, further efforts are still needed to completely introduce these novels materials in all the solid-phase extraction modalities successfully, with real competitive advantages over already commercialized materials.

Acknowledgments

A. G.-S. thanks the Canary Agency of Research and Innovation (ACIISI), co-funded by the European Social Fund, for his FPI PhD fellowship. J.P. thanks the "Agustín de Betancourt" Canary Program for his research associate position at ULL. V.P. thanks the Spanish Ministry of Economy and Competitiveness (MINECO) for the project MAT2017-89207-R.

References

1. Huck, C.W. and Bonn, G.K., Recent developments in polymer-based sorbents for solid-phase extraction. *J. Chromatogr. A*, 885, 51, 2000.
2. Li, Z., Yu, B., Cong, H., Yuan, H., Peng, Q., Recent development and application of solid phase extraction materials. *Rev. Adv. Mater. Sci.*, 48, 87, 2017.
3. Faraji, M., Yamini, Y., Gholami, M., Recent Advances and Trends in Applications of Solid-Phase Extraction Techniques in Food and Environmental Analysis. *Chromatographia*, 82, 1207, 2019.
4. Fontanals, N., Pocurull, E., Borrull, F., Marcé, R.M., Role of solid-phase extraction in wasterwater-based epidemiology. *Curr. Opin. Environ. Sci. Health*, 9, 26, 2019.
5. Hashemi, B., Zohrabi, P., Shamsipur, M., Recent developments and applications of different sorbents for SPE and SPME from biological samples. *Talanta*, 187, 337, 2018.

6. Madikizela, L.M., Ncube, S., Chimuka, L., Recent Developments in Selective Materials for Solid Phase Extraction. *Chromatographia*, 82, 1171, 2019.
7. Poole, C.F., Chapter 12 Principles and practice of solid-phase extraction. *Compr. Anal. Chem.*, 37, 341, 2002.
8. Azzouz, A., Kailasa, S.K., Lee, S.S., Rascón, A.J., Ballesteros, E., Zhang, M., Kim, K.-H., Review of nanomaterials as sorbents in solid-phase extraction for environmental samples. *Trac-Trends Anal. Chem.*, 108, 347, 2018.
9. Wen, Y., Chen, L., Li, J., Liu, D., Chen, L., Recent advances in solid-phase sorbents for sample preparation prior to chromatographic analysis. *Trac-Trends Anal. Chem.*, 59, 26, 2014.
10. Yang, L., Said, R., Abdel-Rehim, M., Sorbent, device, matrix and application in microextraction by packed sorbent (MEPS): A review. *J. Chromatogr. B*, 1043, 33, 2017.
11. Ghorbani, M., Aghamohammadhassan, M., Chamsaz, M., Akhlaghi, H., Pedramrad, T., Dispersive solid phase microextraction. *Trac-Trends Anal. Chem.*, 118, 793, 2019.
12. Socas-Rodríguez, B., Herrera-Herrera, A.V., Asensio-Ramos, M., Hernández-Borges, J., Dispersive Solid-Phase Extraction, in: *Analytical Separation Science*, J.L. Anderson, A. Berthod, V. Pino, A.M. Stalcup (Eds.), pp. 1525–1569, Wiley, Hoboken, 2015.
13. Chisvert, A., Cárdenas, S., Lucena, R., Dispersive micro-solid phase extraction. *Trac-Trends Anal. Chem.*, 112, 226, 2019.
14. Yu, M., Wang, L., Hu, L., Li, Y., Luo, D., Mei, S., Recent applications of magnetic composites as extraction adsorbents for determination of environmental pollutants. *Trac-Trends Anal. Chem.*, 119, 115611, 2019.
15. Jiang, H.-L., Li, N., Cui, L., Wang, X., Zhao, R.-S., Recent application of magnetic solid phase extraction for food safety analysis. *Trac-Trends Anal. Chem.*, 120, 115632, 2019.
16. Capriotti, A.L., Cavaliere, C., La Barbera, G., Montone, C.M., Piovesana, S., Laganà, A., Recent Applications of Magnetic Solid-phase Extraction for Sample Preparation. *Chromatographia*, 82, 1251, 2019.
17. Aguilar-Arteaga, K., Rodriguez, J.A., Barrado, E., Magnetic solids in analytical chemistry: A review. *Anal. Chim. Acta*, 674, 157, 2010.
18. Pawliszyn, J., Theory of Solid-Phase Microextraction, in: *Handbook of Solid Phase Microextraction*, J. Pawliszyn (Ed.), pp. 13–59, Elsevier, Amsterdam, 2012.
19. Reyes-Garcés, N., Gionfriddo, E., Gómez-Ríos, G.A., Alam, Md.N., Boyacı, E., Bojko, B., Singh, V., Grandy, J., Pawliszyn, J., Advances in Solid Phase Microextraction and Perspective on Future Directions. *Anal. Chem.*, 90, 302, 2018.
20. Maciel, E.V.S., de Toffoli, A.L., Neto, E.S., Nazario, C.E.D., Lanças, F.M., New materials in sample preparation: Recent advances and future trends. *Trac-Trends Anal. Chem.*, 119, 115633, 2019.

21. Wang, X. and Ye, N., Recent advances in metal–organic frameworks and covalent organic frameworks for sample preparation and chromatographic analysis. *Electrophoresis*, 38, 3059, 2017.
22. Chen, L., Wu, Q., Gao, J., Li, H., Dong, S., Shi, X., Zhao, L., Applications of covalent organic frameworks in analytical chemistry. *Trac-Trends Anal. Chem.*, 113, 182, 2019.
23. Li, N., Du, J., Wu, D., Liu, J., Li, N., Sun, Z., Li, G., Wu, Y., Recent advances in facile synthesis and applications of covalent organic framework materials as superior adsorbents in sample pretreatment. *Trac-Trends Anal. Chem.*, 108, 154, 2018.
24. Rowsell, J.L.C. and Yaghi, O.M., Metal–organic frameworks: A new class of porous materials. *Micropor. Mesopor. Mat.*, 73, 3, 2004.
25. Safaei, M., Foroughi, M.M., Ebrahimpoor, N., Jahani, S., Omidi, A., Khatami, M., A review on metal–organic frameworks: Synthesis and applications. *Trac-Trends Anal. Chem.*, 118, 401, 2019.
26. Kalmutzki, M.J., Hanikel, N., Yaghi, O.M., Secondary building units as the turning points in the development of the reticular chemistry of MOFs. *Sci. Adv.*, 4, eaat9180, 2018.
27. Deng, L., Zhang, J., Gao, Y., Synthesis, Properties, and Their Potential Application of Covalent Organic Frameworks (COFs), in: *Mesoporous Materials - Properties and Applications*, M. Krishnappa (Ed.), pp. 857–884, IntechOpen, London, 2018.
28. Cao, S., Li, B., Zhu, R., Pang, H., Design and synthesis of covalent organic frameworks towards energy and environment fields. *Chem. Eng. J.*, 355, 602, 2019.
29. Kandambeth, S., Dey, K., Banerjee, R., Covalent Organic Frameworks: Chemistry beyond the Structure. *J. Am. Chem. Soc.*, 141, 1807, 2019.
30. Pacheco-Fernández, I., González-Hernández, P., Pasán, J., Ayala, J.H., Pino, V., The rise of metal–organic frameworks in Analytical Chemistry, in: *Handbook of Smart Materials in Sample Preparation and Analysis*, M. de la Guardia and F.A. Esteve-Turrillas (Eds.), pp. 463–502, Willey, Hoboken, 2019.
31. González-Hernández, P., Gutiérrez-Serpa, A., Rocío-Bautista, P., Pasán, J., Ayala, J.H., Pino, V., Micro-solid-phase extraction using metal–organic frameworks, in: *Metal Organic Frameworks*, V. Mittal (Ed.), pp. 99–136, Central West Publishing, Orange NSW, 2019.
32. Rocío-Bautista, P., González-Hernández, P., Pino, V., Pasán, J., Afonso, A.M., Metal–organic frameworks as novel sorbents in dispersive-based microextraction approaches. *Trac-Trends Anal. Chem.*, 90, 114–134, 2017.
33. Maya, F., Cabello, C.P., Frizzarin, R.M., Estela, J.M., Palomino, G.T., Cerdà, V., Magnetic solid-phase extraction using metal–organic frameworks (MOFs) and their derived carbons. *Trac-Trends Anal. Chem.*, 90, 142, 2017.

34. Gutiérrez-Serpa, A., Pacheco-Fernández, I., Pasán, J., Pino, V., Metal-Organic Frameworks as Key Materials for Solid-Phase Microextraction Devices—A Review. *Separations*, 6, 47, 2019.
35. Rocío-Bautista, P., Pacheco-Fernández, I., Pasán, J., Pino, V., Are metal–organic frameworks able to provide a new generation of solid-phase microextraction coatings?—A review. *Anal. Chim. Acta*, 939, 26, 2016.
36. Rocío-Bautista, P. and Termopoli, V., Metal–Organic Frameworks in Solid-Phase Extraction Procedures for Environmental and Food Analyses. *Chromatographia*, 82, 1191, 2019.
37. Kahkha, M.R.R., Oveisi, A.R., Kaykhaii, M., Kahkha, B.R., Determination of carbamazepine in urine and water samples using amino-functionalized metal–organic framework as sorbent. *Chem. Cent. J.*, 12, 77, 2018.
38. Duo, H., Wang, Y., Wang, L., Lu, X., Liang, X., Zirconium(IV)-based metal-organic framework (UiO-67) as solid-phase extraction adsorbents for extraction of phenoxyacetic acid herbicides from vegetables. *J. Sep. Sci.*, 41, 4149, 2019.
39. Yan, Z., Wu, M., Hu, B., Yao, M., Zhang, L., Lu, Q., Pang, J., Electrospun UiO-66/polyacrylonitrile nanofibers as efficient sorbent for pipette tip solid phase extraction of phytohormones in vegetable samples. *J. Chromatogr. A*, 1542, 19, 2018.
40. Si, T., Liu, L., Liang, X., Duo, H., Wang, L., Wang, S., Solid-phase extraction of phenoxyacetic acid herbicides in complex samples with a zirconium(IV)-based metal–organic framework. *J. Sep. Sci.*, 42, 2148, 2019.
41. Kahkha, M.R.R., Daliran, S., Oveisi, A.R., Kaykhaii, M., Sepheri, Z., The Mesoporous Porphyrinic Zirconium Metal–Organic Framework for Pipette-Tip Solid-Phase Extraction of Mercury from Fish Samples Followed by Cold Vapor Atomic Absorption Spectrometric Determination. *Food Anal. Methods*, 10, 2175, 2017.
42. Zhang, X., Wang, P., Han, Q., Li, H., Wang, T., Ding, M., Metal–organic framework based in-syringe solid-phase extraction for the on-site sampling of polycyclic aromatic hydrocarbons from environmental water samples. *J. Sep. Sci.*, 41, 1856, 2018.
43. Ghiasvand, A.R., Ghaedrahmati, L., Heidari, N., Haddad, P.R., Farhadi, S., Synthesis and characterization of MIL-101(Cr) intercalated by polyaniline composite, doped with silica nanoparticles and its evaluation as an efficient solid-phase extraction sorbent. *J. Sep. Sci.*, 41, 3910, 2018.
44. Li, D., Zhang, X., Kong, F., Qiao, X., Xu, Z., Molecularly Imprinted Solid-Phase Extraction Coupled with High-Performance Liquid Chromatography for the Determination of Trace Trichlorfon and Monocrotophos Residues in Fruits. *Food Anal. Methods*, 10, 1284, 2017.
45. Asiabi, M., Mehdinia, A., Jabbari, A., Spider-web-like chitosan/MIL-68(Al) composite nanofibers for high-efficient solid phase extraction of Pb(II) and Cd(II). *Microchim. Acta*, 184, 4495, 2017.

46. Pang, X., Liu, H., Yu, H., Zhang, M., Bai, L., Yan, H., A metal organic framework polymer monolithic column as a novel adsorbent for on-line solid phase extraction and determination of ursolic acid in Chinese herbal medicine. *J. Chromatogr. B*, 1125, 121715, 2019.
47. Rahimpoor, R., Bahrami, A., Nematollahi, D., Shahna, F.G., Farhadian, M., Facile and sensitive determination of urinary mandelic acid by combination of metal organic frameworks with microextraction by packed sorbents. *J. Chromatogr. B*, 1114–1115, 45, 2019.
48. Wang, X., Ma, R., Hao, L., Wu, Q., Wang, C., Wang, Z., Mechanochemical synthesis of covalent organic framework for the efficient extraction of benzoylurea insecticides. *J. Chromatogr. A*, 1551, 1, 2018.
49. Chang, Q., Zang, X., Wu, T., Wang, M., Pang, Y., Wang, C., Wang, Z., Use of Functionalized Covalent Organic Framework as Sorbent for the Solid-Phase Extraction of Biogenic Amines from Meat Samples Followed by High-Performance Liquid Chromatography. *Food Anal. Methods*, 12, 1, 2019.
50. Liu, J.-M., Hao, J.-L., Yuan, X.-Y., Liu, H.-L., Fang, G.-Z., Wang, S., Spherical covalent organic frameworks as advanced adsorbents for preconcentration and separation of phenolic endocrine disruptors, followed by high performance liquid chromatography. *RSC Adv.*, 8, 26880, 2018.
51. Liu, J.-M., Wang, X.-Z., Zhao, C.-Y., Hao, J.-L., Fang, G.-Z., Wang, S., Fabrication of porous covalent organic frameworks as selective and advanced adsorbents for the on-line preconcentration of trace elements against the complex sample matrix. *J. Hazard. Mater.*, 344, 220, 2018.
52. Li, W., Huang, L., Guo, D., Zhao, Y., Zhu, Y., Self-assembling covalent organic framework functionalized poly (styrene-divinyl benzene-glycidylmethacrylate) composite for the rapid extraction of non-steroidal anti-inflammatory drugs in wastewater. *J. Chromatogr. A*, 1571, 76, 2018.
53. Xu, G., Zhang, B., Wang, X., Li, N., Zhao, Y., Liu, L., Lin, J.-M., Zhao, R.-S., Porous covalent organonitridic frameworks for solid-phase extraction of sulfonamide antibiotics. *Microchim. Acta*, 186, 26, 2019.
54. Ji, W.-H., Guo, Y.-S., Wang, X., Lu, X.-F., Guo, D.-S., Amino-modified covalent organic framework as solid phase extraction adsorbent for determination of carboxylic acid pesticides in environmental water samples. *J. Chromatogr. A*, 1595, 11, 2019.
55. Chen, Z., Yu, C., Xi, J., Tang, S., Bao, T., Zhang, J., A hybrid material prepared by controlled growth of a covalent organic framework on amino-modified MIL-68 for pipette tip solid-phase extraction of sulfonamides prior to their determination by HPLC. *Microchim. Acta*, 186, 393, 2019.
56. Yan, Z., Hu, B., Li, Q., Zhang, S., Pang, J., Wu, C., Facile synthesis of covalent organic framework incorporated electrospun nanofiber and application to pipette tip solid phase extraction of sulfonamides in meat samples. *J. Chromatogr. A*, 1584, 33, 2019.
57. Song, Y., Ma, R., Hao, L., Yang, X., Wang, C., Wu, Q., Wang, Z., Application of covalent organic framework as the adsorbent for solid-phase extraction of

trace levels of pesticide residues prior to high-performance liquid chromatography-ultraviolet detection. *J. Chromatogr. A*, 1572, 20, 2018.
58. Cao, X., Jiang, Z., Wang, S., Hong, S., Li, H., Zhang, C., Shao, Y., She, Y., Jin, F., Jin, M., Wang, J., Metal–organic framework UiO-66 for rapid dispersive solid phase extraction of neonicotinoid insecticides in water samples. *J. Chromatogr. B*, 1077–1078, 92, 2018.
59. Taima-Mancera, I., Rocío-Bautista, P., Pasán, J., Ayala, J.H., Ruiz-Pérez, C., Afonso, A.M., Lago, A.B., Pino, V., Influence of Ligand Functionalization of UiO-66-Based Metal–Organic Frameworks When Used as Sorbents in Dispersive Solid-Phase Analytical Microextraction for Different Aqueous Organic Pollutants. *Molecules*, 23, 2869, 2018.
60. Dou, Y., Guo, L., Li, G., Lv, X., Xia, L., You, J., Amino group functionalized metal–organic framework as dispersive solid-phase extraction sorbent to determine nitrobenzene compounds in water samples. *Microchem. J.*, 146, 366, 2019.
61. Moghaddam, Z.S., Kaykhaii, M., Khajeh, M., Oveisi, A.R., Synthesis of UiO-66-OH zirconium metal–organic framework and its application for selective extraction and trace determination of thorium in water samples by spectrophotometry. *Spectrochim. Acta A*, 194, 76, 2018.
62. Fotovat, H., Khajeh, M., Oveisi, A.R., Ghaffari-Moghaddam, M., Daliran, S., A hybrid material composed of an amino-functionalized zirconium-based metal–organic framework and a urea-based porous organic polymer as an efficient sorbent for extraction of uranium(VI). *Microchim. Acta*, 185, 469, 2018.
63. Tokalıoğlu, S., Yavuz, E., Demir, S., Patat, Ş., Zirconium-based highly porous metal–organic framework (MOF-545) as an efficient adsorbent for vortex assisted-solid phase extraction of lead from cereal, beverage and water samples. *Food Chem.*, 237, 707, 2017.
64. Ji, Z., Cheng, J., Song, C., Hu, N., Zhou, W., Suo, Y., Sun, Z., You, J., A highly sensitive and selective method for determination of phenoxy carboxylic acids from environmental water samples by dispersive solid-phase extraction coupled with ultra high performance liquid chromatography-tandem mass spectrometry. *Talanta*, 191, 313, 2019.
65. González-Sálamo, J., González-Curbelo, M.A., Hernández-Borges, J., Rodríguez-Delgado, M.A., Use of Basolite® F300 metal–organic framework for the dispersive solid-phase extraction of phthalic acid esters from water samples prior to LC-MS determination. *Talanta*, 195, 236, 2019.
66. Jia, X., Zhao, P., Ye, X., Zhang, L., Wang, T., Chen, Q., Hou, X., A novel metal–organic framework composite MIL-101(Cr)@GO as an efficient sorbent in dispersive micro-solid phase extraction coupling with UHPLC-MS/MS for the determination of sulfonamides in milk samples. *Talanta*, 169, 227, 2017.
67. Wang, Y., Zhang, Y., Cui, J., Li, S., Yuan, M., Wang, T., Hu, Q., Hou, X., Fabrication and characterization of metal organic frameworks/polyvinyl

alcohol cryogel and their application in extraction of non-steroidal anti-inflammatory drugs in water samples. *Anal. Chim. Acta*, 1022, 45, 2018.
68. Jia, Y., Zhao, Y., Zhao, M., Wang, Z., Chen, X., Wang, M., Core–shell indium (III) sulfide@metal–organic framework nanocomposite as an adsorbent for the dispersive solid-phase extraction of nitro-polycyclic aromatic hydrocarbons. *J. Chromatogr. A*, 1551, 21, 2018.
69. Nasrollahpour, A., Moradi, S.E., Baniamerian, M.J., Vortex-Assisted Dispersive Solid-Phase Microextraction Using Ionic Liquid-Modified Metal–Organic Frameworks of PAHs from Environmental Water, Vegetable, and Fruit Juice Samples. *Food Anal. Methods*, 10, 2815, 2017.
70. Gao, G., Li, S., Li, S., Wang, Y., Zhao, P., Zhang, X., Hou, X., A combination of computational–experimental study on metal–organic frameworks MIL-53(Al) as sorbent for simultaneous determination of estrogens and glucocorticoids in water and urine samples by dispersive micro-solid-phase extraction coupled to UPLC-MS/MS. *Talanta*, 180, 358, 2018.
71. Rocío-Bautista, P., Pino, V., Pasán, J., López-Hernández, I., Ayala, J.H., Ruiz-Pérez, C., Afonso, A.M., Insights in the analytical performance of neat metal–organic frameworks in the determination of pollutants of different nature from waters using dispersive miniaturized solid-phase extraction and liquid chromatography. *Talanta*, 179, 775, 2018.
72. Gao, G., Li, S., Li, S., Zhao, L., Wang, T., Hou, X., Development and application of vortex-assisted membrane extraction based on metal–organic framework mixed-matrix membrane for the analysis of estrogens in human urine. *Anal. Chim. Acta*, 1023, 35, 2018.
73. Rocío-Bautista, P., Pino, V., Ayala, J.H., Ruiz-Pérez, C., Vallcorba, O., Afonso, A.M., Pasán, J., A green metal–organic framework to monitor water contaminants. *RSC Adv.*, 8, 31304, 2018.
74. González-Hernández, P., Lago, A.B., Pasán, J., Ruiz-Pérez, C., Ayala, J.H., Afonso, A.M., Pino, V., Application of a Pillared-Layer Zn-Triazolate Metal–Organic Framework in the Dispersive Miniaturized Solid-Phase Extraction of Personal Care Products from Wastewater Samples. *Molecules*, 24, 690, 2019.
75. Wang, X., Ma, X., Wang, H., Huang, P., Du, X., Lu, X., A zinc(II) benzenetricarboxylate metal organic framework with unusual adsorption properties, and its application for the preconcentration of pesticides. *Microchim. Acta*, 184, 3681, 2017.
76. Qiang, Y., Wang, W.-F., Dhodary, B., Yang, J.-L., Zeolitic imidazolate framework 8 (ZIF-8) reinforced microporous resin D101 for selective solid-phase extraction of 1-naphtol and 2-npahtol from phenol compounds. *Electrophoresis*, 38, 1685, 2017.
77. Amiri, A., Tayebee, R., Abdar, A., Sani, F.N., Synthesis of a zinc-based metal–organic framework with histamine as an organic linker for the dispersive solid-phase extraction of organophosphorus pesticides in water and fruit juice samples. *J. Chromatogr. A*, 1597, 39, 2019.

78. Mohammadnejad, M., Gudarzi, Z., Geranmayeh, S., Mahdavi, V., HKUST-1 metal–organic framework for dispersive solid phase extraction of 2-methyl-4-chlorophenoxyacetic acid (MCPA) prior to its determination by ion mobility spectrometry. *Microchim. Acta*, 185, 495, 2018.
79. Amiri, A., Ghaemi, F., Maleki, B., Hybrid nanocomposites prepared from a metal–organic framework of type MOF-199(Cu) and graphene or fullerene as sorbents for dispersive solid phase extraction of polycyclic aromatic hydrocarbons. *Microchim. Acta*, 186, 131, 2019.
80. Xia, L., Liu, L., Dou, Y., Guo, L., Li, G., Sun, Z., You, J., A stable mesoporous metal–organic framework as highly efficient sorbent of dispersive micro solid-phase extraction for the determination of polycyclic aromatic hydrocarbons by HPLC. *J. Sep. Sci.*, 41, 4331, 2018.
81. Mohammadnejad, M., Hajiashrafi, T., Rashnavadi, R., Highly efficient determination of malachite green in aquatic product using Tb-organic framework as sorbent. *J. Porous Mat.*, 25, 1771, 2018.
82. Li, W., Chen, N., Zhu, Y., Shou, D., Zhi, M., Zeng, X., A nanocomposite consisting of an amorphous seed and a molecularly imprinted covalent organic framework shell for extraction and HPLC determination of nonsteroidal anti-inflammatory drugs. *Microchim. Acta*, 186, 76, 2019.
83. Salimi, M., Behbahani, M., Sonhi, H.R., Ghambarian, M., Esrafili, A., Dispersive solid-phase extraction of selected nitrophenols from environmental water samples using a zirconium-based amino-tagged metal–organic framework nanosorbent. *J. Sep. Sci.*, 41, 4159, 2019.
84. Lv, Z., Yang, C., Pang, Y., Xie, W., Shen, X., Dispersive solid-phase extraction using the metal–organic framework MIL-101(Cr) for determination of benzo(*a*)pyrene in edible oil. *Anal. Methods*, 11, 3467, 2019.
85. Cai, Q., Zhang, L., Zhao, P., Lun, X., Li, W., Guo, Y., Hou, X., A joint experimental–computational investigation: Metal organic framework as a vortex assisted dispersive micro-solid phase extraction sorbent coupled with UPLC-MS/MS for the simultaneous determination of amphenicols and their metabolite in aquaculture water. *Microchem. J.*, 130, 263, 2017.
86. Cui, Y.-Y., Yang, C.-X., Yang, X.-D., Yan, X.-P., Zeolitic imidazolate framework-8 for selective extraction of a highly active anti-oxidant flavonoid from *Caragana Jubata*. *J. Chromatogr. A*, 1544, 8, 2018.
87. Wang, Y., Dai, X., He, X., Chen, L., Hou, X., MIL-101(Cr)@GO for dispersive micro-solid-phase extraction of pharmaceutical residue in chicken breast used in microwave-assisted coupling with HPLC-MS/MS detection. *J. Pharm. Biomed.*, 145, 440, 2017.
88. González-Rodríguez, G., Taima-Mancera, I., Lago, A.B., Ayala, J.H., Pasán, J., Pino, V., Mixed Functionalization of Organic Ligands in UiO-66: A Tool to Design Metal–Organic Frameworks for Tailored Microextraction. *Molecules*, 24, 3656, 2019.
89. Xia, L., Liu, L., Lv, X., Qu, F., Li, G., You, J., Towards the determination of sulfonamides in meat samples: A magnetic and mesoporous metal–organic

framework as an efficient sorbent for magnetic solid phase extraction combined with high-performance liquid chromatography. *J. Chromatogr. A*, 1500, 24, 2017.

90. Wang, X., Ma, X., Huang, P., Wang, J., Du, T., Du, X., Lu, X., Magnetic Cu-MOFs embedded within graphene oxide nanocomposites for enhanced preconcentration of benzenoid-containing insecticides. *Talanta*, 181, 112, 2018.
91. Liu, G., Huang, X., Lu, M., Li, L., Li, T., Xu, D., Facile synthesis of magnetic zinc metal–organic framework for extraction of nitrogen-containing heterocyclic fungicides from lettuce vegetable samples. *J. Sep. Sci.*, 42, 1451, 2019.
92. Peña-Méndez, E.M., Mawale, R.M., Conde-González, J.E., Socas-Rodríguez, B., Havel, J., Ruiz-Pérez, C., Metal organic framework composite, *nano*-Fe_3O_4@Fe-(benzene-1,3,5-tricarboxylic acid), for solid phase extraction of blood lipid regulators from water. *Talanta*, 207, 120275, 2020.
93. Li, Y., Chen, X., Xia, L., Xiao, X., Li, G., Magnetic metal–organic frameworks-101 functionalized with graphite-like carbon nitride for the efficient enrichment of glucocorticoids in cosmetics. *J. Chromatogr.* A., 1606, 460382, 2019.
94. Wang, T., Liu, S., Gao, G., Zhao, P., Lu, N., Lun, X., Hou, X., Magnetic solid phase extraction of non-steroidal anti-inflammatory drugs from water samples using a metal organic framework of type FeO4/MIL-101(Cr), and their quantitation by UPLC-MS/MS. *Microchim. Acta*, 184, 2981, 2017.
95. Zhang, Q., Yang, Y., Zhi, Y., Wang, X., Wu, Y., Aptamer-modified magnetic metal–organic framework MIL-101 for highly efficient and selective enrichment of ochratoxin A. *J. Sep. Sci.*, 42, 716, 2019.
96. Dargahi, R., Ebrahimzadeh, H., Asgharinezhad, A.A., Hashemzadeh, A., Amini, M.M., Dispersive magnetic solid-phase extraction of phthalate esters from water samples and human plasma based on a nanosorbent composed of MIL-101(Cr) metal–organic framework and magnetite nanoparticles before their determination by GC-MS. *J. Sep. Sci.*, 41, 948, 2018.
97. Jalilian, N., Ebrahimzadeh, H., Asgharinezhad, A.A., Preparation of magnetite/multiwalled carbon nanotubes/metal–organic framework composite for dispersive magnetic micro solid phase extraction of parabens and phthalate esters from water samples and various types of cream for their determination with liquid chromatography. *J. Chromatogr.* A., 1608, 460426, 2019.
98. Lu, N., He, X., Wang, T., Liu, S., Hou, X., Magnetic solid-phase extraction using MIL-101(Cr)-based composite combined with dispersive liquid–liquid microextraction based on solidification of a floating organic droplet for the determination of pyrethroids in environmental water and tea samples. *Microchem. J.*, 137, 449, 2018.
99. Liang, L., Wang, X., Sun, Y., Ma, P., Li, X., Piao, H., Jiang, Y., Song, D., Magnetic solid-phase extraction of triazine herbicides from rice using metal–organic framework MIL-101(Cr) functionalized magnetic particles. *Talanta*, 179, 512, 2018.

100. Jiang, Y., Piao, H., Qin, Z., Li, X., Ma, P., Sun, Y., Wang, X., Song, D., One-step synthesized magnetic MIL-101(Cr) for effective extraction of triazine herbicides from rice prior to determination by liquid chromatography-tandem mass spectrometry. *J. Sep. Sci.*, 42, 1, 2019.
101. Saboori, A., A nanoparticle sorbent composed of MIL-101(Fe) and dithiocarbamate-modified magnetite nanoparticles for speciation of Cr(III) and Cr(IV) prior to their determination by electrothermal AAS. *Microchim. Acta*, 184, 1509, 2017.
102. Huang, Y.-F., Liu, Q.-H., Li, K., Li, Y., Chang, N., Magnetic iron(III)-based framework composites for the magnetic solid-phase extraction of fungicides from environmental water samples. *J. Sep. Sci.*, 41, 1129, 2018.
103. Zhang, S., Han, P., Xia, Y., +Facile extraction of azide in sartan drugs using magnetized anion-exchange metal–organic frameworks prior to ion chromatography. *J. Chromatogr. A*, 1514, 29, 2017.
104. Li, W., Wang, R., Chen, Z., Metal–organic framework-1210(zirconium/cuprum) modified magnetic nanoparticles for solid phase extraction of benzophenones in soil samples. *J. Chromatogr. A.*, 1607, 460403, 2019.
105. Duo, H., Lu, X., Wang, S., Wang, L., Guo, Y., Liang, X., Synthesis of magnetic metal–organic framework composites, Fe_3O_4–NH_2@MOF-235, for the magnetic solid-phase extraction of benzoylurea insecticides from honey, fruit juice and tap water samples. *New J. Chem.*, 43, 12563, 2019.
106. Ma, J., Wu, G., Li, S., Tan, W., Wang, X., Li, J., Chen, L., Magnetic solid-phase extraction of heterocyclic pesticides in environmental water samples using metal–organic frameworks coupled to high performance liquid chromatography determination. *J. Chromatogr. A*, 1553, 57, 2018.
107. Zhou, Q., Lei, M., Li, J., Liu, Y., Zhao, K., Zhao, D., Magnetic solid phase extraction of N- and S-containing polycyclic aromatic hydrocarbons at ppb levels by using a zerovalent iron nanoscale material modified with a metal organic framework of type Fe@MOF-5, and their determination by HPLC. *Microchim. Acta*, 184, 1029, 2017.
108. Wang, T., Zhang, R., Li, D., Su, P., Yang, Y., Application of magnetized MOF-74 to phthalate esters extraction from Chinese liquor. *J. Sep. Sci.*, 42, 1600, 2019.
109. Safari, M., Yamini, Y., Masoomi, M.Y., Morsali, A., Mani-Varnosfaderani, A., Magnetic metal–organic frameworks for the extraction of trace amounts of heavy metal ions prior to their determination by ICP-AES. *Microchim. Acta*, 184, 1555, 2017.
110. Jin, R., Ji, F., Lin, H., Luo, C., Hu, Y., Deng, C., Cao, X., Tong, C., Song, G., The synthesis of Zr–metal–organic framework functionalized magnetic graphene nanocomposites as an adsorbent for fast determination of multi-pesticide residues in tobacco samples. *J. Chromatogr. A*, 1577, 1, 2018.
111. Li, W.-k., Zhang, H.-x., Shi, Y.-p., Selective determination of aromatic amino acids by magnetic hydroxylated MWCNTs and MOFs based composite. *J. Chromatogr. B*, 1059, 27, 2017.

112. Lu, Y., Wang, B., Yan, Y., Liang, H., Wu, D., Silica Protection–Sacrifice Functionalization of Magnetic Graphene with a Metal–Organic Framework (ZIF-8) to Provide a Solid-Phase Extraction Composite for Recognization of Phthalate Easers from Human Plasma Samples. *Chromatographia*, 82, 625, 2019.
113. Parvinizadeh, F. and Daneshfar, A., Fabrication of a magnetic metal–organic framework molecularly imprinted polymer for extraction of anti-malaria agent hydroxychloroquine. *New J. Chem.*, 43, 8508–8516, 2019.
114. Liu, S., Li, S., Yang, W., Gu, F., Xu, H., Wang, T., Sun, D., Hou, X., Magnetic nanoparticle of metal–organic framework with core–shell structure as an adsorbent for magnetic solid phase extraction of non-steroidal anti-inflammatory drugs. *Talanta*, 194, 514, 2019.
115. Li, Y., Zhou, X., Dong, L., Lai, Y., Li, S., Liu, R., Liu, J., Magnetic metal–organic frameworks nanocomposites for negligible-depletion solid-phase extraction of freely dissolved polyaromatic hydrocarbons. *Environ. Pollut.*, 252, 1574, 2019.
116. Deng, Y., Zhang, R., Li, D., Sun, P., Su, P., Yang, Y., Preparation of iron-based MIL-101 functionalized polydopamine@Fe_3O_4 magnetic composites for extracting sulfonylurea herbicides from environmental water and vegetable samples. *J. Sep. Sci.*, 41, 1, 2018.
117. Lian, L., Zhang, X., Hao, J., Lv, J., Wang, X., Zhu, B., Lou, D., Magnetic solid-phase extraction of fluoroquinolones from water samples using titanium-based metal–organic framework functionalized magnetic microspheres. *J. Chromatogr. A*, 1579, 1, 2018.
118. Jalilian, N., Ebrahimzadeh, H., Asgharinezhad, A.A., A nanosized magnetic metal–organic framework of type MIL-53(Fe) as an efficient sorbent for coextraction of phenols and anilines prior to their quantitation by HPLC. *Microchim. Acta*, 186, 597, 2019.
119. Safari, M., Shahlaei, M., Yamini, Y., Shakorian, M., Arkan, E., Magnetic framework composite as sorbent for magnetic solid phase extraction coupled with high performance liquid chromatography for simultaneous extraction and determination of tricyclic antidepressants. *Anal. Chim. Acta*, 1034, 204, 2018.
120. Yamini, Y. and Safari, M., Magnetic Zink-based metal organic framework as advance and recyclable adsorbent for the extraction of trace pyrethroids. *Microchem. J.*, 146, 134, 2019.
121. Yamini, Y., Safari, M., Morsali, A., Safarifard, V., Magnetic frame work composite as an efficient sorbent for magnetic solid-phase extraction of plasticizer compounds. *J. Chromatogr. A*, 1570, 38, 2018.
122. Ren, J.-Y., Wang, X.-L., Li, X.-L., Wang, M.-L., Zhao, R.-S., Lin, J.-M., Magnetic covalent triazine-based frameworks as magnetic solid-phase extraction adsorbents for sensitive determination of perfluorinated compounds in environmental water samples. *Anal. Bioanal. Chem.*, 410, 1657, 2018.

123. Chen, L., He, Y., Lei, Z., Gao, C., Xie, Q., Tong, P., Lin, Z., Preparation of core–shell structured magnetic covalent organic framework nanocomposites for magnetic solid-phase extraction of bisphenols from human serum samples. *Talanta*, 181, 296, 2018.
124. Deng, Z.-H., Wang, X., Wang, X.-L., Gao, C.-L., Dong, L., Wang, M.-L., Zhao, R.-S., A core–shell structured magnetic covalent organic framework (type Fe_3O_4@COF) as a sorbent for solid-phase extraction of endocrine-disrupting phenols prior to their quantitation by HPLC. *Microchim. Acta*, 186, 108, 2019.
125. Yan, Z., He, M., Chen, B., Gui, B., Wang, C., Hu, B., Magnetic covalent triazine framework for rapid extraction of phthalate esters in plastic packaging materials followed by gas chromatography-flame ionization detection. *J. Chromatogr. A*, 1525, 32, 2017.
126. Wang, R. and Chen, Z., A covalent organic framework-based magnetic sorbent for solid phase extraction of polycyclic aromatic hydrocarbons and its hyphenation to HPLC for quantitation. *Microchim. Acta*, 184, 3867, 2017.
127. Wu, F.-F., Chen, Q.-Y., Ma, X.-J., Li, T.-T., Wang, L.-F., Hong, J., Sheng, Y.-H., Ye, M.-L., Zhu, Y., N-doped magnetic covalent organic frameworks for preconcentration of allergenic disperse dyes in textiles of fall protection equipment. *Anal. Methods*, 11, 3381, 2019.
128. Chen, L., Zhang, M., Fu, F., Li, J., Lin, Z., Facile synthesis of magnetic covalent organic framework nanobeads and application to magnetic solid-phase extraction of trace estrogens from human urine. *J. Chromatogr. A*, 1567, 136, 2018.
129. Liu, J.-M., Lv, S.-W., Yuan, X.-Y., Liu, H.-L., Wang, S., Facile construction of magnetic core–shell covalent organic frameworks as efficient solid-phase extraction adsorbents for highly sensitive determination of sulfonamide residues against complex food sample matrices. *RSC Adv.*, 9, 14247, 2019.
130. Shi, X., Li, N., Wu, D., Hu, N., Sun, J., Zhou, X., Suo, Y., Li, G., Wu, Y., Magnetic covalent organic framework material: Synthesis and application as a sorbent for polycyclic aromatic hydrocarbons. *Anal. Methods*, 10, 5014, 2018.
131. Li, Z., Qi, M., Tu, C., Wang, W., Chen, J., Wang, A.-J., Magnetic Metal–Organic Framework/Graphene Oxide-Based Solid-Phase Extraction Combined with Spectrofluorimetry for the Determination of Enrofloxacin in Milk Sample. *Food Anal. Methods*, 10, 4094, 2017.
132. Kalantari, H. and Manoochehri, M., A nanocomposite consisting of MIL-101(Cr) and functionalized magnetite nanoparticles for extraction and determination of selenium(IV) and selenium(VI). *Microchim. Acta*, 185, 196, 2018.
133. Esmaeilzadeh, M., Ultrasound-assisted dispersive magnetic solid phase extraction based on metal–organic framework/1-(2-pyridylazo)-2-naphthol modified magnetite nanoparticle composites for speciation analysis of inorganic tin. *New J. Chem.*, 43, 4929, 2019.

134. Bahrani, S., Ghaedi, M., Dashtian, K., Ostovan, A., Mansoorkhani, M.J.K., Salehi, A., MOF-5(Zn)–Fe2O4 nanocomposite based magnetic solid-phase microextraction followed by HPLC-UV for efficient enrichment of colchicine in root of colchicium extracts and plasma samples. *J. Chromatogr. B*, 1067, 45, 2017.
135. Chen, Y. and Chen, Z., COF-1-modified magnetic nanoparticles for highly selective and efficient solid-phase microextraction of paclitaxel. *Talanta*, 165, 188, 2017.
136. Li, N., Wu, D., Liu, J., Hu, N., Shi, X., Dai, C., Sun, Z., Suo, Y., Li, G., Wu, Y., Magnetic covalent organic frameworks based on magnetic solid phase extraction for determination of six steroidal and phenolic endocrine disrupting chemicals in food samples. *Microchem. J.*, 143, 350, 2018.

10

Anticancer and Antimicrobial MOFs and Their Derived Materials

Nasser Mohammed Hosny

Chemistry Department, Faculty of Science, Port Said University, Port Said, Egypt

Abstract

Metal–Organic Frameworks (MOFs), are considered new promising materials with significant potential for biomedical applications, due to their unique structural features. This chapter describes the possible medical applications of MOFs particularly, the use of MOFs as drug carriers of the cancer drugs and the requirements of good drug carrier. The applications of MOFs as new type of photosensitizers (PSs) in photodynamic therapy (PDT) of tumor were also presented. Also, this chapter focuses on the uses of MOFs as antibacterial and antifungal agents. Some light was shed on mechanism of action of MOFs on the pathogenic cells.

Keywords: Metal–organic frameworks (MOFs), anticancer MOFs, antibacterial MOFs, antifungal MOFs

10.1 Introduction

Metal–organic frameworks (MOFs) are polymeric materials form frameworks from organic bridging linkers and metal ions [1–4]. It is possible to tailor the properties of MOFs by appropriate selection of the metal ion and the ligand (linker) [5]. MOFs are high surface area porous hybrid materials, able to selective uptake small molecules. These unique features gave MOFs characteristics magnetic, optical and catalytic properties [6, 7]. MOFs can be modified and their structures will govern their possible applications. One of the most recent applications of them is the chemotherapy. MOFs

Email: Nasserh56@yahoo.com

Inamuddin, Rajender Boddula, Mohd Imran Ahamed and Abdullah M. Asiri (eds.) Applications of Metal–Organic Frameworks and Their Derived Materials, (263–286) © 2020 Scrivener Publishing LLC

can be used either directly as chemotherapeutic agents in treatment of different diseases or indirectly as drug carriers or biosensors [8, 9].

10.2 Anticancer MOFs

Cancer is the major cause of death worldwide [10]. Chemotherapy therapy and radiation are the traditional methods of treatment of cancer. Due to the side effects of these conventional treatments, searching for new strategies is required. Among the new strategies used in potential treatment of cancer comes the use of MOFs as drug carrier and in photodynamic therapy (PDT).

10.2.1 MOFs as Drug Carriers

The requirements of MOFs to be efficient nanocarriers were discussed by Horcajada *et al.* and they concluded the requirements of the efficient carrier as to be able to trap high load of drugs efficiently and also to control the release of the drug and control matrix degradation. Besides that, the surface of the carrier must be easily engineered to control MOFs *in vivo* and it must be detected by imaging techniques.

It will be important advantage if a nanocarrier could act as a diagnostic agent and a drug carrier in the same time to assist in evaluating drug distribution and treatment efficiency [11] (Figure 10.1).

Horcajada applied these requirements on a series of porous nanoMOFs derived from Fe(III) with different carboxylate compounds labeled (MIL-89) [11–14], (MIL-88A) [12–15], (MIL-88B) [12–15], (MIL-53) [16], (MIL-100) [17], and (MIL-101) [18]. The structures of (MIL-88), (MIL-89), (MIL-100), and (MIL-101_NH_2) MOFs are built up from the assembly oxo-centred trimmers of octahedral iron (Figure 10.2). (MIL-53) exists as chains that share corner octahedra with di- or tri-carboxylate ligands [11]. (MIL-88), (MIL-89), and (MIL-53) are microporous flexible solids, while (MIL-100, MIL-101 _NH_2) are mesoporous rigid frameworks [11].

The drugs delivery in the body efficiently by nontoxic nanocarriers is an important domain. The majority of known carriers as liposomes, nanoparticles, nanoemulsions or micelles [19–23] show poor drug loading (The carrier is poor drug loading when it loads less than 5% (drug/carrier)) and/or when, it releases rapidly the carried drug which is adsorbed on the external surfaces of the nanocarrier or occluded inside the crystals. In this context, porous MOFs with high drug loadings are considered as candidates for delivery applications. From this point, porous MOFs derived from Fe(III)

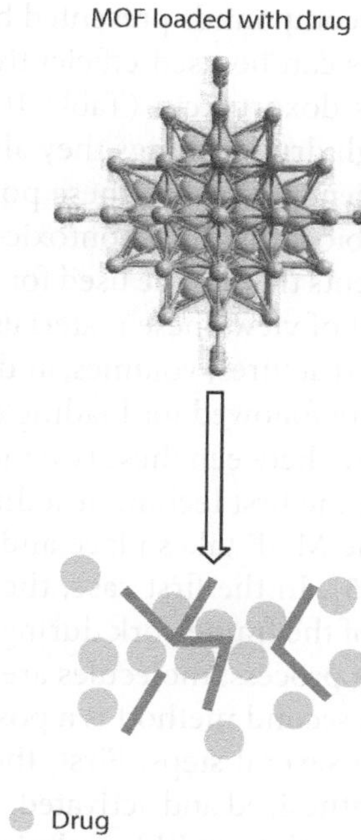

Figure 10.1 Schematic representation of MOFs as drug carriers.

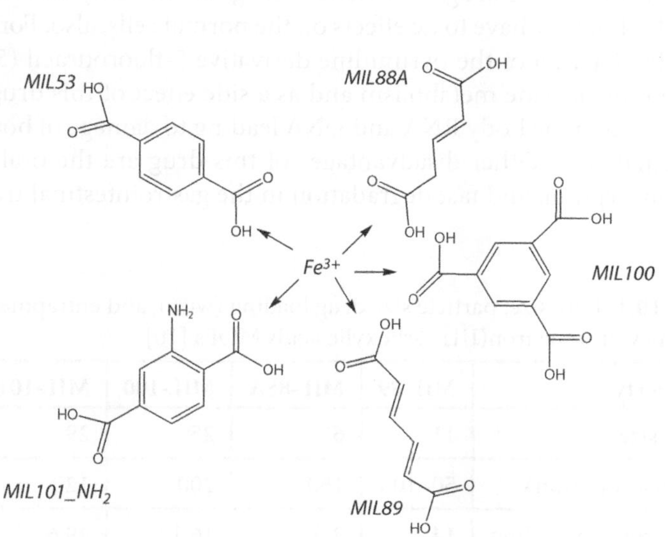

Figure 10.2 Schematic representation of the composition of Iron(III) carboxylic acid MOFs.

with different carboxylate compounds presented by Horcajada are considered superior nanocarriers can be used efficiently for controlled delivery of anti-tumoural drugs as doxorubicin (Table 10.1). These coordination polymers besides their high drug loadings they also exhibit both diagnostics and therapeutics efficiency. Notably, these porous MOFs have several advantages when apply as biocompatible nontoxic drug nanocarriers. Also, they use eco-friendly solvents that can be used for biomedical applications. From the biomedical point of view, these materials are sponges can encapsulate drugs with diverse structures, volumes, and modified groups.

Two main techniques are followed for loading drug into the framework of MOFs [24]. The difference between these two methods depends on when the drug will be added. In the first technique a direct addition of the drug during the synthesis of the MOF takes place and the second method is a post-synthetic method [25]. In the first case, the drug is encapsulated or doped inside the cavities of the framework during the process of synthesis. During this direct loading process, molecules are trapped in the pores, by covalent interactions. The second method is a post-synthetic modification of MOFs which occurs in several steps. First, the metal–organics framework material (host) is synthesized and activated. In a second step the drug is loaded by sorption from solution. This technique involves noncovalent binding of drug (guest) molecules (Figure 10.3).

One of most important problems which face anticancer drugs is the selective action of the drugs. Anticancer drugs act not only on the carcinogenic cells, but they have toxic effects on the normal cells, also. For example the mode of action of the pyrimidine derivative 5-fluorouracil (5- FU), is to prevent nucleoside metabolism and as a side effect of this drug is to be incorporated in the body RNA and DNA leading to damage of both cancer and normal cells. Other disadvantages of this drug are the oral instability, the absorption, and fast degradation in the gastrointestinal tract, short

Table 10.1 Pore size, particle size drug loading (wt%), and entrapment efficiency of some iron(III) carboxylic acids MOFs [10].

Property	MIL-89	MIL-88A	MIL-100	MIL-101_NH_2
Pore size (Å)	11	6	25	29
Particle size (nm)	50–100	150	200	120
Doxorubicin loading	14	2.6	16.1	49.6
Efficiency (%)	81	12	46.2	68.1

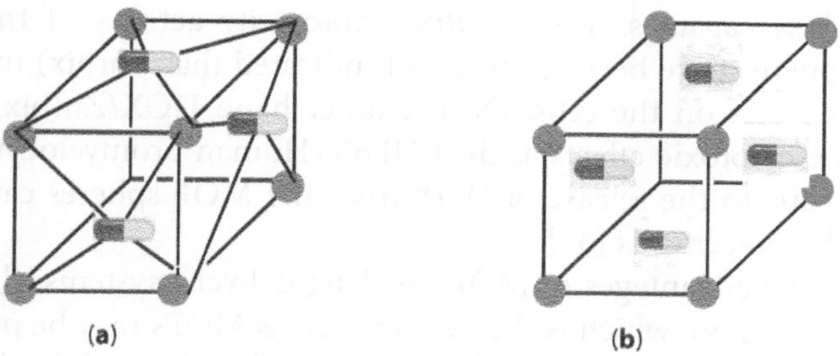

Figure 10.3 (a) The drug is covalently bonded to MOF. (b) The drug is sorped in MOF by some kind of interactions.

contact time with plasma beside the development of drug resistance [24, 26–28]. Another example is the well known antilymphoma drug doxorubicin, which is also restricted by numerous side effects due to its nonspecific bio-distributions, from these adverse effects are the short lifetime in the body and the cause of acute cardio toxicity [28].

The anticancer activity of MOFs made them excellent drug carriers, as they improve the bioavailability and reduce the side effects of these drugs. Besides that, the use of MOFs as carriers of anticancer drugs prevents the untimely decomposition in the biological system and prolongs the drug action without large initial release (burst release) [24]. To control the drug release from the drug carrier the pH must be controlled as the pH of the malignant tumor cells is below the normal. It was noticed that, by using MOF based delivery systems, 5-fluorouracil encapsulated in (Fe-MIL-53-NH_2-FA-5-FAM) releases 32 time faster at pH lower than 7.4 [29]. In case of drug carriers of MIL MOFs, they achieved a prolonged release, which will reduce the frequency administration of the drug [30].

Nano and micro Zn(bix) MOF (bix = 1,4-bis(imidazol-1-ylmethyl) benzene) (Figure 10.4) exists as nanospheres and its ability to encapsulate and release drugs has been tested. Some drugs used in cancer therapy as Doxorubicin (DOX), Camptothecin (CPT), SN-38, and Daunomycin (DAU) were studied. The results demonstrated the release of these drugs

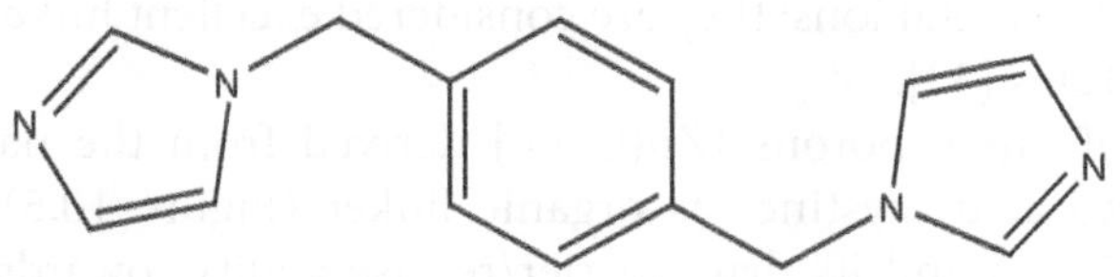

Figure 10.4 1,4-Bis(imidazol-1-ylmethyl)benzene (bix).

from Zn(bix) capsules. The *in vitro* cytotoxicity activity of this DOX/Zn(bix) spheres have been studied and indicated that Zn(bix) matrix has very weak effect on the cells. On the other hand DOX/Zn(bix) showed very strong cytotoxic effects against HL60 (Human promyelocytic leukemia cells) due to the release of DOX from the MOF spheres causing the death of the cancer cells [31].

Despite the advantages of MOFs as drug delivery systems, there is an essential limitation, which is the cytotoxicity, as MOFs may be potentially harmful to humans [32], this problem can be solved by applying biological MOFs.

Biological MOFs are composed of biologically significant safe linker, such as active pharmaceutical ingredients, amino acids, proteins, peptides, carbohydrates, nucleobases, and cyclodextrins and any other safe organic linker. Furthermore, the existed metal in MOFs, must have safe toxicity and high LD50 such as zinc and iron, which exist naturally in the body [32].

Several reviews have discussed the biologically active molecules which have been used as linkers for the design of bioMOFs, their composition and the *in vivo* activity of these MOFs were described also [33, 34].

There are a numerous number of bioactive linkers that allow the development of new biometal–organic frameworks (BioMOFs) which, have additional advantages as: There is no need for the material to be porous, as the release of the drug achieves directly after degradation of the MOF; (ii) Multi-step synthesis of MOF loaded with the drug is avoided as the drug is a constitutive part of the matrix itself; (iii) both, the metal ion and the linker, can be chosen to be bioactive materials to achieve a synergic therapeutic action, (iv) by using porous bioMOF multidrugs can be loaded one as a constituent of the matrix and the other adsorbed in the porous [34].

In bioMOFs, the endogenous metals present in humans as Iron, Zinc, Magnesium, Calcium, or Potassium are important as they can be administered safely in high doses.

One of the most biological linker is the amino acids, which contains two main functional groups both the amino and carboxylic acid groups. They are the main constituents of peptides and proteins, amino acids link together through amide bonds to form peptides and proteins. Amino acids are also excellent chelators to the metal ions. They have different modes of chelation to the metal ions; they are considered excellent linker for formation of bioMOFs [33].

A bioMOF meso-porous [$Zn(Cys)_2$] derived from the naturally presented amino acid (cystine) as organic linker (Figure 10.5) and $ZnCl_2$ has been tailored and its drug carrier/release ability towards Methylene Blue (MB) and Sorafenib was discussed [35]. MB is used in treatment of

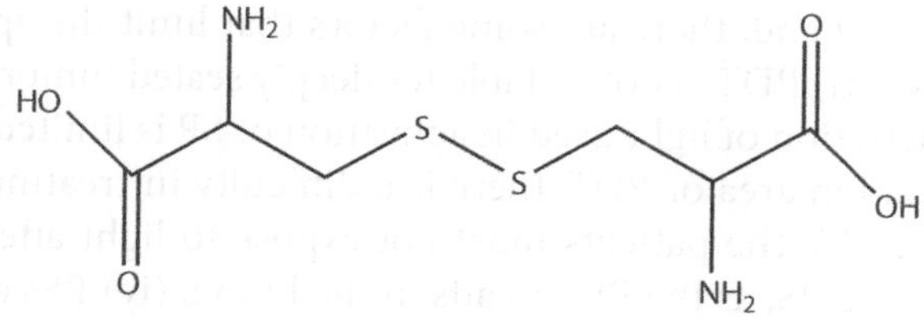

Figure 10.5 Structure of cystine.

colorectal cancer and *Leishmania* in photodynamic therapy and SOR which is used for hepatocellular carcinoma treatment. This synthesized bioMOF when subjected to acidic or reductive environment; it decomposes rapidly releasing the drug.

In conclusion of use MOFs as a drug carriers four common strategies for loading drugs into MOFs are used, in the first strategy the drug is non-covalent encapsulated in MOF by physisorption, in the second, the drug is conjugated to the organic linkers, in the third, the drug itself is used as a linker in building the blocks of the MOF, in the last strategy the drug attaches to the subunits of MOFs [36].

10.2.2 MOFs in Phototherapy

Nanoscale MOFs have recent applications as new type of photosensitizers (PSs) in photodynamic therapy (PDT) of tumor [37]. Guan *et al.* reviewed the strategy of use nanoMOFs in PDT [37]. What is PDT? It is a clinically approved, method for treatment of cancer [38, 39] depends on tumor localization of a photosensetizer, then light activates the photosensitizer to excite tissue oxygen into cytotoxic reactive oxygen species (ROS), which is used in treatment of cancer due to its reactivity in oxidizing many cellular components as nucleic acids, proteins, and phospholipids. The excited PS can transfer electrons directly to DNA of the cell causing cell death [37]. The mechanism of action of PDT to cause cell death [37, 40] takes place through necrosis, apoptosis, and autophagy-associated cell death.

PDT has several advantages over anticancer routine therapies as: (i) It is highly selective with low toxicity to the normal cells, as well as the systemic side effects are rare on activation by local lighting. (ii) Tumor cells cannot acquire resistance to cytotoxicity of ROS. (iii) PDT can be used easily in combination with other cancer treatment regimes as chemotherapy. This advantage will reduce the long-term morbidity. (iv) PDT has little effect on the patient's quality of life as it allows quick recovery with no side effects on long term administration [37].

On the other hand, there are some factors that limit the application of PDT as follows: (i) PDT is not suitable for deeply seated tumor because the depth of penetration of light used in activation of SP is limited. (ii) Due to small illumination area of PDT, there is a difficulty in treatment of metastatic tumors. (iii) The patients must not expose to light after PDT until metabolizing the PS, as the PS spreads in the blood. (iv) PSs are insoluble in water and easily aggregate since the majority of PSs are organic compounds with a high degree of conjugation [41]. (v) Finally, the concentration of PS in tumor cells is generally inadequate [37].

Why MOFs are unique and important in PDT? MOFs can be used as SP or as a carrier of SP, because of their characteristic crystalline porous structures. The PS can also, be encapsulated into MOFs or attached to the nano MOFs frameworks with minimum aggregation of the small-molecules. When PS is activated with light, it excites $^3O_2$ from the tissues to $^1O_2$ (Figure 10.6). Due to their porosity, MOFs facilitate the diffusion of the ROS ($^1O_2$). Nano-MOFs facilitate the solubility of PS and increase their cellular uptake. MOFs generally, assist in adjusting the physicochemical properties of the loaded materials [42]. They are considered excellent nanoplatforms which can be used in combination with other treatments, due to their porous structure and easily modified nature. Besides that, MOFs are biodegradable, biocompatible and have low long-term toxicity to

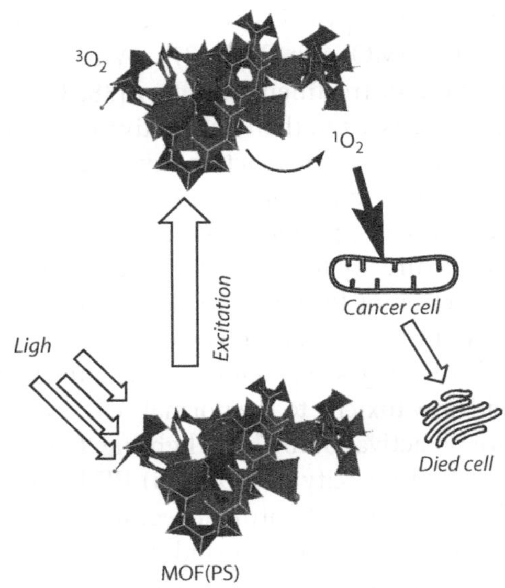

Figure 10.6 Schematic representation of MOFs in PDT.

normal tissues [37, 43]. Based on these unique and characteristic features of MOFs, various strategies as intracellular reactions, self-aggregation, responsive, deep PDT, targeted and combined theranostic strategies were carried out for constructing MOFs-based nanoparticles for PDT [37].

MOFs were first applied in PDT of head and nick cancer treatment by Lu *et al.* [36]. They prepared a Hf–porphyrin nanoMOF of DBP–UiO, (where DBP is 5,15-di(p-benzoato)porphyrin) (Figure 10.7). In DBP–UiO MOF, porphyrin is incorporated as a linker of Hf. An open framework with 1.6 nm triangular channels and tetrahedral–octahedral faces with dimensions 2.0 and 2.8 nm, respectively [36]. The results demonstrated that, nanoDBP–UiO acts as an excellent PDT photosensitizer, as indicated from the efficient generation of $^1O_2$ and its in cytotoxic assay. PDT efficacy demonstrated reduction in the volume of tumor 50 times in half of the investigated mice sample. On the same time, the tumor is completely eradicated in the mice which were treated with DBP–UiO. The results obtained from the group of mice treated with free organic linker H_2DBP, indicated no therapeutic effect of the free linker [36].

Although there are great advantages of MOF based PDT, there are many challenges that still face their clinical applications as The synthesis of these type of compounds on large is not easy and still a problem. Producing simple MOFs-based PDT to be clinically approved is a great challenge. Great efforts are required to enrich the deposition of MOF-based PDT in carcinogenic cells in the same time to reduce the unnecessary systemic toxicity. The presence of tumor marker-triggered for this type of MOF is necessary. It is important to improve the energy transfer efficiency in the deep PDT. The too high power of the used lasers in deep PDT has thermal effects that cause secondary injuries and burns. To study the deep PDT, tumor models are urgently required. However, the methods in determining the therapeutic results of deep PDT are not sufficient, as human tumors are fused with

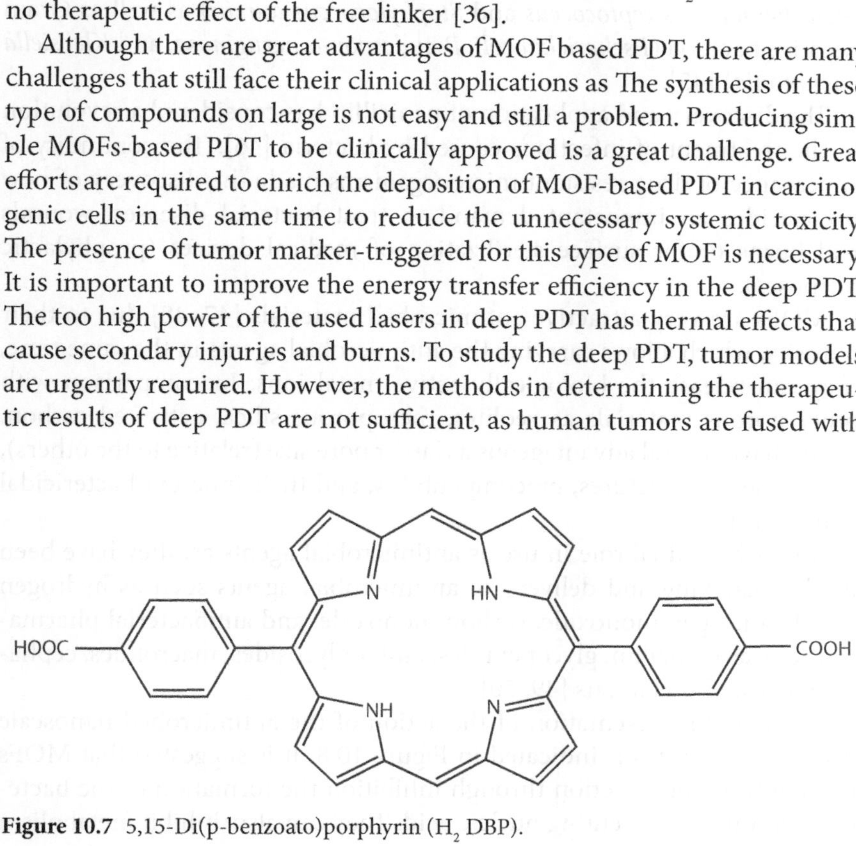

Figure 10.7 5,15-Di(p-benzoato)porphyrin (H_2 DBP).

other tissues within the body. The synergistic mechanism between other tumor therapies and PDT still needs more efforts to be achieved. The biosafety of nanoMOFs still needs further assessments. In the short term, the majority of these compounds show no significant cytotoxicity, however the metabolic process of nanoMOFs is not known clearly. The effect of metal ions accumulation and the metabolic pathways of the organic ligands on the long-term safety in mammals still lacks convincing evidence; where the ligands are not clearly identified and their toxicities are not studied deeply [37].

10.3 Antibacterial MOFs

Millions of nosocomial infections are registered in USA annually [44]. These healthcare-associated infections cause loss of billion dollars annually. These infections are caused by drug-resistant Gram positive bacteria (*Beta-hemolytic streptococcus* and *Staphylococcus aureus*) as well as Gram negative bacteria (*Escherichia coli*, *Pseudomonas aeruginosa*, and *Klebsiella pneumonia*) [45].

The discovery and application of penicillin has considered a revolution in the treatment of infections caused by bacteria [46]. The continuity of appearance of drug resistant microorganisms made the discovery of new bactericides an important demand to treat bacterial diseases, wounds and burns or for surface sterilization of medical devices (prophylactic purposes).

MOFs are very attractive antimicrobial materials [47, 48] due to their porosity, which allows carrying the antimicrobial agents in the same time, they can release the biologically active metal ions. In comparison with other porous materials as zeolites, mesoporous silica, activated carbon, MOFs have several advantageous as larger pore size (relative to the others), their ordered structures, biocompatibility, and their inherent bactericidal nature [47].

MOFs have dual role in use as antimicrobial agents as, they have been used in carrying and delivery of antimicrobial agents such as hydrogen sulfide, nitrogen monoxide, carbon monoxide, and antibacterial pharmaceuticals as penicillin, glycopeptides, aminoglycosides, macrolides, cephalosporin, and metal ions [49, 50].

Schematic representation of the action of the antimicrobial nanoscale drug carrier MOFs is indicated in Figure 10.8. It is suggested that MOFs exert antimicrobial action through inhibition the formation of the bacterial cell wall, or preventing nucleic acid. They can also, inhibit metabolism

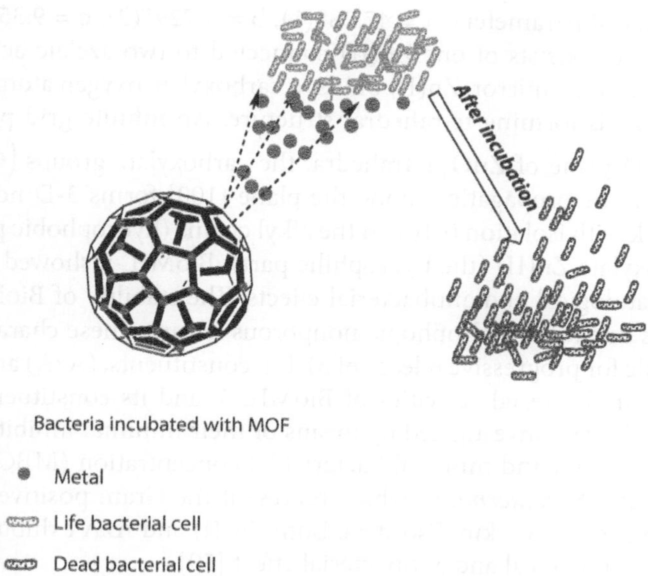

Figure 10.8 Schematic representation of the mechanism of action of antimicrobial MOFs.

of protein or disrupt the membrane structure. On the other hand, MOF is a source of biologically active metal ions such as Zn, Ag, Fe, Cu, Ca, and Co, which release slowly exerting bactericidal effect [48, 51–54].

Zinc based biologically active materials have wide applications as sunscreen lotions, in astringents, moisturizers, and as anti-dandruff agents. Besides that, Zn(II) ion itself has antibacterial activity [45, 55]. However, toxicity analysis indicated the toxicity of Zn(II) toward neuronal cells depends on the concentration of Zn(II) ion [56]. It was found that MOFs have low cytotoxicity, comparing with commercial nanoparticles. Zn containing MOFs were used as carriers of antibacterial drugs or as bactericidal compounds due to their low cytotoxic properties [57, 58].

A bioactive Metal–Organic Framework (BioMIL-5) derived from Zn(II) and azelaic acid (AZA) (Figure 10.9), was synthesized hydrothermally. BioMIL-5 crystallizes in the orthorhombic unit cell with the *Pcca* space

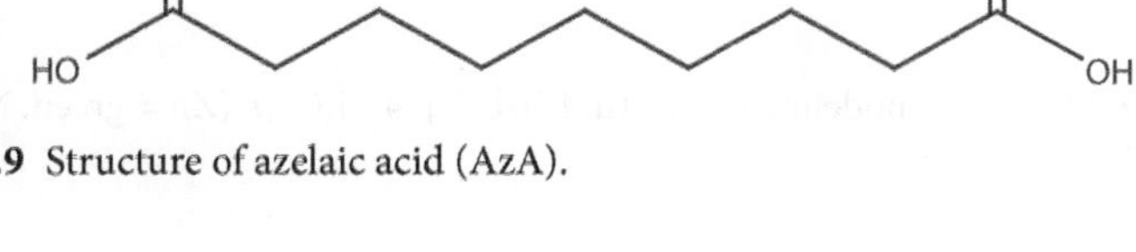

Figure 10.9 Structure of azelaic acid (AzA).

group and cell parameters, a = 47.288(1), b = 4.7297(2), c = 9.3515(3) Å. The unit cell consists of one Zn(II) connected to two azelaic acid molecules located on a mirror. Zn(II) binds to carboxylate oxygen atoms of four azelate ligands forming tetrahedral structure. An infinite grid propagate along (011) plane of ZnO_4 tetrahedra, the carboxylate groups $\left(CO_2^-\right)$ act as bridges. The propagation along the plane [100] forms 3-D nonporous framework with isolation between the alkyl chains (hydrophobic part) and the carboxylate Zn(II) (the hydrophilic part). BioMIL-5 showed interesting dermatological and antibacterial effects. The stability of BioMIL-5 in water was due to its hydrophopic nonporous nature. These characters are responsible for progressive release of MIL-5 constituents, (AzA) and Zn(II) ions. The antibacterial activities of BioMIL-5, and its constituents (AzA and Zn(II)), were investigated by means of their minimal inhibitory concentration (MIC) and minimal bactericidal concentration (MBC) against *S. aureus* and *S. epidermidis*, which represent the Gram positive bacteria responsible for some skin disorders. Both Zn(II) and AzA exhibited interesting dermatological and antibacterial effect [59].

{[Zn(μ-4-hzba)$_2$]$_2$ 4(H_2O)}*n* was prepared from hydrazinebenzoate as linker. Two different stereochemistries are observed around Zn(II) ion (Figure 10.10). Tetrahedral coordination is formed around one Zn(II) while, the other Zn(II) is octahedrally coordinated. The different

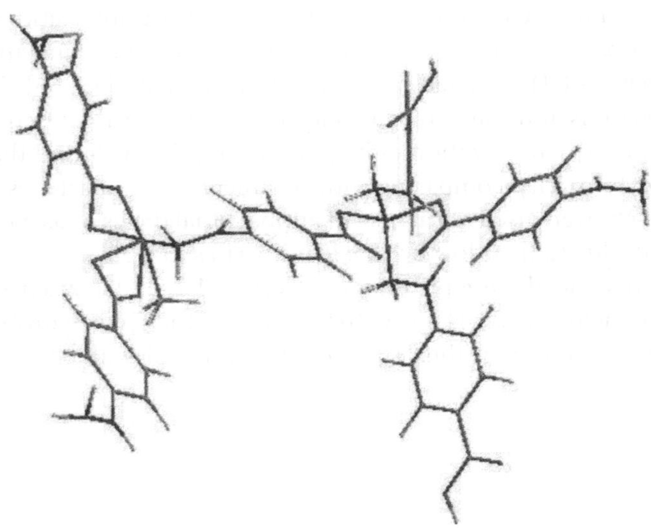

Figure 10.10 Molecular modeling of {[Zn(μ-4-hzba)$_2$]$_2$ 4(H_2O)}*n* (Zn = green, N = blue and O = red).

coordination environments resulted from the mode of chelation of the carboxylate group. It acts as a bidentate chelator in case of the octahedrally Zn(II) center. On the other hand, it acts in a monodentate manner for the other Zn(II) ion. Zn(II) atoms are connected together by the bridging ligand (4-hydrazinebenzoate), constituting extended network. The antimicrobial activity of this MOF was studied against *Staphylococcus aureus.* $\{[Zn(\mu\text{-}4\text{-}hzba)_2]2.4(H_2O)\}n$ inhibited the bacterial growth and metabolic activity. The antibacterial effect was attributed to the release of the linker (4-hydrazinebenzoate) with a low contribution from Zn(II) ion. This Zn based MOF was able to release 4-hydrazinebenzoate for several days [60].

Three Zinc containing nanoMOFs (MOF-5, Zn-BTC, and IRMOF-3) were prepared and their antibacterial efficacies against *Staphylococcus lentus, Escherichia coli, Staphylococcus aureus, and Listeria monocytogenes* were explored.

The antibacterial activity was determined as (MIC) for the free (Zn-MOFs, Ampicillin, and Kanamycin) or mixtures of (MOF + antibiotic), against *L. monocytogenes, S. lentus, S. aureus, and E. coli.* All the tested compounds exhibited bactericidal effect against the tested pathogens. It was observed that, the mixture of antibiotic and MOF has higher activity than the free compounds. Notably, (MOF + Ampicillin) has reduced (MIC) from two- to fourfold (Table 10.2). Also, (IRMOF-3 + Kanamycin)

Table 10.2 Minimum inhibitory concentration (MICs) (μg/mL) of nMOFs and Ampicilline when used alone and as a mixture with the antibiotics [51].

Compound	*S. aureus*	*S. lentus*	*L. monocytogenes*	*E. coli*
Zn-BTC (alone) (1)	200	200	250	150
Zn-BTC + Ampicillin	150	150	100	100
Ampicilline (alone)	32	48	48	32
Ampicilline + (1)	16	24	16	8
IRMOF-3 (alone) (2)	100	150	150	100
IRMOF-3 + Ampicilline	100	50	50	50
Ampicilline + (2)	16	16	16	8
MOF-5 (alone) (3)	200	200	200	200
MOF-5 + Ampicilline	100	150	100	100
Ampicilline +(3)	16	24	24	8

reduced (MIC) from 1.5- to 8-fold when used against *E. coli*. Synergistic antibacterial action of (MOF + antibiotic) had been observed against both Gram positive and Gram-negative bacteria. The significant effect of these tested MOFs was explained as Zn based MOFs are reservoirs of Zn(II) ions, which damage the bacterial cell wall [51].

Silver salts, in the form of Ag(I) have broad spectrum bactericidal effects. One of the most important biological properties of silver is its low toxic effects on the eukaryotic cells [61]. Generally, Ag^+ inhibits DNA replication of the cell or the metabolism of the essential cellular proteins [62].

In general Ag-based antibacterial materials are either elemental silver (which dissolves very slowly in water), or soluble Ag^+ ions or complexes [63–65]. Therefore, great efforts are done to look for new materials with biocidal Ag^+ ion have high antibacterial efficiency. In this concern, silver-based MOFs are considered an alternative approach for preparing interesting antimicrobial materials, as they slowly release biocidal concentrations of metal ions towards wide spectrum of pathogens.

Three Ag-based metal–organoboron frameworks with formulae [(AgL)NO_3]$2H_2O$ (1), [(AgL)CF_3SO_3]$2H_2O$ (2), and [(AgL)ClO_4]$2H_2O$ (3) (where, L = (4-pyridylduryl)borane) were synthesized [66]. The isolated MOFs (1)–(3) were tested against *S. aureus and E. coli*. The test showed that the ligand L had no effect on the tested bacteria. While, MOFs (1)–(3) exhibit significant antibacterial activity against both *E. coli and S. aureus* [67]. The rate of Ag(I) release, was also, studied, MOF (1) showed the slowest rate of release, while, (3) exhibited the fastest rate. All the studied MOFs gave steady release for months of the metal ions and the releasing process relays on changing the network structures of the tested MOFs.

The silver-based MOF Ag_3(3-phosphonobenzoate) (1) (Figure 10.11) in which the linker is 3-phosphonobenzoic acid which is classified as rigid

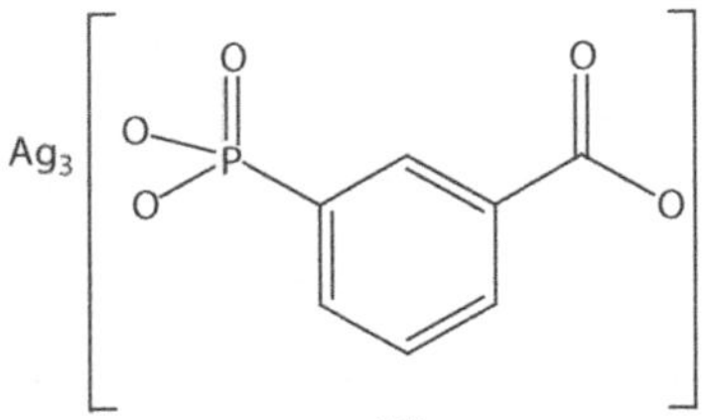

(1)

Ag_3(3-phosphonobenzoate)

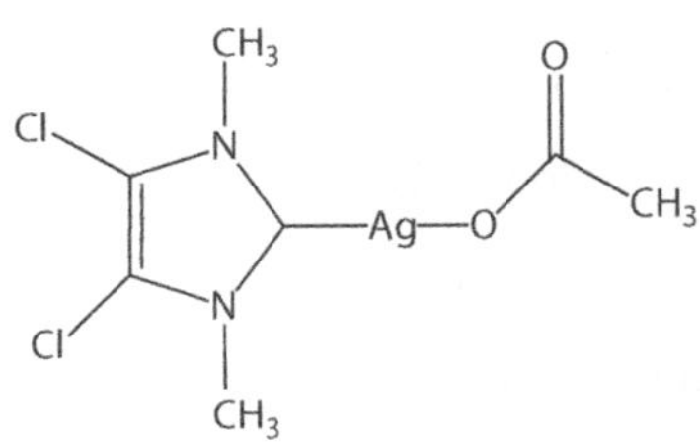

(3)

1,3-dimethyl-4,5-ichloroimidazole-2-ylidene)-silver(I) acetate

Figure 10.11 Structures of MOFs (**1**) and (**3**).

ligand has two hard bases functional groups (phosphonic and carboxylic acids). These structural features of the linker favor the slow release of Ag(I) ions. The antibacterial activity of this MOF was investigated against six bacterial strains. A sustainable release of Ag(I), was followed by cathodic stripping voltammetry. The released Ag(I) is responsible for the activity against the tested microorganisms. It was found that the free phosphonic acid did not show any effect against the tested bacteria. MOF (1) exhibited bactericidal effect even at quite low concentrations due to the release of Ag(I) ions [67].

The microbial activity of Cu(II) ions is well-known due to its ability to destroy the envelope of bacteria [52, 68]. Cu-BTC (MOF-199) was prepared using *in situ* techniques onto cellulosic substrate. The reaction was carried out between copper acetate, the linker (1,3,5-benzenetricarboxylic acid) in presence of carboxymethylated cellulose and triethylamine. It completely eliminated pathogenic bacteria. The antibacterial activity of the cellulose-MOF was attributed to MOF-199 not to cellulose. There were evidences for the presence a strong bonding between MOF-199 and the cellulose substrates forming not detach crystals and allowing the textile modified with MOF to be washed and reused fabricating antibacterial clinical fabrics [69].

The first antibacterial Co-based MOF namely (Co-TDM), where, TDM = tetrakis [(3,5-dicarboxyphenyl)-oxamethyl] methane acid (Figure

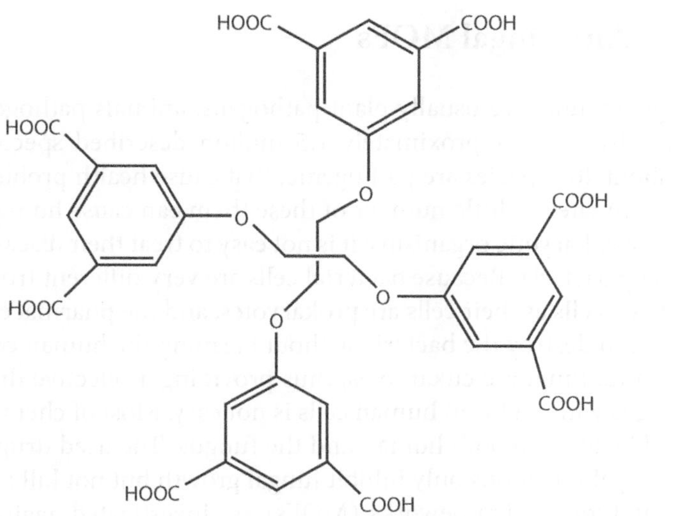

Figure 10.12 Structure of tetrakis[(3,5-dicarboxyphenyl)-oxamethyl] methane acid (H_8TDM).

10.12) was synthesized by Zhuang *et al.* [70] via facile hydrosolvothermal technique. (Co-TDM) crystallizes in the tetragonal space system. An octahedral stereochemistry is formed around each cobalt atom. The ligand $(TDM)^{8-}$ acts as an octa-topic donor coordinating to eight cobalt centers, two Co clusters are bridged by two $(COO)^{-}$ groups from two different (TDM) and one μ_2-H_2O. The Co-TDM was demonstrated to inactivate the Gram negative bacteria (*E. coli*). The mechanism of inactivation may involve the weakening of the cell as a result of several biochemical interactions with the cell wall protein and DNA followed by disruption of iron balance. The unique structure of the Co-TDM which contains octa-topic ligand, enables it to act as a reservoir for the Co ions when interacts with the bacteria cell membrane. The large porosity and the highly available surface of Co play roles in the high antibacterial activity of Co-TDM.

Three different MOFs, namely, Zn (Zn-SIM1), Co(Co-SIM1) (SIMI refers to subfamily of MOF and imidazole ligand) and Ag (Ag-TAZ) (TAZ refers to 1,2,4-triazole), have been prepared by solvothermal procedures and evaluated against two cyanobacteria. The tested bacteria were more sensitive to Co-SIM than Ag (Ag-TAZ), although, all the investigated MOFs can diffuse in the cyanobacteria and inhibit their growth. There is (SAR) structure activity relationship between the structure of the MOFs and their released metal, which is responsible for the variation in their biocidal activities [71].

10.4 Antifungal MOFs

Pathogenic fungi are usually plant pathogens, animals pathogens are few species. There are approximately 1.5 million described species of fungi [72], about 400 species are pathogenic, that cause health problems in animals, fortunately, a little number of these them can cause human diseases. Fungi are eukaryotic organisms; it is not easy to treat their diseases as those caused by bacteria. Because bacterial cells are very different from our own eukaryotic cells as their cells are prokaryotes, and the pharmaceutics (antibiotics) can destroy the bacteria without harming the human cells. On the other hand, fungi are eukaryotes, thus providing a selective drug can kill the fungus without harm human cells is not easy. Most of chemotherapeutics will be toxic to both human and the fungus. The used drugs for treatment fungal infections only inhibit fungal growth but not kill it.

Metal–Organic Frameworks (MOFs) are investigated against fungi as new biologically active materials due to their characteristics structural features.

HKUST-1 (MOF) is an open pore material contains binuclear Cu(II) centers. In this MOF Cu ions are connected to the linker forming three dimensional network. On the other hand, when the network is destroyed Cu ions will liberate causing the biocidal action. HKUST-1 showed strong antifungal activity against *Saccharomyces cerevisiae and Geotrichum candidum* (commonly in the food industries). The antifungal action of HKUST-1 was related to the action of copper ions which releases into the medium after breaking down the crystal of the MOF. Cu ions interact with the fungal cell, leading to inhibiting intracellular enzymes and disrupting the transport of nutrients [73]. Moreover, the variation in the taxonomic group of the walls of *S. cerevisiae and G. candidum*, may cause the variation in sensitivity of them toward HKUST-1 [74]. This copper-based MOF can be used for controlled release of copper ions [75].

A water stable MOF, namely, copper-1,3,5-benzenetricarboxylate (Cu-BTC), was prepared [76] and investigated against several species of fungi. The structure of (Cu-BTC) contains dimeric Cu^{2+} ions locate in the centre of the molecule where, each copper binds with four oxygen atoms from the carboxylate groups of the ligand (BTC) and a coordinated water molecule, giving the general formula $Cu_3(BTC)_2(H_2O)_3$. The crystal structure indicated the formation of a 3D binuclear Cu_2 paddle wheel contains square pores (9Å by 9Å) [77, 78]. This water stable and industrially interesting (Cu-BTC MOF) was investigated against *Aspergillus oryzae, Candida*

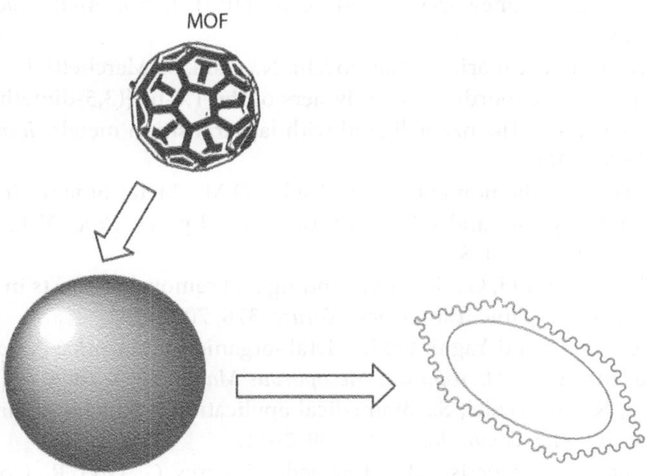

Figure 10.13 Schematic representation of the effect of MOF on the cell wall of the fungi (a) Fungi cell. (b) Fungi cell after treating with MOF.

albicans, Fusarium oxysporum, and *Aspergillus niger*. These pathogens can cause economic and health problems *range* from common to serious infections. *They are* responsible for contamination of foods leading to food spoilage and foodborne diseases or sometimes, serious clinical cases can take place [79] when disseminates through the bloodstream. Besides that, *some of these fungi (Fusarium oxysporum)* present in the soil and has a negative impact on the environment [80]. It was suggested that (Cu-BTC) exhibited a powerful antifungal activity, as it damages the cell causing bleeding of intracellular compounds and complete cell destroying (Figure 10.13). This mechanism was supported by the ability of MOF to reduce the oxygen gas and produces ROS which damage the cell and inhibit the microorganisms.

References

1. Sun, Y., Yu, H., Wang, L., Ding, W., Ji, J., Chen, Y., Ren, F., Tian, Z., Zhao, Y., Huang, L., Ren, P., Tong, R., Synthesis of three coordination polymer microspheres and their application in hydrogen storage. *J. Alloys Compd.*, 585, 667, 2014.
2. Fernandez-Zapico, E., Montejo-Bernardo, J., D'Vries, R., Garcia, J.R., Garcia-Granda, S., Rodriguez, F.J., Pedro, I., Blanco, J.A., Phase stability and magnetic properties of a new cobalt(II) coordination polymer based on 2-carboxyethylphosphonate and 1,10′-phenanthroline. *J. Alloys Compd.*, 536, S507, 2012.
3. Hosny, N.M., Solvothermal synthesis, thermal and adsorption properties of metal-organic frameworks Zn and CoZn (DPB). *Therm. Anal. Calorim.*, 122, 89, 2015.
4. Tabacaru, A., Pettinari, C., Masciocchi, N., Galli, S., Marchetti, F., Angjellari, M., Pro-porous coordination polymers of the 1,4-bis((3,5-dimethyl-1Hpyrazol-4-yl)-methyl)benzene ligand with late transition metals. *Inorg. Chem.*, 50, 11506, 2011.
5. Britt, D., Tranchemontagne, D.J., Yaghi, O.M., Metal-organic frameworks with high capacity and selectivity for harmful gases. *Proc. Natl. Acad. Sci. USA*, 105, 11623, 2008.
6. Yaghi, O.M. and Li, G., Selective binding and removal of guests in a microporous metal–organic framework. *Nature*, 378, 703, 1995.
7. Rowsell, J.L.C. and Yaghi, O.M., Metal–organic frameworks: A new class of porous materials. *Microporous Mesoporous Mater.*, 73, 3, 2004.
8. Keskin, S. and Kızılel, S., Biomedical applications of metal organic frameworks. *Ind. Eng. Chem. Res.*, 50, 1799, 2011.
9. McKinlay, A.C., Morris, R.E., Horcajada, P., Férey, G., Gref, R., Couvreur, P., Serre, C., BioMOFs: Metal–organic frameworks for biological and medical applications. *Angew. Chem. Int. Ed.*, 49, 6260, 2010.

10. Siegel, R.L., Miller, K.D., Jemal, A., Cancer statistics. *Ca-Cancer J. Clin.*, 68, 7, 2018.
11. Horcajada, P., Chalati, T., Serre, C., Gillet, B., Sebrie, C., Baati, T., Eubank, J.F., Heurtaux, D., Clayette, P., Kreuz, C., Porous metal-organic-frameworks nanoscale carriers as a potential platform for drug delivery and imaging. *Nat. Mater.*, 9, 172, 2010.
12. Férey, G. and Serre, C., Large breathing effects in three-dimensional porous hybrid matter: Facts, analyses, rules and consequences. *Chem. Soc. Rev.*, 38, 1380, 2009.
13. Serre, C., Mellot-Draznieks, C., Surblé, S., Audebrand, N., Filinchuk, Y., Férey, G., Role of solvent-host interactions that lead to very large swelling of hybrid frameworks. *Science*, 315, 1828, 2007.
14. Serre, C., Millange, F., Surblé, S., Férey, G., A route to the synthesis of trivalent transition-metal porous carboxylates with trimeric secondary building units. *Angew. Chem. Int. Ed.*, 43, 6285, 2004.
15. Surblé, S., Serre, C., Mellot-Draznieks, C., Millange, F., Férey, G., A new isoreticular class of metal-organic-frameworks with the MIL-88 topology. *Chem. Commun.*, 3, 284, 2006.
16. Whitfield, T.R., Wang, X., Liu, L., Jacobson, A.J., Metal–organic frameworks based on iron oxide octahedral chains connected by benzenedicarboxylate dianions. *Solid State Sci.*, 7, 1096, 2005.
17. Horcajada, P., Surblé, S., Serre, C., Hong, D., Seo, Y., Chang, J., Grenèche, J., Margiolaki, I., Férey, G., Synthesis and catalytic properties of MIL-100(Fe), an iron(III) carboxylate with large pores. *Chem. Commun.*, 27, 2820, 2007.
18. Bauer, S., Serre, C., Devic, T., Horcajada, P., Marrot, J., Férey, G., Stock, N., High-throughput assisted rationalization or the formation of metal–organic framework in the iron(III) aminoterephthalate solvothermal system. *Inorg. Chem.*, 47, 7568, 2008.
19. Peer, D., Karp, J.M., Hong, S., Farokhzad, O.C., Margalit, R., Langer, R., Nanocarriers as an emerging platform for cancer therapy. *Nat. Nanotechnol.*, 2, 751, 2007.
20. Couvreur, P., Gref, R., Andrieux, K., Malvy, C., Nanotechnology for drug delivery: Applications to cancer and autoimmune diseases. *Prog. Solid State Ch.*, 34, 231, 2006.
21. Gref, R., Minamitake, Y., Peracchia, M.T., Trubetskoy, V., Torchilin, V., Langer, R., Biodegradable long-circulating polymeric nanospheres. *Science*, 263, 1600, 1994.
22. Gabizon, A., Stealth liposomes and tumor targeting: One step further in the quest for the magic bullet. *Clin. Cancer Res.*, 7, 223, 2001.
23. Sheikh Hasan, A., Socha, M., Lamprecht, A., El Ghazouani, F., Sapin, A., Hoffman, M., Maincent, P., Ubrich, N., Effect of the microencapsulation of nanoparticles on the reduction of burst release. *Int. J. Pharm.*, 344, 53, 2007.

24. Simagina, A.A., Polynski, M.V., Vinogradov, A.V., Pidko, E.A., Towards rational design of metal-organic framework-based drug delivery systems. *Russ. Chem. Rev.*, 87, 831, 2018.
25. Huxford, R.C., Della Rocca, J., Lin, W., Metal–organic frameworks as potential drug carriers. *Curr. Opin. Chem. Biol.*, 14, 262, 2010.
26. Zhang, N., Yin, Y., Xu, S.J., Chen, W.S., 5-Fluorouracil: Mechanisms of resistance and reversal strategies. *Molecules*, 13, 1551, 2008.
27. Arias, J.L., Ruiz, M.A., LoApez-Viota, M., Delgado, A.A.V., Poly (alkylcyanoacrylate) colloidal particles as vehicles for antitumor drug delivery: A comparative study. *Colloids Surf. B: Biointerfaces*, 62, 64, 2008.
28. Arias, J.L., LoApez-Viota, M., Delgado, A.A.V., Ruiz, M.A., Iron/ethylcellulose (core/shell) nanoplatform loaded with 5-fluorouracil for cancer targeting. *Colloids Surf. B: Biointerfaces*, 77, 111, 2010.
29. Takemura, G. and Fujiwara, H., Doxorubicin-induced cardiomyopathy from the cardiotoxic mechanisms to management. *Prog. Cardiovasc. Dis.*, 49, 330, 2007.
30. Gao, X., Zhai, M., Guan, W., Liu, J., Liu, Z., Damirin, A., Controllable Synthesis of a Smart Multifunctional Nanoscale Metal–Organic Framework for Magnetic Resonance/Optical Imaging and Targeted Drug Delivery. *ACS Appl. Mater. Interfaces*, 9, 3455, 2017.
31. Yang, B., Shen, M., Liu, J., Ren, F., Post-Synthetic Modification Nanoscale Metal-Organic Frameworks for Targeted Drug Delivery in Cancer Cells. *Pharm. Res.*, 34, 2440, 2017.
32. Imaz, I., Rubio-Martinez, M., Garcia-Fernández, L., Garcia, F., Ruiz-Molina, D., Hernando, J., Puntesa, V., Maspoch, D., Coordination polymer particles as potential drug delivery systems. *Chem. Commun.*, 46, 4737, 2010.
33. Imaz, I., Rubio-Martınez, M., An, J., Sole-Font, I., Rosi, N.L., Maspoch, D., Metal-biomolecule frameworks (MBioFs). *Chem. Commun.*, 47, 7287, 2011.
34. Rojas, S., Devic, T., Horcajada, P., Metal organic frameworks based on bioactive components. *J. Mater. Chem. B*, 5, 2560, 2017.
35. Bieniek, A., Wiśniewski, M., Roszek, K., Bolibok, P., Terzyk, A.P., Ferrer, P., da Silva, I., New strategy of controlled, stepwise release from novel MBioF and its potential application for drug delivery systems. *Adsorption*, 25, 383, 2019.
36. Lu, K., Aung, T., Guo, N., Weichselbaum, R., Lin, W., Nanoscale metal–organic frameworks for therapeutic, imaging, and sensing applications. *Adv. Mater.*, 30, 1707634, 2018.
37. Guan, Q., Li, Y., Li, W., Dong, Y., Photodynamic therapy based on nanoscale metal-organic frameworks: From material design to cancer nanotherapeutics. *Chem. Asian J.*, 13, 3122, 2018.
38. Rubin, A.I., Chen, E.H., Ratner, D., Basal-cell carcinoma. *N. Engl. J. Med.*, 353, 2262, 2005.

39. Agostinis, P., Berg, K., Cengel, K.A., Foster, T.H., Girotti, A.W., Gollnick, S.O., Hahn, S.M., Hamblin, M.R., Juzeniene, A., Kessel, D., Korbelik, M., Moan, J., Mroz, P., Nowis, D., Piette, J., Wilson, B.C., Golab, J., Photodynamic therapy of cancer: An update. *Ca-Cancer J. Clin.*, 61, 250, 2011.
40. Zhang, P., Hu, C., Ran, W., Meng, J., Yin, Q., Li, Y., Recent progress in light-triggered nanotheranostics for cancer reatment. *Theranostics*, 6, 948, 2016.
41. Rajora, M.A., Lou, J.W.H., Zheng, G., *Chem. Soc. Rev.*, 46, 6433–6469, 2017.
42. Hynek, J., Ondrusova, S., Buzek, D., Kovar, P., Rathousky, J., Demel, J., Post synthetic modification of a zirconium metal–organic framework at the inorganic secondary building unit with diphenylphosphinic acid for increased photosensitizing properties and stability. *Chem. Commun.*, 53, 8557, 2017.
43. Baati, T., Njim, L., Neffati, F., Kerkeni, A., Bouttemi, M., Gref, R., Najjar, M.F., Zakhama, A., Couvreur, P., Serre, C., Horcajada, P., In depth analysis of the *in vivo* toxicity of nanoparticles of porous iron (III) metal–organic frameworks. *Chem. Sci.*, 4, 1597, 2013.
44. Klevens, R.M., Edwards, J.R., Horan, T.C., Gaynes, R.P., Pollack, D.A., Cardo, D.M., Estimating Health Care-Associated Infections and Deaths in U.S. Hospitals, 2002. *Public Health Rep.*, 122, 160, 2007.
45. Jean, S.S., Hsueh, P.R., Lee, W.S., Chang, H.T., Chou, M.Y., Chen, I.S., Wang, J.H., Lin, C.F., Shyr, J.M., Ko, W.C., Wu, J.J., Liu, Y.C., Huang, W.K., Teng, L.J., Liu, C.Y., Nationwide surveillance of antimicrobial resistance among Enterobacteriaceae in intensive care units in Taiwan. *Eur. J. Clin. Microbiol. Infect. Dis.*, 28, 215, 2009.
46. Kumazawa, J. and Yagisawa, M., The history of antibiotics: The Japanese story. *J. Infect. Chemother.*, 8, 125, 2002.
47. Morris, R.E., *Antimicrobial coatings and modifications on medical devices*, p. 171, Springer, 2017.
48. Quirós, J., Boltes, K., Aguado, S., de Villoria, R.G., Vilatela, J.J., Rosal, R., Antimicrobial metal–organic frameworks incorporated into electrospun fibers. *Chem. Eng. J.*, 262, 189, 2015.
49. McKinley, A.C., Allan, P.K., Renoir, C.L., Duncan, M.J., Wheatley, P.S., War render, S.J., Dawson, D., Ash brook, S.E., Gil, B., Marshaled, B., MultiMate delivery of multiple therapeutic agents from metal–organic frameworks. *APL Mater.*, 2, 124108, 2014.
50. R.E. Morris, Anti-microbial metal organic framework. US 2013/0171228A1, 2013.
51. Santo, C.E., Quaranta, D., Grass, G., Antimicrobial metallic copper surfaces kill Staphylococcus haemolyticus *via* membrane damage. *Microbiologyopen*, 1, 46, 2012.
51. Bhardwaj, N., Pandey, S.K., Mehta, J., Bhardwaj, S.K., Kim, K.H., Deep, A., Bioactive nano-metal-organic frameworks as antimicrobials against Gram-positive and Gram-negative bacteria. *Toxicol. Res.*, 7, 931, 2018.

52. Casey, A., Adams, D., Karpanen, T., Lambert, P., Cookson, B., Nightingale, P., Miruszenko, L., Shillam, R., Christian, P., Elliott, T., Role of copper in reducing hospital environment contamination. *J. Hosp. Infect.*, 74, 72, 2010.
53. Miller, S.R., Alvarez, E., Fradcourt, L., Devic, T., Wuttke, S., Wheatley, P.S., Steunou, N., Bonhomme, C., Gervais, C., Laurencin, D., A rare example of a porous Ca-MOF for the controlled release of biologically active NO. *Chem. Commun.*, 49, 7773, 2013.
54. Plum, L.M., Rink, L., Haase, H., The essential toxin: Impact of zinc on human health. *Int. J. Environ. Res. Public Health*, 7, 1342, 2010.
55. Schwartz, J.R., Marsh, R.G., Draelos, Z.D., Zinc and skin health: Overview of physiology and pharmacology. *Dermatol. Surg.*, 31, 837, 2005.
56. Pavlica, S., Gaunitz, F., Gebhardt, R., Comparative *in vitro* toxicity of seven zinc-salts towards neuronal PC12 cells. *Toxicol. In Vitro*, 23, 653, 2009.
57. Ren, F., Yang, B., Cai, J., Jiang, Y., Xu, J., Wang, S., Toxic effect of zinc nanoscale metal-organic frameworks on rat pheochromocytoma (PC12) cells *in vitro*. *J. Hazard. Mater.*, 271, 283, 2014.
58. Beg, S., Rahman, M., Jain, A., Saini, S., Midoux, P., Pichon, C., Ahmad, F.J., Akhter, S., Nanoporous metal organic frameworks as hybrid polymer–metal composites for drug delivery and biomedical applications. *Drug Discov. Today*, 22, 625–637, 2017.
59. Tamames-Tabar, C., Imbuluzqueta, E., Guillou, N., Serre, C., Miller, S., Elkaïm, E., Horcajada, P., Blanco-Prieto, M., A Zn azelate MOF: Combining antibacterial effect. *CrystEngComm*, 17, 456, 2015.
60. Restrepo, J., Serroukh, Z., Santiago-Morales, J., Aguado, S., Gomez-Sal, P., Mosquera, M.E.G., Rosal, R., *Eur. J. Inorg. Chem.*, 574, 2017.
61. Silver, S., Bacterial silver resistance: Molecular biology and uses and misuses of silver compounds. *FEMS Microbiol. Rev.*, 27, 341, 2003.
62. Russell, A.D. and Hugo, W.B., Antimicrobial activity and action of silver. *Prog. Med. Chem.*, 31, 351, 1994.
63. Sambhy, V., MacBride, M.M., Peterson, B.R., Sen, A., Silver bromide nanoparticle/polymer composites: Dual action tunable antimicrobial materials. *J. Am. Chem. Soc.*, 128, 9798, 2006.
64. Ho, C.H., Tobis, J., Christina, S., Thoman, R., Tiller, J.C., *Adv. Mater.*, 16, 957, 2004.
65. Melaiye, A., Sun, Z., Hindi, K., Milsted, A., Ely, D., Reneker, D.H., Tessier, C.A., Youngs, W.J., Silver(I)-imidazole cyclophane gem–diol complexes encapsulated by electrospun tecophilic nanofibers: Formation of nanosilver particles and antimicrobial activity. *J. Am. Chem. Soc.*, 127, 2285, 2005.
66. Liu, Y., Xu, X., Xia, Q., Yuan, G., He, Q., Cui, Y., Multiple topological isomerism of three-connected networks in silver-based metal–organoboron frameworks. *Chem. Commun.*, 46, 2608, 2010.
67. Berchel, M., Le Gall, T., Denis, C., Le Hir, S., Quentel, F., Elleouet, C., Montier, T., Rueff, J., Salaun, J., Haelters, J., Hix, G.B., Lehn, P., Jaffres, P.,

A silver-based metal–organic framework material as a 'reservoir' of bactericidal metal ions. *New J. Chem.*, 35, 1000, 2011.

68. Santo, C.E., Quaranta, D., Grass, G., Antimicrobial metallic copper surfaces kill *Staphylococcus haemolyticus* via membrane damage. *Microbiologyopen*, 1, 46, 2012.
69. Rodriguez, H.S., Hinestroza, J.P., Ochoa-Puentes, C., Sierra, C.A., Soto, C.Y., Antibacterial activity against *Escherichia coli* of Cu-BTC (MOF-199) metal-organic framework immobilized onto cellulosic fibers. *J. Appl. Polym. Sci.*, 131, 40815, 2014.
70. Zhuang, W., Yuan, D., Li, J.R., Luo, Z., Zhou, H.C., Bashir, S., Liu, J., Highly potent bactericidal activity of porous metal–organic frameworks. *Adv. Healthc. Mater.*, 1, 225–238, 2012.
71. Martín-Betancor, K., Aguado, S., Rodea-Palomares, I., Tamayo-Belda, M., Leganés, F., Rosal, R., Fernández-Piñas, F., Co, Zn and Ag-MOFs evaluation as biocidal materials towards photosynthetic organisms. *Sci. Total Environ.*, 595, 547, 2017.
72. Hawksworth, D.L., Fungi: A neglected component of biodiversity crucial to ecosystem function and maintenance. *Can. Biodivers.*, 1, 4, 1992.
73. Freeman, M.H. and MacIntyre, C.R., A comprehensive review of copper-based wood preservatives with a focus on new micronized or dispersed copper systems. *Forest Prod. J.*, 58, 6, 2008.
74. González, M., de Groot, P.W.J., Klis, F.M., Lipke, P.N., *Microbial glycobiology*, O. Holst, P.J. Brennan, M. von Itzsteim (Eds.), pp. 169–183, Elsevier, Netherlands, 2009.
75. Chiericatti, C., Basilico, J.C., Basilico, M.L.Z., Zamaro, J.M., Novel application of HKUST-1 metal–organic framework as antifungal: Biological tests and physicochemical characterizations. *Microporous Mesoporous Mater.*, 162, 60, 2012.
76. Bouson, S., Krittayavathananon, A., Phattharasupakun, N., Siwayaprahm, P., Sawangphruk, M., Antifungal activity of water-stable copper-containing metal-organic frameworks. *R. Soc. Open Sci.*, 4, 170654, 2017.
77. Castillo, J.M., Vlugt, T.J.H., Calero, S., Understanding water adsorption in Cu-BTC metal-organic frameworks. *J. Phys. Chem. C*, 112, 15934, 2008.
78. Abbasi, A.R., Akhbari, K., Morsali, A., Dense coating of surface mounted CuBTC metal–organic framework nanostructures on silk fibers, prepared by layer-by-layer method under ultrasound irradiation with antibacterial activity. *Ultrason. Sonochem.*, 19, 846, 2012.
79. Fridkin, S.K. and Jarvis, W.R., Epidemiology of nosocomial fungal infections. *Clin. Microbiol. Rev.*, 9, 499, 1996.
80. Nelson, P.E., Dignani, M.C., Anaissie, E.J., Taxonomy, biology and clinical aspects of *Fusarium* species. *Clin. Microbiol. Rev.*, 7, 479, 1994.

11

Theoretical Investigation of Metal–Organic Frameworks and Their Derived Materials for the Adsorption of Pharmaceutical and Personal Care Products

Jagannath Panda[1], Satya Narayan Sahu[1], Tejaswini Sahoo[1], Biswajit Mishra[1], Subrat Kumar Pattanayak[2] and Rojalin Sahu[1]*

[1]*School of Applied Sciences, Kalinga Institute of Industrial Technology, Deemed to be University, Bhubaneswar, India*
[2]*Department of Chemistry, National Institute of Technology Raipur, India*

Abstract

PPCPs, i.e., pharmaceutical and personal care products are the products, which are being used in healthcare, medicated and personal care products. Collectively, these are more than thousands types chemical compounds starting from lotions, perfume, drugs, etc. With the growing population of the world and life style improvement, there is a high demand of PPCPs resulting in adulteration to the surface and also ground water. Recently, the harmful and toxic impact of PPCPs on water bodies both surface and ground water, is serious global problem. Removal of these PPCPs from water bodies is very important and urgent to save aquatic organisms as well as human beings. Different types of porous materials are used to remove PPCPs as adsorbents and also research is going on to find out new porous materials including the new inorganic–organic hybrid Metal organic frameworks (MOFs). In this work, we have focused the general synthesis of MOFs and the adsorptive removal of Ibuprofen, Diclofenac, Macroxen, and Oxybenzone using MIL-100 by adopting *in silico* process. From the study, we have found that MIL-100 have very high binding energy with Diclofenac, Naproxen, and Oxybenzone. The binding energy of MOF (MIL-100) with targeted Pharmaceutical and personal care product (PPCPs) was reported to be negative which signified exothermic adsorption.

**Corresponding author*: rsahufch@kiit.ac.in

Inamuddin, Rajender Boddula, Mohd Imran Ahamed and Abdullah M. Asiri (eds.) Applications of Metal–Organic Frameworks and Their Derived Materials, (287–312) © 2020 Scrivener Publishing LLC

The capacity of adsorption collected from discussed MOF (MIL-100) can be cited as hydrophobic contacts and hydrogen bonding. The binding energy and other parameters like internal energy, Vdw hydrogen bond dissolve energy, electrostatic energy specified that MIL-100 MOF can be used for adsorptive removal of PCPPs.

Keywords: PPCPs, MOFs, MIL-100, adsorption, molecular interaction, synthesis, *in silico*

11.1 Introduction

PPCPs, i.e., pharmaceutical and personal care products are the products which are being used in healthcare, medicated and personal care products [1]. Collectively, these are more than thousands types chemical compounds starting from lotions, perfume, drugs, etc. [2–4]. With the growing population of the world and life style improvement, there is a high demand of PPCPs resulting adulteration surface and also ground water [5–7]. Recently the harmful and toxic impact of PPCPs on water bodies both surface and ground water, is serious global problem [8]. Even, PPCPs are found in fishes and vegetables as well as in water bodies. These are harmful to aquatic organisms as well as to human beings [9–11]. There are some reports on the interference of PPCPs with endocrine system resulting hormonal imbalance. Hence, there is an urgency for the removal of PPCPs from water bodies and now research is being carried globally. Various types of methods have been used for the removal of PPCPs from aqueous medium and these are advance oxidation (AOP), ozonization, photo degradation, chlorination, biodegradation [12–22]. However, no method is proven successful because these methods have their own merits and demerits. For example, AOP is energy consuming and ozonization method produces many side products. Subsequently, other methods have been investigated to overcome these demerits. Adsorption is considered to be the best method for the removal of PPCPs from aqueous medium because it is an economical process, mild operating with little or no energy consumption [23]. Different type of porous materials is used to remove PPCPs as adsorbents and also research is going on to find out new porous materials including the new inorganic–organic hybrid Metal organic frameworks (MOFs) [24].

MOFs (metal organic framework) are available in 1D (one dimensional), 2D (two dimensional) as well as 3D (three dimensional) structure in which clusters or metal ions are linked with different types of organic ligands [25]. These MOFs are one of the subgroups of coordination polymer compounds which exhibit unique characteristics, i.e., porosity. BDC (1,4-benzene dicarboxylic acid) one of the organic linker (struts) used in preparation of coordination polymer [26]. The coordinating networks in metal organic framework which are integrated with these organic ligands consist of voids or holes. Materials which are porous have application in the storing of gas, separation of gas using adsorption method, catalytic reaction, which is size selective, storing, and delivery of drug, etc. [27–37]. Conventionally, these porous materials can be either inorganic or organic material. Activated carbon is the most general porous material synthesized via pyrolysis of materials which are rich in carbon (high adsorption capacity and high surface area bearing), comes under organic material category [38, 39]. But these carbons based porous organic material are not ordered. Zeolites are the porous inorganic materials having structures which are highly ordered [40, 41].

Synthesis of these frameworks often requires an inorganic or organic template with strong interactions forming between the inorganic framework and the template during the synthesis [42]. On merging the pros of properties of inorganic as well as organic porous material, Metal organic framework (porous hybrid) is resulted which possess stability, highly ordered and high surface area [43]. MOFs are identified by number of names such as porous coordination networks or polymers, etc.

As MOFs are known for their various industrial applications, other possibilities to employ these promising materials are also looked for. MOFs are very attractive because of their tunability and a remarkable degree of diversity in the inorganic and organic units they can possess in their structures [44].

MOFs are categorized in three generations: The stability of first-generation metal organic framework is dependent on presence of guest molecules i.e. in the absence of guest molecules irreversible breakdown of structure takes place. MOF-5, an example of second-generation metal organic framework unlike first generation metal organic framework exhibit stability even in the absence of guest molecule and also possess constant porous behavior [45].

The third-generation metal organic framework exhibit flexibility and dynamic network, which react to external factors like electric fields, light, pressure, guest molecules, etc., and the outcome is reflected in modification of void size reversibly [46].

MOFs have been generally synthesized from isolated metal ions and organic linkers under hydrothermal or solvothermal conditions via conventional electrical heating on small scales. Recently, the development of the precursor approach and kinetically tuned dimensional augmentation strategy provides more possibilities to obtain novel MOFs with new structures and interesting properties. There are different synthesis methods (spray drying, electrochemical, mechanochemical, sonochemical, etc.), which quicken and speed up the process of crystallization and production of reduced size even crystal within short duration and better quality ,thereby making it suitable for various industrial applications. The schemes of formation of diversified networks of metal organic framework use the concept of reticular synthesis. Due to flexible coordination numbers a large category of metal ions belong to class of transition metal class with octahedron, tetrahedron, and square planar geometry. The organic linkers or ligands are classified as cyanide, halide, anionic, or neutral organic molecules like 4,4′-bipyridine, benzene dicarboxylic acid, etc. [47–52].

MOFs have been applied for the adsorptive removal of PCPPs from aqueous medium. The PCPPs which are removed by MOFs are naproxen, bis-phenol-A(BPA), diclofenac and clofibric acid. Among the MOFs, MIL-101 has been used widely for PCPPs adsorption. In this work, the adsorptive removal of Ibuprofen, Diclofenac, Macroxen and Oxybenzone have been studied by adopting in silico process which we are reporting for the first time as no *in silico* study have been adopted before. MOFs used for adsorptive removal of pharmaceutical products from waste water are given in Table 11.1.

11.2 General Synthesis Routes

Hydrothermal or solvothermal methods are exclusively used for the production of MOFs and zeolite. In this process the crystals slowly develops from hot solution. Unlike zeolites, metal organic frameworks are created using bridging organic ligand which remains unchanged during the entire process of synthesis. The ions which sway the developing inorganic framework structure are called as templates. Secondary building unit (SBU) and organic ligands are used as templates bin metal organic framework. Metal organic frameworks applied for storing of gas is templated via solvents (metal binding) like water and N,N dimethyl formamide.

The method of solvothermal synthesis is very vital for developing crystals appropriate for elucidation of structure because in this process growth of crystal occurs within time period of hours to days. Using metal organic framework as storage materials requires vast scale up for their synthesis process in order to meet demands of consumer products. One such scale up technique is microwave assistance which is utilized to nucleate crystals of metal organic framework from the solution rapidly. Synthesis of wide

Table 11.1 MOFs used for adsorptive removal of pharmaceutical products from waste water.

Sl no	MOF	Adsorbate	Dominant mechanism	Reference
1	MIL-53(Cr)	2,4-Dichlorophenoxyacetic acid	Electrostatic interaction, π–π interaction	[53]
2	Mesoporous ZIF-8	*p*-Arsalinic acid	Electrostatic interaction	[53]
3	ZIF-8	*p*-Arsalinic acid	Electrostatic interaction	[53]
4	MIL-100(Fe)	*p*-Arsalinic acid	CUSs	[54]
5	NH_2-CAU-1	Bentazon	n.a.	[55]
6	MOF-235(Fe)	Bentazon	n.a.	[55]
7	MIL-53(Cr)	Bisphenol A	π–π interaction, hydrogen bonding	[56]
8	MIL-101(Cr)	Bisphenol A	π–π interaction, hydrogen bonding	[57]
9	ED-MIL-101(Cr)	Clofibric acid	Acid-base interaction	[58]
10	MIL-101(Cr)	Clofibric acid	Electrostatic interaction	[58]
11	AMSA-MIL-101(Cr)	Clofibric acid	Electrostatic interaction	[58]
12	MOF-235(Fe)	Clopyralid	n.a.	[55]
13	NH_2-CAU-1	Clopyralid	n.a.	[55]

(*Continued*)

Table 11.1 MOFs used for adsorptive removal of pharmaceutical products from waste water. (*Continued*)

Sl no	MOF	Adsorbate	Dominant mechanism	Reference
14	Zn(bdc)(tib)] • $3H_2O$	Congo red	Pore size, CUSs	[59]
15	MIL-53(Al)*	Dimethyl phtalate	π–π interaction, pore size	[60]
16	[Cu(INA)2]	Fluorescein	Electrostatic interaction	[61]
17	MIL-100(Cr)	Furosemide	n.a.	[62]
18	UiO-67(Zr)	Glufosinate	Chemisorption	[63]
19	UiO-67(Zr)	Glyphosate	Chemisorption	[63]
20	ZIF-8	Humic acid	Electrostatic, π–π interaction	[64]
21	NH_2-CAU-1	Isoproturon	n.a.	[55]
22	MOF-235(Fe)	Isoproturon	n.a.	[55]
23	MIL-100(Fe)	Methyl blue	Electrostatic interaction, π–π interaction	[65]
24	NH_2-MIL101(Al)	Methyl orange	Electrostatic interaction	[66]
25	[Cd6(L)2(bib)2(DMA)4]	Methyl orange	n.a.	[67]

(*Continued*)

Table 11.1 MOFs used for adsorptive removal of pharmaceutical products from waste water. (*Continued*)

Sl no	MOF	Adsorbate	Dominant mechanism	Reference
26	NH_2-TMU-16(Zn)@silk	Methyl orange	Hydrogen bonding	[68]
27	UiO-66(Zr)	Methylchlorophenoxypropionic acid	Electrostatic, π–π interaction	[69]
28	NH_2-MIL101(Al)	Methylene blue	Electrostatic interaction	[66]
29	MIL-101(Al)	Methylene blue	Electrostatic interaction	[66]
30	MIL-101(Cr)–SO_3H	Fluoroquinolone	Electrostatic interaction	[70]
31	[bbpy][Bi_4I_{16}]	Methylene blue	Hydrogen bonding	[71]
32	NH_2-TMU-16(Zn)@silk	Morphine	Hydrogen bonding	[68]
33	ED-MIL-101(Cr)	Naproxen	Acid–base interaction	[72]
34	MIL-101(Cr)	Naproxen	Electrostatic interaction	[72]
35	MIL-100(Fe)	Naproxen	Electrostatic interaction	[73]
36	AMSA-MIL-101(Cr)	Naproxen	Electrostatic interaction	[73]
37	MIL-68(Al)	Nitrobenzene	Hydrogen bonding	[74]

(*Continued*)

Table 11.1 MOFs used for adsorptive removal of pharmaceutical products from waste water. *(Continued)*

Sl no	MOF	Adsorbate	Dominant mechanism	Reference
38	CAU-1	Nitrobenzene	Hydrogen bonding	[74]
39	MIL-53(Al)	Nitrobenzene	Hydrophobic/hydrophilic, π–π interactions	[75]
40	NH_2-MIL-101(Al)	*p*-Nitrophenol	Hydrogen bonding	[76]
41	NH_2-MIL-101(Al)	Phenol	n.a.	[76]
42	MIL-68(Al)	Phenol	π–π interaction, hydrogen bonding	[77]
43	ZIF-8	Phthalic acid	Acid–base, electrostatic interaction	[78]
44	NH_2-UiO-66(Zr)	Phthalic acid	Acid–base, electrostatic interaction	[78]
45	UiO-66(Zr)	Phthalic acid	π–π interaction	[78]
46	[Cd6(L)2(bib)2(DMA)4]	Rhodamine B	n.a.	[67]
47	MIL-100(Fe)	Roxarsone	CUSs	[54]
48	MIL-100(Cr)	Sulfasalazine	n.a.	[62]

range of metal organic framework in the absence of solvent is discussed. In general organic pro ligand and metal acetate are grounded and mixed via ball mill. This synthetic route performed in the absence of solvent resulted that in quick quantities production of Cu3(BTC)2 and it was found that structure of the product is similar to that of Basolite C300 which is an industrial product. It was believed that the collision energy in the ball mill assisted in completion of reaction by melting the components. During the reaction acetic acid was produced as side product in the ball mill and it played the role of solvent in the ball mill. CVD or chemical vapor deposition is another advancement in the field of solvent free synthesis of metal organic framework. The above process based on CVD was first applied for MOF-ZIF-8 consisting of dual methods. In initial step there was deposition of metal oxide parent layer and in next step these parent layers are uncovered to fine ligand molecules which finally cause induction of phase change in metal organic framework crystal lattice. Release of water during this process acted as vital role in guiding the process of phase transformation [47–52].

11.2.1 Hydrothermal Synthesis

This synthesis method involves production of single crystal depending on minerals solubility under high pressure in hot water. The process of development of crystal takes place in an autoclave (steel pressure vessel) where there is supply of a nutrient via water. At the end of the growth chamber there is a temperature gradient. The solute nutrient gets dissolved at the end which is hotter and it gets deposited at the cooler end leading to growth of required crystal.

Hydrothermal synthesis includes the crystallization of substances at high-temperature and at high vapor pressures which are generated due to solvent synthesis. High temperature range below 200°C is needed to perform this technique for organic and aqueous solvent by using autoclave which lined by Teflon under pressure. These conditions enable the solvent temperature to increase beyond its boiling point and atmospheric pressure while the solvent dielectric constant and viscosity gets reduced. The decrease in dielectric constant and viscosity causes enhancement in the process of diffusion and thereafter growth of crystal. Material having high vapor pressure close to their melting points can be synthesized by hydrothermal methods. The main hurdle associated with this technique is requirement of costly autoclaves and observation of growing crystal [79–81].

Recently it was revealed that this technique can also be used in the preparation of MOFs, which undergo *in situ* ligand synthesis.

11.2.2 Solvothermal Synthesis of MOFs

Generally metal organic framework produced via solvothermal route by electrical heating.

Collaboration process in the formation of metal organic framework is performed between organic ligand and isolated metal ions [82]. Solvothermal synthesis is a method of producing crystalline MOFs, which is very similar to the hydrothermal route. The only difference being that the precursor solution is usually not aqueous. The solvothermal synthesis route gains the benefit of both the sol–gel and hydrothermal routes, permitting the accurate hold over distribution in shape, size and crystallinity of MOFs by varying experimental parameters like reaction temperature and time, solvent type, surfactant type, precursor type, etc. Solvothermal synthesis has been used in the laboratory to make nanostructured titanium dioxide, graphene, carbon, and other materials [83, 84].

11.2.3 Room Temperature Synthesis

It is a unique category of solvothermal process in which heating is not mandatory to get metal organic framework crystals. The reaction mixture combined with solvent is kept for crystallization at room temperature. During the process of synthesis bases like triethylamine are added which results in organic ligands deprotonation in order to precipitate metal organic framework [85].

11.2.4 Microwave Assisted Synthesis

The process of microwave irradiation is utilized for the development of metal organic frameworks. This category of synthesis depends on electromagnetic waves interaction with mobile electric charge. The system temperature can be increased by application of required frequency which in initiate collision between the molecules, thereby increasing the kinetic energy. This is a very efficient procedure of heating because of the direct faceoff between solution and radiation. The pros of this method enlist control in morphology, reduction of particle size, and selectivity of phase [86].

11.2.5 Mechanochemical Synthesis

This process can cause induction of a large number of chemical as well as mechano physical phenomena during the availability of solvents. Here intermolecular bonds undergoes mechanical breakdown and later on chemical modifications occurs. Reaction of precursor materials for few minutes inside a reactor made up of steel ball results in the formation of single phased highly crystalline metal organic frameworks having guest particles in its voids. In order to obtain guest free compound activating the MOF thermally will give the desired result [87, 88].

11.2.6 Electrochemical Synthesis

In large scale process of production this method helps in obstructing the entry of unwanted anions like halides, perchlorate, or nitrate. Hence, metal organic framework can be synthesized on large scale by utilizing metal ions and organic ligands. During this process, highly crystalline and pristine products are expected [89].

11.3 Postsynthetic Modification in MOF

The term PSM was derived from the post translational modification of proteins, whereby chemical functionality is introduced by chemical modification of an intact polypeptide. Basically, PSM is the structural modifications of the synthesized MOFs. By this way, we can synthesize a number of MOFs modifying the structure. The main advantage of PSM is incorporation of different functional groups in the MOF matrix [90]. Different derivatives of a single MOF can be achieved by using varieties of reagents so that the modified MOFs are topologically identical but functionally diversified. This method is already established in solid state materials such as CNT, organosilicas, zeolites, and meso porous silicas. Various methods for the synthesis of MOFs are shown in Figure 11.1.

11.4 Computational Method

In this report, we have studied computationally how the pharma products can be removed from the pharmaceutical waste water to make the water bodies PPCPs free to be reuse or recycle by using MOFs. The

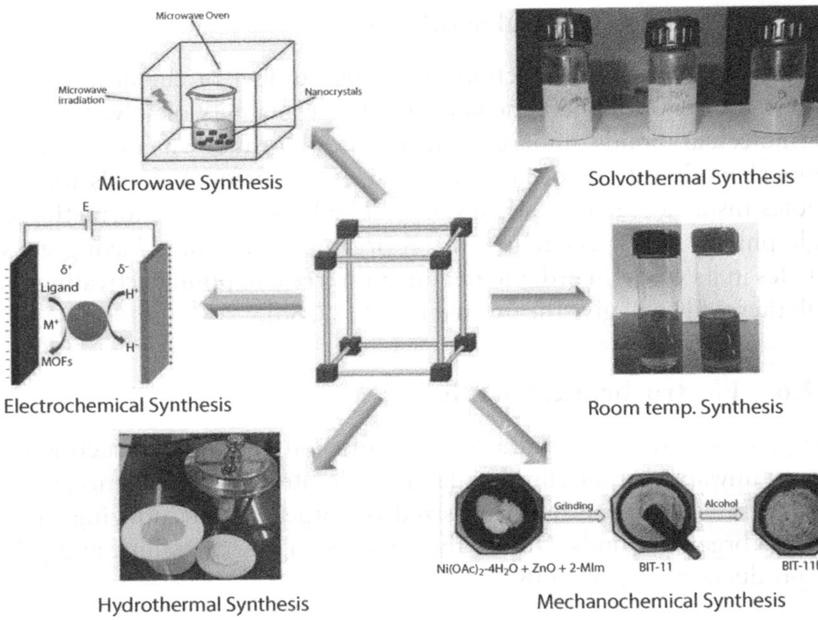

Figure 11.1 Various methods for the synthesis of MOFs.

theoretical analysis was taken for the present study. The study was performed by using only computational tools and algorithm. Various properties of molecular interaction of MOF and organic hazardous molecules of pharmaceutical waste water are studied by using computational methods. In this study molecular docking approach were used to find out the adsorbent capacity of MOF with organic hazardous molecules of pharmaceutical waste water. To predict the binding orientation of small molecules, the frequently used computational method is docking [91–96]. The molecular docking study were performed by using Auto dock 4.2 tool in which Auto docking software is used [97–101]. The docking study was performed with the flexible ligand and the rigid receptor by tacking grid size 126, 126, and 126 points in x, y, and z directions was created with center_x = 14.48, center_y = 11.81 and center_z = 7.12 for MIL-100. For docking study the input file should be in .pdb file format, and to converted the structure into .pdbqt file format we have to remove bad contacts and added hydrogens on polar contacts of the receptor by using Autodock tool where both the internal geometry of the receptor and ligand is kept fixed and docking is performed or may be flexible docking (induced fit), which allows conformational changes. Two approaches are particularly popular within the molecular docking [102]. One approach uses for searching algorithms that describes the receptor and the ligand binding modes. Another approach is based on to calculate the interaction energies between ligand-receptor in pair wise. Both approaches are based on mathematical methods to find out the different interactions, including non-covalent interaction. During docking process, the intermolecular interactions were evaluated by scoring functions. The predicted grid size and coordinates of MOFs were investigated by molecular docking method. The initial coordinates of MIL-100 was taken from the Cambridge structural database and other structures are taken from pubchem database. The visualization and image representation were performed by using Discovery Studio [101]. The molecular interaction between MIL-100 MOF and hazardous molecules present in waste water would be obtained from inter molecular interactions. Hence docking plays an important role in the molecular recognition process.

11.5 Results and Discussion

In this report, we have used molecular docking approach to show the binding coordination between MIL-100-Fe with different molecules present in pharmaceutical waste water. Specifically, we have studied the interaction between MIL-100-Fe and the pharmaceutical products, Diclofenac, Ibuprofen, Naproxen, and Oxybenzone, which are present in pharmaceutical waste water like to analyze the binding and adsorption performance over the MIL-100 MOF.

11.5.1 Binding Behavior Between MIL-100 With the Adsorbates (Diclofenac, Ibuprofen, Naproxen, and Oxybenzone)

When the MIL-100 MOF binds with Diclofenac, the binding energy is found to be −5.8 kcal/mol. The hydrogen bonds are found within the range 2.02 to 3.45 Å. Generally, the hydrogen bonds are present between oxygen and hydrogen and polar hydrogen atoms bind to chlorine and oxygen, whereas the hydrophobic interactions are maintained between chlorine and π orbitals. The pictorial representation of docking between MIL-100 with Diclofenac is shown in Figure 11.2(a). The binding energy of Ibuprofen, Naproxen, and Oxybenzone with MIL-100 were −5.9, −6.7, and −6.4 kcal/mol, respectively, as shown in Table 11.2.

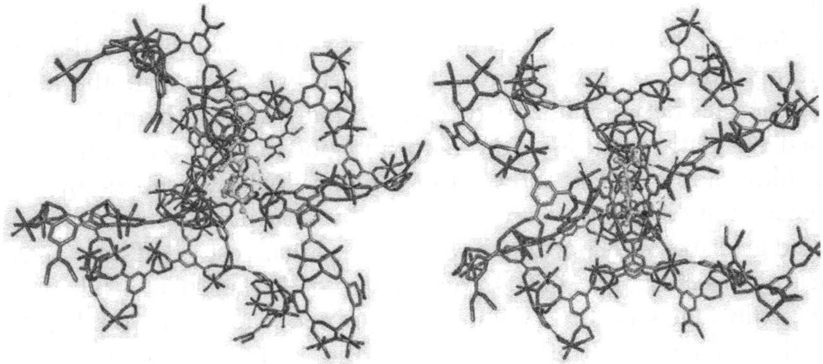

Figure 11.2 Schematic representation of binding coordinate between MIL-100-Fe with (a) Diclofenac and (b) Ibuprofen. The yellow colored ball and stick model represented as adsorbents.

In case of Ibuprofen there is only one hydrogen bond formed and the bond length is 2.72 Å and three hydrophobic interactions are present with bond lengths varies from 3.92 to 4.54 Å. The pictorial representation of docking between MIL-100-Fe with Ibuprofen was shown in Figure 11.2(b).

When we have analyzed the binding performance between MIL-100-Fe and Naproxen, we have found that there are three hydrogen bonds formed with bond lengths 2.76, 3. 87, and 4.10 Å and five hydrophobic interactions are present with variable bond lengths in the range of 4.17 to 5.21 Å. The binding coordinates are shown in Figure 11.3(a).

In case of Oxybenzone, there are four numbers of hydrogen bonds present which are formed between MIL-100 and Oxybenzone and the bond lengths are in the ranges of 2.86 to 3.57 Å. In addition to this, there are three numbers of hydrophobic interactions are present with varying bond lengths. The schematic representation of binding behavior between MIL-100 and Oxybenzone is shown in Figure 11.3(b). From our study, we have confirmed that the adsorption behavior of MIL-100-Fe with Naproxen and Oxybenzone have shown best binding performance due to the high binding energy [103–105].

From the computational study, we found that MIL-100-Fe MOF have very high binding energy with respect to Diclofenac, Naproxen, and Oxybenzone. The negative value of binding energy of MIL-100-Fe with all the studied adsorbates signifies the exothermic adsorption. The predominant mechanism of adsorption of these pharma products over this MOF is

Table 11.2 Binding and dominant mechanisms of MIL-100 with various organic molecules, Diclofenac, Ibuprofen, Naproxen, and Oxybenzone present in pharmaceutical waste water.

Adsorbates	Binding energy (kcal/mol)	Binding atoms	Nature of the bond	Bond length (Å)
Diclofenac	−5.8	:UNK0:O - :UNK0:Cl	Hydrogen bond	3.261
		:UNK0:O - :UNK0:O	Hydrogen bond	2.903
		:UNK0:H - :UNK0:O	Hydrogen bond	2.796
		:UNK0:H - :UNK0:O	Hydrogen bond	2.024
		:UNK0:O - :UNK0	Hydrogen bond	3.455
		:UNK0:Cl - :UNK0	Hydrophobic contact	4.195
		:UNK0:Cl - :UNK0	Hydrophobic contact	4.963
		:UNK0 - :UNK0	Hydrophobic contact	4.148
Ibuprofen	−5.9	:UNK0:O - :UNK0:O	Hydrogen bond	2.725
		:UNK0 - :UNK0	Hydrophobic contact	3.966
		:UNK0:C - :UNK0	Hydrophobic contact	4.546
		:UNK0 - :UNK0	Hydrophobic contact	3.921

(Continued)

Table 11.2 Binding and dominant mechanisms of MIL-100 with various organic molecules, Diclofenac, Ibuprofen, Naproxen, and Oxybenzone present in pharmaceutical waste water. (*Continued*)

Adsorbates	Binding energy (kcal/mol)	Binding atoms	Nature of the bond	Bond length (Å)
Naproxen	−6.7	:UNK0:O - :UNK0:O	Hydrogen bond	2.763
		:UNK0:O - :UNK0	Hydrogen bond	4.106
		:UNK0:O - :UNK0	Hydrogen bond	3.874
		:UNK0 - :UNK0	Hydrophobic contact	4.146
		:UNK0 - :UNK0	Hydrophobic contact	5.216
		:UNK0 - :UNK0	Hydrophobic contact	4.178
		:UNK0 - :UNK0	Hydrophobic contact	4.997
		:UNK0 - :UNK0	Hydrophobic contact	4.566
Oxybenzone	−6.4	:UNK0:O - :UNK0:O	Hydrogen bond	2.867
		:UNK0:O - :UNK0:O	Hydrogen bond	2.723
		:UNK0:O - :UNK0	Hydrogen bond	3.575
		:UNK0:O - :UNK0	Hydrogen bond	3.565
		:UNK0 - :UNK0	Hydrophobic contact	4.962
		:UNK0 - :UNK0	Hydrophobic contact	4.408
		:UNK0 - :UNK0	Hydrophobic contact	4.251

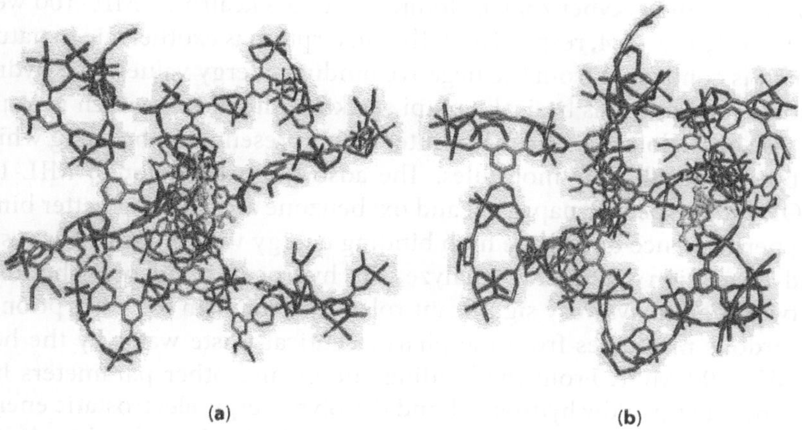

Figure 11.3 Schematic representation of binding coordinate between MIL-100 with (a) Naproxen and (b) Oxybenzone. The yellow colored ball and stick model represented the adsorbents.

mainly due to hydrophobic and hydrogen bonding interactions. The binding energy and other parameters like internal energy, Vdw hydrogen bond dissolve energy, electrostatic energy specified that, this studied MOF can be a better adsorbent for the PPCPs present in water medium. Hydrogen bonding is the active site of interaction present in between MOF and the pharmaceutical products. In this study H-bonding plays an important role in case of adsorption. Hydrogen bonding is maintaining the interactions between MOF and organic molecules during adsorption. From our result and discussion section we analyze, the hydrogen bond or H-bonding networks are plays very significant role in mechanism for adsorption of hazardous molecules from the pharmaceutical waste water by the help of MIL-100 MOF. The detail of binding energy by docking analysis of MOF with organic molecules present in pharmaceutical waste water is given in the Table 11.1. Visualization and image preparation were performed by using Discovery Studio.

11.6 Conclusion

In conclusion, we have studied the interaction of PPCPs with MIL-100-Fe by adopting *in silico* process. We have found that MIL-100-Fe have very high binding energy with Diclofenac, Naproxen, and Oxybenzone. The binding energy between MIL-100-Fe and Diclofenac, Ibuprofen,

Naproxen, and Oxybenzone is found to be -5.8 kcal/mol, MIL-100 were −5.9, −6.7, and −6.4, respectively. This adsorption is exothermic in nature which is confirmed from the negative binding energy values. The hydrophobic interaction is basically pi–pi stacking interaction which is found in case of Diclofenac. The major interaction present is H-bonding which is present in all four molecules. The adsorption behavior of MIL-100 MOF with respect to naproxen and oxybenzone are showing better binding performance due to the high binding energy values. From our result and discussion section we analyze, the hydrogen bond or H-bonding networks are plays very significant role in mechanism for adsorption of hazardous molecules from the pharmaceutical waste water by the help of MIL-100 MOF. From the binding energy and other parameters like internal energy, Vdw hydrogen bond dissolve energy, electrostatic energy specified that MIL-100-Fe can be a very good adsorbent for the adsorptive removal of PPCPs from aqueous medium.

References

1. Liu, J.L. and Wong, M.H., Pharmaceuticals and personal care products (PPCPs): A review on environmental contamination in China. *Environ. Int.*, 59, 208–224, 2013.
2. Sharma, B.M., Bečanová, J., Scheringer, M., Sharma, A., Bharat, G.K., Whitehead, P.G., Klánová, J., Nizzetto, L., Health and ecological risk assessment of emerging contaminants (pharmaceuticals, personal care products, and artificial sweeteners) in surface and groundwater (drinking water) in the Ganges River Basin, India. *Sci. Total Environ.*, 646, 1459–1467, 2019.
3. Papageorgiou, M., Zioris, I., Danis, T., Bikiaris, D., Lambropoulou, D., Comprehensive investigation of a wide range of pharmaceuticals and personal care products in urban and hospital wastewaters in Greece. *Sci. Total Environ.*, 694, 133565, 2019.
4. Yang, Y.Y., Zhao, J.L., Liu, Y.S., Liu, W.R., Zhang, Q.Q., Yao, L., Hu, L.X., Zhang, J.N., Jiang, Y.X., Ying, G.G., Pharmaceuticals and personal care products (PPCPs) and artificial sweeteners (ASs) in surface and ground waters and their application as indication of wastewater contamination. *Sci. Total Environ.*, 616, 816–823, 2018.
5. Jjemba, P.K., *Pharma-ecology: The occurrence and fate of pharmaceuticals and personal care products in the environment*, John Wiley & Sons, 2018.
6. Richardson, B.J., Lam, P.K., Martin, M., Emerging chemicals of concern: Pharmaceuticals and personal care products (PPCPs) in Asia, with particular reference to Southern China. *Mar. Pollut. Bull.*, 50, 9, 913–920, 2005.

7. Ebele, A.J., Abdallah, M.A.E., Harrad, S., Pharmaceuticals and personal care products (PPCPs) in the freshwater aquatic environment. *Emerg. Contam.*, 3, 1, 1–16, 2017.
8. Kallenborn, R., Brorström-Lundén, E., Reiersen, L.O., Wilson, S., Pharmaceuticals and personal care products (PPCPs) in Arctic environments: Indicator contaminants for assessing local and remote anthropogenic sources in a pristine ecosystem in change. *Environ. Sci. Pollut. Res.*, 25, 33, 33001–33013, 2018.
9. Reiss, R., Lewis, G., Griffin, J., An ecological risk assessment for triclosan in the terrestrial environment. *Environ. Toxicol. Chem.: Int. J.*, 28, 7, 1546–1556, 2009.
10. Harada, A., Komori, K., Nakada, N., Kitamura, K., Suzuki, Y., Biological effects of PPCPs on aquatic lives and evaluation of river waters affected by different wastewater treatment levels. *Water Sci. Technol.*, 58, 8, 1541–1546, 2008.
11. Ashfaq, M., Li, Y., Rehman, M.S.U., Zubair, M., Mustafa, G., Nazar, M.F., Yu, C.P., Sun, Q., Occurrence, spatial variation and risk assessment of pharmaceuticals and personal care products in urban wastewater, canal surface water, and their sediments: A case study of Lahore, Pakistan. *Sci. Total Environ.*, 2019.
12. Wang, J. and Wang, S., Removal of pharmaceuticals and personal care products (PPCPs) from wastewater: A review. *J. Environ. Manage.*, 182, 620–640, 2016.
13. Yang, Y., Ok, Y.S., Kim, K.H., Kwon, E.E., Tsang, Y.F., Occurrences and removal of pharmaceuticals and personal care products (PPCPs) in drinking water and water/sewage treatment plants: A review. *Sci. Total Environ.*, 596, 303–320, 2017.
14. Carballa, M., Omil, F., Alder, A.C., Lema, J.M., Comparison between the conventional anaerobic digestion of sewage sludge and its combination with a chemical or thermal pre-treatment concerning the removal of pharmaceuticals and personal care products. *Water Sci. Technol.*, 53, 8, 109–117, 2006.
15. Sui, Q., Huang, J., Deng, S., Chen, W., Yu, G., Seasonal variation in the occurrence and removal of pharmaceuticals and personal care products in different biological wastewater treatment processes. *Environ. Sci. Technol.*, 45, 8, 3341–3348, 2011.
16. He, Y.J., Chen, W., Zheng, X.Y., Wang, X.N., Huang, X., Fate and removal of typical pharmaceuticals and personal care products by three different treatment processes. *Sci. Total Environ.*, 447, 248–254, 2013.
17. Grassi, M., Rizzo, L., Farina, A., Endocrine disruptors compounds, pharmaceuticals and personal care products in urban wastewater: Implications for agricultural reuse and their removal by adsorption process. *Environ. Sci. Pollut. Res.*, 20, 6, 3616–3628, 2013.
18. Oulton, R.L., Kohn, T., Cwiertny, D.M., Pharmaceuticals and personal care products in effluent matrices: A survey of transformation and removal

during wastewater treatment and implications for wastewater management. *J. Environ. Monit.*, 12, 11, 1956–1978, 2010.
19. Paucar, N.E., Kim, I., Tanaka, H., Sato, C., Ozone treatment process for the removal of pharmaceuticals and personal care products in wastewater. *Ozone: Sci. Eng.*, 41, 1, 3–16, 2019.
20. Rajapaksha, A.U., Premarathna, K.S.D., Gunarathne, V., Ahmed, A., Vithanage, M., Sorptive removal of pharmaceutical and personal care products from water and wastewater, in: *Pharmaceuticals and Personal Care Products: Waste Management and Treatment Technology*, pp. 213–238, Butterworth-Heinemann, 2019.
21. Wang, Y. and Huang, H., Carbon nanotube composite membranes for microfiltration of pharmaceuticals and personal care products, in: *Advanced Nanomaterials for Membrane Synthesis and its Applications*, pp. 183–202, Elsevier, 2019.
22. Phonsiri, V., Choi, S., Nguyen, C., Tsai, Y.L., Coss, R., Kurwadkar, S., Monitoring occurrence and removal of selected pharmaceuticals in two different wastewater treatment plants. *SN Appl. Sci.*, 1, 7, 798, 2019.
23. de Andrade, J.R., Oliveira, M.F., da Silva, M.G.C., and Vieira M.G.A., Adsorption of pharmaceuticals from water and wastewater using nonconventional low-cost materials: A review. *Ind. Eng. Chem. Res.*, 57, 9, 3103–3127, 2018.
24. Hasan, Z., Khan, N.A., Jhung, S.H., Adsorptive removal of diclofenac sodium from water with Zr-based metal–organic frameworks. *Chem. Eng. J.*, 284, 1406–1413, 2016.
25. Bhadra, B.N., Ahmed, I., Kim, S., Jhung, S.H., Adsorptive removal of ibuprofen and diclofenac from water using metal–organic framework-derived porous carbon. *Chem. Eng. J.*, 314, 50–58, 2017.
26. Perry Iv, J.J., Perman, J.A., Zaworotko, M.J., Design and synthesis of metal–organic frameworks using metal–organic polyhedra as supermolecular building blocks. *Chem. Soc. Rev.*, 38, 5, 1400–1417, 2009.
27. Kurmoo, M., Magnetic metal–organic frameworks. *Chem. Soc. Rev.*, 38, 5, 1353–1379, 2009.
28. Chaudhari, A.K., Nagarkar, S.S., Joarder, B., Ghosh, S.K., A continuous π-stacked starfish array of two-dimensional luminescent MOF for detection of nitro explosives. *Cryst. Growth Des.*, 13, 8, 3716–3721, 2013.
29. Wang, B., Xie, L.H., Wang, X., Liu, X.M., Li, J., Li, J.R., Applications of metal–organic frameworks for green energy and environment: New advances in adsorptive gas separation, storage and removal. *Green Energy Environ.*, 3, 3, 191–228, 2018.
30. Kundu, T., Shah, B.B., Bolinois, L., Zhao, D., Functionalization-Induced Breathing Control in Metal–Organic Frameworks for Methane Storage with High Deliverable Capacity. *Chem. Mater.*, 31, 8, 2842–2847, 2019.

31. Yoon, M., Srirambalaji, R., Kim, K., Homochiral metal–organic frameworks for asymmetric heterogeneous catalysis. *Chem. Rev.*, 112, 2, 1196–1231, 2011.
32. Wang, L., Zheng, M., Xie, Z., Nanoscale metal–organic frameworks for drug delivery: A conventional platform with new promise. *J. Mater. Chem. B*, 6, 707–717, 2018.
33. Chen, Z., Hanna, S.L., Redfern, L.R., Alezi, D., Islamoglu, T., Farha, O.K., Reticular chemistry in the rational synthesis of functional zirconium cluster-based MOFs. *Coord. Chem. Rev.*, 386, 32–49, 2019.
34. Tanabe, K.K. and Cohen, S.M., Postsynthetic modification of metal–organic frameworks—A progress report. *Chem. Soc. Rev.*, 40, 2, 498–519, 2011.
35. Kitagawa, S., Kitaura, R., Noro, S.I., Functional porous coordination polymers. *Angew. Chem. Int. Ed.*, 43, 18, 2334–2375, 2004.
36. Ma, S. and Zhou, H.C., A metal–organic framework with entatic metal centers exhibiting high gas adsorption affinity. *J. Am. Chem. Soc.*, 128, 36, 11734–11735, 2006.
37. Cychosz, K.A., Wong-Foy, A.G., Matzger, A.J., Liquid phase adsorption by microporous coordination polymers: Removal of organosulfur compounds. *J. Am. Chem. Soc.*, 130, 22, 6938–6939, 2008.
38. Faria, P.C.C., Orfao, J.J.M., Pereira, M.F.R., Adsorption of anionic and cationic dyes on activated carbons with different surface chemistries. *Water Res.*, 38, 2043–2052, 2004.
39. Rafatullah, M., Sulaiman, O., Hashim, R., Ahmad, A., Adsorption of methylene blue on low-cost adsorbents: A review. *J. Hazard. Mater.*, 177, 70–80, 2010.
40. Alpat, S.K., Ozbayrak, O., Alpat, S., Akcay, H., The adsorption kinetics and removal of cationic dye, Toluidine Blue O, from aqueous solution with Turkish zeolite. *J. Hazard. Mater.*, 151, 213–220, 2008.
41. Karadag, D., Akgul, E., Tok, S., Erturk, F., Kaya, M.A., Turan, M., Basic and reactive dye removal using natural and modified zeolites. *J. Chem. Eng. Data*, 52, 2436–2441, 2007.
42. Cheetham, A.K., Férey, G., Loiseau, T., Open-framework inorganic materials. *Angew. Chem. Int. Ed.*, 38, 22, 3268–3292, 1999.
43. Bučar, D.K., Papaefstathiou, G.S., Hamilton, T.D., Chu, Q.L., Georgiev, I.G., MacGillivray, L.R., Template-Controlled Reactivity in the Organic Solid State by Principles of Coordination-Driven Self-Assembly. *Eur. J. Inorg. Chem.*, 2007, 29, 4559–4568, 2007.
44. Kuppler, R.J., Timmons, D.J., Fang, Q.R., Li, J.R., Makal, T.A., … Young, M.D., Zhou, H.C., Potential applications of metal–organic frameworks. *Coord. Chem. Rev.*, 253, 23–24, 3042–3066, 2009.
45. Horike, S., Shimomura, S., Kitagawa, S., Soft porous crystals. *Nat. Chem.*, 1, 9, 695, 2009.

46. Sanda, S., Parshamoni, S., Konar, S., Third-generation breathing metal–organic framework with selective, stepwise, reversible, and hysteretic adsorption properties. *Inorg. Chem.*, 52, 22, 12866–12868, 2013.
47. Liang, Y., Yuan, W.G., Zhang, S.F., He, Z., Xue, J., Zhang, X., ... Qin, D.B., Hydrothermal synthesis and structural characterization of metal–organic frameworks based on new tetradentate ligands. *Dalton Trans.*, 45, 4, 1382–1390, 2016.
48. Sun, Y. and Zhou, H.C., Recent progress in the synthesis of metal–organic frameworks. *Sci. Technol. Adv. Mater.*, 16, 5, 054202, 2015.
49. Clausen, H.F., Poulsen, R.D., Bond, A.D., Chevallier, M.A.S., Iversen, B.B., Solvothermal synthesis of new metal organic framework structures in the zinc–terephthalic acid–dimethyl formamide system. *J. Solid State Chem.*, 178, 11, 3342–3351, 2005.
50. Chen, D., Zhao, J., Zhang, P., Dai, S., Mechanochemical synthesis of metal–organic frameworks. *Polyhedron*, 162, 59–64, 2019.
51. Furukawa, H., Cordova, K.E., O'Keeffe, M., Yaghi, O.M., The chemistry and applications of metal–organic frameworks. *Science*, 341, 6149, 1230444, 2013.
52. O'Keeffe, M. and Yaghi, O.M., Deconstructing the crystal structures of metal–organic frameworks and related materials into their underlying nets. *Chem. Rev.*, 112, 2, 675–702, 2011.
53. Jung, B.K., Hasan, Z., Jhung, S.H., Adsorptive removal of 2, 4-dichlorophenoxyacetic acid (2, 4-D) from water with a metal–organic framework. *Chem. Eng. J.*, 234, 99–105, 2013.
54. Jun, J.W., Tong, M., Jung, B.K., Hasan, Z., Zhong, C., Jhung, S.H., Effect of central metal ions of analogous metal–organic frameworks on adsorption of organoarsenic compounds from water: Plausible mechanism of adsorption and water purification. *Chem. Eur. J.*, 21, 1, 347–354, 2015.
55. De Smedt, C., Spanoghe, P., Biswas, S., Leus, K., Van Der Voort, P., Comparison of different solid adsorbents for the removal of mobile pesticides from aqueous solutions. *Adsorption*, 21, 3, 243–254, 2015.
56. Park, E.Y., Hasan, Z., Khan, N.A., Jhung, S.H., Adsorptive removal of bisphenol-A from water with a metal–organic framework, a porous chromium-benzenedicarboxylate. *J. Nanosci. Nanotechnol.*, 13, 4, 2789–2794, 2013.
57. Qin, F.X., Jia, S.Y., Liu, Y., Li, H.Y., Wu, S.H., Adsorptive removal of bisphenol A from aqueous solution using metal–organic frameworks. *Desalination Water Treat.*, 54, 1, 93–102, 2015.
58. Song, J.Y., and Sung H.J., Adsorption of pharmaceuticals and personal care products over metal-organic frameworks functionalized with hydroxyl groups: Quantitative analyses of H-bonding in adsorption. *Chem. Eng. J.*, 322, 366–374, 2017.
59. Zhang, X., Gao, Y., Liu, H., Liu, Z., Fabrication of porous metal–organic frameworks via a mixed-ligand strategy for highly selective and efficient dye adsorption in aqueous solution. *CrystEngComm*, 17, 31, 6037–6043, 2015.

60. Li, Z., Wu, Y.N., Li, J., Zhang, Y., Zou, X., Li, F., The Metal–Organic Framework MIL-53 (Al) Constructed from Multiple Metal Sources: Alumina, Aluminum Hydroxide, and Boehmite. *Chem. Eng. J.*, 21, 18, 6913–6920, 2015.
61. Tella, A.C., Owalude, S.O., Ojekanmi, C.A., Oluwafemi, O.S., Synthesis of copper–isonicotinate metal–organic frameworks simply by mixing solid reactants and investigation of their adsorptive properties for the removal of the fluorescein dye. *New J. Chem.*, 38, 9, 4494–4500, 2014.
62. Cychosz, K.A. and Matzger, A.J., Water stability of microporous coordination polymers and the adsorption of pharmaceuticals from water. *Langmuir*, 26, 22, 17198–17202, 2010.
63. Zhu, X., Li, B., Yang, J., Li, Y., Zhao, W., Shi, J., Gu, J., Effective adsorption and enhanced removal of organophosphorus pesticides from aqueous solution by Zr-based MOFs of UiO-67. *ACS Appl. Mater. Interfaces*, 7, 1, 223–231, 2014.
64. Lin, K.Y.A. and Chang, H.A., Efficient adsorptive removal of humic acid from water using zeolitic imidazole framework-8 (ZIF-8). *Water Air Soil Pollut.*, 226, 2, 10, 2015.
65. Jia, Y., Jin, Q., Li, Y., Sun, Y., Huo, J., Zhao, X., Investigation of the adsorption behaviour of different types of dyes on MIL-100 (Fe) and their removal from natural water. *Anal. Methods*, 7, 4, 1463–1470, 2015.
66. Haque, E., Lo, V., Minett, A.I., Harris, A.T., Church, T.L., Dichotomous adsorption behaviour of dyes on an amino-functionalised metal–organic framework, amino-MIL-101 (Al). *J. Mater. Chem. A*, 2, 1, 193–203, 2014.
67. Yi, F.Y., Li, J.P., Wu, D., Sun, Z.M., A series of multifunctional metal–organic frameworks showing excellent luminescent sensing, sensitization, and adsorbent abilities. *Chem. Eng. J.*, 21, 32, 11475–11482, 2015.
68. Abbasi, A.R., Azadbakht, A., Morsali, A., Safarifard, V., Synthesis and characterization of TMU-16-NH 2 metal–organic framework nanostructure upon silk fiber: Study of structure effect on morphine and methyl orange adsorption affinity. *Fibers Polym.*, 16, 5, 1193–1200, 2015.
69. Seo, Y.S., Khan, N.A., Jhung, S.H., Adsorptive removal of methylchlorophenoxypropionic acid from water with a metal–organic framework. *Chem. Eng. J.*, 270, 22–27, 2015.
70. Guo, X., Kang, C., Huang, H., Chang, Y., Zhong, C., Exploration of functional MOFs for efficient removal of fluoroquinolone antibiotics from water. *Microporous Mesoporous Mater.*, 286, 84–91, 2019.
71. Du, H., Wang, C., Li, Y., Zhang, W., Xu, M., Li, S., … Hou, H., A supramolecular metal–organic framework derived from bismuth iodide and 4, 4′-bipyridinium derivative: Synthesis, structure and efficient adsorption of dyes. *Microporous Mesoporous Mater.*, 214, 136–142, 2015.
72. Hasan, Z., Choi, E.J., Jhung, S.H., Adsorption of naproxen and clofibric acid over a metal–organic framework MIL-101 functionalized with acidic and basic groups. *Chem. Eng. J.*, 219, 537–544, 2013.

73. Hasan, Z., Jeon, J., Jhung, S.H., Adsorptive removal of naproxen and clofibric acid from water using metal–organic frameworks. *J. Hazard. Mater.*, 209, 151–157, 2012.
74. Xie, L., Liu, D., Huang, H., Yang, Q., Zhong, C., Efficient capture of nitrobenzene from waste water using metal–organic frameworks. *Chem. Eng. J.*, 246, 142–149, 2014.
75. Patil, D.V., Rallapalli, P.B.S., Dangi, G.P., Tayade, R.J., Somani, R.S., Bajaj, H.C., MIL-53 (Al): An efficient adsorbent for the removal of nitrobenzene from aqueous solutions. *Ind. Eng. Chem. Res.*, 50, 18, 10516–10524, 2011.
76. Liu, B., Yang, F., Zou, Y., Peng, Y., Adsorption of phenol and p-nitrophenol from aqueous solutions on metal–organic frameworks: Effect of hydrogen bonding. *J. Chem. Eng. Data*, 59, 5, 1476–1482, 2014.
77. Han, T., Xiao, Y., Tong, M., Huang, H., Liu, D., Wang, L., Zhong, C., Synthesis of CNT@ MIL-68 (Al) composites with improved adsorption capacity for phenol in aqueous solution. *Chem. Eng. J.*, 275, 134–141, 2015.
78. Khan, N.A., Jung, B.K., Hasan, Z., Jhung, S.H., Adsorption and removal of phthalic acid and diethyl phthalate from water with zeolitic imidazolate and metal–organic frameworks. *J. Hazard. Mater.*, 282, 194–200, 2015.
79. Choi, E.-Y., Park, K., Yang, C.-M., Kim, H., Son, J.-H., Lee, S.W., Lee, Y.H., Min, D., and Kwon Y.-U., Benzene - templated hydrothermal synthesis of metal–organic frameworks with selective sorption properties. *Chem. Eur. J.*, 10, 21, 5535–5540, 2004.
80. Wang, X., Li, Q., Yang, N., Yang, Y., He, F., Chu, J., … Xiong, S., Hydrothermal synthesis of NiCo-based bimetal-organic frameworks as electrode materials for supercapacitors. *J. Solid State Chem.*, 270, 370–378, 2019.
81. Peh, S.B., Cheng, Y., Zhang, J., Wang, Y., Chan, G.H., Wang, J., Zhao, D., Cluster nuclearity control and modulated hydrothermal synthesis of functionalized Zr 12 metal–organic frameworks. *Dalton Trans.*, 48, 21, 7069–7073, 2019.
82. Esrafili, L., Tehrani, A.A., Morsali, A., Carlucci, L., Proserpio, D.M., Ultrasound and solvothermal synthesis of a new urea-based metal–organic framework as a precursor for fabrication of cadmium (II) oxide nanostructures. *Inorganica Chim. Acta*, 484, 386–393, 2019.
83. Wang, X., Yang, N., Li, Q., He, F., Yang, Y., Wu, B., … Xiong, S., Solvothermal synthesis of flower-string-like NiCo-MOF/MWCNT composites as a high-performance supercapacitor electrode material. *J. Solid State Chem.*, 277, 575–586, 2019.
84. Sanada, T., Tominaka, S., Kojima, K., Cheetham, A.K., Violet Luminescence from Zinc-Based Metal–Organic Frameworks Prepared by Solvothermal Synthesis. *Bull. Chem. Soc. Jpn.*, 92, 2, 427–434, 2019.
85. Shi, J., Zhang, J., Tan, D., Cheng, X., Tan, X., Zhang, B., … Xiang, J., Rapid, Room-Temperature and Template-Free Synthesis of Metal-Organic Framework Nanowires in Alcohol. *ChemCatChem*, 11, 8, 2058–2062, 2019.

86. Zhang, H., Gao, X.W., Wang, L., Zhao, X., Li, Q.Y., Wang, X.J., Microwave-assisted synthesis of urea-containing zirconium metal–organic frameworks for heterogeneous catalysis of Henry reactions. *CrystEngComm*, 21, 9, 1358–1362, 2019.
87. Stock, N. and Biswas, S., Synthesis of metal–organic frameworks (MOFs): Routes to various MOF topologies, morphologies, and composites. *Chem. Rev.*, 112, 2, 933–969, 2011.
88. Samal, M., Panda, J., Biswal, B.P., Sahu, R., Kitchen grinder: A tool for the synthesis of metal–organic frameworks towards size selective dye adsorption. *CrystEngComm*, 20, 18, 2486–2490, 2018.
89. de Lima Neto, O.J., de Oliveira Frós, A.C., Barros, B.S., de Farias Monteiro, A.F., Kulesza, J., Rapid and efficient electrochemical synthesis of a zinc-based nano-MOF for Ibuprofen adsorption. *New J. Chem.*, 43, 14, 5518–5524, 2019.
90. Yin, Z., Wan, S., Yang, J., Kurmoo, M., Zeng, M.H., Recent advances in post-synthetic modification of metal–organic frameworks: New types and tandem reactions. *Coord. Chem. Rev.*, 378, 500–512, 2019.
91. Pattanayak, S.K. and Chowdhuri, S., Effect of water on solvation structure and dynamics of ions in the peptide bond environment: Importance of hydrogen bonding and dynamics of the solvents. *J. Phys. Chem. B*, 115, 45, 13241–13252, 2011.
92. Pattanayak, S.K., Prashar, N., Chowdhuri, S., Effect of temperature and pressure on the structure, dynamics, and hydrogen bond properties of liquid N-methylacetamide: A molecular dynamics study. *J. Chem. Phys.*, 134, 15, 154506, 2011.
93. Pattanayak, S.K. and Chowdhuri, S., A molecular dynamics simulations study on the behavior of liquid N-methylacetamide in presence of NaCl: Structure, dynamics and H-bond properties. *J. Mol. Liq.*, 172, 102–109, 2012.
94. Pattanayak, S.K. and Chowdhuri, S., Size dependence of solvation structure and dynamics of ions in liquid n-methylacetamide: A molecular dynamics simulation study. *J. Theor. Comput. Chem.*, 11, 02, 361–377, 2012.
95. Pattanayak, S.K. and Chowdhuri, S., Effects of concentrated NaCl and KCl solutions on the behaviour of aqueous peptide bond environment: Single-particle dynamics and H-bond structural relaxation. *Mol. Phys.*, 111, 3297–3310, 2013.
96. Pattanayak, S.K. and Chowdhuri, S., Effects of methanol on the hydrogen bonding structure and dynamics in aqueous N-methylacetamide solution. *J. Mol. Liq.*, 194, 141–148, 2014.
97. Bissoyi, A., Singh, A.K., Pattanayak, S.K., Bit, A., Sinha, S.K., Patel, A., Patra, P.K., Understanding the molecular mechanism of improved proliferation and osteogenic potential of human mesenchymal stem cells grown on a polyelectrolyte complex derived from non-mulberry silk fibroin and chitosan. *Biomed. Mater.*, 13, 1, 015011, 2017.

98. Bissoyi, A., Pattanayak, S.K., Bit, A., Patel, A., Singh, A.K., Behera, S.S., Satpathy, D., Alphavirus nonstructural proteases and their inhibitors, in: *Viral proteases and their inhibitors*, pp. 77–104, Academic Press, 2017.
99. Hiremath, S.M., Suvitha, A., Patil, N.R., Hiremath, C.S., Khemalapure, S.S., Pattanayak, S.K., Armaković, S., Synthesis of 5-(5-methyl-benzofuran-3-ylmethyl)-3H-[1, 3, 4] oxadiazole-2-thione and investigation of its spectroscopic, reactivity, optoelectronic and drug likeness properties by combined computational and experimental approach. *Spectrochim. Acta Part A: Mol. Biomol. Spectrosc.*, 205, 95–110, 2018.
100. Meng, X.Y., Zhang, H.X., Mezei, M., Cui, M., Molecular docking: A powerful approach for structure-based drug discovery. *Curr. Comput.-Aided Drug Des.*, 7, 2, 146–157, 2011.
101. D.S. Visualizer, Accelrys software inc. Discovery Studio Visualizer 2, 2005.
102. Panda, J., Sahoo, J.K., Panda, P.K., Sahu, S.N., Samal, M., Pattanayak, S.K., Sahu, R., Adsorptive behavior of zeolitic imidazolate framework-8 towards anionic dye in aqueous media: Combined experimental and molecular docking study. *J. Mol. Liq.*, 278, 536–545, 2019.
103. Gao, G., Xing, Y., Liu, T., Wang, J., Hou, X., UiO-66 (Zr) as sorbent for porous membrane protected micro-solid-phase extraction androgens and progestogens in environmental water samples coupled with LC-MS/MS analysis: The application of experimental and molecular simulation method. *Microchem. J.*, 146, 126–133, 2019.
104. Yang, Q., Wiersum, A.D., Jobic, H., Guillerm, V., Serre, C., Llewellyn, P.L., Maurin, G., Understanding the thermodynamic and kinetic behavior of the CO2/CH4 gas mixture within the porous zirconium terephthalate UiO-66 (Zr): A joint experimental and modeling approach. *J. Phys. Chem. C*, 115, 28, 13768–13774, 2011.
105. Rodrigues, M.O., de Paula, M.V., Wanderley, K.A., Vasconcelos, I.B., Alves, S., Jr., Soares, T.A., Metal organic frameworks for drug delivery and environmental remediation: A molecular docking approach. *Int. J. Quantum Chem.*, 112, 20, 3346–3355, 2012.

12

Metal–Organic Frameworks and Their Hybrid Composites for Adsorption of Volatile Organic Compounds

Shella Permatasari Santoso[1,2*], Artik Elisa Angkawijaya[3], Vania Bundjaja[1], Felycia Edi Soetaredjo[1,2] and Suryadi Ismadji[1,2]

[1]*Chemical Engineering Department, Widya Mandala Surabaya Catholic University, Surabaya, East Java, Indonesia*
[2]*Department of Chemical Engineering, National Taiwan University of Science and Technology, Taiwan, Taipei*
[3]*Graduate Institute of Science and Technology, National Taiwan University of Science and Technology, Taiwan, Taipei*

Abstract

Environmental pollution caused by anthropogenic emissions is a growing concern throughout the world; volatile organic compounds (VOCs) are the main constituents that make up the emissions. In handling VOCs, adsorption is considered the most efficient method to date. Engineers have been intensively studied the synthesis and usage of metal–organic frameworks (MOF) as versatile adsorbents for many types of VOC. Improving adsorption efficiency and performance is a routine agenda in the development of the MOF; prior to achieving an efficient MOFs as VOCs adsorbents, insight to MOF-key features such as structure, pore, and the functional group is very crucial. To this end, several topics related to adsorption of VOCs by MOF is discussed; specifically, the adsorption performance of some MOFs against VOCs, the effect of some key-features of MOF to the adsorption performance, the development of MOF composite for improvement of adsorption performance, analytical method for modeling the adsorption, and factors influencing the adsorption performance.

Keywords: Metal–organic framework, MOF, volatile organic compound, adsorption, porous adsorbent, dynamic sorption, breakthrough curve

**Corresponding author*: shella_p5@yahoo.com

Inamuddin, Rajender Boddula, Mohd Imran Ahamed and Abdullah M. Asiri (eds.) Applications of Metal–Organic Frameworks and Their Derived Materials, (313–356) © 2020 Scrivener Publishing LLC

12.1 Introduction

The deployment of air pollution caused by toxic and hazardous substances worsens as a result of the rapid increase in anthropogenic emissions. Chemicals combustion, evaporation, reaction, and purification are the predominant sources of anthropogenic emissions and are growing concerns of worldwide. A substantial amount of hazardous gases such as hydrocarbons, volatile organic compounds, sulfuric compounds, nitrogen compounds, and greenhouse compounds are produced from those emissions. The volatile organic compounds (VOCs) are considered as major air pollutants since they always produced in every activity involving chemicals [1, 2]. As reported by the United States Environmental Protection Agency (US EPA), 50% of VOCs is contained in the emission of fossil fuel combustion [3]. Moreover, given the continuous development of industry throughout the world, it is prominent that VOCs emissions will continue to increase in the future.

VOCs are those organic compounds that quickly evaporate at standard ambient conditions. Acetone, benzene, chloroform, ethanol, formaldehyde, isopropanol, naphthalene, propane, and toluene are several types of VOCs. VOCs are dangerous health threats that are often underestimated since they are commonly not physically visible. The entry of VOCs into the respiratory tract leads to various health problems ranging from mild to acute, such as irritation, headache, nausea, tissue or organ damage, and cancer. Unwittingly, VOCs pollutions may come from daily items and/or various surrounding materials; i.e., natural gas, paint, adhesive, candle, room freshener, ink, household cleaner, cigarette, varnish, disinfectant, rubbing alcohol, pesticide, cosmetic, etc. VOCs have a peculiar strong odor that signifies high concentration, at low concentration the odor may remain unnoticeable; nevertheless, the damaging effects may potentially fatal. The presence of VOCs as an airborne contaminant can aggravate global warming since they cause stratospheric depletion, greenhouse effect, and smog accumulation. Thus VOCs considered as one of the root of climate problems [1–4].

Numerous VOCs removal techniques have been introduced, including photo-catalysis, filtration, oxidation, biological degradation, condensation, and adsorption [5, 6]. Adsorption by using natural or synthetic adsorbent is the most frequently performed since it offers various advantages such as flexible (can be used in removal of specific or non-specific VOCs), low cost, simple, and controllable. The development of engineered-adsorbents has been substantially examined with priority to acquire adsorbent that effectively eliminates VOCs. Out of various engineered-adsorbents, metal–organic frameworks (MOFs) are being intensively developed as adsorbent of VOCs; MOFs offer a large specific surface area that relates to the adsorption ability. MOFs can be easily constructed, designed, and fabricated by coordinating organic compound (ligand) and metal at the particular condition. Some MOFs possess semiconducting nature that makes them capable of self-regeneration by photo-degrading the VOCs. Furthermore, recyclability is also one of the superiority of MOFs. Provided with those superior properties, the emergence of utilization and development of MOFs to eliminate VOCs is obviously visible.

12.2 VOCs and Their Potential Hazards

VOC is an epithet for carbon-containing compounds, which possess distinct characteristics such as high vapor pressure, low boiling point, evaporate freely, and reactive at a standard temperature and pressure. There are certain usual extents, related to the characteristics of VOCs, between each country that causes a slight difference in the definition of VOCs. Several agencies issuing the standard of VOCs are National Institute for Occupational Safety and Health (NIOSH), Environmental Protection Agency (EPA), Air Quality Management (AQM), Environmental Protection Department (EPD), Department of Occupational Safety and Health (DOSH), etc. In general, the standards over the countries gives the following specific characteristics of VOCs; *viz.* boiling point of 50°C to 250°C and vapor pressure equal to 0.01 kPa (9.87×10^{-5} atm) or higher (measured at 293.15 K and 101.325 kPa). VOCs sources increase every decade due to increasingly diverse type of emissions that may come both from natural sources and anthropogenic sources. Undoubtedly anthropogenic sources, which mainly due to the rapid development of industrial sections in various parts of the world, contribute the highest VOCs emissions. Daily activities such as cooking, smoking, driving, printing, and burning, are also pay a role for this VOCs emissions [3, 7].

VOCs come in a variety of structures ranging from aromatic, alcohol, aldehyde, alkene, ketone, polycyclic, and halogenated forms, as listed in Table 12.1. The concentration index of VOCs that is likely to cause dangerous threat is proposed by NIOSH as immediately dangerous to life and health (IDLH). The index can be expressed as recommended exposure limit (REL) from adverse health effects over a working lifetime, and 10% of the lower explosive limit (LEL) for safety considerations related to the potential for explosion hazards. Health and/or life threats caused by VOCs vary depending on the length of exposure time, number of toxicants, and

Table 12.1 Properties of some VOCs and their IDLH levels as proposed by NIOSH [8].

Compound	Formula	b.p. (K)	p (kPa)	IDLH	
				REL (mg/m^3)	LEL (ppm)
Alcohols					
Ethyl alcohol	C_2H_5OH	351.39	7.91	1,900[b]	3,300
Methyl alcohol	CH_3OH	337.85	16.93	260[b]	6,000
Isopropyl alcohol	$CH_3CHOHCH_3$	355.45	6.05	980[b]	2,000
Phenol	C_6H_5OH	454.90	0.05	19[a]	1,800
Aldehydes					
Acetaldehyde	CH_3CHO	293.95	120.26	n.e.	4,000
Formaldehyde	H_2CO	253,65	518.62	n.e.	7,000
Benzaldehyde	C_6H_5CHO	451.85	0.17	NA	1,400
Alkanes					
n-Hexane	C_6H_{14}	341.88	20.40	180[a]	1,100
n-Heptane	C_7H_{16}	371.53	6.13	350[a]	1,050
Octane	C_8H_{18}	398.77	1.88	350[a]	1,000

(Continued)

Table 12.1 Properties of some VOCs and their IDLH levels as proposed by NIOSH [8]. (*Continued*)

Compound	Formula	b.p. (K)	*p* (kPa)	IDLH	
				REL (mg/m^3)	LEL (ppm)
Aromatics					
Aniline	$C_6H_5NH_2$	457.25	0.09	n.e.	1,300
Benzene	C_6H_6	353.23	12.64	3.2[a]	1,200
Toluene	C_7H_8	383.75	3.79	375[a]	1,100
Naphthalene	$C_{10}H_8$	491.05	0.01	50[a]	900
Chlorobenzene	C_6H_5Cl	404.75	1.60	NA	1,300
Carboxylics					
Acetic acid	CH_3COOH	391.05	2.09	25[a]	4,000
Formic acid	HCOOH	374.15	5.68	9[a]	18,000
Esters					
Ethyl acetate	$CH_3COOC_2H_5$	350.25	12.43	1,400[a]	2,000
Diethyl phthalate	$C_6H_4(COOC_2H_5)_2$	568.15	0.0003	5[a]	700*
Halogenated					
Benzyl chloride	$C_6H_5CH_2Cl$	447.15	0.16	5[c]	1,100

(*Continued*)

Table 12.1 Properties of some VOCs and their IDLH levels as proposed by NIOSH [8]. (*Continued*)

Compound	Formula	b.p. (K)	*p* (kPa)	IDLH	
				REL (mg/m^3)	LEL (ppm)
Chlorine	Cl_2	239.11	777.27	1.45[c]	n.f.
Chloroform	$CHCl_3$	334.27	26.26	9.78[b]	n.f.
Ketones					
Acetone	$(CH_3)_2CO$	329.23	30.80	590[a]	2,500
2-Butanone	$CH_3C(O)CH_2CH_3$	352.74	12.08	590[a]	1,400
Cyclohexanone	$(CH_2)_5CO$	428.75	0.67	100[a]	1,100
Peroxides					
Hydrogen peroxide	H_2O_2	423.35	0.26	1.4[a]	n.f.
Pnictogens					
Ammonia	NH_3	239.80	999.92	18[a]	15,000
Dimethylamine	$(CH_3)_2NH$	280.45	202.65	18[a]	2,800
Miscellaneous					
Hydrogen sulfide	H_2S	212.82	2026.50	15[a]	4,000

n.e. none established (potential occupational carcinogenic); n.f. nonflammable; n.a. not available.
[a]8-h time-weighted average (TWA); [b]60-min short-term exposure limit (STEL); [c]Ceiling value
*measured at 368°F = 459.8 K

route of exposure. Age and pre-existing health conditions are also determining factor in level of hazard.

Acute health effect usually can be seen soon after exposure includes irritation to the eyes, skin, upper respiratory—the nose, nasal cavity, and throat, and lower respiratory—the larynx, trachea, bronchi, and lungs, and other short-term effects such as headache, grogginess, drowsiness, fatigue, nausea, and vomiting. Long-term exposure of some VOCs is associated with persistent diseases and even life-threatening diseases, such as damage of liver, kidneys, lungs, nerves, and other organs [8]. Known and probable human carcinogenic properties were also observed for several VOCs as reported by the International Agency for Research on Cancer (IARC) and the US National Toxicology Program (NTP). Health and/or life threats caused by VOCs vary depending on their reactivity.

12.2.1 Other Sources of VOCs

Air emissions by VOCs are traditionally only associated with various anthropogenic activities. However recent observations revealed that biomass also contributed to the VOCs emission. The VOCs emitted by biomass is known as biogenic volatile organic compounds (*b*VOCs); with isoprene, methanol, acetone, acetaldehyde, limonene, and myrcene are several types of *b*VOCs. Soil is commonly linked to the emission of *b*VOCs due to biotic or abiotic activities of living organisms in soil. *b*VOCs produced by the living organisms mostly end up buried and accumulated in the soil; then, after certain (high) concentration, the *b*VOCs are released from soil. In this context, soil may acts as a complex-adsorbent material that restrains the immediate release of *b*VOCs [9].

The litter of diverse pine trees can produce *b*VOCs including acetone, methanol, acetaldehyde, ethane, hexane, also some groups of alkanes, and terpenes. The emission rates of those *b*VOCs are relatively low, between 936 to 0.35 $\mu g/m^2$ in an hour. This emission rate normally increased at hot season due to increased temperature and solar irradiance. Some studies revealed that plant-roots commonly emit *b*VOCs in the form of aldehydes, alcohols, and ketones [9, 10]. Microorganisms, such as bacteria and fungi, dwell in the soil also contributed to the formation of *b*VOCs as a result of their metabolism. *Bacillus subtilis* metabolism is known to produce 2,3-butanediol. *Saccharomyces cerevisiae* and *Lactobacillus* produce ethanol in their fermentation. *Clostridium* species fermentation produces acetone and butanol. Biodegradation of organic matter done by microorganisms is known to produce methane, one of the greenhouse gases.

Some flavoring compounds, such as pyrazines, are produced from the biosynthesis of some bacteria. A similar type of *b*VOCs can also be produced by fungi [9, 10].

Although *b*VOCs are usually not being considered as hazardous substances due to their relatively low emission rate and their natural characteristics, it is worth to mention to broaden insight about VOCs sources. Moreover, some *b*VOCs are known to have antifungi activity and plant-growth promoting effect. For instance, bacteria-VOCs can inhibit the growth of phytopathogenic fungi *Rhixoctonia solani*. Trimethylamine, dimethyloctylamine, and benzaldehyde produced by bacteria are known to have strong antifungi activity. Ryu *et al.* stated that 2,3-butanediol produced by bacteria could promote the growth of plant [11]. Several other *b*VOCs and their producing organism are listed in Table 12.2.

12.3 VOCs Removal Techniques

A considerable number of VOCs handling techniques are mandatory due to the noticeable harmful impact of VOCs on environmental sustainability and human health. There are two fashions in the handling of VOCs that is recovery-based and destruction-based techniques [6]. Recovery technique is based on the idea of capturing VOCs using a transfer media to be reused later; the method is including adsorption, evaporation–condensation, absorption, and membrane separation. Whilst, destruction technique is a process of degrading VOCs into air constituents such as water vapor and carbon dioxide by combustion, oxidation, biodegradation, or plasma treatment [6]. Some studied removal techniques of VOCs are listed in Table 12.3. So far, recovery-based methods are more appealing than destruction-based methods from many viewpoints such as:

- VOCs recovery allow for reuse which is considered beneficial economically
- Absence of environmentally toxic by products such as reactive oxygen species and greenhouse gases
- Low energy necessity
- Comparable removal efficiency with the destruction techniques
- A relatively shorter process

Table 12.2 Some biogenic volatile organic compounds and their producing organisms.

Biomass	Produced VOCs	Ref.
Plant species		
Arabidopsis thaliana (mouse-ear cress)	Limonene, ethyl acetate, rhizathalene, ethanol, aldehydes, ketones	[12]
Artemisia tridentate (sage brush)	Limonene, nerol, camphor, caryophyllene, neryl isovalerate	[13]
Citrus aurantium (bitter orange)	Monoterpenes (e.g., limonene, terpinolene, sabinene, camphene, camphor, carvone, geranyl acetate), monoterpenoid (e.g., linalool, citronellol, geraniol, nerol, citronellal, perillaldehyde), 1-octanol, octanal, dodecanal, decanal	[14]
Pinus sylvestris (Scots pine)	Acetone, isoprene, tricyclene, α- and β-pinene, camphene, limonene, careen, γ-lerpinene	[15]
Zea mays (corn)	Hexadecanal, tetradecanal, β-macrocarpene, β-bisabolene, (*E*)-β-caryophyllene	[16]
Bacteria species		
Saccharomyces cervisiae	Ethanol, ethyl acetate, 3-methyl-1-butanol, 2-methyl-1-butanol, ethyl octanoate, phenylethyl acohol	[17]

(*Continued*)

Table 12.2 Some biogenic volatile organic compounds and their producing organisms. (*Continued*)

Biomass	**Produced VOCs**	**Ref.**
Wickerhamomyces anomalus	Acetaldehyde, ethyl acetate, ethyl butyrate, isoamyl acetate, ethyl hexanoate, phenylethyl acetate 2-phenyl-ethanol, isobutanol, amylic alcohol, isoamyl alcohol	[18]
Lactobacillus helveticus	Ketones (e.g., 2-propanone, 2-butanone, diacetyl), esters (e.g., ethyl acetate, ethyl butanoate, isoamyl acetate), benzaldehyde, phenylacetaldehyde, phenylethanol, ethanol, 2- and 3-menthylbutanal	[19]
Staphylococcus aureus	3-Methyl-butanal, 3-methyl-butanoic acid	[20]
Clostridium beijerinckii	acetone, butanol, ethanol	[21]
Fungi species		
Muscodor vitigenus	Naphtalene	[22]
Phomopsis sp.	Sabinene, 1-butanol, 3-methyl, benzeneethanol, 1-propanol, 2-propanone	[23]
Phoma sp.	Trans-caryophyllene, naphthalenes, alcohols	[24]
Trametes versicolor	δ-cadinene, β-guaiene, γ-patchoulene, isoledene	[25]
Metarhizium anisopliae	*n*-tetradecane, alkenes	[26]

Table 12.3 Several studied VOCs removal techniques and their performance.

Technique	Material	VOC	Operating condition	Removal efficiency	Ref
Destruction-based techniques					
Catalytic oxidation	Au/MnO_x/3DOM SiO_2	Toluene	225°C	90%	[31]
Thermal oxidation	Porous cordierite packed bed	Mix: methanol, xylene, ethyl acetate	400–975°C	90%	[32]
Biofiltration	*Candida utilis* biofilm on sugar cane bagasse	Ethanol	30°C, pH 6, moisture 63%	70%	[33]
Recovery-based techniques					
Absorption	Polypropylene bioscrubber	Mix: oxygenated, aromatic, halogenated	Room temp.	12–36%	[34]
Adsorption	Diatomite stellerite vitric tuff	butyl propionate	25°C	9.2% 59.3% 20.5%	[30]
Membrane separation	Poly-dimethyl-siloxane/α-alumina membrane	Mix: toluene, propylene, butadiene	21°C, 3–5 mbar	95%	[35]

Adsorption is acknowledged as the most favorable recovery-based VOCs elimination techniques because of its notable effectiveness (selective for each type of adsorbent and adsorbate), practical operation, and economic feasibility [27–30]. The characteristic of the solid adsorbent (porosity, selectivity, and activity) play the most vital role in the performance of adsorption. Metal organic framework (MOF) is a rising adsorbent material that is recognized for its large surface area, high porosity, ease fabrication, modifiable selectivity, high efficiency, etc. Many engineering processes are being developed to synthesize MOF with high efficiency in adsorption. The specificity of MOFs for adsorption of certain VOC is very dependent on their active adsorption sites. Many studies stated that high porosity and surface area of MOFs is not adequate for achieving efficient adsorption towards specific VOC gases/vapors. In fact, the strong interaction between host adsorbent and guest adsorbate is depended on each other compatibility. For instance, some interactions (i.e., van der Waals interactions, π–π interaction, and coordination bond) can increase the adsorption efficiency and/or selectivity when unsaturated (open sites) metal centers and organic ligand with certain functional groups interact.

12.4 Fabricated MOF for VOC Removal

Beside the potential of VOCs adsorption, the development of novel MOFs has been extensively carried out in the basis of their extent of freedom and easiness in fabrication, pore tunability, and strong durability. MOFs are constructed from the coordination of organic compounds as the primary building blocks to metal cluster (or ion) as the secondary building blocks. Substantial choices of building blocks are available to synthesize MOFs which offers distinct structural diversity of MOFs. However, a proper understanding of the occurring coordination/interaction between MOF and VOC is crucial prior to gain insight in their adsorption mechanism.

Physically, the adsorption potential of MOFs arises from the porous nature that grants them with large surface area and deep pore volume. The large surface area of MOFs provides large active adsorption sites, while the deep pore volume allows the adsorbate to be entrapped in the pore of MOFs. The choice of building blocks is important in designing the molecular structure of MOFs; it is usually based on the size and chemical features of the building blocks. However, along with progress, more and more cases disclose that the merit of MOFs is not enough for efficient adsorptive removal of VOCs. Instead, effective adsorption requires specific interactions and coordination between the VOCs adsorbate and MOFs adsorbent.

In this section, the adsorption performance of several analogs of MOF, such as MIL series MOFs, the isoreticular MOFs, and some MOF derivatives, is gathered. Only dynamic adsorption of some common VOCs such as acetaldehyde, acetone, benzene, ethyl acetate, formaldehyde, etc, is discussed.

12.4.1 MIL Series MOFs

MIL, which represents Materials of the Institute Lavoisier, series of MOFs consist of MIL-53, MIL-47, MIL-100, and MIL-101. MIL-53 and MIL-47 share the same geometrical structure, while MIL-100 shares the same structure with MIL-101. The difference in MIL-53 and MIL-47 is mostly based on their properties which caused by the different metal used for the synthesis. In the synthesis of MIL-100 and MIL-101 different type of organic linker (ligand) is used, that is trimesic acid for MIL-100 and terephthalic acid for MIL-101. The building blocks and the shape of MIL series MOFs are presented in Figure 12.1. Here we can see only several metals that often used for the MIL synthesis is given.

MIL-53 consists of three structural analogues that were synthesized with different metal ions, namely Cr, Sc, Fe, or Al, through a hydrothermal reaction between metal nitrate salt and terephthalic acid (H_2BDC) in water (1:0.5:80 in molar ratio) at a temperature of 180°C. MIL-53 arranged of one

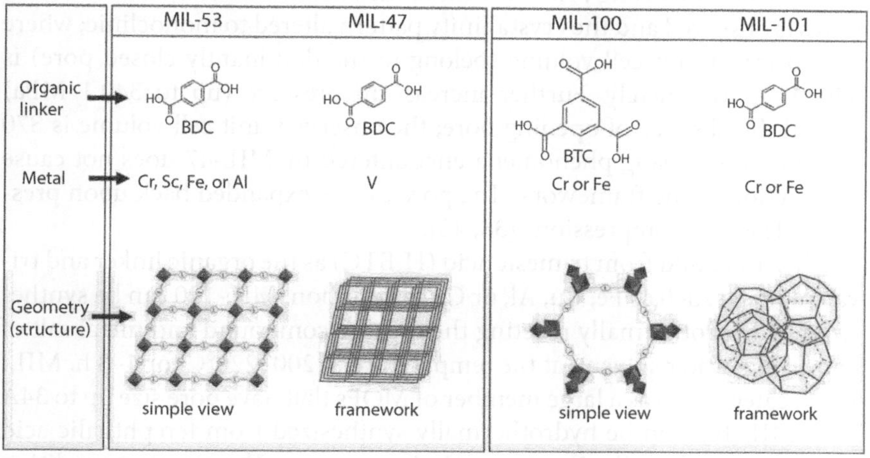

Figure 12.1 Organic and metal building blocks of MIL-53, MIL-47, MIL-100, and MIL-101, and their structures. MIL-53 and MIL-47 built from octahedral metal cluster coordinated to organic linkers to form a 3D network with 1D diamond-shaped pores. MIL-100 and MIL-101 built from trimer of metal octahedral coordinated with organic linkers to form a 3D dodecahedron with the large pore. Adapted from Ref. [36–42].

dimensional parallel chains of the octahedral metal cluster (that is $M(O_4)(OH)_2$, where M = Cr^{3+}, Fe^{3+}, Sc^{3+}, or Al^{3+}) coordinated to terephthalic organic linkers into a three dimensional network with diamond-shaped pores (up to 8.5 Amstrong in size) [43–45]. MIL-53 is known to have unique behavior, namely breathing behavior which portrays MIL-53 reversible structural change from an open pore to a closed pore structure. The structural transformation of MIL-53 is induced by heating. An exposure of MIL-53 to high temperature caused pore opening and change in crystalline structure into the orthorhombic crystal system, while low temperature condition caused pore closing and monoclinic crystal arrangement [38, 43–45]. Breathing behavior is explained further in subsection 12.8.1.

Constructed from the same organic linker as the MIL-53, MIL-47 possesses a similar geometry with MIL-53. MIL-47 is constructed with VO_6 octahedral metal cluster, it has a one dimensional pores which shaped like diamond and diameter of 7.5 Å. MIL-47 is as stable as MIL-53, where no structural collapse observed after the adsorption process. MIL-47 has a rigid structure which not allowing the occurrence of breathing behavior. The rigidity of MIL-47 can be preserved up to mechanical pressure of 178.1 MPa; which also confirmed by the X-ray diffraction (XRD). The XRD analysis shows no (or slight) orthorhombic pattern change upon exposure to a pressure up to 178.1 MPa; where the unit cell volume maintained approximately 1500 $Amstrong^3$. At 178.1 MPa, the coexistence of open and closed pores is observed and the crystallinity pattern altered to monoclinic; where the observed unit cell volume (belong to the dominantly closed pore) is 950 $Å^3$ approximately. Further increase in pressure (up to 340.1 MPa) caused the absence of opening pore; the observed unit cell volume is 870 $Å^3$. The pore closing phenomena encountered by MIL-47 does not cause destruction to the framework. The pore can be expanded back upon pressure release (decompression) [38, 42].

MIL-100 build from trimesic acid (H_3BTC) as the organic linker and trivalent metal such as Fe, Cu, Al, or Cr, in common. MIL-100 can be synthesized by hydrothermally reacting the organic compound and metal salt of choice in a water solvent, at the temperature of 200–220°C for 4–8 h. MIL-100 is categorized as a large member of MOFs that have pore size up to 34Å [37]. MIL-101 can be hydrothermally synthesized from terephthalic acid with metal such as Fe, Cu, Al, or Cr, in common. The reacting condition is similar to that of the synthesis of MIL-100. Earlier synthesis of MIL-101 (also MIL-100) was done by using hydrofluoric acid as solvent, but recent studies showed that either MOF-100 or MOF-101 can be obtained by using water as solvent without addition of hydrofluoric acid [36, 40].

The summarized adsorption capacity of MIL series MOF is available in Table 12.4. Among the MIL series of MOFs, MIL-53(Al or Cr) and MIL-47(V) are the one known to possess reversible adsorption behavior owing to their breathing capability. The reversible properties of MIL-53 and MIL-47 have been extensively studied against hydrogen sulfide (H_2S) gas under different pressure. While the pore structure of MIL-47 remains the same after H_2S adsorption at a pressure below 1.8 MPa; MIL-53 encounter a sequence of pore opening and closing. At high pressure of H_2S, MIL-53 encounters pore opening while at low pressure of H_2S MIL-53 encounters pore closing. Both MIL-53 and MIL-47, are able to maintain their structural stability after the adsorption process. The adsorption in MIL-53 and MIL-47 happened trough the H-bonding between the adsorbate and carboxylic groups of the MOFs. In the case of MIL-53, especially with Cr building block, H_2S acts as the H-acceptor. On the contrary, H_2S acts as H-donor against MIL-47(V) [38].

In the similar adsorption process against H_2S as adsorbate, MIL-100(Cr) and MIL-101(Cr) exhibit irreversible desorption due to the partial structural collapse of the building blocks which initiated by the strong adsorptive interactions between H_2S and the MOFs. This phenomenon also indicated the lack regenerative ability of MIL-100 and MIL-101, thus limit its reusability and practical application [38]. The adsorption performance of breathable MOFs (MIL-100 and MIL-101) against other VOCs also has been investigated in other studies. They show different adsorption capacity for each studied VOC. It is emphasized that different type of VOC has a different affinity towards the adsorbent, depending on their dominant bonding characteristics (such as π or σ bond, proton, or H bond, etc.).

12.4.2 Isoreticular MOFs

Isoreticular is a term given for MOFs with similar topological structure. In this section, the dynamic adsorption capacities of six isoreticular MOFs (Figure 12.2), namely, MOF-177, MOF-5, MOF-74, IRMOF-3, MOF-199, IRMOF-62, are discussed. MOF-177 has been widely synthesized and used for the adsorption of gas and VOCs [29, 46–49]. A MOF-177 was synthesized from zinc nitrate hexahydrate as the metal source and synthesized-organic-ligand namely 4,4′,4″,-benzene-1,3,5-triyl-trisbenzoic acid (H_3BTB) and are used for the preparation of. H_3BTB is a well-known building block for producing MOFs with a high surface area up to 5000 m^2/g; this property also serves it as a good gas storage material. As previously introduced by many researchers, the

Table 12.4 Studied adsorption performance of MIL series MOFs against some VOCs.

MOF	BET (m^2/g)	Pore volume (cm^3/g)	Adsorption type	Adsorption model	Adsorption capacity	Ref
MIL-47 (V)	~1000	–	Static	–	14.6 mmol/g (H_2S).	[38]
MIL-53(Cr)	~1000	–	Static	–	13.12 mmol/g (H_2S).	[38]
MIL-53(Fe)	~1000	–	Static	–	11.77 mmol/g (H_2S).	[38]
MIL-53(Al)	~1000	–	Static	–	8.53 mmol/g (H_2S).	[38]
MIL-100(Cr)	1695	–	Static	SL	0.38 (phenol), 0.20 (*p*-nitrophenol). Value in mmol/g.	[27]
MIL-100(Cr)	1900	–	Static	–	16.7 mmol/g (H_2S).	[38]
MIL-100(Fe)	1675	–	Static	SL	0.38 (phenol), 0.23 (*p*-nitrophenol). Value in mmol/g.	[27]
MIL-101(Cr)	4293	2.43	Dynamic	FL-PFO	83.44 (benzene), 97.74 (toluene), 7.34 (xylene), 98.47 (*n*-hexane), 81.97 (*n*-heptane). Value in wt%.	[41]
MIL-101(Cr)	2736	1.50	Static	DA	0.16 (*n*-hexane), 0.53 (toluene), 0.83 (butanone), 0.99 (dichloromethane), 1.47 (methanol), 12.8 (*n*-butylamine). Value of limiting adsorption capacity in mmol/g.	[39]
MIL-101(Cr)	2600	–	Static	–	38.4 mmol/g (H_2S).	[38]

FL-PFO: fractal-like pseudo-first-order, DA: Dubinin-Astakhov, SL: Sips and Langmuir.

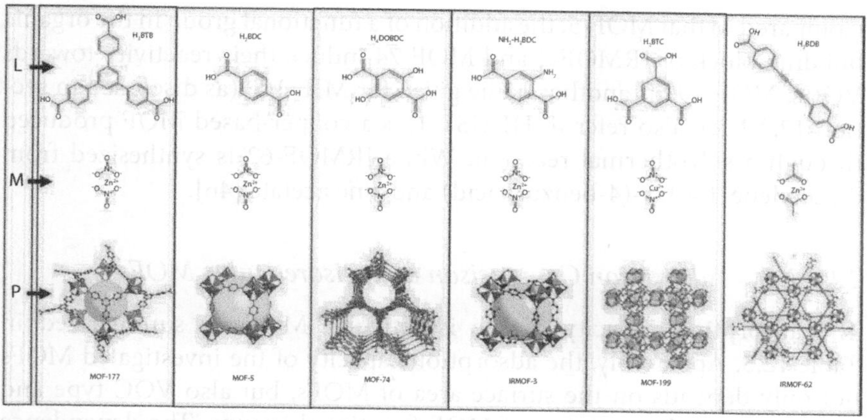

Figure 12.2 Organic and metal building blocks of the isoreticulars MOF-177, MOF-5, MOF-74, IRMOF-3, MOF-199, and IRMOF-62. L = organic ligand, M = metal ion source, and P = MOF product. Adapted from Ref. [46].

preparation of the H_3BTB ligand involve two step reactions; first step by electrophilic reaction involving $AlCl_3$, and second step by Hofmann rearrangement involving NaOBr. The preparation of MOF-177 itself is conducted solvothermally, in *N, N*-diethylformamide solvent (common solvent for synthesizing cubic-shaped zinc oxide). Characteristics of the reported MOF-177 have slight differences from each other, which might be caused by aberrations in the concentration and purity of the starting materials, also differences in heating and cooling rate. The relative characteristics of MOF-177 are Langmuir surface area of 4,170–5,640 m^2/g, BET surface area of 2,970–4,630 m^2/g, pore volume of 1.11–1.69 cm^3/g, and pore diameter 0.94–1.06 nm. The synthesized MOF-177 has a crystal clear appearance with a slightly yellow color and thermally stable in the air up to a temperature of 350°C [46].

Combination of terephthalic acid (1,4-benzenedicarboxylate, abbreviated as H_2BDC) and zinc nitrate as building blocks are used to synthesize MOF-5. The as-synthesized MOF-5 (also called as IRMOF-1) have a chemical composition of $Zn_4O(BDC)_3$ which have a three-dimensional cubic shape. MOF-5 is reportedly having a high surface area of 2,500 to 3,000 m^2/g owing to its scaffolding-like nature [46]. Both MOF-74 and IRMOF-3 have similar building blocks to that of MOF-5 but with the addition of functional groups in the organic linker structure. MOF-74 is synthesized from 2,5-dihydroxyterephthalic acid (H_4DOBDC) that has a similar structure to that of H_2BDC but with additional of two hydroxyl groups while in IRMOF-3, the functional group comes from its NH_2- site.

Compared to that MOF-5, the addition of a functional group in the organic building block of IRMOF-3 and MOF-74 induce their reactivity towards VOCs. MOF-199 is another name given for MIL-100 (as discussed in section 12.4.1.3.), also refer to HKUST-1, is a copper-based MOF produced through a solvothermal reaction. While IRMOF-62 is synthesized from diacetylene-1,4-bis-(4-benzoic acid) and zinc acetate [46].

12.4.2.1 Adsorption Comparison of the Isoreticular MOFs

The adsorption capacity of the isoreticular MOFs is summarized in Table 12.5. Apparently, the adsorption capacity of the investigated MOFs not only depends on the surface area of MOFs, but also VOC type and reactivity, and the presence of MOF functional groups. The dependence of adsorption performance of MOFs toward the type of VOCs is apparent, in which a MOF might be effective in adsorbing one type of adsorbent but ineffective for other types of VOC. For instance, MOF-177 can adsorb ammonia 42-times higher than the other tested VOCs. Most of the investigated MOFs failed to adsorb chlorine; this is because chlorine cannot act as a ligand that binds to the metal cluster; unlike other VOCs. IRMOF-3 shows successful adsorption towards chlorine; this is ascribed to the reactive nature of the gas and also the effect of open metal sites as Lewis acids that interact with the adsorbate gases as Lewis base [46]. In the case of aromatic VOC (i.e., benzene), the adsorption may improve due to π–π interaction [29].

Based on Table 12.5, it is noted that a larger surface area not always in accordance with a higher adsorption capacity. MOF-177 and MOF-5 have larger BET surface area than other investigated MOFs, but they have the worst dynamic adsorption performance. MOFs with a large surface area may have a high capacity for thermodynamic adsorption; however, they are less effective in kinetic adsorption due to the absence of functional group that essential for the adsorbate-adsorbent interactions. The high surface area of IRMOF-62 seems to assist in enhancing the adsorption capacities; however, the lack of functional group caused IRMOF-62 to be less effective against certain VOC (i.e., sulfur dioxide) [46].

The role of MOF-functional groups in improving the adsorption performance can be noticed by comparing MOF-5, MOF-74, and IRMOF-3. Addition of NH_2– functional group to the organic building block of MOF-5 produces IRMOF-3 which have better dynamic adsorption capacity; i.e., IRMOF-3 possesses adsorption capacity towards ammonia 18-times higher than MOF-5. Addition of OH– functional group (i.e., in producing MOF-74) also accentuate the adsorption performance. Such improvement

Table 12.5 Dynamic adsorption performance of some isoreticular MOFs against VOCs at gas flow 25 ml/min [46] and 600 ml/min [29].

Properties	MOF-177	MOF-5	MOF-74	IRMOF-3	MOF-199	IRMOF-62
BET surface area (m^2/g)	3,875	2,205	632	1,568	1,264	1,814
Adsorption capacity (mg/g) at gas flow rate of 25 ml/min						
Ammonia	42	6	93	105	87	23
Chlorine	<1	–[a]	–[a]	335	36	92
Tetrahydrothiophene	<1	1	90	7	351	84
Sulfur dioxide	<1	1	194	6	32	<1
Benzene	1	2	96	56	176	109
Dichloromethane	<1	<1	32	1	55	19
Ethylene oxide	<1	1	110	2	95	11
BET surface area (m^2/g)	2,970					
Adsorption capacity (mg/g) at gas flow rate of 600 ml/min						
Benzene	800					

[a]Experiments not conducted due to apparatus corrosion by chlorine.

in adsorption capacity is due to the additional functional groups, such as amine, which function as electron rich groups that readily form hydrogen bonding with the adsorbate-molecules [46].

The vacant coordinated metal sites of MOF act as Lewis acids that is deficient of electron hence generate a stronger interaction with adsorbate molecules that can act as Lewis base, such as ammonia. The role of metal as Lewis acids can be observed in MOF-199 and MOF-74. Both MOF-199 and MOF-74 have open copper sites which can readily adsorb VOCs [46]. As presented in Table 12.5, apparently a larger gas purge flow may stimulate adsorption driving force thus promote penetration of VOC gas molecules into the porous matrix of the MOF. This insight can be drawn from adsorption capacity of MOF-177 at 600 ml/min gas flow which gave 800-times-higher benzene adsorption capacity compared to the studies done at 25 ml/min gas flow [29, 46].

12.4.3 NENU Series MOFs

A group in NENU (Northeast Normal University) synthesized another series of isoreticular MOF namely NENU-X. Prior to the synthesis of NENU-X MOFs, the organic linker used in a series of NENU MOFs was synthesized by integrating thiophene-containing bidentate ligand into benzene-1,3,5-tribenzoate (BTB). The MOFs were then synthesized through a solvothermal method with *N,N*-diethyl-formamide as a solvent (85°C, 2 days), and zinc nitrate was used as the metal source. The, as synthesized NENU-511 was built from coordination of six BTB to a six-connected node and possess meso- and micro-sized pore. Meanwhile, NENU-512, NENU-513, and NENU-514 have more disorderly structure due to their large-size pores [50].

The NENU-X MOFs have an enormous BET surface area up to 4240 m^2/g, which play an important role in the adsorption of VOC, namely benzene. The adsorption capacity of the NENU-X MOFs against benzene is shown in Table 12.6. Adsorption of benzene onto NENU-511 and NENU-512, which do not possess a specific functional group on their inner matrix, is ascribed from their spacious matrix that can accommodate significant amount of benzene. Their spacious matrix also postponed their saturation to $P/P_0 = 0.20$. Meanwhile, NENU-513 and NENU-514 showed earlier saturation at $P/P_0 = 0.10$. This is due to the presence of phenyl-rings on the inner surfaces. While these phenyl-rings helps to enhance the adsorption towards benzene by raising the π–π interaction, excessive occurrence of phenyl-rings on the inner surface may reduce MOFs porosity and cause

Table 12.6 Surface area and adsorption capacity of NENU series toward benzene [50].

	NENU-511	NENU-512	NENU-513	NENU-514
Surface area (m^2/g)				
BET	4240	3648	3554	3457
Langmuir	6491	5640	5438	5320
Adsorption capacity (mg/g)	1556	1519	1687	1311

repulsion effect between each adsorbate molecules thus adversely affect adsorption capacity [50].

Another NENU-X MOF, namely NENU-520, has been synthesized by using a tetrazolate-containing ligand. However, no work has been done to study its adsorption ability against VOCs. Adsorption data of NENU-520 for CO_2 uptake has been reported by Lan *et al.*, where the uptake capacity is 79.9 cm^3/cm^3 at standard temperature and pressure [51].

12.4.4 MOF-5, Eu-MOF, and MOF-199

A research study by Vellingiri *et al.* demonstrated the superiority of MOF as adsorbent of VOC and semi-VOC, which is comparable to some commercial adsorbents (Tenax TA, Carbopack-X, and Carboxen-1000) [28]. The adsorption ability of MOF-5, Eu-MOF, and MOF-199 against hydrocarbons, carboxylics, phenolics, and indoles VOC are investigated in their study. Some of the dynamic adsorption data from their research are shown in Table 12.7. MOF-199 show better adsorption performance compared to the other investigated MOFs; which was attributed from the effect of vacant copper–metal sites of MOF-199 that also act as Lewis sites. The lower adsorption capacities of Eu-MOF may be in consequence to the absence of vacant coordination sites and low adsorption energy of Eu metal. Meanwhile, the poor adsorption performance of MOF-5 is suggested to be the effect of its cubic-pore that eliminates the functionality of the frameworks.

Density functional theory (DFT) estimation revealed some specific interaction between the six-member ring compounds and MOFs that occurs during the adsorption of VOCs. The AA-stacking phenomena can be observed in the case of adsorption by using MOF-5, while higher

Table 12.7 The equilibrium adsorption capacity of MOF-5, Eu-MOF, and MOF-199 against several types of VOCs (selected from the Ref. [28]).

Properties	MOF		
	MOF-5	**Eu-MOF**	**MOF-199**
Surface area (m^2/g)[a]	535	281	1591
Pore volume (cm^3/g)	0.22	0.95	0.46
SEM morphology	Cubic crystal	Grass-like crystal	Octahedral crystal
Adsorption capacity (mg/g)			
Aromatic VOCs mixture (p_i[b])			
Benzene (8.6 mPa)	n. m.[c]	1.00	>1.1[d]
Toluene (6.8 mPa)	n. m.	0.95	>2.6[d]
p-Xylene (6.0 mPa)	n. m.	0.76	>5.2[d]
Styrene (5.5 mPa)	n. m.	0.85	>4.9[d]
Phenol VOCs mixture			
Phenol (12 mPa)	n. m.	1.50	13.0
p-Cresol (9.8 mPa)	1.50	3.00	15.0

[a]Langmuir surface area; [b]partial pressure of each species in VOCs mixture in parenthesis; [c]n.m. refers to not measurable; [d]saturation of those adsorbates were not achieved within 15 L sample loading.

energy AB-stacking phenomena are observed for MOF-199. Moreover, the Cu-metal sites of MOF-199 can attract the compound containing polar -OH group thus increasing the adsorption capacity; this phenomenon does not occur in MOF-5 with Zn-ion as the central [28].

12.4.5 Amine-Impregnated MIL-100

MIL-100(Al) with amine functional group, coded as NH_2-MIL-101(Al) can be synthesized by reacting aluminum chloride salt with 2-amino terephthalic acid in DMF solvent (403K, 3days). The as synthesized NH_2-MIL-101 has BET surface area of 1943 m^2/g. In static (liquid-phase) adsorption experiment, by using phenol and *p*-nitrophenol as adsorbate, NH_2-MIL-101(Al) can adsorb up to 1.2 mmol/g which is higher than some

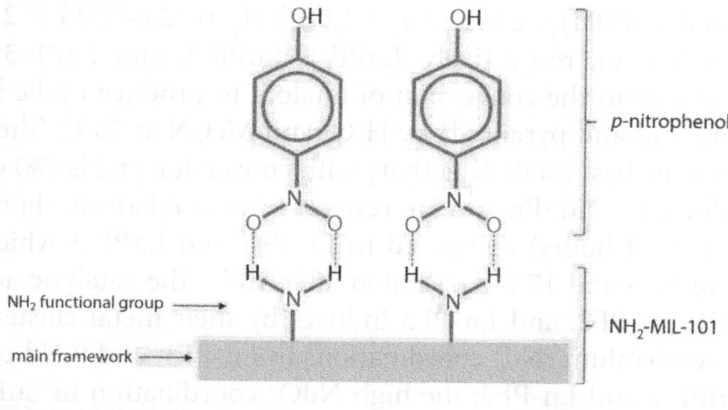

Figure 12.3 Adsorption of *p*-nitrophenol solution onto functionalized MOF. Amine group of NH_2–MIL-101(Al) provides proton donor to the oxygen atoms of nitro group of *p*-nitrophenol. Redrawn form Ref. [27].

highly crosslinked-polymer such as HJ-1 (1.0 mmol/g) and Amberlite XAD (0.4 mmol/g). Hydrogen bonding plays an important role in promoting the adsorption of *p*-nitrophenol onto the amine-containing adsorbent as illustrated in Figure 12.3 [27].

12.4.6 Biodegradable MOFs MIL-88 Series

The sorption of nitric oxide (NO) onto a series of biocompatible and biodegradable Fe-based MOFs, namely MIL-88, has been studied. Various organic ligands, for instance fumarate-acetic acid, 1,4-benzenedicarboxylate (1,4-bdc), 2-nitroterephthalate (2-NO_2-bdc), and 2,5-dihydroxyterephthalate (2,5-dhbdc), have been used in the synthesized of MIL-88A, MIL-88B, MIL-88B-NO_2, and MIL-88B-2OH, respectively. It is observed that the MIL-88 MOFs can adsorb a relatively low amount of NO approximately 1–2.5 mmol/g. The adsorption performance of MIL-88 series is affected by the accessibility of the gas penetration to the pore matrix which is quite low due to the absence of breathing ability; as a result, the gas cannot easily diffuse to interact with the central Fe metal cluster [52].

12.4.7 Catalytic MOFs

A certain metal cluster can produce MOFs with catalytic activity. A series of rare earth arenedisulfonate MOFs have been produced and tested for their catalytic activity. The tested MOFs are coded as LnPF-1 (Ln = La, Nd, Pr), LnPF-2 (Ln = Nd), and LnPF-3 (Ln = Nd) from the Ln(OH)(1,5-NDS)H_2O (1,5-NDS = 1,5-naphthalenedisulfonate), [Ln(OH)(1,5-NDS)

$(H_2O)]_4$, and $Ln(OH)_{4.5}(2,6\text{-NDS})_{1.5}(H_2O)_2](H_2O)$ (2,6-NDS = 2,6-naphthalenedisulfonate), respectively. LnPF-1, LnPF-2, and LnPF-3 showed catalytic activity in the conversion of linalool to produce cyclic hydroxyl ethers (furanoid and pyranoid) by H_2O_2 and MeCN at 70°C. The Ln-PF1 MOFs show the best catalytic activity with conversion yields (%) of 76, 94, and 100 (for Ln = Nd, Pr, and La, respectively) in relatively shorter reaction time (22-24 hours) compared to Ln-PF2 and LnPF-3 which barely can achieve 51% and 17% conversion after 48 h. The catalytic activity of the Ln-PF1, Ln-PF2, and Ln-PF3 induces by their metal clusters. There are eight neodymium (Nd) coordinations in Ln-PF1, and 9 Nd coordination in LnPF-2 and Ln-PF3; the high NdO_9 coordination in LnPF-2 and LnPF-3 makes it difficult for guest molecules to reach the active centers and thereby reducing the actuality of the catalyst. Better catalytic activity is shown by using praseodymium (Pr) and lanthanum (La) as the metal clusters in Ln-PF1, which is related to the decreasing of their radii [53].

Some other studied catalytic MOFs is the ytterbium (Yb) based MOFs, namely Yb-LRH with the cationic formula $[Yb_4(OH)_{10}(H_2O)_4]_n^{2+}$ and Yb-RPF-5 with formula $[Yb(OH)(2,6\text{-AQDS})(H_2O)]$ (AQDS = anthraquinone-2,6-disulfonate). The Yb-LRH is shown a selective-catalytic activity for the sulfide oxidation, while Yb-RPF-5 can catalyze hydrodesulfurization of thiophene (7 bar, 70°C) [54]. The use of Au as a metal cluster produced a MOF (Au@ZIF-8) that can catalyze the oxidation of carbon monoxide into CO_2 [55]. The $NH_4[Cu_3(OH)(capz)]$-MOF, together with *t*-butyl hydroperoxide, show catalytic activity in the oxidation of cyclohexane and cyclohexene [56]. Also, it is worth to mention, a lot of molybdenum (Mo)-based MOFs show a catalytic behavior in the epoxidation of alkenes [57].

12.4.8 Photo-Degradating MOFs

Some MOFs possess adsorption and photo-degradation ability toward VOCs. The hybrid of TiO_2 nanoparticles within a typical NH_2-UiO-66 MOF, namely TiO_2@NH_2-UiO-66, have an excellent interface contact that allow the penetration of visible light and consequently accelerate the photogenerated electron–hole separation. The hybrid MOFs, which contain 5% TiO_2, show a remarkable photo-degradation activity against styrene gas into CO_2; with removal efficiency >99% within 10 hours [58]. The Ti-Zr based MOFs, NH_2-MIL-125 (TiZr15), shows a high photo-degradation activity of acetaminophen (with a pseudo-first order rate of 0.0121/min); the activity was found to be better compare to the bare Ti MOF. A better photo-degradation activity was achieved at lower Zr proportions. The efficiency of photocatalytic activity of TiZr15 remains above 90% upon three successive acetaminophen conversion; which indicate its stability and reusability [59].

12.4.9 Some Other Studied MOFs

Various divalent metal ions have been used in the preparation of M-MOF-74 (where M = Zn, Co, Ni, Mg) and each metal produced MOF with different surface area, specifically 1206, 835, 599, 496 m^2/g for M-MOF-74 with M = Mg, Co, Ni, Zn, respectively. The adsorptive ability of M-MOF-74 is tested against ammonia, cyanogen chloride, and sulfur dioxide. It is noted that the dynamic adsorption performance was different when conducted in dry and humid condition. As shown in Table 12.8, all MOF-74 analogs show better adsorption performance at dry condition. The adsorption capacity for all type of adsorbate was reduced when exposed to the humid gas stream; which suggests that there is a competition between the water molecules and adsorbate molecules to attach on the MOF surface [60].

Pyrazolate ligands and nickel hydroxo metal cluster have been used to synthesized some hydrophobic isoreticulars MOFs, namely, $[Ni_8(OH)_4(H_2O)_2(L)_6]$ or $[Ni_8(L)_6]$; where L symbolized various ligands such as H_2L1 (1H-pyrazole-4-carboxylic acid; H_2L2 (4-(1H-pyrazole-4-yl)benzoic acid), H_2L3 (4,4′-benzene-1,4-diylbis(1H-pyrazole)), H_2L4

Table 12.8 Dynamic adsorption data of MOF-74 analogues [60].

	Adsorption capacity (mmol/g)							
	Co-MOF-74		Mg-MOF-74		Ni-MOF-74		Zn-MOF-74	
	Dry	Wet	Dry	Wet	Dry	Wet	Dry	Wet
Ammonia	6.70	4.30	7.60	1.70	2.30	1.90	3.70	2.80
Cyanogen chloride	5.60	0.05	1.20	0.08	2.40	0.17	3.60	0.10
Sulfur dioxide	0.63	0.03	1.60	0.72	0.04	0.02	0.26	0.04
Amount of adsorbed water on adsorbent at different relative humidity (RH)								
15% RH	0.46		0.25		0.08		0.19	
50% RH	0.49		0.31		0.11		0.24	
80% RH	0.50		0.39		0.13		0.32	

(4,4′-buta-1,3-diyne-1,4-diylbis(1H-pyrazole)), H_2L5 (4,4′-(benzene-1,4-diyldiethyne-2,1-diyl)bis(1H-pyrazole)), or H_2L5-R (R = methyl, trifluoromethyl). The geometry of the resulting MOFs is the combination of an octahedral and two tetrahedral cavities which analogues to that of UiO series MOFs. The BET surface area of the MOFs increases with the longer ligand chain; where the BET ranges between 205–2215 m²/g with $[Ni_8(L1)_6]$ have the lowest BET (205 m²/g), while $[Ni_8(L5)_6]$ have the largest BET (2215 m²/g). The $[Ni_8(L)_6]$ series have been proven to be stable against hydrolysis with their bipyrazolate linkers are unaltered even after 24 h soaking in water. The hydrophobic nature of the $[Ni_8(L)_6]$ series allows performing adsorption under humid condition. Increase in hydrophobicity was observed by the presence of a fluorine functional group in $[Ni_8(L5\text{-}CF_3)_6]$. As reported, at 80% relative humidity, $[Ni_8(L5\text{-}CF_3)_6]$ is not suitable for water molecule adsorbent, in contrast to non-fluorine MOFs, e.g., $[Ni_8(L4)_6]$ and $[Ni_8(L5)_6]$ which possess high water adsorption capacity. The $[Ni_8(L5\text{-}CF_3)_6]$ show efficient adsorption against diethylsulfide under a very humid condition [61].

12.5 MOF Composites

12.5.1 MIL-101 Composite With Graphene Oxide

Combination of graphene oxide (GrO) with MIL-101 produced a composite (GrO@MIL-101) which suggest its potential in CO_2 and CH_4 removal. GrO@MIL-101 has the higher surface area and larger pore volume than that of MIL-101 prepared in the study; where the BET surface are (m²/g), and pore volume (cm³/g) for MIL-101 are 2670 and 1.28, while for GrO@MIL-101 are 2950 and 1.42. As revealed by SEM, the addition of GrO induced irregular particle shape of originally octahedral shaped MIL-101 [62]. The static adsorption study of the composite revealed that GrO@MIL-101 composite could remove CO_2 and CH_4 more effectively than MIL-101, as shown in Table 12.9. Also, at 1.5 bar, the adsorption selectivity of GrO@MIL-101 against CO_2/CH_4 was up to 32.

12.5.2 MIL-101 Composite With Graphite Oxide

Composites of MIL-101(Cr) and graphite oxide (GO), coded as MIL-101@GO-*n*, have been synthesized via the hydrothermal method by Sun *et al.* In their study, the composites were prepared by introducing *n*-amount of GO (*n* = 2, 5, 10, and 15 wt%). As characterized by XRD, the addition of GO gives a reduction to the crystallinity of MIL-101(Cr). The SEM images

Table 12.9 Static adsorption performance of MIL-101 and its composite against CO_2 [62].

MOF composite	Adsorption model	Pressure (bar)	Adsorption capacity (mmol/g)		Desorption efficiency
			CO_2	CH_4	
MIL-101(Cr)	DSLF[a]	25	50.39	26.17	–
GrO@MIL-101(Cr)	DSLF[a]	25	66.12	37.21	>95%

[a]Dual Site Langmuir Freundlich (DSLF) model equation.

revealed that the addition of GO causes the alteration of octahedral MIL-101(Cr) into irregular-shape particles. All of the composites possess a Type I isotherm-sorption as analyzed by N_2-sorption. The addition of GO also affecting the surface area of the prepared MOF-composites; where the BET surface area (m^2/g) and pore volume (cm^3/g, value in parenthesis) value are 2881.1 (1.36), 2988.6 (1.54), 3502.2 (1.75), 3318.6 (1.72), 1742.7 (1.00) for MIL-101, MIL-101@GO-2, MIL-101@GO-5, MIL-101@GO-10, and MIL-101@GO-15, respectively [63]. There is no certain decrementing or incrementing trend, caused by the addition of GO, observed for the surface area and pore volume. Reportedly, the addition of GO has been proven to increase the adsorption capacity toward *n*-hexane, as shown in Table 12.10. As shown in Table 12.10, MIL-101@GO-5 exhibit the highest adsorption of *n*-hexane compared to the other investigated MOF. The excellent adsorption performance of MIL-101@GO-5 is not only related to the high surface

Table 12.10 Static (vacuumed) adsorption performance of MIL-101 and its composite against *n*-hexane [63].

MOF composite	Adsorption model	Pressure (mbar)	Adsorption capacity (mg/g)	Desorption efficiency
MIL-101	L-F[a]	175	540.3	–
MIL-101@GO-5	L-F[a]	175	1042.1	–
MIL-101@GO-10	L-F[a]	175	952.8	96.78

[a] Langmuir–Freundlich (L–F) model equation.

are but also due to the addition of GO in the proper ratio which in turns affect their surface dispersive force [63].

12.6 Generalization Adsorptive Removal of VOCs by MOFs

MOFs with the adsorptive surface that contain certain functional groups or vacant metal sites are more likely to have an enhanced adsorption efficiency toward adsorbate compounds. The organic building block of MOF which contain certain functional site can generate an electrostatic interaction or hydrogen bonding that can enhance their adsorption activity; meanwhile, vacant metal sites can promote coordination bond or π-interaction against the VOCs adsorbate. Adsorption transpires through continuous pore filling by VOCs to the inner and outer pore surfaces of the MOF. Pore filling by VOCs is possible since VOCs are vapors which have verge properties to non-ideal gases, with some deviations due to their gas-liquid phase equilibrium. The VOCs detention by MOF is consolidated through the non-covalent coupling involving the π-system of the metal cluster of MOF and VOCs. The π-interaction is triggered due to the different charge loading between the metal and VOCs. The metal cluster of MOF is an electron-poor system; meanwhile, the VOCs are the electron-rich system. The π-interaction is common, especially for the aromatic VOCs.

The adsorption of VOCs containing saturated chain, nonpolar property, and unbranched sites (e.g., hexane, heptane, and octane) are more likely occur due to physical interactions of the adsorbate molecules with the adsorbent lattice [64]. For aromatic molecules, a combination of physical and chemical adsorption may occur from the adsorbate and adsorbent interactions. Liu *et al.* demonstrated that the binding of benzene to the Lewis site of metal emerges due to the van der Waals and covalent bonding [65]. In adsorption of VOCs which behave as Lewis base (e.g., ammonia), the adsorption usually occurs as the effect of Lewis acid and Lewis base interaction. Ammonia molecules act as the Lewis base, while the metal sites serve as Lewis acid.

12.7 Simple Modeling the Adsorption

12.7.1 Thermodynamic Parameters

The Clausius–Clapeyron equation can be applied to estimate the heat of adsorption. The variation of retention volume (V_g) with the change of temperature (T) can be translated into a mathematical equation to estimate the heat of adsorption ΔH_{ads}, Eq. (12.1) [61].

$$\Delta H_{ads} = -R\frac{\delta(\ln V_g)}{\delta(1/T)} \tag{12.1}$$

where R is the ideal gas constant. The V_g is directly related to Henry's law constant (H_{lc}), the constant that stated the proportionality between the partial pressure of the gas phase to the quantity of gas dissolved [61]. In adsorption, the value of H_{lc} is directly related to that of the amount of gas adsorbed. The calculation of H_{lc} can be translated into a mathematical formula, as shown in Eq. (12.2).

$$H_{lc} = \frac{n_{ads}}{m \times P} \tag{12.2}$$

where, n_{ads} is the amount of gas adsorbed, m is the mass of adsorbent used, and p is the outlet pressure of gas.

A virial-type expansion, with virial coefficient A_i and B_i, also can be applied to estimate the heat of adsorption.

$$\ln P = \ln n_{ads} + \frac{1}{T}\sum_{i=0}^{m} A_i n_{ads}^i + \sum_{i=0}^{n} B_i n_{ads}^i \tag{12.3}$$

$$\Delta H_{ads} = -R\sum\nolimits_{i=0}^{m} A_i n_{ads}^i \tag{12.4}$$

where m and n are the number of coefficients used to determine the isotherms [48].

12.7.2 Dynamic Sorption Methods

The determination of adsorption capacity of porous materials to adsorb gas adsorbate mostly done by using dynamic sorption methods since, in real practice, the gas adsorption is always done dynamically. Static volumetric measurements are one of (semi)-dynamic sorption methods in which the process conducts in a closed chamber, then the amount of gas adsorbed upon saturation at a given pressure (p / p_0) is recorded. The schematic diagram of the static volumetric and its resulting plot are shown in

Figure 12.4. The measurement only can be done for a single component, which is become the shortage of this method [66–69]. The amount of adsorbed gas can be calculated by treated the gas component as an ideal gas, as shown by the Eq. (12.5) below.

$$n_{\text{adsorbed}} = n_{\text{dosed}} - n_{\text{free}} = \left(\frac{pV}{RT}\right)_{\text{dose}} - \left(V_{\text{dose}} + V_{\text{chamber}}\right)\left(\frac{p}{RT}\right)_{\text{chamber}} \tag{12.5}$$

where n is the mol of gas, p is pressure, V is volume, T is temperature, and R is the ideal gas constant. The adsorption isotherm curve can be obtained by plotting amount of gas adsorbed vs. the given pressure.

A dynamic method proceeds under forced-flow conditions. The resulting data (that is adsorption time (duration of the test) vs. concentration of adsorbate in the effluent stream) is depicted as a breakthrough curve. The breakthrough curve (dynamic method) process is similar to that separation process in the industry, where the adsorbent is a gas mixture. The measurement itself is done in an open system while maintaining the pressure of the feed. The schematic diagram of the experimental setup and its corresponding breakthrough curve are depicted in Figure 12.5. Control of the gas flow rate is an important parameter that affected the amount of VOC introduced to the adsorption system, the acceleration of gas flow rate also directly related to the contact time between the VOC and MOF-adsorbent. The dynamic adsorption capacity can be calculated through breakthrough data by using the following series of equations [60, 66–69].

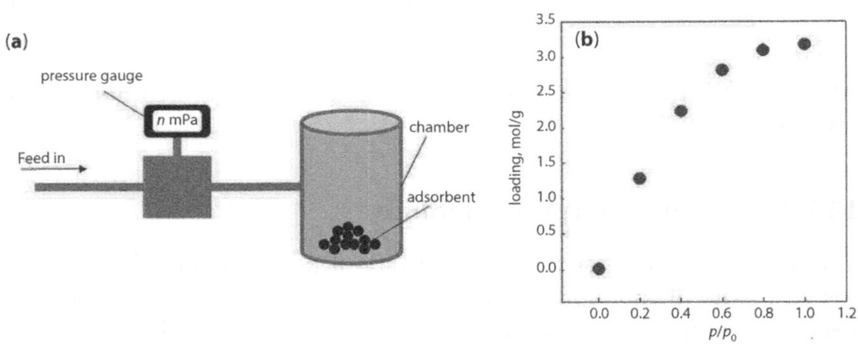

Figure 12.4 (a) Schematic diagram of static volumetric measurement; a single component of VOC gas is pressurized to the desired pressure and then introduced into a closed chamber filled with a certain amount of adsorbent. (b) Adsorption curve; obtained by plotting amount of gas adsorbed at saturation per given pressure.

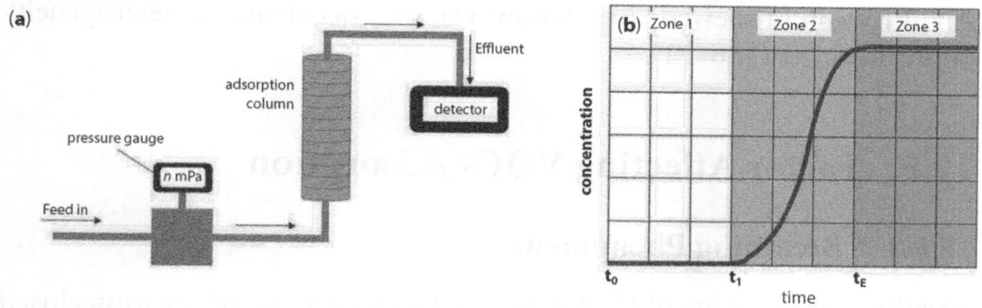

Figure 12.5 (a) Schematic diagram of dynamic method; the pressurized mixture of VOC gases continuously flowed into the adsorption column (filled with adsorbent) and the gas effluent is measured by using analytical device/detector. (b) Ideal S-shape breakthrough curve; zone 1—unsaturated zone, zone 2—mass transfer zone, zone 3—saturated zone.

The concentration of VOC over time (Ct, mg·min/m^3) is calculated by multiply concentration of the feed (C_{feed}, mg/m^3) with the length of time that the feed passed (t_f, min), as shown in Eq. (12.6).

$$Ct_{\text{feed}} = t_f C_{\text{feed}} \tag{12.6}$$

The subsequent eluted concentration of VOC per time (Ct_{elution}, mg·min/m^3) done by the adsorbent until the flow stopped can be calculated by integrating under the elution curve, as shown by Eq. (12.7); which also can be applied for desorption.

$$Ct_{\text{elution}} = \sum\nolimits_{t=0}^{t_s} \frac{C_n + C_{n-1}}{2} \times (t_{n-1} - t_n) \tag{12.7}$$

in this case, t_s is saturation time (min), t_n is the time n (min), and C_n is the concentration at time n (mg/m^3) [60]. Note that there are many options of a modified form of the equations that can be used to modeling the adsorption experiment.

Zone 1, unsaturated zone, cannot be obtained by using the static volumetric measurements. Zone 2 is a mass transfer zone (also abbreviated as MTZ), the approximation of the effective-amount of adsorbent can be determined by analyzing the steepness of this zone. A very narrow zone 2 indicates that the adsorbent has a low adsorption capacity. The saturated zone, zone 3, is similar to that of the equilibrium value of adsorption isotherm in which the adsorption capacity can be determined. An ideal breakthrough curve would have an S-shaped curve, as shown in Figure 12.5.

An alteration (a skewed curve), however, may occur due to heterogeneity of the adsorbent [66–71].

12.8 Factor Affecting VOCs Adsorption

12.8.1 Breathing Phenomena

Breathing phenomena of MOF refers to the turnover structure from closed pores (non-porous) to open pores (porous). The phenomenon is pioneered by Ferey and Kitagawa *et al.* This dynamic phenomenon in adsorption is stimulated by atomic displacement (dipole moment interaction, in specific) which is caused by the presence of adsorbate. The phenomena are introduced by Kitagawa *et al.*, in their study on pillared layer MOFs. Kitagawa *et al.* declared that MOF is undergoing a locking and unlocking phenomena before and after adsorption, respectively [72, 73]. The distortion of the molecular structure of MOFs and cell-dimensions expansion occurred along with the breathing phenomena.

A similar breathing behavior observed for MIL-53. A reversible change in the pore volume happened when the MIL-53 (Al or Cr) is being air-cooled after heating at high temperature. As illustrated in Figure 12.6, a decrease in pore volume occurred when MIL-53 is at low temperature due to absorption of water molecules and reversed back to original volume when water molecules are removed by treating in high temperature. Upon high temperature treatment, the H-bond between the water molecules is getting weaker an

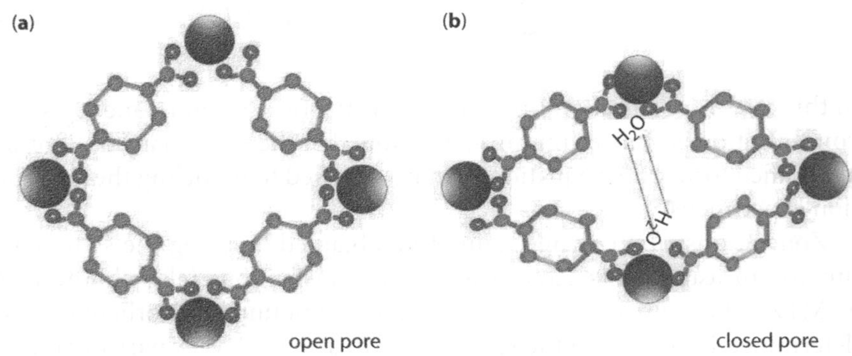

Figure 12.6 Illustrated breathing phenomena of MIL-53. (a) Open pore after high temperature treatment (unit cell volume 1522 Amstrong^3); (b) closed pore (unit cell volume 1013 Amstrong^3) after air cooling due to H-bond interactions between water molecules. Adapted from Ref. [74].

eventually disappeared along with the dehydration process; subsequently, the pores of MIL-53 convert back to large open pores [74].

A similar breathing phenomena also observed upon pressurized-CO_2 exposure to the hydrated MIL-53 at low temperature. CO_2 exposure at low pressure (<5 bar) is not sufficient to cause pore opening to MIL-53. The co-existence of open and closed pores is observed when the more CO_2 molecules are introduced, and consequently, the pressure is increased. Further increase of CO_2 pressure, > 15 bar (approximately 7.2 mmol/g), allow the presence of MIL-53 with the whole open pores. In this case, there is a slight different dehydrated phenomenon; that is, a two steps pore transition (open–closed–specific) is observed. Upon dehydration, a plateau (steady pore size) happens when 3.0 mmol/g CO_2 introduced. Hereafter, the pores start to expand as the CO_2 added (up to 7.7 mmol/g) before another plateau occurs. The plateau is suggested to be the effect of steric hindrance between the water molecules. The structural change due to the breathing effect also caused alteration to the sorption–isotherm behavior of MOFs. At dehydrated state, the N_2-sorption of MIL-53 (both with Al or Cr building blocks) shows a Type-I-sorption-isotherm which indicate the dominance in microporous structure. A behavioral discrepancy is observed for MIL-53 with ferric (Fe) and scandium (Sc) metal center. Both dehydrated MIL-53(Fe) and (Sc) show nonporous structure; this difference is due to the two MOFs is initially have a closed or very-narrow pores unlike MIL-53(Al) or (Cr). When CO_2 introduced, instead of N_2, a Type-VI-sorption-isotherm is observed. MOF attributed with breathing ability reportedly is chemically more stable than other MOFs without this ability. MIL-53, for instance, remains indestructible either in the presence of water molecules or heat up to 500°C [44, 75, 76].

12.8.2 Activation of MOFs

Many studies reported that the enormous surface area and porosity of MOFs can be achieved by removing the guest molecules from the framework matrix without destroying the integrity of the structure and pores [77]. This process is called activation or sometimes referred to as purification. The presence of guest molecules is originated by the excess reactant, solvent, and reaction intermediate ion or molecules. Often, those guest molecules cause pore blocking, thus reducing the surface area as well as pore volume of MOFs; subsequently, they may also demote the adsorption performance of MOFs. For instance, the presence of dimethylformamide (DMF) residual-guest molecules in the structural of BioMIL-3

([Ca_2(azbz-TC)-(H_2O)(DMF)], azbz-TC = 3,3′,5,5′-azobenzenetetracarboxylate) cause steric hindrance and diffusion blocking thus constraint the interaction of the gas adsorbate (nitric oxide) with the metal sites [78]. In some cases, activation of MOFs may also grant breathing ability to MOFs [79]. Several activation processes have been developed by engineers, such as heat treatment, vacuum treatment, chemical treatment, solvent exchange, supercritical treatment, and freeze drying.

Ideally activation process by using heat treatment can only be successfully applied to MOFs with high thermal stability. However, in most cases, the implementation of heat activation fails to purify the pore and may destruct the crystallinity of the MOFs in response to the demolition of the coordination bond of MOFs by a large amount of heat energy. A subclass of MOFs namely ZIF-8 (ZIF = zeolitic imidazolate framework) with surface area 874 m^2/g show an improvement in adsorption activity upon heat treatment. The adsorption capacity of ZIF-8 towards toluene increased from 5.4 to 41.1 mg/g after heating treatment at 300°C [80]. Similarly, vacuum treatment is also known to cause loss of crystallinity due to the large capillary forces and surface tension (as the liquid phase transforms into gas phase) that result in the destruction of MOF coordination bonds. In chemical treatment, a strong acid solution is commonly used to rigorously wash the guest molecules out of the pore of MOF. Activation by the solvent is done by exchanging the high boiling point solvent with a lower boiling point solvent, this is to minimize the molecular forces caused by the solvent during activation. Supercritical treatment is basically similar to that solvent exchange, but the treatment is done at supercritical condition by using liquid CO_2 as the common solvent. The implementation of freeze drying activation is preceded by the introduction of benzene into MOFs followed by several freezing-thawing cycles [77, 81–84].

12.8.3 Applied Pressure

Application of pressure play an important role in promoting the adsorbate-adsorbent interaction. Hamon *et al.* show that adsorption of H_2S onto MIL-53(Cr) at low pressure (4.5 kPa) resulting in low adsorption capacity of 3.02 mmol/g. At higher applied pressure of 1.6 MPa, 13.12 mmol H_2S can be adsorbed per gram of MIL-53(Cr). This behavior attributed to the breathing of MIL-53(Cr). At low pressure, closing pore phenomena occurs due to the strong interaction of OH-group of MOF with H_2S. At higher pressure, there is a complete pore vacancy as the interaction of OH-group of MOF with H_2S getting weaker. The pore opening is allowing the total filling of the pores, which leads to higher adsorption capacity [38].

12.8.4 Relative Humidity

Adsorption which performed in humid gas flow or in the presence of water molecules, may have a decreased adsorption capacity. As shown in Table 12.5, the adsorption performs in a dry gas flow have better adsorption capacities than that of wet flow. A similar phenomenon also occurs in the adsorption study of MOF-177 by Yang *et al.* The adsorption of toluene and trichloroethylene on MOF-177 is decreased steeply as the relative humidity altered from 0% to 50%. The water molecules initiate competition against adsorbate to bind to the open metal sites of MOF, thus cause suppressions to the adsorption performance. The other possibility is that the exposure of water to MOF may disrupt its structure [29, 60]. In particular, [Ni(bpb)] MOF may effectively adsorb VOCs in a humid condition. In a dynamic adsorption experiment with 60% humidity, 1 g of [Ni(bpb)] (which synthesized from organic linker namely 1,4-(4-bispyrazolyl)benzene (bpb)) can adsorb 340 mg of thiopene [85].

12.8.5 Breakthrough Conditions

As so called breakthrough curve give a full depiction of adsorption dynamic. The curve shows the change of VOC concentration in the effluent throughout adsorption time. The sorption capacity, thermodynamic (heat transfer or release), kinetics (adsorption rate), selectivity, and gas flow rate are the crucial parameters that influence the curve. An adsorbent with large sorption capacity will have longer breakthrough times since it can retain more adsorbate molecules. However, if the sorption kinetics proceed very slowly, there is a possibility of the spontaneous breakthrough, this will cause alteration in the shape of the curve. A rapid mass transfer (high flow rate) produces a steep breakthrough curve and narrow mass transfer zone. Broadening of the mass transfer zone may occur in the case of exothermic adsorption. The temperature increase due to heat release by adsorption leads to flatter and more asymmetric curve. An evenly mass dispersion may also lead to broadening of the mass transfer zone [69].

12.8.6 Functional Group of MOFs

The presence of functional groups of MOF in enhancing the adsorption has been demonstrated. A particular functional group may also help to prevent competition between adsorbate molecules and water molecules; consequently, the adsorption still performs excellently in the humid condition. Addition of fluorine in the building block of FMOF-1 (prepared by reacting monovalent silver and 3,5-bis(trifluoromethyl)-1,2,4-triazolate) cause it to be hydrophobic, therefore the water molecules cannot enter the pore of FMOF-1. The hydrophobicity of FMOF-1 also makes it be selective against certain VOCs (benzene, *p*-xylene, cyclohexane, toluene, and *n*-hexane) [86]. A similar phenomenon is observed for $[Ni_8(OH)_4(H_2O)_2(L5\text{-}CF_3)_6]$, which build from 4,4'-(benzene-1,4-diyldiethyne-2,1-diyl)bis(1*H*-pyrazole) abbreviated as L5. The presence of coordination of the metal with the pyrazole group induces the stability of the MOF against water molecules, and the MOF still able to adsorb diethylsulfide at 80% relative humidity [61].

The presence of the ionic group in the MOF structure may induce selectivity against certain VOC. The NH_4^+ in the framework of MOF namely $NH_4[Cu_3(OH)(capz)]$, where capz = 4-carboxypyrazolato, causes alteration in the pore structure of the MOF and also lead to adsorption selectivity toward cyclohexane–benzene mixtures, which related to the size-exclusion of the adsorbate molecules.

12.8.7 Concentration, Molecular Size, and Type of VOCs

Adsorption can occur effectively when there is compatibility between VOCs and MOFs as the guest and host molecules. The key features of the compatibility, in this case, are the molecular size, molecular orientation, and the number of molecules of the VOC adsorbate. Adsorption can occur efficiently when all key features of VOCs are consistent with the pore structure and crystal symmetry of MOF as adsorbent. For instance, MIL-53(Al) can adsorb doubled amount of *p*-xylene when the pore is in the vacant form. The availability of space in the vacant pore is adequate to accommodate two molecules of *p*-xylene as illustrated in Figure 12.7. Also, the length of *p*-xylene is compatible with the size of MIL-53 open pore [75]. If the amount of adsorbate molecules is too many, the active adsorption sites of the adsorbent may not enough to accommodate all of the desired adsorbate molecules (saturation of the adsorbent). Also, there might be repulsive force between the adsorbate molecules thus decrease the efficiency of the adsorption.

The type of VOC adsorbents determines the adsorption performance. The strength of inter- or intra-molecular interaction of each type of VOC is determined by their affinity; the interaction can be polarization and/or electrostatic, acid-base, or physisorption. A density functional theory has been applied to investigate the adsorption energies between several small organic molecules with a copper metal cluster of HKUST-1. It is revealed that only the adsorption energy of NH_3 can surpass the

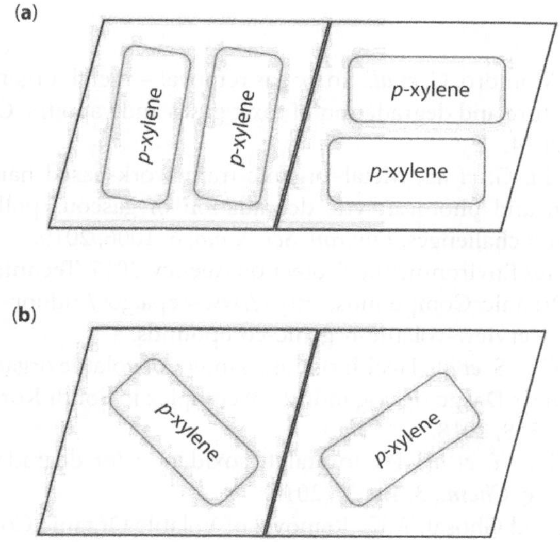

Figure 12.7 Adsorption of *p*-xylene by MIL-53 at open pore condition. (a) There is double amount of *p*-xylene adsorbed when the molecular orientation is compatible with pore structure and symmetry, (b) a lesser amount of *p*-xylene can be adsorbed since the orientation is not compatible. Adapted from Ref. [75].

water molecules; with the value of adsorption energies is decreased in the following order: $NH_3 > H_2O > PH_3 > H_2S > SO_2 > CO \sim OCS \sim CO_2 \sim N_yO_x > N_2 > O_2$ [87].

12.9 Future Perspective

The superiority and efficacy of MOF in overcoming VOCs has been proven by many engineers. MOF can be used as an efficient adsorbent for removing and/or storing VOCs, some MOFs also have catalytic activity against VOCs. In many lab scale experiments, it has been shown that the MOF adsorption capacity is very high; i.e., can reach >1 g VOC per g MOF. But there are still very few experiments that show the feasibility of VOC adsorption by MOF on an industrial scale. The feasibility of MOF application in the industry can be said to be far from ideal, both in terms of economics and synthesis. This is because in industry, VOC is always produced in large quantities so it requires large amounts of MOF as an adsorbent; this will cause an increase in total production costs due to the high cost of MOF synthesis.

References

1. Barea, E., Montoro, C. *et al.*, Toxic gas removal – metal–organic frameworks for the capture and degradation of toxic gases and vapours. *Chem. Soc. Rev.*, 43, 5419, 2014.
2. Wen, M., Li, G. *et al.*, Metal–organic framework-based nanomaterials for adsorption and photocatalytic degradation of gaseous pollutants: Recent progress and challenges. *Environ. Sci. Nano*, 6, 1006, 2019.
3. United States Environmental Protection Agency, 2017. Technical Overview of Volatile Organic Compounds, https://www.epa.gov/indoor-air-quality-iaq/technical-overview-volatile-organic-compounds.
4. Shuai, J., Kim, S. *et al.*, Health risk assessment of volatile organic compounds exposure near Daegu dyeing industrial complex in South Korea. *BMC Public Health*, 18, 528, 2018.
5. Lin, L., Chai, Y. *et al.*, Photocatalytic oxidation for degradation of VOCs. *Open J. Inorg. Chem.*, 3, 14–25, 2013.
6. Khan, F.I. and Ghosal, A.K., Removal of Volatile Organic Compounds from polluted air. *J. Loss Prevent. Proc.*, 13, 527–545, 2000.
7. U.S. National Library of Medicine, 2017. Volatile Organic Compounds (VOCs), https://toxtown.nlm.nih.gov/chemicals-and-contaminants/volatile-organic-compounds-vocs.
8. The National Institute for Occupational Safety and Health (NIOSH), Centers for Disease Control and Prevention, The NIOSH Pocket Guide to Chemical Hazards, https://www.cdc.gov/niosh/npg/pgintrod.html.
9. Loreto, F. and Fares, S., Biogenic Volatile Organic Compounds and Their Impacts on Biosphere–Atmosphere Interactions, in: *Developments in Environmental Science*, Matyssek, R., *et al.* (Eds.), pp. 57–75.
10. Peñuelas, J., Asensio, D. *et al.*, Biogenic volatile emissions from the soil. *Plant Cell Environ.*, 37, 1866–1891, 2014.
11. Ryu, C.-M., Farag, M.A. *et al.*, Bacterial volatiles promote growth in Arabidopsis. *PNAS*, 100, 4927–4932, 2003.
12. Naznin, H.A., Kiyohara, D. *et al.*, Systemic Resistance Induced by Volatile Organic Compounds Emitted by Plant Growth-Promoting Fungi in *Arabidopsis thaliana*. *PLoS One*, 9, e86882, 2014.
13. Jaeger, D.M., Runyon, J.B. *et al.*, Signals of speciation: Volatile organic compounds resolve closely related sagebrush taxa, suggesting their importance in evolution. *New Phytol.*, 211, 1393–1401, 2016.
14. González-Mas, M.C., Rambla, J.L. *et al.*, Volatile Compounds in *Citrus* Essential Oils: A Comprehensive Review. *Front. Plant Sci.*, 10, 12, 2019.
15. Shao, M., Czapiewski, K.V. *et al.*, Volatile organic compound emissions from Scot pine: Mechanisms and description by alogarithms. *J. Geophys. Res.*, 106, 20483–20491, 2001.
16. Rasmann, S., Köllner, T.G. *et al.*, Recruitment of entomopathogenic nematodes by insect-damaged maize roots. *Nature*, 434, 732–737, 2005.

17. Fialho, M.B., Toffano, L. *et al.*, Volatile organic compounds produced by *Saccharomyces cerevisiae* inhibit the *in vitro* development of *Guignardia citricarpa*, the causal agent of citrus black spot. *World J. Microbiol. Biotechnol.*, 26, 925–932, 2010.
18. Oro, L., Feliziani, E. *et al.*, Volatile organic compounds from *Wickerhamomyces anomalus*, *Metschnikowia pulcherrima* and *Saccharomyces cerevisiae* inhibit growth of decay causing fungi and control postharvest diseases of strawberries. *Int. J. Food Microbiol.*, 265, 18–22, 2018.
19. Cuffia, F., Bergamini, C.V. *et al.*, Characterization of volatile compounds produced by *Lactobacillus helveticus* strains in a hard cheese model. *Food Sci. Technol. Int.*, 24, 67–77, 2017.
20. Chen, J., Tang, J.N. *et al.*, The production characteristics of volatile organic compounds and their relation to growth status of Staphylococcus aureus in milk environment. *J. Dairy Sci.*, 101, 4983–4991, 2018.
21. Ezeji, T.C., Qureshi, N. *et al.*, Production of acetone, butanol and ethanol by *Clostridium beijerinckii* BA101 and *in situ* recovery by gas stripping. *World J. Microbiol. Biotechnol.*, 19, 595–603, 2003.
22. Daisy, B.H., Strobel, G.A. *et al.*, Naphthalene, an insect repellent, is produced by *Muscodor vitigenus*, a novel endophytic fungus. *Microbiology*, 148, 3737–3741, 2002.
23. Singh, S.K., Strobel, G.A. *et al.*, An endophytic Phomopsis sp. possessing bioactivity and fuel potential with its volatile organic compounds. *Microb. Ecol.*, 61, 729–739, 2011.
24. G. Strobel, D. Manker *et al.*, Endophytic fungi and methods of use. U.S. Patent 7754203, assigned to Agra Quest, Inc. 2010.
25. Drilling, K. and Dettner, K., Electrophysiological responses of four fungivorous coleoptera to volatiles of *Trametes versicolor*: Implications for host selection. *Chemoecology*, 19, 109–115, 2009.
26. Hussain, A., Tian, M.Y. *et al.*, Differential fluctuation in virulence and VOC profiles among different cultures of entomopathogenic fungi. *J. Invertebr. Pathol.*, 104, 166–171, 2010.
27. Liu, B., Yang, F. *et al.*, Adsorption of Phenol and *p*Nitrophenol from Aqueous Solutions on Metal–Organic Frameworks: Effect of Hydrogen Bonding. *J. Chem. Eng. Data*, 59, 1476–1482, 2014.
28. Vellingiri, K., Szulejko, J.E. *et al.*, Metal organic frameworks as sorption media for volatile and semi-volatile organic compounds at ambient conditions. *Sci. Rep.*, 6, 27813, 2016.
29. Yang, K., Xue, F. *et al.*, Adsorption of volatile organic compounds by metal–organic frameworks MOF-177. *J. Environ. Chem. Eng.*, 1, 713–718, 2013.
30. Zhang, G., Liu, Y. *et al.*, Adsorption of volatile organic compounds onto natural porous minerals. *J. Hazard. Mater.*, 364, 317–324, 2019.
31. Yang, H., Deng, J. *et al.*, Au/MnOx/3DOM SiO2: Highly active catalysts for toluene oxidation. *Appl. Catal., A*, 507, 139–148, 2015.

32. Hao, X., Li, R. *et al.*, Numerical simulation of a regenerative thermal oxidizer for volatile organic compounds treatment *Environ. Eng. Res.*, 2018.
33. Christen, P., Domenech, F. *et al.*, Biofiltration of volatile ethanol using sugar cane bagasse inoculated with *Candida utilis*. *J. Hazard. Mater.*, 89, 253–265, 2002.
34. Lalanne, F., Malhautier, L. *et al.*, Absorption of a mixture of volatile organic compounds (VOCs) in aqueous solutions of soluble cutting oil. *Bioresour. Technol.*, 99, 1699–1707, 2008.
35. Rebollar-Perez, G., Carretier, E. *et al.*, Volatile Organic Compound (VOC) Removal by Vapor Permeation at Low VOC Concentrations: Laboratory Scale Results and Modeling for Scale Up. *Membranes*, 1, 80–90, 2011.
36. Alivand, M.S., Shafiei-Alavijeh, M. *et al.*, Facile and high-yield synthesis of improved MIL-101(Cr) metal–organic framework with exceptional CO_2 and H_2S uptake; the impact of excess ligand-cluster. *Microporous Mesoporous Mater.*, 279, 153–164, 2019.
37. Du, P.D., Thanh, H.T.M. *et al.*, Metal–Organic Framework MIL-101: Synthesis and Photocatalytic Degradation of Remazol Black B Dye. *J. Nanomat.*, 2019, Article ID 6061275, 15 pages, 2019.
38. Hamon, L., Serre, C. *et al.*, Comparative Study of Hydrogen Sulfide Adsorption in the MIL-53(Al, Cr, Fe), MIL-47(V), MIL-100(Cr), and MIL-101(Cr) Metal–Organic Frameworks at Room Temperature. *J. Am. Chem. Soc.*, 131, 8775–8777, 2009.
39. Huang, C.-Y., Song, M. *et al.*, Probing the Adsorption Characteristic of Metal–Organic Framework MIL-101 for Volatile Organic Compounds by Quartz Crystal Microbalance. *Environ. Sci. Technol.*, 45, 4490–4496, 2011.
40. Leng, K., Sun, Y. *et al.*, Rapid Synthesis of Metal–Organic Frameworks MIL-101(Cr) Without the Addition of Solvent and Hydrofluoric Acid. *Cryst. Growth Des.*, 16, 1168–1171, 2016.
41. Shafiei, M., Alivand, M.S. *et al.*, Synthesis and adsorption performance of a modified micro-mesoporous MIL-101(Cr) for VOCs removal at ambient conditions. *Chem. Eng. J.*, 341, 164–174, 2018.
42. Yot, P.G., Ma, Q. *et al.*, Large breathing of the MOF MIL-47(V^{IV}) under mechanical pressure: A joint experimental–modelling exploration. *Chem. Sci.*, 3, 1100–1104, 2012.
43. Boutin, A., Coudert, F.-X. *et al.*, The Behavior of Flexible MIL-53(Al) upon CH_4 and CO_2 Adsorption. *J. Phys. Chem. C*, 114, 22237–22244, 2010.
44. Alhamami, M., Doan, H. *et al.*, A Review on Breathing Behaviors of Metal–Organic-Frameworks (MOFs) for Gas Adsorption. *Materials*, 7, 3198–3250, 2014.
45. Finsy, V., Kirschhock, C.E. *et al.*, Framework breathing in the vapour-phase adsorption and separation of xylene isomers with the metal–organic framework MIL-53. *Chemistry*, 15, 7724–7731, 2009.

46. Britt, D., Tranchemontagne, D. *et al.*, Metal–organic frameworks with high capacity and selectivity for harmful gases. *Proc. Natl. Acad. Sci.*, 105, 11623–11627, 2008.
47. Chae, H.K., Siberio-Pérez, D.Y. *et al.*, A route to high surface area, porosity and inclusion of large molecules in crystals. *Let. Nature*, 427, 523–527, 2004.
48. Furukawa, H., Miller, M.A. *et al.*, Independent verification of the saturation hydrogen uptake in MOF-177 and establishment of a benchmark for hydrogen adsorption in metal–organic frameworks. *J. Mater. Chem.*, 17, 3197–3204, 2007.
49. Tranchemontagne, D.J., Hunt, J.R. *et al.*, Room temperature synthesis of metal–organic frameworks: MOF-5, MOF-74, MOF-177, MOF-199, and IRMOF-0. *Tetrahedron*, 64, 8553–8567, 2008.
50. He, W.-W., Yang, G.-S. *et al.*, Phenyl Groups Result in the Highest Benzene Storage and Most Efficient Desulfurization in aSeries of Isostructural Metal–Organic Frameworks. *Chem. Eur. J.*, 21, 9784–9789, 2015.
51. Bao, S.-J., Krishna, R. *et al.*, A stable metal–organic framework with suitable pore sizes and rich uncoordinated nitrogen atoms on the internal surface of microspores for highly efficient CO_2 capture. *J. Mater. Chem. A*, 3, 7361–7367, 2015.
52. McKinlay, A.C., Eubank, J.F. *et al.*, Nitric Oxide Adsorption and Delivery in Flexible MIL-88(Fe) Metal–Organic Frameworks. *Chem. Mater.*, 25, 1592–1599, 2013.
53. Gándara, F., García-Cortés, A. *et al.*, Rare Earth Arenedisulfonate Metal–Organic Frameworks: An Approach toward Polyhedral Diversity and Variety of Functional Compounds. *Inorg. Chem.*, 46, 3475–3484, 2007.
54. Gándara, F., Puebla, E.G. *et al.*, Controlling the Structure of Arenedisulfonates toward Catalytically Active Materials. *Chem. Mater.*, 21, 655–661, 2009.
55. Jiang, H.L., Liu, B. *et al.*, Au@ZIF-8: CO oxidation over gold nanoparticles deposited to metal–organic framework. *J. Am. Chem. Soc.*, 131, 11302–11303, 2009.
56. Procopio, E.Q., Linares, F. *et al.*, Cation-Exchange Porosity Tuning in Anionic Metal–Organic Frameworks for the Selective Separation of Gases and Vapors and for Catalysis. *Angew. Chem. Int. Ed.*, 49, 7308, 2010.
57. Shen, Y., Jiang, P. *et al.*, Recent Progress in Application of Molybdenum-Based Catalysts for Epoxidation of Alkenes. *Catalysts*, 9, 31, 2019.
58. Yao, P., Liu, H. *et al.*, Enhanced visible-light photocatalytic activity to volatile organic compounds degradation and deactivation resistance mechanism of titania confined inside a metal–organic framework. *J. Colloid Interface Sci.*, 522, 174–182, 2018.
59. Gómez-Avilés, A., Peñas-Garzón, M. *et al.*, Mixed Ti-Zr metal–organic-frameworks for the photodegradation of acetaminophen under solar irradiation. *Appl. Catal., B*, 253, 253–262, 2019.
60. Glover, T.G., Peterson, G.W. *et al.*, MOF-74 building unit has a direct impact on toxic gas adsorption. *Chem. Eng. Sci.*, 66, 163–170, 2011.

61. Padial, N.M., Procopio, E.Q. *et al.*, Highly Hydrophobic Isoreticular Porous Metal–Organic Frameworks for the Capture of Harmful Volatile Organic Compounds. *Angew. Chem. Int. Ed.*, 52, 1–6, 2013.
62. Zhou, X., Huang, W. *et al.*, Enhanced separation performance of a novel composite material GrO@MIL-101 for CO_2/CH_4 binary mixture. *Chem. Eng. J.*, 266, 339–344, 2015.
63. Sun, X., Xia, Q. *et al.*, Synthesis and adsorption performance of MIL-101(Cr)/graphite oxide composites with high capacities of n-hexane. *Chem. Eng. J.*, 239, 226–232, 2014.
64. Hercigonja, R. and Rakić, V., Enthalpy–entropy Compensation for n-hexane Adsorption on Y Zeolite Containing Transition Metal Cations. *Sci. Sinter.*, 47, 83–88, 2015.
65. Liu, W., Carrasco, J. *et al.*, Benzene Adsorbed on Metals: Concerted Effect of Covalency and van der Waals Bonding. *Phys. Rev. B*, 86, 245405, 2012.
66. Möller, A., Eschrich, R. *et al.*, Dynamic and equilibrium-based investigations of CO_2-removal from CH_4-rich gas mixtures on microporous adsorbents. *Adsorption*, 23, 197–209, 2017.
67. Grande, C.A., Blom, R. *et al.*, High-Pressure Separation of CH_4/CO_2 Using Activated Carbon. *Chem. Eng. Sci.*, 89, 10–20, 2013.
68. Bastos-Neto, M., Moeller, A. *et al.*, Adsorption Measurements of Nitrogen and Methane in Hydrogen-Rich Mixtures at High Pressures. *Ind. Eng. Chem. Res.*, 50, 10211–10221, 2011.
69. INSTRUMENTS GmbH & Co. KG, 2019. Dynamic Sorption, https://www.dynamicsorption.com/dynamic-sorption-method/breakthrough-measurement/.
70. Hori, H., Tanaka, I. *et al.*, Breaktrough Time on Activated Carbon Fluidized Bed Adsorbers. *J. Air Pollut. Control Assoc.*, 38, 269–271, 1988.
71. Wang, W.-Z., Brusseau, M.L. *et al.*, Nonequilibrium and Nonlinear Sorption during Transport of Cadmium, Nickel, and Strontium through Subsurface Soils. *Adsorpt. Met. Geomedia*, 427–443, 1998.
72. Kitagawa, S., Kitaura, R. *et al.*, Functional Porous Coordination Polymers. *Angew. Chem. Int. Ed.*, 43, 2334, 2004.
73. Foo, M.L., Matsuda, R. *et al.*, An Adsorbate Discriminatory Gate Effect in a Flexible Porous Coordination Polymer for Selective Adsorption of CO_2 over C_2H_2. *J. Am. Chem. Soc.*, 138, 3022–3030, 2016.
74. Serre, C., Millange, F. *et al.*, Very Large Breathing Effect in the First Nanoporous Chromium(III)-Based Solids: MIL-53 or $Cr^{III}(OH)\cdot\{O_2C–C_6H_4–CO_2\}\cdot\{HO_2C–C_6H_4–CO_2H\}_x\cdot H_2O_y$. *J. Am. Chem. Soc.*, 124, 13519–13526, 2002.
75. Wu, H., Gong, Q. *et al.*, Commensurate adsorption of hydrocarbons and alcohols in microporous metal organic frameworks. *Chem. Rev.*, 112, 836–868, 2012.

76. Llewellyn, P.L., Bourrelly, S. *et al.*, How hydration drastically improves adsorption selectivity for CO_2 over CH_4 in the flexible chromium terephthalate MIL-53. *Angew. Chem. Int. Ed.*, 45, 7751–7754, 2006.
77. Mondloch, J.E., Karagiaridi, O. *et al.*, Activation of metal–organic framework materials. *Cryst. Eng. Comm.*, 15, 9258–9264, 2013.
78. Miller, S.R., Alvarez, E. *et al.*, A rare example of a porous Ca-MOF for the controlled release of biologically active NO. *Chem. Commun.*, 49, 7773, 2013.
79. Engel, E.R., Jouaiti, A. *et al.*, Activation-Dependent Breathing in a Flexible Metal–Organic Framework and the Effects of Repeated Sorption/Desorption Cycling. *Angew. Chem. Int. Ed.*, 56, 8874–8878, 2017.
80. Jafaria, S., Ghorbani, F. *et al.*, Removal of Toluene from Air by Zeolitic Imidazolate Framework-8: Synthesis, Characterization, and Experimental Breakthrough Curve. *Int. J. Sci. Stud.*, 5, 1073–1082, 2017.
81. Abraham, B.F., Hoskins, B.F. *et al.*, Assembly of porphyrin building blocks into network structures with large channels. *Nature*, 369, 727–729, 1994.
82. Bedia, J., Muelas-Ramos, V. *et al.*, A Review on the Synthesis and Characterization of Metal Organic Frameworks for Photocatalytic Water Purification. *Catalysts*, 9, 52, 2019.
83. Férey, G., Mellot-Draznieks, C. *et al.*, A chromium terephthalate-based solid with unusually large pore volumes and surface area. *Science*, 23, 2040–2042, 2005.
84. Cavka, J.H., Jakobsen, S. *et al.*, A new zirconium inorganic building brick forming metal organic frameworks with exceptional stability. *J. Am. Chem. Soc.*, 130, 13850–13851, 2008.
85. Galli, S., Masciocchi, N. *et al.*, Adsorption of Harmful Organic Vapors by Flexible Hydrophobic Bis-pyrazolate Based MOFs. *Chem. Mater.*, 22, 1664–1672, 2010.
86. Yang, C., Kaipa, U. *et al.*, Fluorous metal–organic frameworks with superior adsorption and hydrophobic properties toward oil spill cleanup and hydrocarbon storage. *J. Am. Chem. Soc.*, 133, 18094–18097, 2011.
87. Supronowicz, B., Mavrandonakis, A. *et al.*, Interaction of Small Gases with the Unsaturated Metal Centers of the HKUST-1 Metal Organic Framework. *J. Phys. Chem. C*, 117, 14570–14578, 2013.

13

Application of Metal–Organic Framework and Their Derived Materials in Electrocatalysis

Gopalram Keerthiga[1]*, Peramaiah Karthik[2] and Bernaurdshaw Neppolian[2]

[1]*Chemical Engineering Department, SRM Institute of Science and Technology, Kattankulathur, Chennai, Tamil Nadu, India*
[2]*SRM Research Institute, SRM Institute of Science and Technology, Kattankulathur, Chennai, Tamil Nadu, India*

Abstract
Metal organic framework and its group are foreseen as the next-generation leading porous materials and find enormous potential in energy and environmental applications. Exclusive properties such as large surface area, high crystallinity, well-defined pore properties, easily tunable and tailorable structures, and chemical functionality make MOFs unique. This book chapter summarizes the related contribution of metal organic framework and its application pertaining to its usage as electrocatalyst in diverse applications. Recent developments of MOF in energy and environmental application such as water splitting, hydrogen evolution reaction, oxygen evolution reaction, carbon dioxide reduction, and electrochemical sensing has been summarized. The theme of the book chapter focuses on designing of nanostructured heterogeneous catalyst modified with size effect, with added second element (metal, metal oxide), atom scaled catalyst, addition of ionic liquid, and alternation of ligand from literature has been abridged.

Keywords: Electrocatalyst, MOF, HER, ORR, OER, CO2RR, electrochemical sensing

**Corresponding author*: keerthig@srmist.edu.in

Inamuddin, Rajender Boddula, Mohd Imran Ahamed and Abdullah M. Asiri (eds.) Applications of Metal–Organic Frameworks and Their Derived Materials, (357–376) © 2020 Scrivener Publishing LLC

List of Abbreviations

MOF—Metal Organic Framework
HER—Hydrogen Evolution Reaction
ORR—Oxygen Evolution Reaction
OER—Oxygen Reduction Reaction
CO_2RR—Carbon dioxide Reduction Reaction

13.1 Introduction

The innovative ideas on energy and environmental problems demand the development and discovery of novel materials. Although there is no good technology available in hand to solve the energy and environmental problems, synergistic independent methods like photoelectrochemical, photochemical, and electrochemical methods are some of the promising emerging fields. Being chosen with electrochemical reduction of CO_2 (ECR), the merits of this technique is not restricted to the following, (i) directs the intermediates toward reduction, (ii) quick industrial upgradation, (iii) less formation of byproducts, (iv) avoids complex start up processes, (v) avoids sophisticated costly reactors [1, 2].

The surge in the development of metal organics framework is due to its inherent merits of large surface area and enormous pore volume. Structurally MOFs has porous networks made up of a connecting a metal with the linker (Figure 13.1) [3]. Any modification in the structure can be related to its activity and hence the efforts to develop new MOF focus on increasing chemical stability, conductivity, and superior efficiency. The change in its properties can be brought by altering or introducing organic linkers or inorganic metal nodes.

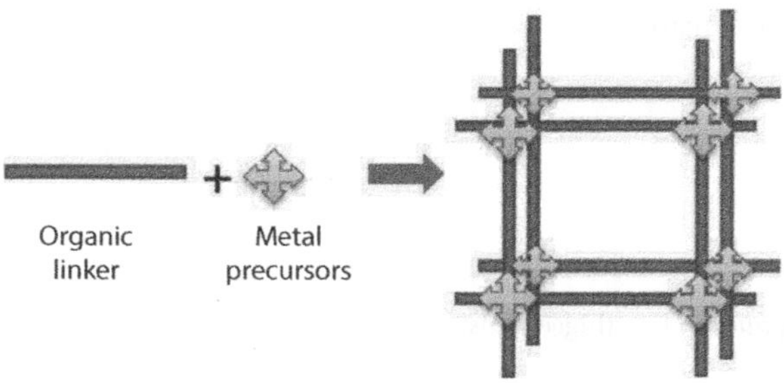

Figure 13.1 Illustration of metal organic framework with metal node and organic linker.

The virtual applications of MOF as electrocatalyst has been established in the field of Hydrogen Evolution Reaction (HER), Oxygen Evolution Reaction (OER), Oxygen Reduction Reaction (OER), Carbon dioxide Reduction Reaction (CO_2RR), and electrochemical sensing, energy storage for its embarked structure and properties with its half-cell reactions (Table 13.1). Tuning the pore size of structure building units and design of appropriate functional groups in the metal sites helps to suit the particular application with improved kinetics of targeted reaction.

The choice in the method of synthesis, templates aids helps in designing MOF with morphologies relating to hollow spheres, nanocages, nanostructures, and polyhedrons. Care should be taken to prevent the decomposition of organic linker and kinetics of crystallization to allow nucleation and growth with sufficient open pore structure. Fabricated MOF should

Table 13.1 Electrochemical half-cell reaction for ORR, OER, HER, and CO_2RR [4].

Electrochemical half-cell reactions	
Oxygen reduction reaction *Acidic aqueous solution* $O_2+4H^++4e^- \rightarrow 2H_2O$	1
Alkaline aqueous solution $O_2+2H_2O+4e^- \rightarrow 4OH^-$	2
Oxygen evolution reaction *Acidic aqueous solution* $2H_2O \rightarrow O_2+4e^-+4H^+$	3
Alkaline aqueous solution $4OH^- \rightarrow O_2+4e^-+2H_2O$	4
Hydrogen evolution reaction *Acidic aqueous solution* $2H^++2e^- \rightarrow H_2$	5
Alkaline aqueous solution $2H_2O+2e^- \rightarrow H_2+2OH^-$	6
Carbon dioxide reduction reaction *Cathodic reaction* $CO_{2(g)} + 8H^+ + 8e^- \rightarrow CH_4+2H_2O$	7
Anodic Reaction $2H_2O \rightarrow O_2 + 4H^++4e^-$	8

exhibit the characteristics of uncoordinated metal sites, specific heteroatoms, covalent functionalization, other building unit interactions, hydrophobicity, porosity, defects. Addition of nanoscale metal catalysts is crucial for the development of higher-performance MOFs [5].

The active metal centers and organic linker can be tuned to attain the following: 1. highly porous structure for enhanced diffusion, 2. high surface area, 3. improved adsorption of reactants, 4. chemical stability, 5. thermal stability, and 6. high conductivity. MOF functionality as electrocatalyst has been realized in many vital applications, inspired by its merits and its developmental scope in catalysis. This chapter summarizes the recent developments and future directions of applications of MOF and their derived materials as an electrocatalyst.

13.2 Perspective Synthesis of MOF and Their Derived Materials

Basic characteristics of MOF to exhibit electrocatalytic property is to possess three essential electrochemical factors of onset potential, current density, and redox-active metal sites for catalytic activity [6]. The preparation was influenced by the addition of surface directing agents, ionic liquid, choice of precursor, and method of synthesis to reap high yield at a large-scale synthesis of MOF.

Preparation in the form of sheets instead of powder finds wide use in energy and environmental application, whereby the strategies by sonication, exfoliation, interfacial synthesis, three-layer synthesis, and surfactant-assisted synthesis methods are on the verge to exhibit new structure and functions [7]. Any surfactant (structure-directing agents) with charged linker helps in development of new MOF with tuning of size and morphology of MOF crystals [8]. The addition of surfactants like PVP, formic acid for preparation of 2D MOF sheets aids to stabilize the growth of MOF and inhibits growth in the vertical direction. The thickness of the MOF-2 nanosheets obtained is 1.5–6.0 nm and the lateral dimension varies from 100 nm to 1 μm.

MOF for the catalytic application was first illustrated by Yaghi in the 1990s, since then the development of MOF attributes to the change in either linker or the metal to suit any desire applications [3]. The surface area of MOFs is larger than compared to zeolites due to catenation. It has a good ability to adsorb any desired molecule and store in its space available. Framework catenation of MOF happens by self-assembly of zwitterionic linker molecules of opposite charge with each other. This interpenetration and interweaving phenomena of catenation build rigidity to MOF at the expense of its porosity with high surface area compared to that of a playground (≌10,000 m^2). The catenation can be controlled by varying solvent, SBU, and its linkers, adjusting pH of the solution, doping of active sites on to the linker [8–10]. Figure 13.2 shows the choice of various metal ions and organic linkers used for MOF synthesis.

Electrosynthesis has attracted good attention in the directional formation of MOF. ITO was first electrodeposited from simple salt solutions where the metal surface can be taken as cation source and also as initial nucleation sites for film growth [11]. Continuous salt free electrochemical production of MOF employs an electrolyte comprising of organic ligand, conduction salt, metal ion, and MOF building block for versatile and effective production of MOF thin films [12]. The continuous synthesis can be varied by the application of an applied potential (2.5 V to 25 V), the concentration of the metal salts and also by the addition of water during synthesis.

The continuous electrochemical synthesis of MOF (HKUST-1) has been attempted using room temperature ionic liquids (ILs) as a solvent and a suitable structure-directing agent [8]. Beside ionic liquids, the eutectic solvents (DESs) having lower melting point have been scaled up to 100 g per batch. Due to their high stability and specific application, the MOF films have been prepared, tested for energy and environmental applications. Figure 13.3 shows the schematic of an H-type cell for electrochemical synthesis and for being used as electrocatalyst for numerous applications.

Realizing its potential application in various fields of electro and photocatalysis, the criteria for industrial-scale synthesis of MOF-2, MOF-5, and IRMOF-8 with the yield of 87%, 91%, and 91% have been established [8]. Among various precursors, copper acetate gives better yield than compared

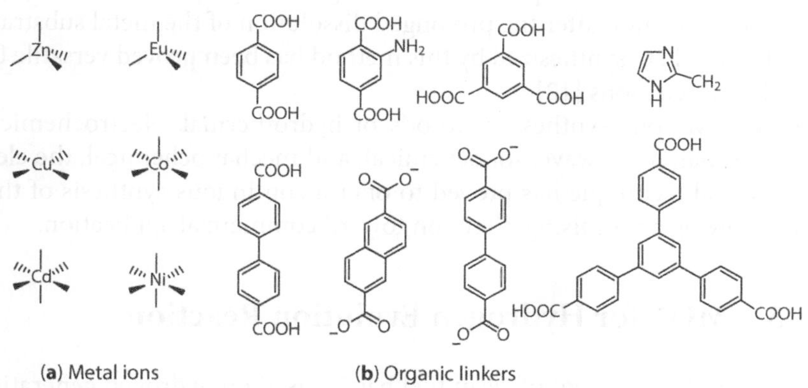

Figure 13.2 Examples of metal ions and organic linkers used in MOF synthesis [8].

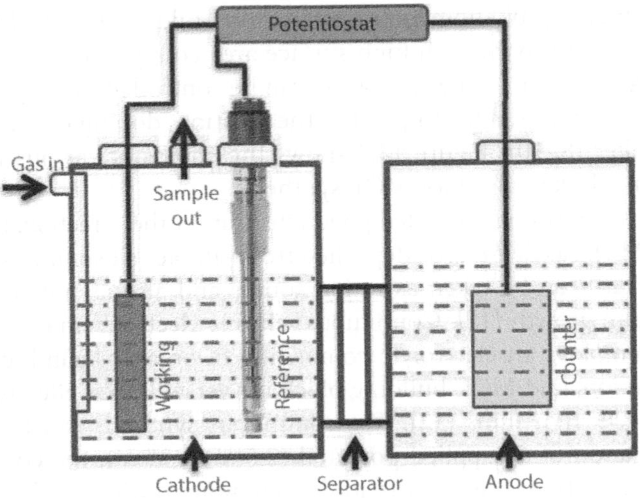

Figure 13.3 Schematic of H cell for electrochemical based applications and synthesis.

to copper nitrate and large-scale synthesis is established in zinc nitrate. The role of counterions has been realized to establish charge balance and to act as templates for MOF synthesis. The industrial scale yield was influenced by the choice of precursor, reaction medium, reaction temperature, and the overall concentration [8].

The combination of precursor and the linker can also be tuned by its choice on the anodic and cathodic deposition. In the anodic deposition, MOF develops at the anode surface and cathodic deposition occurs in the presence of metal salt and organic linker present in solution. It offers an inexpensive and large-scale production of MOF films on conducting films which occurs in steps of nucleation, growth, progressive nucleation, and film exfoliation after the prolonged dissolution of the metal substrate. Cu-ITO, Zn-ITO synthesized by this method has been proved versatile for HER based reactions [12].

Out of various synthesis methods of hydrothermal, electrochemical, solvothermal, microwave, sonochemical, and mechanochemical, the electrochemical technique has proved to offer a continuous synthesis of thin films showing a promising direction toward commercial application.

13.3 MOF for Hydrogen Evolution Reaction

The recent development of clean fuel has focused on hydrogen generation where commercialization is limited due to the high cost and low efficiency of the catalyst. MOF is targeted for lowering reaction potentials and to speed up the reaction rates. Moreover, the high porosity and surface area of MOFs based materials facilitate effective H_2 production, which could lead to the devolvement of stable catalyst for HER [13]. Theoretical and experimental literature-based evidence of MOF for HER has promised new directions for itscommercial development.

The practical application of HER using MOF can be achieved by proper design strategies. Main strategies focused by Wen *et al.* (2019) are (1) the particle size of the metal/metal oxide embedded onto the carbon-based support, (2) its high conductivity for HER applications, (3) energy barrier of reduction by the design of bimetallic and polymetallic MOF, (4) non-metal doped strategy to increase the HER activity and stability [14].

Hybridization of metal with carbon linked to MOF (MOF derived Cu/Carbon) shows cauliflower-like morphology of about 20 nm. The smaller micro-mesostructured catalyst shows good surface area with more positive onset potential and higher current density. XRD results showed Cu^{2+} and Cu^{+} states and offered single electron transfer with Tafel slope of 0.34 and high exchange current density (1.2 mA/cm^2) than compared to bare GCE [15].

Nickel foam (NiMo/$NiMoO_4$@NC/NF) based bimetallic MOF is found to be a low-cost, non-Pt electrode MOFs at a lower overpotential of 80 mV for HER. The catalyst shows long-term stability in an acidic medium for a duration of 24 h. The high catalytic activity is due to its structural merits and the synergy between the MOF and the guest species [16].

A computational screening of a two-dimensional metal–organic framework, to provide an optimum single-atom catalyst for HER and their configuration as a potential catalyst was established by Ji *et al.*, 2019 [17]. Post synthetic modification of MOF, MOF derived and MOF supported catalyst will offer a new avenue in industrial synthesis of MOF for hydrogen evolution reaction.

13.4 MOF for Oxygen Evolution Reaction

Oxygen evolution reaction was improved with MOF based catalyst where a bimetal addition, synthesis strategy, linker change, addition of plasmonic nanoparticles, and progeny of self-supporting MOF is found to influence the current density and overpotential of the desired reaction.

The bimetallic MOFs of Co, Ni with tunable pillar linkers is found to be better than conventional bimetal oxide catalyst. The first metal occupies the random nodal position with metal content varying from 0 to 1 (Wt %) and the addition of second metal alters the stability and band position of MOF and hence influences the properties of MOF for OER. The metal ratios of Co and Ni are varied to study the performance of oxygen evolution reaction [18] where the lowest Tafel slope 55.6 mV dec^{-1} at an overpotential (335 mV) for a current density of 10 mA cm^2 is observed.

The hetero doped atom, metal oxides, phosphides, sulfides, and selenides based pristine MOF and their derivatives for OER based reaction offers tunable compositions, high surface area, and large large pore volume-based catalyst. The intrinsic mechanism behind the actual active sites for OER enhancement was also discussed [19]. Similarly, imidazolium linker (ZIF-67-Co) promises relatively lower onset potential, lower Tafel slope, and long-term stability in acidic media for OER when performed under phosphorylation conditions [20].

In situ grown ultrathin nanosheet arrays of Fe doped Co based 2D MOFs is found to exhibit for quick and feasible oxygen evolution reaction exhibiting lower overpotential and higher turnover frequency [21]. AuNPs based plasmonic nanoparticles doped MOF and ultrathin semiconductor-like Co/Ni-MOF nanosheets is reported to offer a 10-fold increase in oxidation current density than compared to conventional MOF [22]. The additional advantage of illumination of light stimulates the plasmonic nanoparticles with hot holes generation to elevate OER performance.

Maruthapandian *et al.* (2018) have investigated Ni MOF for OER and Urea Oxidation Reaction (UOR) in alkaline medium [23]. The Ni MOF exhibited low overpotential (346 mV) and high current density (10 mA cm^{-2}) than were compared with NiO and RuO_2 for OER with 1 M KOH. The metal and organic moiety coexisted with enhanced actives sites and porous structure. The material offers large specific mass activity and appreciable stability for 12 h [23].

The presence of an amine-based linker (2-aminobenzene-1,4-dicarboxylic acid) finds potential application in electrocatalyst for fuel cells and batteries. Amine linker with metal-doped MOF significantly influences the morphology and performance of catalysts for ORR and OER in alkaline and acidic media. The low-cost doping of Co and Cu based MOF formed petal-like morphology with a halfway potential 0.76 V versus RHE for ORR in 0.1 M KOH and an overpotential of 0.31 V at 10 mA/cm^2 versus RHE for OER in 1 M KOH, respectively [23].

Another novel approach uses self-sacrificial template for building MOF. Ultra-small Fe-rich Fe(Ni)-MOFs, cluster-decorated ultrathin Ni-rich Ni(Fe)-MOFs nanosheets with thickness of 1.56 nm are tightly anchored on the substrate for oxygen evolution reaction. It offers high current density than compared to benchmark RuO_2 catalyst at low overpotential for overall water splitting reaction [24]. Hence the futuristic development in MOF could lead to high performance self-supporting electrocatalyst.

Chen *et al.*, (2017) have devised Co embedded carbon structures as a bifunctional electrocatalyst for OER and ORR in the alkaline medium where Co/nanoporous N doped carbon composites have resulted to have ORR activity at onset potential of 0.92 V and half-wave potential of 0.82 V than compared to benchmark Pt/C catalyst. Optimized Co@C-800 at −1.61 V showed enhanced water splitting activity at 10 mA/cm^2 current density [25].

Developments of MOF as electrocatalyst has been remarkable for OER where new strategies of nanoparticles addition, self-supporting template, linker and metal combination has brought significant development.

13.5 MOF for Oxygen Reduction Reaction

Oxygen reduction reaction and its development lead to better performance of fuel cells which demands commercial demonstration in the current scenario. The sluggish reaction kinetics of fuel cell demands the invasion of new catalyst like MOF where its recent developments in ORR have been summarized.

Tripathy *et al.*, (2019) worked on Co-MOF with accessible surface area, robust stability and availabilities of more active sites for DMFC. The turnover frequency was found to be 93.21 s^{-1} at the overpotential of 350 mV than compared to RuO_2 at a higher current density of 10 mA cm^{-2}. The MOF shows higher durability for long hours of reaction and tolerance toward methanol [24].

Hierarchical bimetallic MOF nanostructures and trimetallic MOF helps in the promotion of OER activity. During the controlled heat treatment Ni, Co, and Fe, the step by step addition of metals at its corresponding temperature leads to trimetallic NiCoFe-MOF-74. Similarly, a hexadentate carboxylate ligand with a (6,6)-connected via the network has shown a promising result for OER as a trimetallic catalyst [25].

Gong *et al.* (2019) discovered annealing as the versatile and inexpensive approach for the preparation of MOF. OER analysis of Co complexed annealed MOF in KOH with high surface area and wider pore size distribution offered an overpotential of 420 mV @10 mA/cm^2 at a Tafel slope of 55 mV/dec [26].

Bucci *et al.*, (2018) have synthesized three imidazolate-based Co-MOFs, IFP-5, -8, and 10 with a different peripheral group –R (–Me, –OMe, and –OEt, respectively) by a solvothermal method and tested toward oxygen evolution reaction (OER). The imidazolate works on the formation of Co(O)OH phase of MOF during the reaction and facilitates OER along with the altered phase of peripheral group R [27].

Varying the bimetallic content of Ni^{2+}/Zn^{2+} ratio in metal–organic framework nanomaterials as an electrocatalyst regulated the morphology and function for OER. The Urchin-like microspheres works for facilitating the diffusion of gas and reducing the transport resistance of ions by its features of the large interfacial area and convenient diffusion channels. Higher the Ni ratio, higher was the performance of bimetallic MOF [28].

Li *et al.* (2018) have worked on the preparation of CoP nanosheets with Co as center by a facile one-step low-temperature phosphidation process for bifunctional hydrogen and oxygen evolution electrocatalyst. It has a large specific surface area and rich catalytic active sites in acidic and alkaline environments. The CoP-NSC works as a robust catalyst than compared to CoP/C, CoP particles, and comparable to those of commercial noble-metal catalysts [19].

Thus, amendments focus on the preparation, doping of bimetallic MOF, annealing, change in linker for lower overpotential based oxygen evolution reaction.

13.6 MOF for CO_2 Electrochemical Reduction Reaction

The craving demand to combat atmospheric concentrations of CO_2 emerges hastily to survive the planet and its inhabitants. CO_2 is the most stable molecule whose reduction targets at multi electron transfer kinetics and drives our attention in activating CO_2 molecule for its utilization. Though Jaramillo's group reports 16 different multicarbon products (like ethylene glycol, glycolaldehyde, hydroxyacetone, acetone, and glyoxal) on polycrystalline Cu surface, commercialized demonstration on continuous flow set up targeting higher hydrocarbon products by a novel material remains to be one of the biggest scientific challenge [29]. The initial formation of the CO_2 radical intermediate before reduction offers a high activation energy barrier and hence the need for suitable catalyst is on demand [30]. The catalyst amendments by novel hybrids, nanostructure and alloys is limited by CO formation with traces of hydrocarbon and hence there is shifting paradigm in the research toward ECR of CO_2 targeting at C_{n+1} value added products [31].

13.6.1 Electrosynthesis of MOF for CO_2 Reduction

The opportunities in MOF for CO_2 reduction is huge where the materialistic behavior ranges from choice of bimetals to electrolytes and its preparation methods. Electrosynthesis of Cu MOF at room temperature employs two Cu electrodes in ethanol with BTC as linker and TBATFB as supporting electrolyte under a potential of 30 V for 2.5 h in nitrogen atmosphere [32]. The catalyst ($Cu_3(BTC)_2$) was washed, dried followed by activation at 150°C for 3 h to yield a coordination state of Cu^{4+}. It shows high absorption capacities than the MOF prepared by solvothermal method and yields CO and CH_4 [32].

Re-based MOF thin film has been deposited onto a conductive FTO electrode by liquid-phase epitaxy with high Faradic efficiency of 93% for CO. The monolithic coatings showed good results with one order of magnitude higher than electrocatalytically active MOF thin films [33]. This proves the fact that metal–organic frameworks (MOFs) are regarded as promising material for CO_2 adsorption, where the mass transfer limitation near the electrode is eliminated for ECR of CO_2.

13.6.2 Composite Electrodes as MOF for CO_2 Reduction

The challenges in catalyst design are therefore to select water- and acid-stable catalyst that can operate with a low overpotential to reduce the undesirable hydrogen byproduct and to enhance selectivity for a particular product from CO_2. Recently, Cu-MOF [34], Zn-BTC-MOF [35], Cu-HKUST [33, 36] has been reported for ECR of CO_2 in a standard three electrode set-up for ionic liquids.

Zr incorporated Fe porphyrin units in MOF 525 involves to achieve a high turnover of 1520 over 3.2 h to convert CO_2 to CO in DMF but again with high H_2 production. The active center was found to be Fe (0) at more positive potential than Fe redox couple. Ti/TiO_2–ZIF-8 has been investigated as photoelectrochemical reduction of CO_2. MOF helps in the absorption of CO_2 in the form of carbamate with a further reduction to methanol and ethanol due to the presence of the nitrogenous group in ZIF-8 and also due to alternation in energy levels of the materials. The selectivity of methanol can be increased with bias potential, which in turn generates photoelectrons. The recombination of photoelectrons is prevented by linkers and more availability of photoelectrons reduces methanol to ethanol [3].

Inorganic such as graphite oxide, silica, alumina, and their combinations, organics, polymers was used in the formation of composites [8]. The chemistry of metal–organic framework has been reviewed by Trickett *et al.*, where Al-based MOF with a Cu-containing porphyrin active center shows methanol as the predominant product. Ti-based MIL-125-NH_2 with gold nanoparticles yields methane as a product [5]. Core–shell MOF and metal nanoparticle doped MOF has been studied for CO_2RR where core–shell material with TiO_2 nanoparticles as the shell and HKUST-1 as the core offers high CO_2 uptake and hence improves CO_2 conversion, reduction of

H_2 production. Zinc imidazolate framework synthesized with various zinc sources showcased 65% CO yield with NaCl being the best electrolyte [37].

Preparation of nitrogen-doped carbon materials via pyrolysis of ZIF-8 followed by acid treated MOF for ECR of CO_2 yields CO and H_2. The high activity can be attributed to the presence of a large amount of pyridinic-N and quaternary-N species in the carbon structure, which are known to lower the energy barrier for the formation of COOH*, an intermediate to produce CO. In addition, the well-developed porosity further promotes the activity by making more active sites accessible for reduction [37].

Recently, copper nanoparticles are embedded into a solvothermally grown thin film of a zirconium metal–organic framework (MOF) and demonstrated to yield formate as the major products for $NaClO_4$ electrolyte. Some of the Cu sites in the film are subsequently reduced to metallic Cu by applying negative potential in the aqueous electrolyte, resulting in electroactive nanoparticles of Cu deposited near the MOF/FTO interface [38]. The authors have claimed simultaneous capture and reduction of CO_2 using $Cu_3(BTC)_2$ (Cu-MOF) on carbon paper-based gas diffusion electrode to yield CH_4 with two- to three-fold higher than GDE without Cu MOF addition with reduced H_2 evolution.

In the preview of electrocatalytic reduction, the challenges remain to achieve a water and acid-stable catalyst which can operate with low overpotential to reduce the hydrogen evolution and to enhance CO_2 reduction. The aqueous CO_2 reduction with Cu MOF yields formic acid at a higher yield than bare Cu where non-aqueous solvents yields oxalic acid followed by dimerization of CO_2. A hydric scheme with MOF as electrocatalyst for CO_2 reduction has been depicted in Figure 13.4.

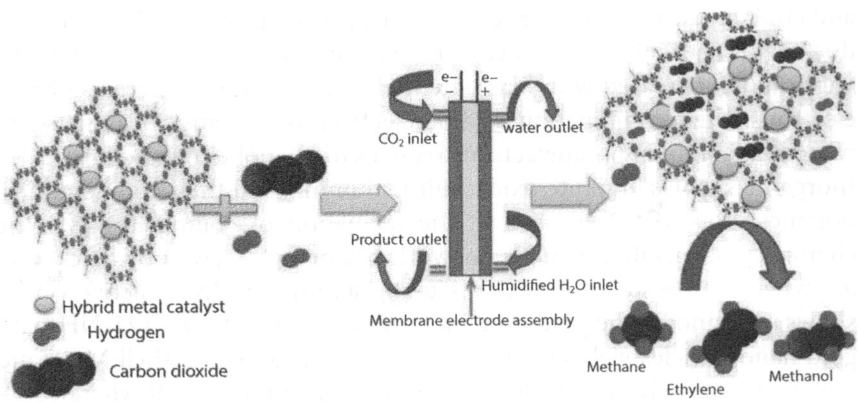

Figure 13.4 Scheme of concept of hybrid MOF for CO_2 reduction.

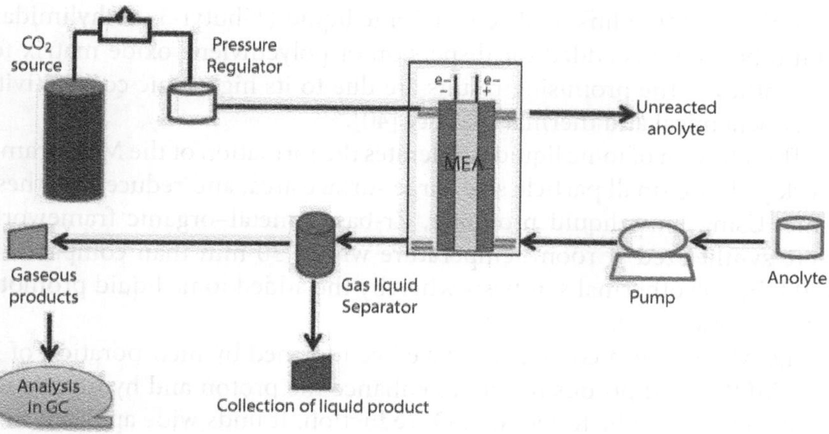

Figure 13.5 Schematic assembly of continuous-flow ECR CO_2 reduction.

13.6.3 Continuous Flow Reduction of CO_2

For commercialization of any processes, the reaction feasibility and scale up plays a major role. As an electrocatalyst, fuel cell has been reached a platform of commercialized device where MOF finds application in such similar device. In the continuous system, the choice of electrolyte also influences the solubility of CO_2 and hence the Faradaic efficiency and current density. The configuration of an electrochemical cell also plays a vital role in catalytic output. Continuous flow fuel cell like configuration as depicted in Figure 13.5 has been studied for CO_2 reduction with an attempt to commercialize the product formation and to work on hurdles faced in a single cell set up. On a membrane electrode assembly set up, the electroreduction of CO_2 happens on the cathode and and O_2 evolution reaction occurs at anode (Table 13.1). Gas diffusion electrodes (GDEs) reduce the internal resistance and promotes mass transfer of reactants [3].

13.6.4 CO_2 Electrochemical Reduction in Ionic Liquid

Ionic liquid plays a significant role in CO_2 electrochemical reduction. The merits of using IL are for its unique physio-chemical properties, altered reaction's conditions, CO_2^- stabilization, reduced overpotential, and improved current density and Faradaic efficiency [39]. The IL can be varied for its cation, anion, and its functional groups to tune an electrochemical reaction.

Polymer dispersed IL based MOF composites are found to enhance the structural and electrochemical properties where metal precursor

($Cu(NO_3)_2$), MOF linker (BTC), an ionic liquid (1-butyl-3 methylimidazolium bromide) is added for dispersion of polyethylene oxide matrix for its synthesis. The promising results are due to its high ionic conductivity, electrochemical and thermal stability [40].

The addition of ionic liquid accelerates the formation of the MOF framework yielding small particle size, large surface area, and reduced synthesis time. Using ionic liquid promoter, Zr-based metal–organic frameworks were synthesized at room temperature within 30 min than compared to 120 h by solvothermal synthesis whereby the added ionic liquid promotes crystallization [41].

IL@MOF hybrid composites have been formed by incorporation of IL into MOF based porous matrix to enhance the proton and hydroxide ion diffusion and can be tested for CO_2 reduction. It finds wide application in solvent-free devices and prevents liquid leakage. The preferred synthesis approach is to start with well-established structures of IL and MOF and hence to establish the desirable properties [42].

Designing of nanostructured heterogeneous catalyst reviews on inorganic heterogeneous along with its computational aspects [43]. Development of protocol for assessing the properties of continuous flow EC cells for CO_2 reduction shows a clear indication on the progress of this field. The opportunities and focus of MOF for CO_2 reduction are numerous where the recent approaches focus on the addition of bimetals, inorganic metal addition on a substrate with ionic liquid as a synthesis strategy for activation of CO_2.

13.7 MOF for Electrocatalytic Sensing

Redemption of pollutants and its derivatives is important to break the biomagnification of the human and animal food chain. Detection, quantification of pollutants and its biomarkers are gaining importance in health and environment sector. The development of sensors is emphasized for its accuracy, limits of detections, the viability in the presence of other limiting ions, cost, and portability.

Recently, Zr (IV)-based UiO-66 MO, a fluorescent based MOF sensor is reported for sensing of 4-nitrobenzaldehye. The sensitivity of probe is about 4.7 μM and has been tested for tap water, lake water, human urine, and human blood serum. The material offers selective and sensitive fluorimetric detection of phosphate ions in the buffer and in aqueous medium [44].

Karthik *et al.* (2017) synthesized metal Fe added amino-functionalized MOF of type NH_2–MIL-101(Fe) for sensing nitroaromatic compounds. The feasible transfer of electrons is due to intermolecular H bond formation. The detection of nitro compounds by the material is in the order of 4-nitrobenzene (32 ppm) > 4-nitrophenol (17 ppm) > 1,3-dinitrobenzene (11.5 ppm) > 4-nitrotoluene (10 ppm) [45].

Indium metal–organic framework (MOF), namely MFM-300 has been tested for SO_2 gas sensing within the limits as low as 5 ppm to 75 ppm [46]. Gaseous SO_2 at low concentrations ranging from 25 to 500 p.p.m was studied in MOF coated quartz crystal microbalance electrode [47].

A new class of hybrid MOF for sensing various target components, including small molecules, solvents, pesticides, explosives, and biological markers in various research fields, such as catalysis, energy storage, drug-delivery systems, non-linear optics, and gas storage was reviewed by by Pawan kumar *et al.*, (2015) [48].

The nanocomposite of HKUST-1 (MOF) and electroreduction graphene oxide (ERGO) prepared by one-step electrodeposition promise high stability in aqueous solution. High sensitivity, good selectivity, and excellent stability have been observed for the determination of paracetamol and dopamine in test solution [49].

Given the significant diversity and tailorability of MOFs, a facile strategy for fast synthesis of efficient sheet-like electrocatalysts based on MOF nanosheets, metal incorporated MOF, and luminescent based MOF quenching is progressing as a strategy for electrochemical sensing of MOF.

13.8 Electrocatalytic Features of MOF

MOFs act as a host material for variety of guest molecules, the framework itself acts as a catalyst, in which its activity depends on metal centers, organic or pseudo-organic linkers and other functional groups from post-synthetic modification which act as active metal sites. MOF combine the inherent advantages of homogenous and heterogeneous catalyst [49] and has reported for is high adsorption and deterred electron hole recombination properties. Synergistic activity between highly dispersed metal with porous MOF matrix will pave an integrated platform for modular ECR reduction. Some of the key insights for the functionality of MOF are not limited to the following. (i) Hybrid MOF offers high surface area; crystallinity and porosity needed for the facile electron transfer of ECR of CO_2. (ii) MOF has been regarded as a good medium for gas storage where mass transfer limitation at the surface of the electrode can be ignored in electrochemical applications. (iii) Encapsulation of metal nanoparticles, metal oxides, and its hybrids on MOF will help to tune selectivity as understood from literatures summarized in this book chapter.

13.9 Conclusion

The immeasurable potential application of metal organic framework as electrocatalyst has been realized for energy and environment related application. Recent work on MOF as electrocatalyst emphasizes on metal, bimetal incorporation, change of linker, and alternative synthesis technology like one-shot electrodeposition, sonochemical, and addition of ionic liquid in synthesis rather than conventional solvothermal and hydrothermal synthesis. Utmost care is taken to maintain the stability of MOF. Development of MOF and its specific applications from the recent literature has been summarized in this book chapter.

Acknowledgment

The authors acknowledge SRM institute of Science and Technology for its research facilities.

References

1. Shibata, H., *Electrocatalytic CO_2 reduction catalysis engineering and reaction mechanism*, Ph.D. thesis, Technical University of Delft, 2009.
2. Peterson, A.A., Pedersen, F.A., Studt, F., Rossmeisl, J., Nørskov, J.K., How copper catalyzes the electroreduction of carbon dioxide into hydrocarbon fuels, *Energy Environ. Sci.*, 3, 1311, 2010.
3. Rowaili, F.A., Jamal, A., Ba-Shammakh, M.S., Rana, A., A review on recent advances for electrochemical reduction of carbon dioxide to methanol using MOFs and non- MOFs Catalysts; Challenges and Future Prospects. *ACS Sustain. Chem. Eng.*, 12, 15895–15914, 2018.
4. Yap, M.H., Fow, K.L., Chen, G.Z., Synthesis and applications of MOF-derived porous nanostructures. *Green Energy Environ.*, 2, 218–245, 2017.
5. Trickett, C.A., Helal, A., Al-Maythalony, B.A., Yamani, Z.H., Cordova, K.E., Yaghi, O.M., The chemistry of metal–organic frameworks for CO_2 capture, regeneration and conversion. *Nat. Rev. Mater.*, 2, 17045, 2017.

6. Zhang, H., Li, J., Tan, Q., Lu, L., Wang, Z., Wu, G., Metal–Organic Frameworks and Their Derived Materials as Electrocatalysts and Photocatalysts for CO_2 Reduction: Progress, Challenge, and Perspective. *Chem. Eur. J.*, 24, 18137–18157, 2018.
7. Zhao, M., Lu, Q., Ma, Q., Zhang, H., Two-Dimensional Metal–Organic Framework Nanosheets. *Small Methods*, 1, 1600030–1600038, 2017.
8. Stock, N. and Biswas, S., Synthesis of Metal–Organic Frameworks (MOFs): Routes to various MOF topologies, morphologies, and composites. *Chem. Rev.*, 112, 933–969, 2012.
9. Yagi, O.M., Li, G., Li, H., Selective Binding and Removal of Guests in a Microporous Metal–organic Framework, *Nature*, 378, 703, 1995.
10. Sachdeva, S., Pustovarenko, B., Sudholter, B.J.R., Louis C.P.M De Smeta, F., Gascon, J., Control of Interpenetration of Copper-Based MOFs on Supported Surfaces by Electrochemical Synthesis. *Cryst. Eng. Comm.*, 18, 4018–4022, 2016.
11. Hauser, J.L., Tso, M., Fitchmun, K., Oliver, S.R.J., Anodic Electrodeposition of Several Metal Organic Framework Thin Films on Indium Tin Oxide Glass. *Cryst. Growth Des*, 19, 2358–2365, 2019.
12. Ameloot, R., Stappers, L., Fransaer, J., Alaerts, L., Sels, B.F., De Vos, D.E., Patterned Growth of Metal–Organic Framework Coatings by Electrochemical Synthesis. *Chem. Mater.*, 21, 2580–2582, 2009.
13. Zhu, B., Zou, R., Xu, Q., Metal–Organic Framework Based Catalysts for Hydrogen Evolution. *Adv. Energy. Mater.*, 27, 1801193, 2018.
14. Raoof, J.B., Hosseini, S.R., Ojani, R., Mandegarza, S., MOF-derived Cu/nanoporous carbon composite and its application for electro-catalysis of hydrogen evolution reaction. *Energy*, 90, 1075–1081, 2015.
15. Wen, X. and Guan, J., Recent progress on MOF-derived electrocatalysts for hydrogen evolution reaction. *Appl. Mater. Today*, 16, 146–168, 2019.
16. Karuppasamy, K., Jothi, V.R., Vikraman, D., Prasanna, K., Maiyalagan, T., Sang, B.I., Yi, S.C., Kim, H.S., Metal–organic framework derived NiMo polyhedron as an efficient hydrogen evolution reaction electrocatalyst. *Appl. Surf. Sci.*, 478, 916–923, 2019.
17. Ji, Y., Dong, H., Liu, C., Li, Y., Two-dimensional π-conjugated metal–organic nanosheets as single-atom catalysts for the hydrogen evolution reaction. *Nanoscale*, 11, 454–458, 2019.
18. Zhang, H., Liu, X., Wu, Y., Guan, C., Cheetham, A.K., Wang, J., MOF-derived nanohybrids for electrocatalysis and energy storage: Current status and perspectives. *Chem. Commun.*, 54, 5268–5288, 2018.
19. Shi, Q., Fu, S., Zhu, C., Song, J., Du, D., Lin, Y., Metal–organic frameworks-based catalysts for electrochemical oxygen evolution. *Mater. Horiz.*, 6, 684, 2019. 10.1039/C8MH01397G.
20. Li, H., Ke, F., Zhu, J., MOF-Derived Ultrathin Cobalt Phosphide Nanosheets as Efficient Bifunctional Hydrogen Evolution Reaction and Oxygen

Evolution Reaction Electrocatalysts. *Nanomaterials*, 8, 89, 2018. doi:10.3390/nano8020089.

21. Chen, W., Zhang, Y., Chen, G., Huang, R., Zhou, Y., Wu, Y., Hu, Y., Ostrikov, K., Mesoporous cobalt–iron–organic frameworks: Plasma-enhanced oxygen evolution electrocatalyst References CO_2 reduction. *J. Mater. Chem. A*, 2019, doi: 10.1039/C8TA10952D.
22. Wang, X., Xiao, H., Li, A., Li, Z., Liu, S., Zhang, Q., Gong, Y., Zheng, L., Zhu, Y., Chen, C., Wang, D., Peng, Q., Gu, L., Han, X., Li, J., Li, Y., Constructing $NiCo/Fe_3O_4$ Heteroparticles within MOF-74 for Efficient Oxygen Evolution Reactions. *J. Am. Chem. Soc.*, 45, 140, 15336–15341, 2018.
23. Maruthapandian, V., Kumaraguru, S., Mohan, S., Saraswathy, V., Muralidharan, S., An Insight on the Electrocatalytic Mechanistic Study of Pristine Ni MOF (BTC) in Alkaline Medium for Enhanced OER and UOR. *ChemElectroChem*, 19, 2795–2807, 2018.
24. Tripathy, R.K., Samantara, A.K., Behera, J.N., A cobalt metal–organic framework (Co-MOF): A bi-functional electro active material for the oxygen evolution and reduction reaction. *Dalton Trans.*, 48, 10557–10564, 2019.
25. Zhou, W., Dan Huang, D., Wu, Y.P., Zhao, J., Wu, T., Zhang, J., Li, D.S., Sun, C., Feng, P., Bu, X., Stable Hierarchical Bimetal–Organic Nanostructures as High Performance Electrocatalysts for the Oxygen Evolution Reaction. *Angew. Chem. Int. Ed.*, 58, 1–6, 2019.
26. Gong, Y., Xu, Z., Pan, H., Facile Synthesis and Characterization of MOF-Derived Porous Co_3O_4 Composite for Oxygen Evolution Reaction. *ChemistrySelect*, 4, 1131–1137, 2019.
27. Bucci, A., Mondal, S.S., M.-Diaconescu, V., Shafir, A., Fillo, J.L., Highly Active Oxygen Evolution Electrocatalysts Based on Cobalt Amide Imidate Imidazolate Frameworks. *ChemRxiv*, 2019. preprint, doi.org/10.26434/chemrxiv.8273240.v1.
28. Tang, Y., Zheng, S., Xue, H., Pang, H., Regulation of the Ni^{2+} Content in a Hierarchical Urchin-Like MOF for High-Performance Electrocatalytic Oxygen Evolution. *Front. Chem.*, 7, 411, 2019, doi: 10.3389/fchem.2019.00411.
29. Kuhl, K.P., Cave, E.R., Abram, D.N., Jaramillo, T.F., New insights into the electrochemical reduction of carbon dioxide on metallic copper surfaces. *Energy Environ. Sci.*, 5, 7050–7075, 2012.
30. Keerthiga, G. and Chetty, R., Electrochemical Reduction of Carbon Dioxide on Zinc-Modified Copper Electrodes. *J. Electrochem. Soc.*, 164, H164–H169, 2017.
31. Vesborg, P.C.K. and Seger, B., Performance Limits of Photoelectrochemical CO_2 Reduction Based on Known Electrocatalysts and the Case for Two-Electron Reduction Products. *Chem. Mater.*, 24, 8844–8850, 2016.
32. Pirzadeh, K., Ghoreyshi, A.A., Rahimnejad, M., Mohammadi, M., Electrochemical Synthesis, Characterization and Application of a Microstructure $Cu_3(Btc)_2$ Metal Organic Framework for CO_2 and CH_4 Separation. *Korean J. Chem. Eng.*, 35, 974–983, 2017.

33. Hinogami, R., Yotsuhashi, S., Deguchi, M., Zenitani, Y., Hashiba, H., Yamada, Y., Electrochemical reduction of carbon dioxide using a copper rubeanate metal organic framework. *ECS Electrochem. Lett.*, 1, 4, H17–H19, 2012.
34. Senthil Kumar, R., Senthil Kumar, S., Anbu Kulandainathan, M., Highly selective electrochemical reduction of carbon dioxide using Cu based metal organic framework as an electrocatalyst. *Electrochem. Commun.*, 25, 70–73, 2012.
35. Kang, X., Zhu, Q., Sun, X., Hu, J., Zhang, J., Liu, Z., Han, B., Highly efficient electrochemical reduction of CO_2 to CH_4 in an ionic liquid using a metal–organic framework cathode. *Chem. Sci.*, 7, 266–273, 2016.
36. Zhao, K., Liu, Y., Quan, X., Chen, S., Yu, H., CO_2 Electroreduction at Low Overpotential on Oxide-Derived Cu/Carbons Fabricated from Metal Organic Framework. *ACS Appl. Mater. Interfaces*, 9, 6, 5302–5311, 2017.
37. Wang, Y., Hou, P., Wang, Z., Kang, P., Zinc Imidazolate Metal–Organic frameworks (ZIF-8) for Electrochemical Reduction of CO_2 to CO. *Chemphyschem*, 18, 3142–3147, 2014.
38. Kung, C.W., Audu, C.O., Peters, A.W., Noh, H., Farha, O.K., Hupp, J.T., Copper Nanoparticles Installed in Metal–Organic Framework Thin Films are Electrocatalytically Competent for CO_2 Reduction. *ACS Energy Lett.*, 2, 2394–2401, 2017.
39. Faggion, D., Gonçalves, W.D.G., Dupont, J., CO_2 Electroreduction in Ionic liquids. *Front. Chem.*, 2019. https://doi.org/10.3389/fchem.2019.00102.
40. Dutta, R. and Kumar, A., Structural and Dielectric Properties of Ionic Liquid Doped Metal Organic Framework based Polymer Electrolyte Nanocomposites. *J. Phys. Conf. Ser.*, 765, 01, 2016. 24th Condensed Matter Days National Conference (CMDAYS2016).
41. Sang, X., Zhang, J., Xiang, J., Cui, J., Zheng, L., Zhang, J., Wu, Z., Li, Z., Mo, G., Xu, Y., Song, J., Liu, C., Tan, X., Luo, T., Zhang, B., Han, B., Ionic liquid accelerates the crystallization of Zr-based metal–organic frameworks. *Nat. Commun.*, 8, 175, 2017, 1848. DOI:10.1038/s41467-017-00226-y.
42. Yoshida, Y. and Kitagawa, H., Ionic Conduction in Metal–Organic Frameworks with Incorporated Ionic Liquids. *ACS Sustain. Chem. Eng.*, 7, 70–81, 2019.
43. Vickers, J.W., Alfonso, D., Kauffman, D.R., Electrochemical Carbon Dioxide Reduction at Nanostructured Gold, Copper, and Alloy Materials. *Energy Technol.*, 5, 1–22, 2017.
44. Das, A., Das, S., Trivedi, A., Biswas, S., A Dual Functional MOF-Based Fluorescent Sensor for Intracellular Phosphate and Extracellular 4-Nitrobenzaldehyde. *Dalton Trans.*, 48, 1332–1343, 2019.
45. Karthik, P., Pandikumar, A., Preeyangha, M., Kowsalya, M., Neppolian, B., Amino-Functionalized Mil-101(Fe) Metal–Organic Framework as a Viable Fluorescent Probe for Nitroaromatic Compounds. *Microchim. Acta*, 184, 2265–2273, 2017.

46. Ghosh, D., Pal, A., Ghosh, S., Gayen, A., Motin Seikh, Md., Mahata, P., Metal Ion Sensing and Electrochemical Behavior of MOF Derived $ZnCo_2O_4$. *Eur. J. Inorg. Chem.*, 26, 3076–3083, 2019.
47. Kumar, P., Deep, A.B., Kim, K.-H., Metal Organic Frameworks for Sensing Applications. *Trends Anal. Chem.*, 73, 39–53, 2015.
48. Ma, B., Guo, H., Wang, M., Li, L., Jia, X., Chen, H., Xue, R., Yang, W., Electrocatalysis of Cu-MOF/Graphene Composite and its Sensing Application for Electrochemical Simultaneous Determination of Dopamine and Paracetamol. *Electroanalysis*, 31, 1002–1008, 2019.
49. Kornienko, N., Zhao, Y., Kley, C.S., Zhu, C., Kim, D., Lin, S., Chang, C.J., Yaghi, O.M., Yang, P., *J. Am. Chem. Soc.*, 137, 14129–14135, 2015.

14

Applications of MOFs and Their Composite Materials in Light-Driven Redox Reactions

Elizabeth Rojas-García[1]*, José M. Barrera-Andrade[2], Elim Albiter[2], A. Marisela Maubert[3] and Miguel A. Valenzuela[2]

[1]*Área de Ingeniería Química, Universidad Autónoma Metropolitana-Unidad, Iztapalapa, Vicentina, CDMX, México*
[2]*Laboratorio de Catálisis y Materiales, ESIQIE—Instituto Politécnico Nacional, Zacatenco, CDMX, México*
[3]*Área de Química de Materiales, Universidad Autónoma Metropolitana-Unidad Azcapotzalco, Reynosa Tamaulipas, CDMX, México*

Abstract

Metal–organic frameworks (MOFs) also known as porous coordination polymers (PCP) are a type of hybrids porous crystalline systems formed from metallic clusters and organic ligand. Due to their capacity for structural and functional adjustment, they have had a growing interest in chemistry and related areas in the last decade. MOFs have found a promising niche in photocatalysis, because of their excellent tunable structure and optical properties. Even though their applications in photocatalysis are in a childhood stage, but in the last five years, they have shown an exponential increase in the number of publications, very close to that in heterogeneous catalysis. Their applications in photocatalysis can be grouped into three categories: energy (hydrogen production, CO_2 conversion), environment (degradation of organic/inorganic pollutants), and organic synthesis. This chapter reviews recent investigations in the synthesis of MOFs pristine and their modifications and the impact they have had when applied in the photocatalytic reactions mentioned above. Following a brief introduction on the fundamentals of MOFs and their applications, the synthesized new MOFs were classified according to the metal cluster belonging to groups 4, 8 and 9–12 of the periodic table. Subsequently, MOFs-composites including metals, semiconductors, and multicomponent were

**Corresponding author*: erg@xanum.uam.mx

Inamuddin, Rajender Boddula, Mohd Imran Ahamed and Abdullah M. Asiri (eds.) Applications of Metal–Organic Frameworks and Their Derived Materials, (377–462) © 2020 Scrivener Publishing LLC

analyzed in terms of preparation methods, properties, and reaction mechanisms involved in the selected photocatalytic reactions.

Keywords: MOFs, light-driven applications, photocatalysis, PCP, MOFs-composites

14.1 Introduction

Metal–organic frameworks (MOFs) also known as porous coordination polymers (PCP) are a type of hybrid porous crystalline materials formed from metallic ions or clusters called as secondary construction units (SBUs) and multidentate organic linkers, connected through coordination bonds of moderate force [1, 2]. The organic ligands act as "buffers" or bridges for the metallic centers or SBUs, which in turn act as "joints" in the frameworks of the resulting MOF [3]. The first MOF with permanent microporosity was synthesized by O. Yaghi in 1998 called MOF-5 or IRMOF-1, with which he coined the name of "Metal-Organic Frameworks" to this type of materials [4]. Due to their capacity for structural and functional adjustment, they have had a growing interest in chemistry and related areas in the last decade (Figure 14.1) [5]. Figure 14.1 shows an exponential growth in the study of MOFs as photocatalysts and their composites.

A large number of articles have demonstrated that MOFs are materials with excellent physicochemical and textural properties, such as moderated

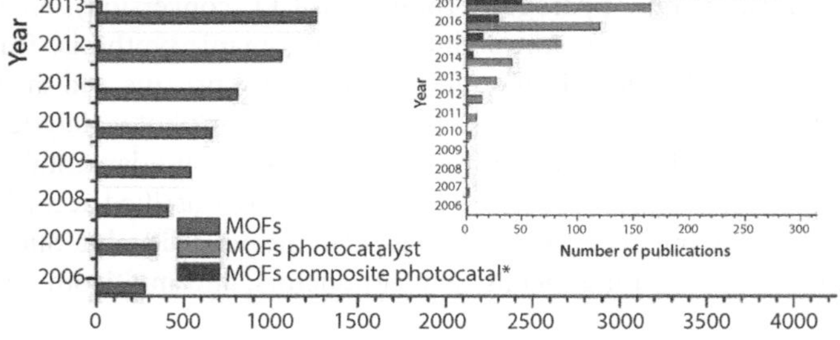

Figure 14.1 Year wise publication status from 2006 to 2019 of various aspects of MOFs. Obtained from Scopus database. Keywords: MOFs, MOF photocatalyst, MOF composite photocatal*.

thermal stability, large internal specific surface area (up to 7,000 m^2/g), large pore volumes (up to 1 cm^3/g) and ultra-low densities, which can reach up to 90% of the volume of the crystal, among others [6–10]. If compared with the conventional and well-known materials (e.g., metal oxides, hydrotalcites, mesoporous materials, etc.), MOFs present other structural features, which can be classified into four categories: rigid structure, flexible/dynamic structure, functionalized surface, and open metallic sites. For example, MOFs with a rigid structure usually have stable and robust porosity, while MOFs with flexible structure show significant changes when inserting or eliminating host molecules and responding to external stimuli, such as pressure and temperature, a characteristic that is absent in traditional adsorbents, such as zeolites and activated carbons [11]. On the other hand, MOFs have crystalline networks of distinct topologies depending of the type or inorganic/organic molecular building blocks and the synthesis conditions, therefore more than 20,000 different structures have been synthesized, so far.

In recent years, MOFs have been rapidly developed due to the variety of structures that might be formed as well as their easy adaptation finding different applications in gas storage and separation, chemical sensors, proton conduction, biomedicine, catalysis and photocatalysis, among others [12]. For instance, several applications have been reported in energy, environmental remediation, and organic synthesis using MOFs as photocatalysts, as shown in Figure 14.2.

Due to their high porosity, specific surface, and molecular nature, design of MOFs as photocatalysts is an area of great interest with relevant differences to conventional semiconductors. As is well known conventional semiconductors (e.g., metal oxides) present a low specific surface, so that their surface available to carry out the reaction is very limited.

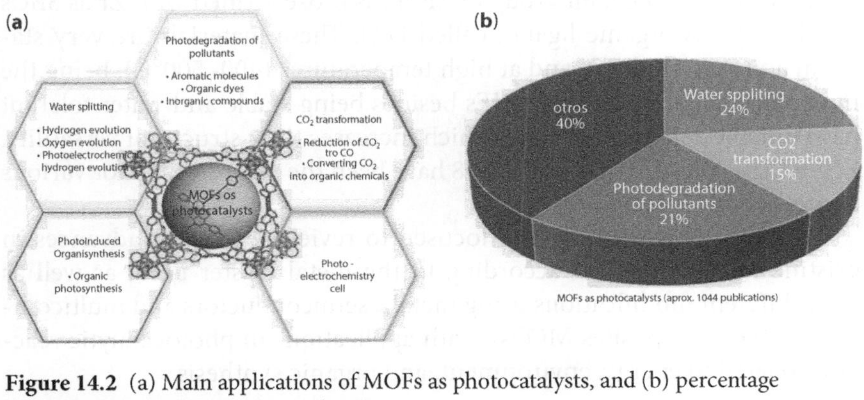

Figure 14.2 (a) Main applications of MOFs as photocatalysts, and (b) percentage according to main applications of MOFs.

In addition, although MOFs present a porous crystalline nature, light scattering and absorption is not a problem [2]. Despite the good textural and structural properties of MOFs their photoactivity is still low compared to conventional semiconductors [13]. However, the high porosity in MOFs facilitates the diffusion of reagents and products, where the active catalytic sites are located allowing to increase the efficiency of the photocatalytic reaction [14].

Recent studies have shown that MOFs are not only a new class of photocatalysts applicable in the phodegradation of organic contaminants under UV-Visible light irradiation [8], but they can also be used in hydrogen evolution reactions (e.g., water splitting) [15] and photocatalytic reduction of CO_2 [16]. This is because some MOFs can behave like semiconductors, so these can be used as transport systems for load carriers through the excitation of organic ligand or metal ions. In addition to that, they present photocatalytic activity due to their unsaturated metal atoms or catalytically active organic bonds. Compared to commonly used inorganic semiconductors, MOFs can be easily adjusted by modifying their organic bonds or metal centers. Theoretical calculations suggest that MOFs as semiconductors have band-gap values in the range of 1.0–5.5 eV.

Different strategies are being developed for potentialize their catalytic performance, including functionalization and modification of ligands, sensitization, decoration with a co-catalyst, and coupling with a semiconductor [17]. Among these materials we can find a subclass of MOFs, the MIL-ns, which have received special attention for their resistance to water, common solvents and temperatures above 300°C. MIL-125 (responsive to UV-light irradiation), and MIL-101, MIL-100, and MIL-53, which are visible light-responsive MOFs, have been tested in several photocatalytic reactions. Another important group of MOFs is those formed with Zr as SBUs and H_2BDC as organic ligand, called UiO. These materials are very stable in aqueous medium, and at high temperatures (400–500°C), being the most representative UiO-66. ZIFs besides being stable and water tolerant have tunable zeotype topology, which increases their structural versatility. ZIF-8 (Zn), ZIF-67 (Co), and ZIF-8 have been the most studied for various photocatalytic reactions [18].

Therefore, this work will be focused to review the recent advances in pristine MOFs classified according to the metal cluster used, as well as their different modifications using metals, semiconductors and multicomponent (i.e., composites MOFs), with applications in photocatalytic reactions related to energy, environment, and organic synthesis.

14.1.1 MOFs as Photocatalysts

So far, TiO_2 has been the most studied semiconductor for photocatalytic reactions, it has a wide band gap (Eg = 3.0–3.2 eV), so only UV light can be used to generate to load carriers, which limits its use in photocatalysis under visible light irradiation, so as well as solar energy. In addition to the above, TiO_2 has a high recombination of photogenerated (electron-hole) charge carriers, so several efforts have been made to increase the photocatalytic activity of TiO_2, however it is still far from use this system in an economically viable process. Therefore, the development of new photocatalytic materials with adjustable function remains a major challenge in the field of photocatalysis [19]. Thus, compared to traditional semiconductors, MOFs can be modified to molecular level and perform photocatalytic reactions [20, 21].

Nowadays, MOFs are of great interest because they have shown optical properties allowing the formation of charge carriers (holes, h^+, and electrons, e^-) [22]. However, these materials, due to their particular structure and organic nature, must take into account some additional and complementary considerations to traditional metal semiconductors [23]: 1) both, organic ligands and metal centers can absorb photons of light incident and convert a photogenerated excited state where transfer electrons from ligand to metal center, and 2) both, organic ligand and metal center can also be carriers of chromophore groups, which change considerably the optical properties as gap band [24].

In the synthesis of MOFs, the most used ligands are aromatic polycarboxylates, due to their excess of electrons they can transfer to the metallic centers attached to them, after proper light excitation. In addition, organic ligands have intense absorption bands centered on a wavelength of 250 nm and that can travel up to 300 nm and even to the visible region, depending on the type of ligand used. This means that with a proper ligand selection, it is possible to design a responsive visible-light MOF [25]. A clear example of the above mentioned is the introduction of functional groups ($-NH_2$, $-OH$, $-CH_3$, or $-Cl$) in the organic linker, which can shift the band gap to a lower value. The mechanism consists in the 2p electron donation from the functional groups to the organic linker; for example, amino group ($-NH_2$) introduces inter-band states increasing the visible light absorption without modifying the crystalline structure of the MOF. In addition, the presence of functional groups linked to MOFs, improves their photocatalytic properties during irradiation with visible light.

Currently, MOFs have had great interest in the field of photoredox catalysis due mainly to the advantages of having porous nanostructures and controllable semiconductor properties. The high crystallinity in the

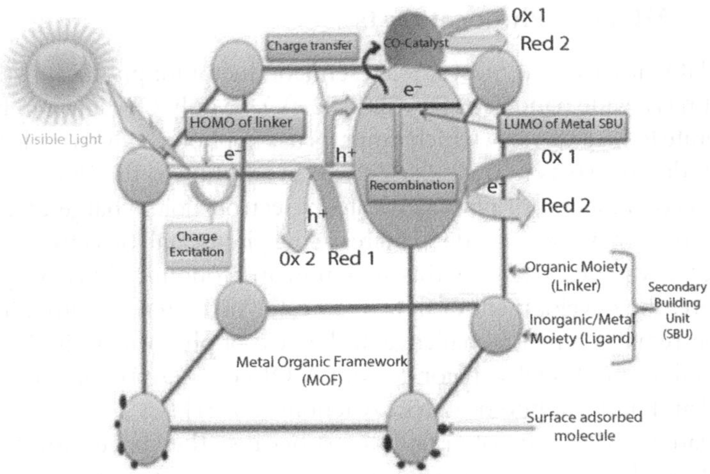

Figure 14.3 Functionality of MOFs as photocatalysts. Reprinted with permission from Ref. [26].

MOFs and their adjustable porosity facilitates the transfer of light from the organic ligand to the metal cluster, as well as the mass transfer of reactants and products in the photocatalytic reaction.

The functionality of MOFs as photocatalysts is possible due to the excellent textural properties that allow the incorporation of co-catalysts on their structure such as metals, metal oxides or metal and metal oxide, among other, modifying optical and textural properties of MOF pristine which in most cases improves considerably their photocatalytic activity (Figure 14.3).

14.1.2 Charge Transfer Mechanisms

A difference relevant between the classical heterogeneous catalysis and photocatalysis is the mode of activation of the catalysts where the thermal activation is replaced by a photonic activation making use of light in the visible and UV region generating to the active sites or charge carriers (electron, e^-, and holes, h^+). In traditional periodic-structured semiconductors, valence and conduction bands are separated for a prohibited band or gap band, which is the minimum energy necessary for generating the charge carriers (electron–hole). The band model is a fundamental method that allows to determine as are formed the valence and conduction bands based on bonding and anti-bonding orbitals. Therefore in these organic semiconductors (or MOFs) the model band is not enough to explain the electronic transitions and absorption of light, so it is necessary to use the theory of molecular orbitals, which the difference in energy between the highest occupied molecular orbital of more energy (HOMO) and lowest unoccupied molecular orbital (LUMO) determine the energy of the gap band of these photocatalysts. Is important note, MOFs are composed of organic ligands and metal clusters, both components co-participate in the formation of charge carriers, useful for several redox reactions occurring on their surface, which recently has led to propose various charge transfer mechanisms for MOFs.

a) Ligand-to-metal charge-transfer (LMCT) mechanism

In this mechanism, organic ligand that is part of the MOF structure absorbs light and excited electrons transfer to the metal cluster reducing to metal. Commonly, organic ligands are polycarboxylates rich in electrons (with aromatic ring in their structure) without/with chromophore groups, which act as an antenna absorbing light energy and directing excited electrons to metal sites. Fu *et al.* studied an example very representative of this mechanism using MOF NH_2–MIL-125 (Ti) for CO_2 photo-reduction [27] (Figure 14.4). They proposed that the amino group of organic ligand absorbs visible light, which could act as an antenna. The photogenerated π–electron of benzene ring of organic ligand is transferred at the site of O^{2-} where later a band is formed by the transfer of electron of site of O^2 to Ti-oxo cluster with the reduction of Ti^{4+} to Ti^{3+}, transforming it into a reducing agent for the molecules reduction and then oxidized to its Ti^{4+} oxo-structure state.

b) Ligand-to-ligand charge transfer mechanism (LLCT)

In this case, the organic ligand of MOF forms two energy bands that affecting the global conductive properties of MOF, which forms a dual photocatalytic route that contributes getting better photocatalytic properties. An example of this mechanism is shown for NNU-28 photocatalyst used

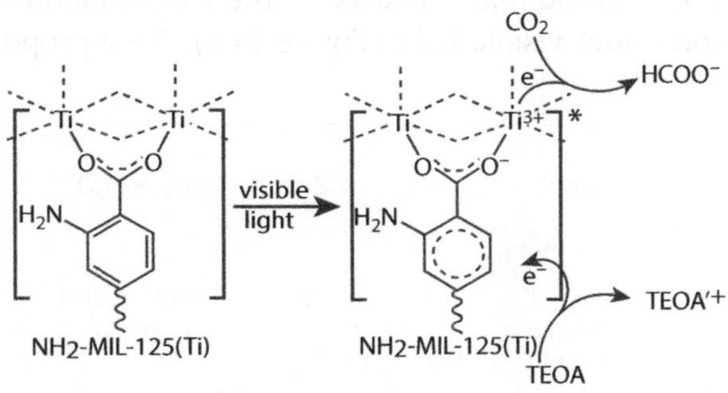

Figure 14.4 Ligand-to-metal charge-transfer (LMCT) mechanism in MOF NH_2–MIL-125 (Ti). Reprinted with permission from Ref. [27].

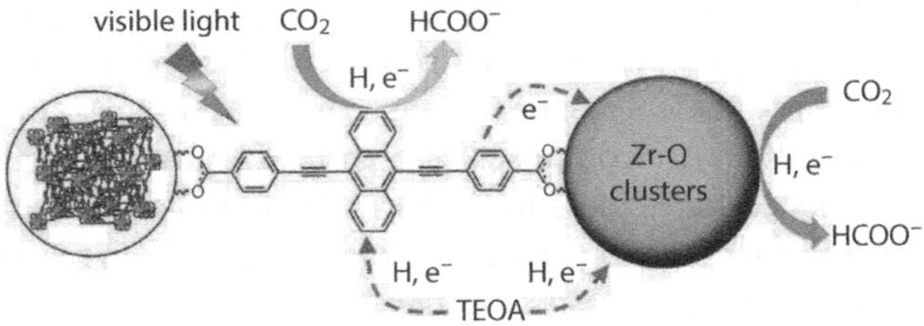

Figure 14.5 An example of ligand-to-ligand charge transfer (LLCT) mechanism for NNU-28 photocatalysts reported by Chen *et al.* Reprinted with permission from Ref. [28].

in the visible-light-driven CO_2 photoreduction (Figure 14.5) [28]. Chen *et al.* demonstrated that dual photocatalytic route due to anthracene-based ligand contributes highly to efficient visible-light-driven CO_2 photoreduction. The LLTC mechanism starts with the photoexcitation of organic ligand, and subsequent delocalization of photogenerated π-electron.

c) Metal-to-ligand charge transfer (MLCT) or metal-to-metal charge transfer (MMCT) mechanism

The use of metalloligands-based photosensitizer such as dyes or (Ru, Ir)-based complex initially are integrated to MOF, changing its semiconductor properties and forming a visible light responsible photocatalyst [29]. In this case, the photosensitizer absorbs light emitting an electron, followed the excited electron migrated to the LUMO of the organic ligand or metal node of MOF, called MLCT or MMCT mechanism, respectively. In addition, both mechanisms have also been observed when new synthesis strategies are used for the modification of MOFs, such as post-synthetic exchange (PSE) strategy used for the metal and/or ligands substitution [30]. For example, Sun *et al.* prepared Ti-substituted NH_2–UiO-66(Zr/Ti) material by PSE method and was used in the CO_2 reduction and hydrogen production under visible light (Figure 14.6). They proposed that the

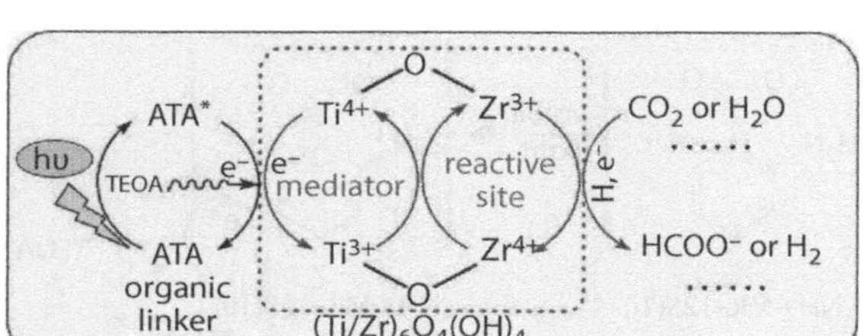

Figure 14.6 Proposed mechanism for the photocatalytic reactions over NH_2–UIO-66 (Zr/Ti). Reprinted with permission from Ref. [30].

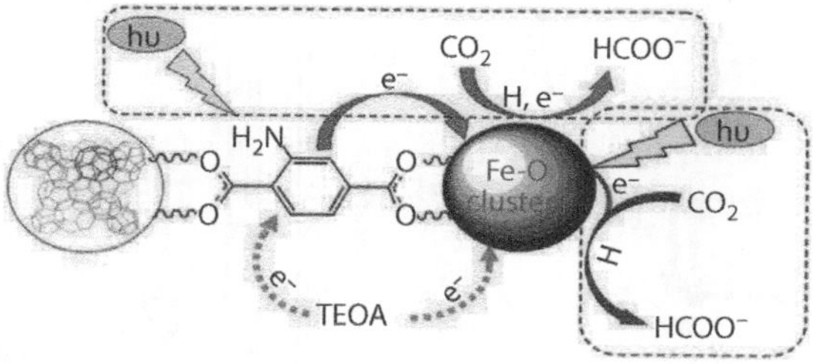

Figure 14.7 Dual excitation pathway for Fe-containing MOFs. Reprinted with permission from Ref. [31].

enhanced photocatalytic activity *via* Ti-mediated electron transfer mechanism, where electron charge transfer of excited (ATA) organic ligand or linker to Zr-oxo clusters is promoted by Ti-mediator species present in the structure of MOF. The electron charge transfer of excited Ti-mediator to Zr-oxo clusters can be carry out by the MMCT mechanism.

d) Dual excitation pathways

MOFs are composed of organic ligands and metal oxo-clusters, and if both are capable of absorbing incident light and promoting excited electrons to the LUMO orbital of the metal clusters, it would increase the efficiency of photocatalytic activity. Wang *et al.* studied examples of iron-based MOFs very representative of this mechanism (Figure 14.7) [31]. Iron-based MOFs such as NH_2–MIL-101 (Fe), NH_2–MIL-53 (Fe), and NH_2–MIL-88B (Fe) showed to be able of efficiently CO_2 reduction under visible light, *via* dual excitation pathways, i.e., the excitation of a NH_2-group present in the organic ligand and excited electron transfer to the Fe-metal center in addition to the direct excitation of the Fe-O clusters.

14.1.3 Methods of Synthesis

The main objective of making a variation in the synthesis methods is to be able to establish conditions that allow obtaining well-defined structures, without a possible decomposition of the organic ligand as well as finding the appropriate kinetics of crystallization that allows nucleation and growth of the desired phase. Commonly, MOFs are synthesized by the solvothermal or hydrothermal method, although other synthesis methods have been investigated to produce smaller and more uniform crystals with very short synthesis times, such as microwave assisted, sonochemistry, mechanochemistry and electrochemistry methods [32] (Figure 14.8).

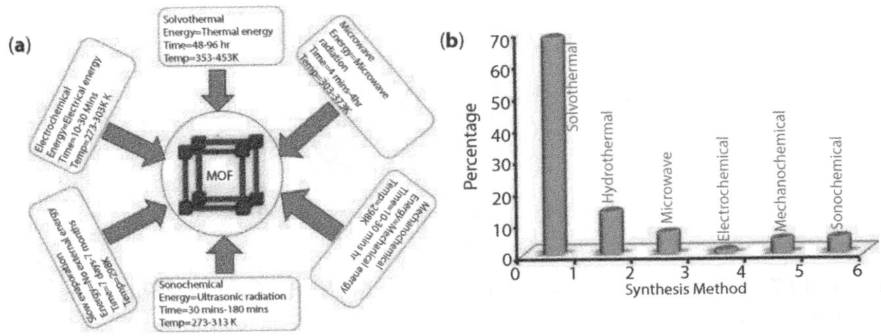

Figure 14.8 (a) Conventional synthesis methods of MOFs and (b) percentage of MOFs synthesized using the various preparation routes. Reprinted with permission from Ref. [32].

MOFs synthesis frequently takes place in a solvent and at temperatures in the range from room temperature to approximately 500 K. The energy can be introduced through an electric potential, electromagnetic radiation, mechanical waves, or mechanically. Synthesis time plays a very important role on the final properties to be obtained in the MOF, for instance, hydro (solvo) thermal, diffusion, and solvent evaporation methods need very long reaction times, while microwave assisted, sonochemical, electrochemical, and mechanochemical methods use very short reaction times.

However, nowadays there are several strategies that are being used for MOFs modification:

a) *Post-synthetic modification* (PSM or PSE), is being an excellent synthesis strategy that allows to obtain topologically identical structures but with diverse functionality structures. The functionality of the structure allows to modify its textural properties, as well as optical for MOF as semiconductors. So that, it is possible to design new MOFs with desired properties, which gives this strategy a great versatility.
b) *Solvent-assisted linker exchange method (SALE)*, is a synthesis route that modifies the structure and properties of MOFs [33, 34]. This strategy allows to obtain MOFs with different textural, morphological and structural properties. In this case, the substitution of organic ligand of MOF-structure could create defects at metal centers due to the missing of linker.

c) *Supermolecular building blocks (SBB) and supermolecular building layer (SBL) strategy*: these two strategies facilitate the design and deliberate construction of metal–organic frameworks or made-to-order MOFs. Both strategies have been described and illustrated by Guillerm *et al.* [35].
d) *Dual-ligand or mixed-ligand strategy*, where one ligand contributes to form the layer, and the other acts as the pillar [36].

14.2 Pristine MOFs and Their Application in Photocatalysis

14.2.1 Group 4 Metallic Clusters

Group 4 (or IVB group of ancient periodic table) comprise transition metals such as titanium (Ti), zirconium (Zn), and hafnium (Hf). It is well known that photocatalysts based on titanium and zirconium transition elements (i.e., metal oxides) present good photocatalytic properties, and in MOFs is not the exception. MOFs composed of metals of group 4 have shown to be very attractive as photocatalysts for several chemical process such as water splitting, pollutant degradation, and CO_2 reduction, among others (Table 14.1). It is important to note that two of the MOFs more studied (UIO-66 (Zr) and MIL-125 (Ti) in photocatalysis are in this group. However, several strategies are being studied to enhance the photocatalytic activity of these MOFs.

Li *et al.* observed that the incorporation of different electron-donor groups in MOF UiO–66(Zr)–X (X = –OH, $-NH_2$, –COOH, $-NO_2$, and –H) modified considerably the optical properties and its photocatalytic efficiency in the degradation of phenanthrene under visible light irradiation [19] (see Figure 14.9A). XRD and XPS studies showed that the structure of the UiO-66 (Zr) remained unchanged after the incorporation of electron-donor groups and photocatalytic reaction. The photocatalytic efficiency showed the following order of –OH > $-NH_2$ > –COOH > $-NO_2$ > –H, as shown in Figure 14.9A. Therefore, the introduction of electron donor groups in MOFs to regulate their electronegativity represents an alternative to obtain more efficient photocatalysts under irradiation with visible light.

Post-synthetic modification strategy (PSE) of UIO-66 (Zr) have been used for modified the optical properties and electrons excited charge transfer mechanism. The cation-exchange strategy using the microwave-assisted method was developed to incorporate Ti on substitution of Zr in

Table 14.1 MOFs with group 4 metallic clusters.

MOF	Metallic cluster	Organic ligands	Gap band (eV)	Application/ degraded compound	Reaction conditions	References
UiO-66(Zr/Ti)	Zr–O Ti–O	H_2BDC	3.75	Reduction of Se(VI)	2.5 mg of MOF; Se (VI) (200 ppb); UV lamp (Philips 15 W/T8, 254 nm); 5 mL mixture (4:1, v:v) of water and formic acid. *60% reduction after 8 days	[37]
NH_2–MIL-125(Ti)	Ti–O	NH_2–ATA	2.75	Water splitting	30 mg of a MOF; 500 W Xe/Hg lamp (385 nm); 23.5 mL of CH_3CN, 4.7 mL of TEA and 0.5 mL of H_2O; 40°C. *49.3 μmol/h·g_{MOF}	[38]
VNU-1	Zr–O	H_2CPEB	2.88	Degradation of MB and MO	10 mg of MOF; 15 mL of dye (100 ppm); visible-light (300 W xenon lamp) *100% (MB) and 83% (MO) degradation at 3 h	[39]
NNU-36	Zr–O	H_2L	2.28	Reduction of Cr(VI) to Cr(III)	15 mg of MOF; 40 mL of Cr(VI) aqueous solution (10 ppm); 0.2 M H_2SO_4; visible-light (300 W xenon arc lamp) *95.3% of degradation at 60 min	[36]
PCN-133	Zr–O	BTB and DCDPS	–	Reduction of Cr(VI) to Cr(III)	Cr (VI) aqueous solution (50 ppm) *95% reduction at 10 min	[40]

(Continued)

Table 14.1 MOFs with group 4 metallic clusters. (*Continued*)

MOF	Metallic cluster	Organic ligands	Gap band (eV)	Application/ degraded compound	Reaction conditions	References
Zr–SDCA–NH_2	Zr–O	H_2SDCA–NH_2	2.27	CO_2 reduction	40.0 mg of MOF; CH_3CN/TEOA (50 ml, v/v = 30/1); visible light (300 W Xe lamp (420 to 800 nm)) *96.2 µmol ($HCOO^-$)/h·mmol MOF	[41]
NH_2–UiO-66	Zr–O	H_2BDC–NH_2	2.8	Phenanthrene	50 mg of MOF; 10 mL of phenanthrene–methanol solution (10 ppm); UV light (175 W Hg lamp, $\lambda > 330$ nm). *90% degradation at 159 min	[19]
Zr–MOFs	Zr–O	THPP and THBPP	1.78	CO_2 reduction	20 mg of MOF; 5 mL (acetonitrile and triethanolamine, v/v, 4/1); visible light (300 W Xe lamp, 420 nm) *14 mmol CO/h·g_{cat}	[42]
NH_2–UiO-66(Zr)	Zr–O	NH_2–ATA	2.92	Water splitting	30 mg of MOF; 23.5 mL of CH_3CN; 4.7 mL of TEA and 0.5 mL of H_2O; T =25°C, UV light (500 W Xe/Hg lamp, 385 nm) *1.7 µmol H_2/h·g_{cat}	[38]

(*Continued*)

Table 14.1 MOFs with group 4 metallic clusters. (*Continued*)

MOF	Metallic cluster	Organic ligands	Gap band (eV)	Application/ degraded compound	Reaction conditions	References
VNU-2	Hf–O	H_2CPEB	3.36	MB and MO	10 mg of MOF; 15 mL of dye (100 ppm); visible-light (300 W xenon lamp) *53 (MB) and 72 (MO)% degradation at 3 h	[39]
NH_2–UiO-66(Hf)	Hf–O	NH_2–ATA	2.92	Water splitting	30 mg of MOF; 23.5 mL of CH_3CN, 4.7 mL of TEA, and 0.5 mL of H_2O; T = 25°C, UV light (500 W Xe/Hg lamp, 385 nm). *1.7 μmol H_2/h·g_{cat}	[38]

2-Aminoterephthalic acid (NH_2–ATA), 4,4′-bipyridine (4,4′-bipy), 4′-(1H-tetrazol-5-yl)-[1,1′-biphenyl]-3,5-dicarboxylic acid (H_3L), 1,4-bis((1H-1,2,4-triazol-yl)methyl)benzene (btx), dichloroterephthalic acid (H_2DCTP), 2-methylimidazole (2MIM), bis[5-(2-pyridyltetrazolato)]diaquazinc(II) ($(2\text{-PTZ})_2Zn(H_2O)_2$), trimesic acid ($H_3BTC$), Phenanthroline (phen), 4,4′-oxybisbenzoic acid (H_2-oba), 1,4-bis(4-pyridyl)-2,3-diaza-1,3-butadiene (4-bpdb), 2,5-bis(4-pyridyl)-3,4-diaza-2,4-hexadiene (4-bpdh), terphenyl-3,3″,5,5″-tetracarboxylic acid (H_4L), 9-phenylcarbazole-3,6-dicarboxylic acid (H_2PDA), 9,10-bis(4′-pyridylethynyl)-anthracene (BPEA), 4,4′-biphenyldicarboxylate (BPDC), camphoric acid (D-camH_2), 2-(1-hydroxyethyl) benzimidazole) (OH-bim), thiophene-2,5-dicarboxylic acid (H_2TDC), 1,4-naphthalene dicarboxylic acid (H_2NDC), 3,5-bis(pyridin-4-ylmethoxy) benzoic acid (Hbpba), 3,3′,5,5′-biphenyltetracarboxylate (H_4bptc), pyridine (py), and 2,2′-bipyridine (bpy).

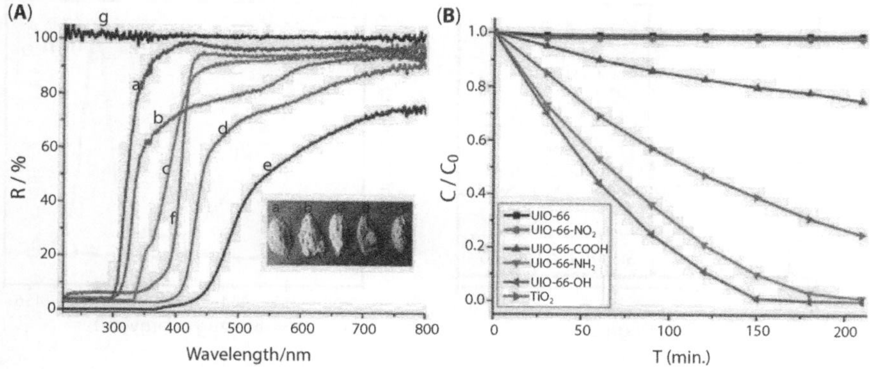

Figure 14.9 (A) UV-Vis DRS of (a) UiO-66, (b) UIO-66–NO_2, (c) UiO-66–COOH, (d) UiO-66–NH_2, (e) UiO-66–OH, (f) TiO_2 and (g) $BaSO_4$ and (B) photocatalytic degradation of phenanthrene by UiO-66–X. Reprinted with permission from Ref. [19].

MOF UiO-66 (UiO-66 (Zr/Ti)-M) (Figure 14.10) [37]. The crystallinity of the solid was maintained and the catalytic activities during a PCVG process for the reduction of Se (VI), were greatly improved. This method was more efficient, and energy saving compared to the traditional solvothermal methods, which makes it very promising for practical applications.

On the other hand, the use of complex ligands as H2L (2,2′-diamino-4,4′-stilbenedi-carboxylic acid, H_2SDCA-NH_2) to allow obtained MOFs with a response to the visible light irradiation. Sun *et al.* [41] synthesized and characterized a Zr-based-MOF and was evaluated in the CO_2 photocatalytic reduction to CO ($[Zr_6O_4(OH)_4(L)_6]\cdot 8DMF$, denoted as Zr–SDCA–$NH_2$) (Figure 14.11). They showed that is MOF present a good chemical stability, a narrow band gap that absorbs in the visible region and a formation rate of 96.2 μmol h^{-1} $mmol^{-1}_{MOF}$, which was higher than that of Zr–MOF functionalized with amine. Mott-Schottky and photoluminescence measurements as well as with the results of photocatalytic activity, demonstrated a charge transfer mechanism through the Zr-O cluster (LMCT), therefore, the two entities (i.e., organic ligand and metal cluster) contribute to the reduction of CO_2 to the anion formate, as shown in Figure 14.11. They demonstrated that the combination of amino groups and highly conjugated molecules, can extend absorption of light in the visible region, increase stability, and better the photocatalytic properties under visible light irradiation.

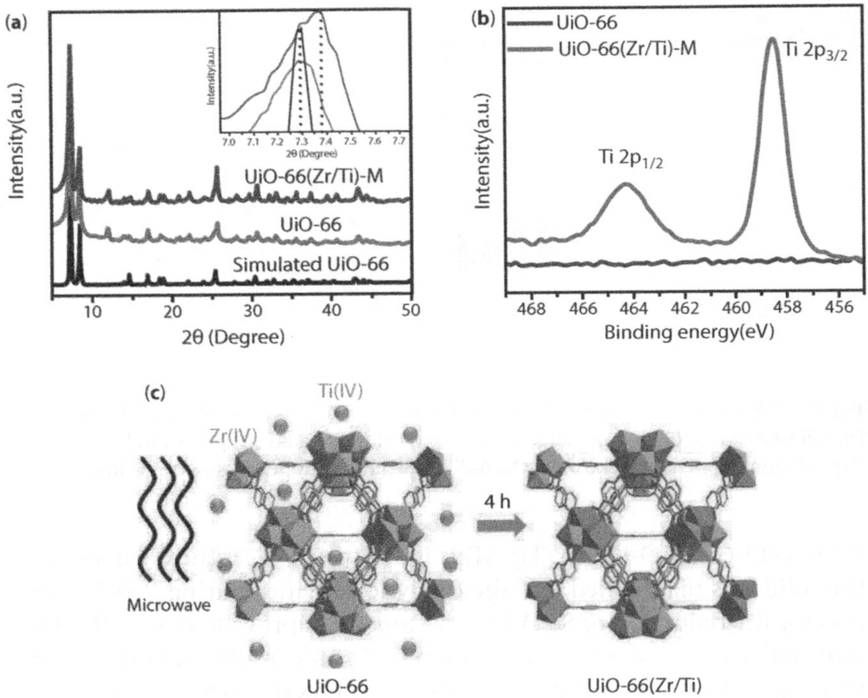

Figure 14.10 (a) PXRD patterns of the simulated UiO-66, prepared UiO-66, and UiO-66(Zr/Ti)-M; (b) XPS spectra of UiO-66 and UiO-66(Zr/Ti)–M in the Ti 2p region; and (c) Schematic illustration of the synthesis of UiO-66(Zr/Ti) using the microwave-assisted method. Reprinted with permission from Ref. [37].

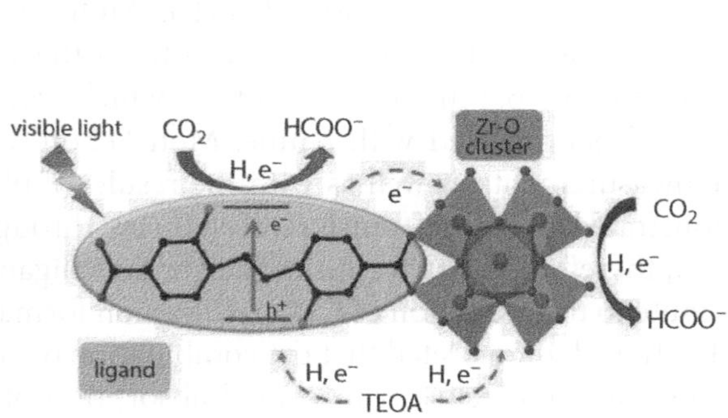

Figure 14.11 Proposed mechanism of photocatalytic CO_2 reduction to formate over Zr–SDCA–NH_2 under visible light irradiation. Reprinted with permission from Ref. [41].

14.2.2 Groups 8, 9, and 10 Metallic Clusters

Table 14.2 shows the MOFs synthesized with transition metals of Groups 8, 9, and 10 (or VIIIB group of ancient periodic table). Is important to note that in this case only MOFs synthesized with iron, cobalt, and nickel have been investigated so far. It is a common practice to use photosensitizers to increase the photocatalytic activity by absorbing light and directing the photogenerated electrons to the reaction sites, where the reduction reactions are carried out. Zhu *et al.* [43] studied the effect of Tris(2,2′-bipyridyl) dichlororuthenium (II) hexahydrate ($[Ru(bipy)_3]Cl_2 \cdot 6H_2O$) as photosensitizer and TEOA as electron donor and a 2D-MOF $Ni_3(HITP)_2$ conductive nanosheet co-catalyst in the selective photoreduction of CO_2 to CO. The co-catalyst showed a well-defined honeycomb structure with a Ni–N4 planar coordination motif. The active Ni–N4 sites and 2D nanosheet morphology in the MOF allow a good synchrony to improve photocatalytic activity. In addition, without photosensitizers it was not possible to observe photocatalytic activity, so that, this is key point to carry out the CO_2 reduction, since the photogenerated electrons come from the absorption of photons by the photosensitizer. The good conductivity of MOF $Ni_3(HITP)_2$ co-catalyst allows electrons to move through the well-defined MOF structure to the Ni–N4 sites, where the reduction of CO_2 to CO takes place, as shown in Figure 14.12.

The ligand-exchange strategy has been used, among other things, to obtain core-shell structures with MOFs. Guo *et al.* [44] using this strategy to synthesize ZIF-67 @ Co–MOF-74 core-shell catalysts and their use in the photocatalytic oxidation of water under visible light. The ZIF-67 @ Co–MOF-74 core-shell material was synthesized by exchanging ZIF-67 surface ligands with DHTP molecules (H_2BDC, H_3BTC, and NH_2–BDC). This exchange is possible since the organic ligand that is conforming to ZIF-67 2-methylimidazole has less coordination than DHTP molecules. SEM studies showed the growth of the Co–MOF-74 MOF (shell) on the surface of the ZIF-67 (core) forming layers of different sizes, making it possible to obtain materials with different thicknesses of the controllable layer, as can be seen in Figure 14.13. The results of the photocatalytic activity experienced an increase in the activity in the core-shell catalyst ZIF-67 @ Co-MOF-74 compared to ZIF-67 and Co-MOF-74.

14.2.3 Group 11 Metallic Clusters

Group 11 comprises the transition metals of copper (Cu), silver (Ag), gold (Au), and roentgenium (Rg). A few studies in the use of these metals for the

Table 14.2 MOFs with groups 8, 9, and 10 metallic clusters.

MOF	Metallic cluster	Organic ligands	Gap band (eV)	Application/ degraded compound	Reaction conditions	References
NH_2–MIL-101(Fe)	Fe–O	NH_2–ATA	2.59	CO_2 reduction	50 mg of MOF; MeCN and TEOA solution (60 mL, 5/1 v/v); visible light (300 W Xe lamp). *178 μmol $HCOO^-$ at 8 h	[31]
MIL-53(Fe)	Fe–O	H_2BDC	2.88	Rhodamine B	10 mg of MOF; 25 mL of RhB (10 ppm); visible light (500 W halogen tungsten lamp at 420 nm); pH 5; H_2O_2 (20 mM) *100% degradation at 50 min	[45]
MIL-88A (Fe)	Fe–O	Fumaric acid	2.05	Methylene blue	40 mg of MOF; 100 mL of MB (1×10^{-4} mol/L); visible light (300 W Xe lamp at 420 nm); pH = 7 *100% degradation in 20 min	[46]
NH_2–MIL-101(Fe) hexagonal microspindles	Fe–O	NH_2–ATA	1.32	Toluene	20 mg of MOF; 4 mL toluene; UV light (500 W Xe lamp) *79.4% degradation after 10 h	[47]

(Continued)

Table 14.2 MOFs with groups 8, 9, and 10 metallic clusters. (*Continued*)

MOF	Metallic cluster	Organic ligands	Gap band (eV)	Application/ degraded compound	Reaction conditions	References
Fe-based MOF	Fe–O	H_2BDC	1.61	Water splitting	5 mg of MOF; acetonitrile/ water solution v/v = 1:1); Eosin Y (EY) and triethylamine (TEA); white-light LED (400–750 nm) *125 µmol of H_2	[48]
MA-MOF 235	Fe–O	H_2BDC	1.94	Rhodamine B	10 mg of MOF; 50 mL of RhB (40 ppm); visible light (300 W Xe lamp); 1 mmol H_2O_2 *100% degradation at 20 min	[49]
MIL-101(Fe)–OH	Fe–O	H_2BDC–OH	3.54	Phenanthrene	50 mg of MOF; 10 mL of phenanthrene (10 ppm); UV light (175 W Hg lamp, λ > 330 nm) *99.98% of degradation at 120 min	[19]

(*Continued*)

Table 14.2 MOFs with groups 8, 9, and 10 metallic clusters. (*Continued*)

MOF	Metallic cluster	Organic ligands	Gap band (eV)	Application/ degraded compound	Reaction conditions	References
NH_2–MIL-53	Fe–O	NH_2–ATA	2.59	Water oxidation	0.70 mM of MOF, 1.0 mM $[Ru(bpy)_3]Cl_2$ (sensitizer), 0.08 M $Na_2S_2O_8$ (electron acceptor) and 20 mM buffered water (pH = 8.5, Na_2HPO_4/NaH_2PO_4), LED lamp (450–550 nm, 4 W) *120 μmol (TON: 51.1) of O_2 in 140 min	[50]
MIL-53(Fe)-1:2–150	Fe–O	H_2BDC	2.69	Rhodamine B	20 mg of MOF; 50 mL RhB (10 PPM); visible light (500 W Xe lamp, 420 nm); 20 mM H_2O_2 *k = 0.0286 min^{-1}	[51]

(*Continued*)

Table 14.2 MOFs with groups 8, 9, and 10 metallic clusters. (*Continued*)

MOF	Metallic cluster	Organic ligands	Gap band (eV)	Application/ degraded compound	Reaction conditions	References
Co6–MOF	Co6(μ3-OH)6	NTB and 4,4′-bpy	4.98	CO_2 Reduction	0.005 mmol of MOF; 0.01 mmol of $[Ru(bpy)_3]Cl_2{\cdot}6H_2O$; MeCN (4 mL), H_2O (1 mL), and TEOA (1 mL); visible light (150 W Xe lamp at 420 nm); T = 35°C *39.36 μmol of CO and 2.81 μmol of H_2	[52]
Core–shell ZIF-67@ Co–MOF-74	Co–O	H_2BDC, NH_2–ATA, and H_3BTC		Water splitting	10 mL of borate buffer solution (80 mM); pH = 8.0–10.0; 1.0 mM of $[Ru(bpy)_3](ClO_4)_2$; 50 mM $Na_2S_2O_8$; pH = 9.0; visible light (300 W Xe lamp at 420 nm) *122 μmol of O_2	[44]
[Co(4,4′-bipy)·$(HCOO)_2$]n	Co–O	4,4′-Bipy	2.14	Methylene blue	4 mg of MOF; 40.0 mL MB (10.0 ppm); visible light (500 W Xe lamp); 400 μL H_2O_2 *54.7% degradation in 150 min	[53]

(*Continued*)

Table 14.2 MOFs with groups 8, 9, and 10 metallic clusters. (*Continued*)

MOF	Metallic cluster	Organic ligands	Gap band (eV)	Application/ degraded compound	Reaction conditions	References
{[Co(bpba)(bdc)$_{1/2}$]}n	Co–O(N)	Hbpba and H_2BDC	2.63	Safranine T	10 mg of MOF; 12 mL of 40 ppm ST; UV light (500 W Hg lamp); 1.0 mL H_2O_2 *99% degradation in 180 min	[54]
$N_{i3}(HITP)_2$ nanosheets	Ni–O	HITP·6HCl	–	CO_2 reduction	2 mg of MOF; 80 mg of $[Ru(bpy)_3]Cl_2·6H_2O$; TEOA/H_2O/MeCN (4 mL/2 mL/10 mL); T = 4°C; LED light (100 W 420 nm) *207 µmol CO and 7.49 µmol H_2 in 3 h	[43]
[Ni(azp)(ppa)$(H_2O)_2$]n	Ni–O	4,4′-Azp and H_2-ppa	3.27	Methylene blue	60 mg of MOF; 100 mL of MB (40 ppm); UV light (400 W high-pressure Hg lamp) *96.8% of degradation in 90 min	[55]

2-Aminoterephthalic acid (NH_2–ATA), terephthalic acid (H_2BDC), 2-hydroxy terephthalic acid (H_2BDC–OH), 4,4′,4″-nitrilotribenzoic acid (NTB), 4,4′-bipyridine (4,4′-bpy), trimesic acid (H_3BTC), 4,4′-bipyridine (4,4′-bipy), 3,5-bis(pyridin-4-ylmethoxy)benzoic acid (Hbpba), 11-hexa-aminotriphenylene hexahydrochloride (HITP·6HCl), 1,4-phenylenedipropionic acid (H_2-ppa), 4,4′-azodipyridine (4,4′-azp).

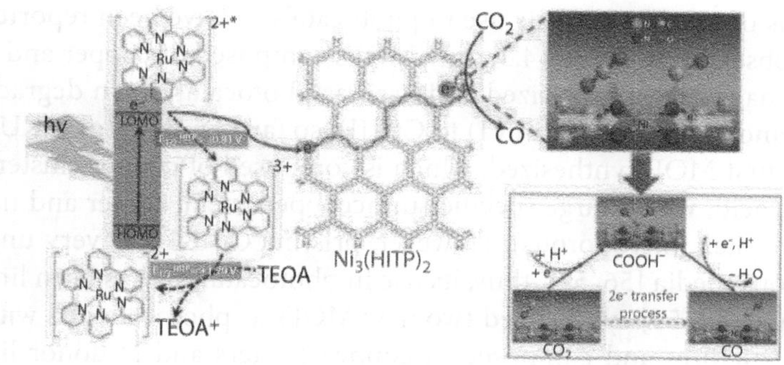

Figure 14.12 Proposed mechanism of photocatalytic CO_2 reduction to CO with $Ni_3(HITP)_2$ under visible-light irradiation. Reprinted with permission from Ref. [1].

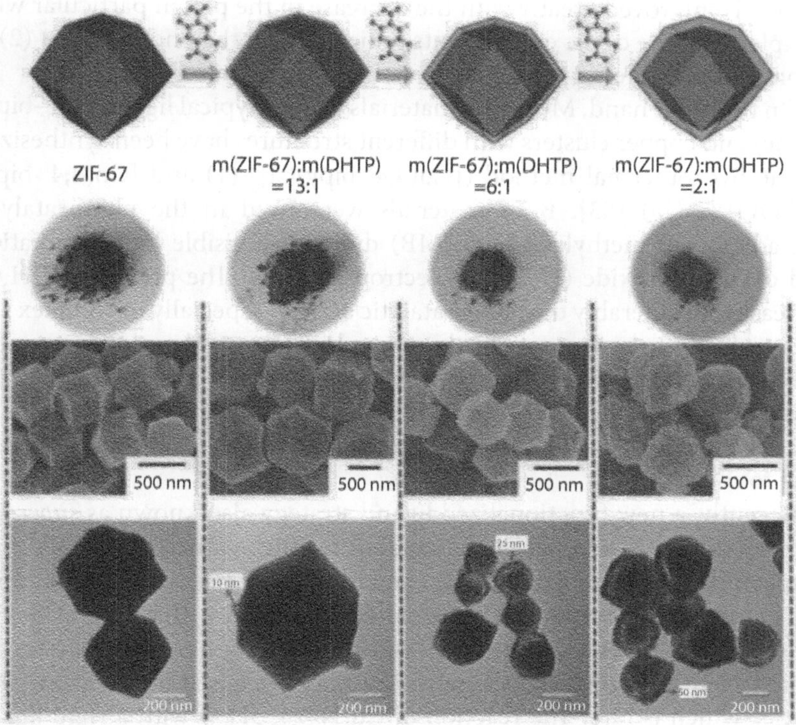

Figure 14.13 SEM/TEM images of core–shell ZIF-67 @Co-MOF-74 material with controllable shell thickness. Reprinted with permission from Ref. [44].

synthesis of MOFs and their use in photocatalysis have been reported. As can be observed in Table 14.3, only MOFs composed of copper and silver clusters have been synthesized and used as photocatalysts in degradation of dyes and reduction of Cr (VI) to Cr(III), so far. Cu–BTC (or HKUST-1) was the first MOF synthesized, which is composed of copper clusters and trimesic acid, with a large specific surface, open metal center and narrow pore size (i.e., microporous). However, pristine Cu-BTC is very unstable in aqueous media [56, 57], thus, its use in photocatalysis has been limited.

Wang *et al.* [59] synthetized two new MOFs as photocatalysts with different structures and composed of copper clusters and N-donor ligands (btx) ($[Cu(btx)_2(ClO_4)_2]_n$ (1) and $[Cu(btx)(ClO_4)]_n$ (2)) (Figure 14.14). These materials were tested in the degradation of methylene blue (MB) dye and the reduction of Cr(VI) to Cr(III) in aqueous solution under UV light irradiation. They demonstrated that both photocatalyst are efficient in MB degradation, besides, they found that the photo-reduction efficiency of Cr(VI) improved greatly with the decrease in the pH, in particular with complex (2). The cyclic experiments indicated that the photocatalyst (2) is stable and reusable.

On the other hand, MOFs 3D materials using a typical ligand (4,4′-bipyridine) and copper clusters with different structures have been synthesized by the hydrothermal method ($[Cu(4,4'\text{-bipy})Cl]_n$ (1) and $[Co(4,4'\text{-bipy})(HCOO)_2]_n$ (2)) [53]. Both materials were used in the photocatalytic degradation of methylene blue (MB) dye under visible light irradiation and oxygen peroxide (H_2O_2) as electron acceptor. The presence of H_2O_2 increases considerably the photocatalytic activity especially in complex (1), due probably to a decrease in the recombination of the photogenerated (electron–hole) charge carrier. Complex (1) was better than complex (2) in the photocatalytic activity of MB under the same conditions, it can be seen in Figure 14.14. In addition, complex (1) showed a higher stability after four cycles.

Recently, a new functionalized ligand strategy also known as *supermolecular building blocks* have been studied for the synthesis of MOFs with semiconductor properties [58]. This method generates two types of binding groups (carboxylate and pyrazole functionalities) that interacts with the metal centers. Also, from the perspective of a topological analysis, can alternatively be described as a network (3, 3, 5)-c based on basic building blocks, which is rare. The reaction of $Cu(NO_3)_2{\cdot}3H_2O$ with a rigid ligand 4′-(1H-tetrazol-5-yl)-[1,1′-biphenyl]-3,5-dicarboxylic acid (H3L) produces a new metal-organic complex $[Cu_2(\mu_3\text{-}OH)(L)(H_2O)_2]_n$. This complex has a truncated cuboctahedron connected by $Cu_3O(N_4CR)_3$ trimers trigonal using tetrazolate (N4CR). Also, this complex contains a large pore

Table 14.3 MOFs with group 11 metallic clusters.

MOF	Metallic cluster	Organic ligands	Gap band (eV)	Application/degraded compound	Reaction conditions	Ref.
$[Cu(4,4'\text{-bipy})Cl]_n$	Cu–O	4,4′-Bipy	2.14	Methylene blue	4 mg of MOF; 40.0 mL of MB (10 ppm); visible light (420 nm); H_2O_2 oxidant agent (400 mL). *93.93% degradation at 150 min	[53]
$[Cu_2(\mu_3\text{–}OH)(L)(H_2O)_2]_n$	Cu–O	H_3L	Not show	Methyl violet	50 mg of MOF; 50 mL of MV (10 ppm); UV-light irradiation (Hg lamp, 250 W). *64.2% of degradation at 100 min	[58]

(*Continued*)

Table 14.3 MOFs with group 11 metallic clusters. (*Continued*)

MOF	Metallic cluster	Organic ligands	Gap band (eV)	Application/degraded compound	Reaction conditions	Ref.
$[Cu(btx)_2(ClO_4)_2]_n$	Cu–O	Btx	2.59	Methylene blue dye and reduction of Cr(VI)	Degradation of MB dye: 7 mg of MOF; 40 mL of MB (10 ppm); UV-light irradiation (125 W Hg lamp, 365 nm); *14.6% degradation at 100 min. reduction of Cr(VI); 7 mg of MOF; 40 mL of Cr(VI) solution (10 ppm); UV-light irradiation (125 W Hg lamp, 365 nm); 6 M H_2SO_4 was used to adjust the acidity; 500 mL of methanol. *82.92% reduction at 60 min.	[59]
$[Ag(btx)_{0.5}(DCTP)_{0.5}]n$	Ag–O	H_2DCTP and btx	3.45	Rhodamine	0.01 mmol of MOF; 50 mL of RhB (10 ppm); UV light irradiation (high-pressure Hg lamp, 300 W). *93.1% of degradation at 120 min.	[60]

4,4′-Bipyridine (4,4′-bipy), 4′-(1H-tetrazol-5-yl)-[1,1′-biphenyl]-3,5-dicarboxylic acid (H_3L), 1,4-bis((1H-1,2,4-triazol-yl)methyl)benzene (btx), dichloroterephthalic acid (H_2DCTP).

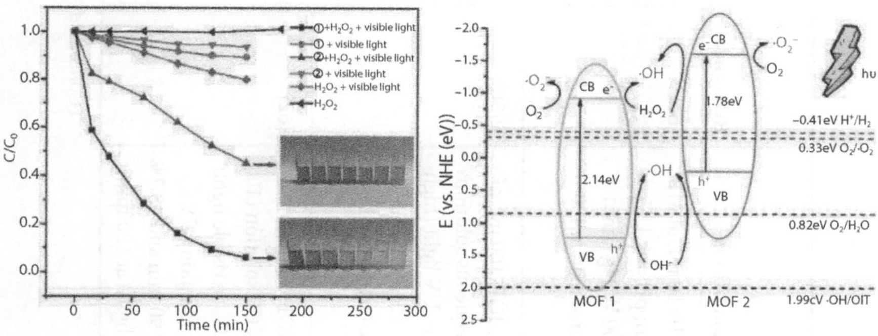

Figure 14.14 Photocatalytic degradation of MB under different photocatalytic conditions (left) and schematic illustration of the energy position and MB degradation over two complexes (right). Reprinted with permission from Ref. [53].

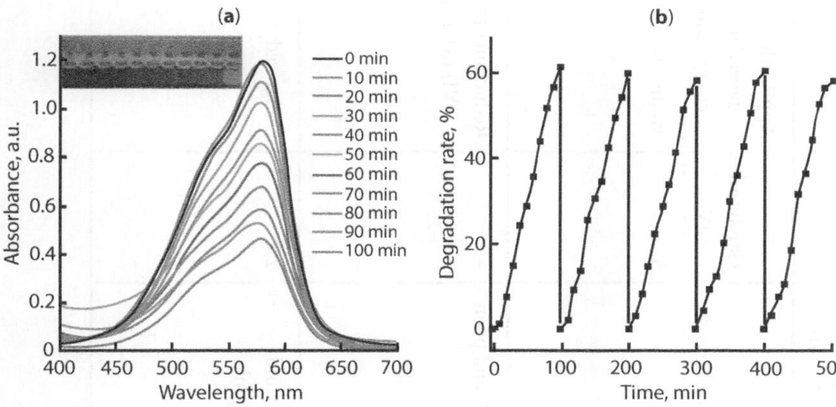

Figure 14.15 (a) UV-Vis absorption spectra of the Methyl violet dye degradation vs time, and (b) recycling tests, of a new metal-organic complex $[CU_2(\mu_3\text{-}OH)(L)(H_2O)_2]_n$. Reprinted with permission from Ref. [58].

size, the above is detrimental to the ability of gas adsorption, which was tested for H_2, CO_2 and CH_4 adsorption. However, the material showed excellent catalytic activity the for photo-degradation of methyl violet (MV) dye in aqueous medium, as it can be seen in Figure 14.15.

14.2.4 Group 12 Metallic Clusters

MOFs synthetized with Group 12 (or group IIB of ancient version of the periodic table) comprises transition metals of zinc (Zn) and cadmium (Cd) and their use as photocatalysts in different reactions, as shown in

Table 14.4 MOFs pristine with group 12 metallic clusters.

MOF	Metallic cluster	Organic ligands	Gap band (eV)	Application/ degraded compound	Reaction conditions	References
ZIF-8	Zn–O	2MIM	5.16	Methylene blue	25 mg of ZIF; 50 mL of MB (10 ppm); pH = 6; UV light (500 W Hg lamp) *82.3 % degradation in 120 min	[61]
HUT-11	Zn–O	H_3BTC and phen	–	Reduction of Cr (VI) to Cr (III)	100 mg of MOF; 100 mL of Cr(VI) aqueous solutions (10 ppm); UV light (high-pressure Hg lamp, 125 W) * 99 % of reduction in 315 min	[65]
TMU-6	Zn–O	H_2-oba and 4-bpmb	2.2	Congo red	25mg of MOF; 50 mL of aqueous CR solution (100 ppm); UV or visible light; *98.3% degradation (UV light) at 90 min and 48.2% (visible light) at 120 min	[62]

(*Continued*)

Table 14.4 MOFs pristine with group 12 metallic clusters. (*Continued*)

MOF	Metallic cluster	Organic ligands	Gap band (eV)	Application/ degraded compound	Reaction conditions	References
$\{[Zn_2(L)(DMF)_3]\cdot 2DMF\cdot 2H_2O\}$	Zn–O	H_4L	3.87	Methyl violet and Rhodamine B	UV light irradiation (250 W Hg lamp) *72.5% (MV) and 92.8% (RhB)	[66]
TMU-6′	Zn–O	4-bpmb and H2-oba	2.5	Reduction of Cr (VI) to Cr (III)	25 mg of MOF; 50 mL of a aqueous Cr(VI) (5 ppm); UV irradiation (high-pressure Hg lamp, 80 W) *90.8 % reduction at 120 min	[63]
Zn–PDA2	Zn–O	H_2PDA	3.48	Water splitting	T = 25°C; 5% TEA in pure water solution (pH = 10); visible light (Xe lamp 500 W) *TOF: 930 mol of $H_2/mol_{cat}\cdot h$	[67]

(*Continued*)

Table 14.4 MOFs pristine with group 12 metallic clusters. (*Continued*)

MOF	Metallic cluster	Organic ligands	Gap band (eV)	Application/ degraded compound	Reaction conditions	References
NNU-36	Zn–O	BPEA and BPDC	–	Rhodamine B, Methylene blue and reactive black 6 and reduction of Cr(VI) to Cr (III)	Reduction of Cr (VI): 15 mg of MOF; 40 mL of Cr(VI) aqueous solution (10 ppm), visible light (Xenon arc lamp, 300 W), *33.2% within 60 min Degradation of dye: 15 mg of MOF; 40 mL of dye solution (10 ppm); 30% H_2O_2; visible light (Xenon arc lamp, 300 W), *96.2% of RhB degradation at 70 min, 94.2% of MB after 80 min and 93.5% of R6G after 90 min	[36]
PL-MOF	$[Zn_2(COO)_3]^+$	D-camH_2 and OH-bim	1.52	Crystal violet	Visible light (589 nm), *100% of degradation after 20 min	[68]

(*Continued*)

Table 14.4 MOFs pristine with group 12 metallic clusters. (*Continued*)

MOF	Metallic cluster	Organic ligands	Gap band (eV)	Application/ degraded compound	Reaction conditions	References
Cd(TDC)(bix)(H_2O)]n	Cd–O	H_2TDC and bix	3.31	Methyl orange	50 mg of MOF; 50 mL of MO (20 ppm); UV light irradiation (height-pressure Hg lamp, 300 W) *90% degradation MO in 2.5 h	[69]
{Cd[Ru-L_2]·3(H_2O)}$_n$	Cd–O	Ru(4,4′-H_2dcbpy)$_2Cl_2$ Metalloligand		CO_2 reduction	40 mg of MOF; MeCN/TEOA (20/1 v/v); UV light (500 W Xe lamp) *71.7 μmol ($HCOO^-$)/gcat*h	[70]
Ru-MOF(Cd) (nanoflowers)	Cd–O	Ru(H_2dcbpy)$_3^2$ metalloligand	–	CO_2 reduction	40 mg of MOF; 60 mL of MeCN/TEOA (20/1), visible-light (500 W Xe lamp) *77.2 μmol/gcat*h	[64]
[Cd(NDC)(biim-4)]·0.5H_2O	Cd–O	Biim-4 and H_2NDC	4.42	Methyl orange	*92% degradation after	[71]

(*Continued*)

Table 14.4 MOFs pristine with group 12 metallic clusters. (*Continued*)

MOF	Metallic cluster	Organic ligands	Gap band (eV)	Application/ degraded compound	Reaction conditions	References
$[Cd_2(BPTC)(solvent)_3]$	Cd–O	H_4BPTC	3.72	Rhodamine B	10 mg of MOF; 20 mL of RhB (32.0 ppm); UV light *72% of degradation after 10 h	[72]
$\{[Cd_3(\mu_4\text{-dbba})_2(phen)_3]\cdot H_2O\}n$	Cd–O	H_3cpta, phen and py	3.20	Methylene blue	50 mg of MOF; 100 mL of MB (10 ppm); UV light (Hg lamp, 125 W) *78.3% of degradation after 150 min	[73]
Cd-TCAA	Cd–O	H_3TCA with abp	1.89	Methylene blue	5 mg of MOF, 50 mL of 28 μM of an aqueous solution of MB; 500 W Xenon lamp. *81% at 175 min	[74]

(*Continued*)

Table 14.4 MOFs pristine with group 12 metallic clusters. (*Continued*)

MOF	Metallic cluster	Organic ligands	Gap band (eV)	Application/ degraded compound	Reaction conditions	References
Cd(4-Hptz)$_2$ (H$_2$O)$_2$Cl$_2$]	Cd–O	4-Hptz	4.12	Methylene blue and reduction of Cr (VI) to Cr (III)	Degradation of MB; 7 mg of MOF; 40.0 mL of MB (10.0 ppm); UV-irradiation (125 W Hg lamp) *93.43% at 100 min Reduction of Cr(VI); 7 mg of MOF; 40 mL of Cr(VI) aqueous solution (10 ppm), 6 M H2SO4,UV light (125 W Hg lamp) *100% of reduction at 50 min	[59]
[Cd(NiL) (PTA)]·DMAC$_3$	Cd–O	NiL and PTA	2.43	Methyl orange	50 mL of MB (20 ppm); visible light (1000 W xenon lamp) *85% of the MO	[75]

2-Methylimidazole (2MIM), trimesic acid (H$_3$BTC), Phenanthroline (phen), 4,4′-oxybisbenzoic acid (H$_2$-oba), Hbpmb = 3,5-bis(4-pyridylmethoxy)benzoic acid, terphenyl-3,3″,5,5″-tetracarboxylic acid (H$_4$L), 9-phenylcarbazole-3,6-dicarboxylic acid (H$_2$PDA), 9,10-bis(4′-pyridylethynyl)-anthracene (BPEA), 4,4′-biphenyldicarboxylate (BPDC), camphoric acid (D-camH$_2$), 2-(1-hydroxyethyl) benzimidazole) (OH-bim), thiophene-2,5-dicarboxylic acid (H$_2$TDC), biim-4 = 1,1′-(1,4-butanediyl)bis(imidazole), 1,4-naphthalene dicarboxylic acid (H$_2$NDC), 3,3′,5,5′-biphenyltetracarboxylate (H4bptc), 2(-5yl)terephthalic acid (H$_3$cpta), tricarboxytriphenyl amine (H$_3$TCA), 3,5-dicarboxylatobenzyloxy)benzoic (H$_3$dbba), 1,4-bis(4-pyridyl)-2,3-diaza-1,3-butadiene (4-bpdb), 1,4-bis(4-pyridyl)-2,3-diaza-1,3-butadiene (4-bpdb).

Table 14.4. MOFs composed of these transition metals have been synthetized using organic ligands from a very simple compound such as, 2-methylimidazole (2MIM) to very complexes as H4L ligand and metalloligand or using a secondary organic ligand (phen, bix, etc). The configuration and flexibility of the secondary ligands play a key role in directing the related properties of MOFs-complexes.

For example, ZIF-8 belongs to a new subfamily of MOFs conformed of N-donor bridging ligands and Zn-O clusters where interactions nitrogen-metal are formed and their use as photocatalysts is attractive due to their excellent stability and physicochemical properties. Jing *et al.* [61] showed that ZIF-8 is an efficient photocatalyst for the photocatalytic degradation of methylene blue (MB) under UV light irradiation. Also, they demonstrated that ZIF-8 present a good photocatalytic performance in a strong alkaline medium due to an elevated concentration of hydroxyl radicals (·OH) formed from water oxidation ($\cdot OH + H^+$) (Figure 14.16).

However, ZIF-8 presents a broad band gap (ca. 5.16 eV) that limits its use under visible light. Other MOFs composed of zinc clusters have been synthesized looking for a decrease in its band gap. Masoomi *et al.* [62, 63] synthesized three MOFs based on Zn(II) clusters *via* mechanosynthesis as a solvent free, rapid and green process. During the synthesis procedure, they used different organic ligands (oba, 4–bpdb and 4–bpmb) which leads to formation of structures of double (TMU-4 and TMU-5) or threefold (TMU-6) interpenetration that showed significant differences in the band gap, thermal stability, pore size, and BET surface area (Figure 14.17). It should be noted that all materials were stable in water, and then, they were soaked in water for 24 h. The three MOFs were used in the photocatalytic degradation of Congo red (CR) dye without oxidizing agents under visible

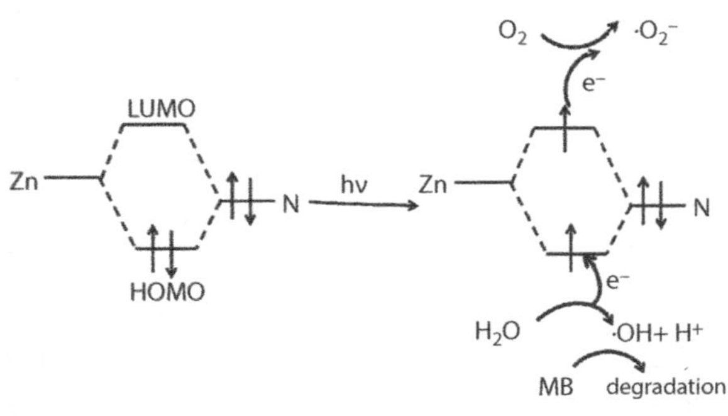

Figure 14.16 A simplified model of photocatalytic reaction mechanism of MB on ZIF-8(Zn). Reprinted with permission from Ref. [61].

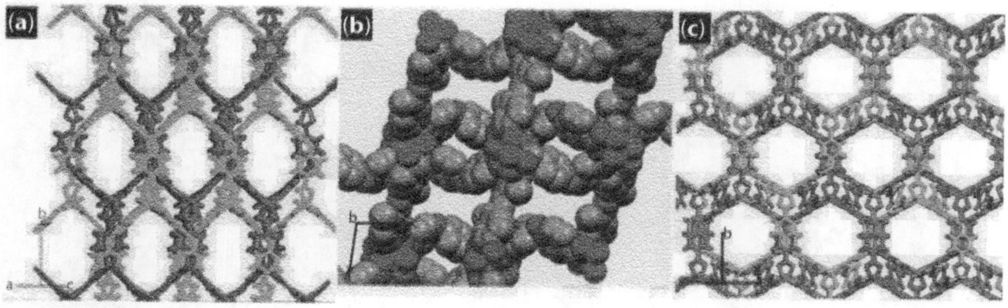

Figure 14.17 (a) Representation of the pore channels and the doubly interpenetrated structure in TMU-4, (b) schematic illustration of the network and pores in TMU-5, and (c) representation of the pore channels and of the network threefold interpenetrated (in red, blue, and brown). Reprinted with permission from Ref. [62].

and UV light irradiation. All materials showed good photocatalytic performance, however, MOF TMU-6 observed a degradation of 98.3% (UV light) at 90 min and 48.2% (visible light) at 120 min, being this material the highest active, compared to TMU-4 and TMU-5 samples. The high photocatalytic activity shown by MOF TMU-6 was due to its lower band gap and a better electron donating ability of N-donor ligands to metal centers. Later, same authors used solvent-assisted linker exchange method (SALE) for exchange the organic ligands (4-bpdb and 4-bpmb) in MOFs TMU-4 and TMU-6, which they called TMU-4′ ($[Zn_2(oba)_2(4\text{–}bpdb)_n{\cdot}(DMF)_2]$) and TMU-6′ ($[Zn(oba)(4\text{–}bpmb)_{0.5}]_n{\cdot}(DMF)_{1.5}$). These materials showed a significant improvement in their photocatalytic activity in the photocatalytic reduction of Cr(VI) under visible light irradiation with respect to that shown by TMU-4 and TMU-6 materials [63].

Other strategy to improve the stability and optical properties in MOFs was shown by Zhao *et al.* [36]. They synthetized a photocatalyst-MOF (NNU–36) with a pillared-layer framework thought pillared-layer synthetic method using two organic ligands (BPEA and BPDC). MOF NNU-36 structure is formed by "in-layer" zinc dimers and BPDC, and "interlayer" bipyridine pillars (BPEA) (Figure 14.18). This MOF presented a high efficiency in the photocatalytic reduction of Cr(VI) in the presence of methanol as hole scavenger and degradation of dyes using hydrogen peroxide as oxidizing agent under visible light irradiation. They demonstrated that the use of hole scavenger or oxidizing agents increase considerably the photocatalytic performance by decreasing or by suppressing the recombination of photogenerated electron–hole pairs in MOFs photocatalysts. It is important to highlight that pillared-layer strategy is a very useful method for synthetizing MOFs with high stability and enhanced photocatalytic performance.

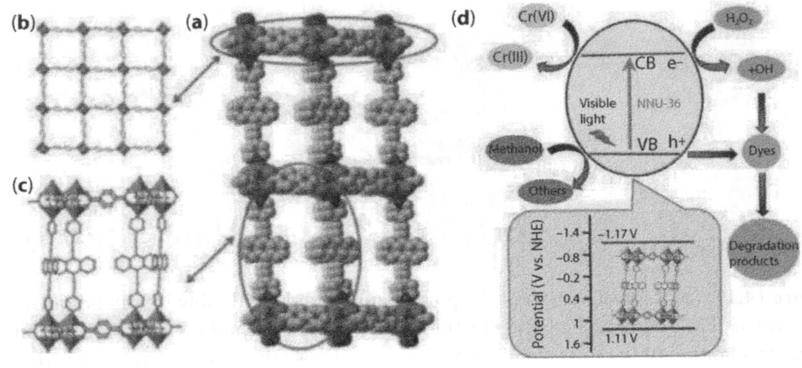

Figure 14.18 (a) Schematic illustration of pillared-layer structure of NNU-36, (b) layer structure constituted by BPDC ligand and zinc dimer; (c) schematic illustration of the pillaring ligand coordinated to layer in the structure, and (d) proposed mechanism of photocatalytic reduction Cr(VI) and degradation of dyes over NNU-36 under visible light irradiation. Reprinted with permission from Ref. [36].

So far, the use of metalloligand in the synthesis of MOFs is an effective strategy to obtain materials with excellent properties and high stability. Zhang *et al.* [64] obtained the first MOF using interpenetration structural strategy and a Ru-metalloligand. They synthetized two MOFs (with non-interpenetrated and interpenetrated structures, respectively) by solvothermal method. Both MOFs are composed of Ru–Cd–polypyridine and were used in the CO_2 photoreduction to CO. They observed an increase in the durability, recyclability, and thermal stability in the interpenetrated material that than non-interpenetrated count. This can be seen in Figure 14.19.

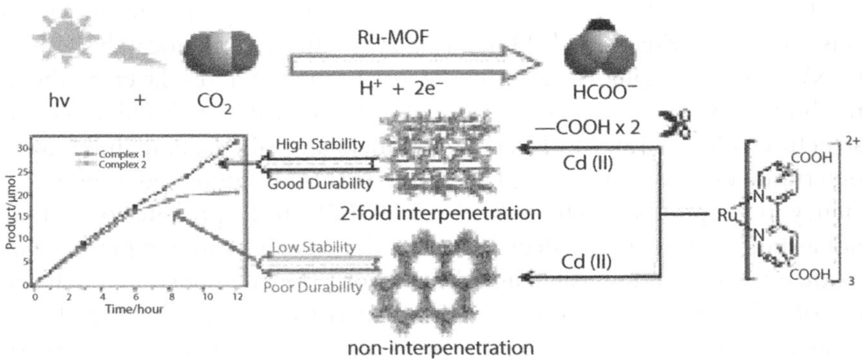

Figure 14.19 Photocatalytic CO_2 reduction over two Ru-MOFs. Reprinted with permission from Ref. [64].

14.3 Metal Nanoparticles–MOF Composites and Their Application in Photocatalysis

The use of metallic nanoparticles (MN) deposited on semiconductor materials (SM) has been a common practice to improve their photocatalytic performance in several reactions involving pollutant degradation, hydrogen production, CO_2 photoreduction, and organic synthesis, among others [76, 77]. This overperform compared to that of the semiconductor alone, has been explained in terms of an improved charge separation at the MN/SM interface (Schottky effect) and increased absorption of visible light due to the localized surface plasmon resonance (LSPR effect) of metals [78]. Indeed, when the MN/SM composite is irradiated with light, it causes the flow of electrons from SM to the MN until the equilibrium is reached in the Fermi levels of both constituents. In addition, this process creates positive charges (holes) on the surface of the SM, while the MNs are negatively charged. Subsequently, a barrier is formed that prevents electron back flow into the semiconductor [79]. It is worth noting that to form the Schottky barrier, both, the work function and electronic affinity of the metals must be higher than those of the semiconductors [80]. On the other hand, the LSPR effect is associated to the collective oscillations of conduction electrons in metals having dimensions smaller than that of wavelength of excitation light [81]. One of the most important properties of MNs presenting LSPR effect is their strong light absorption efficiency, which depends on the metal type, shape, and dimensions [82]. In fact, only certain MNs, such as Au, Al, Ag, and Cu present evident LSPR effect, also showing photothermal effects that can be modulated depending on their morphology and the way they are integrated into the composites [82]. In summary, incorporating plasmonic MNs on SM photocatalysts is a potential way to enlarge the light absorption, charge generation, and separation during the photocatalytic process [77].

As mentioned before, MOFs are crystalline porous materials with high specific surface area, structural diversity, and durability, among others, which are much appreciated in catalysis. Due to their permanent porosity allows the confinement of MNs generating a synergy that leads to better catalytic performance. The use of MNs/MOFs composites has already been reviewed in the literature for photocatalytic reactions, addressing topics of synthesis, characterization, and synergy on different reactions and conditions [83–86].

Although a wide variety of synthesis strategies for MNs/MOFs composites have been investigated, to date they have been classified into three categories [87–90]: (i) the "ship-in-a-bottle" approach, (ii) the "bottle-around-ship" approach, and (iii) the "one-pot" approach, as shown in Figure 14.20. The first route it is intended to control the size of the MN inside the MOF and preventing their agglomeration by adding the MN precursor together with the selected MOF and subsequently effect a careful reduction. In the second route, the pre-synthesized MNs are dispersed in the reaction medium, in which the MOF-forming precursors are found. This procedure ensures a well-defined size, shape, and morphology of MNs, however, it could hinder the MOFs growth because of the high interfacial energy barrier between the two kinds of materials [91]. The third route refers to the simultaneous formation of the two components MNs/MOF, which could represent the easy and versatile way of synthesis. However, due to the lack of synchronization of the self-assembly of the two entities, the final composite could result in low uniformity and homogeneity. Several preparation methods have been used for immobilizing MNs in MOFs, such as, solution infiltration technique, chemical vapor deposition, solid grinding technique, and dual solvent technique [83, 92]. Nonetheless, due to the great importance of MOFs in commercial applications, new methods of preparation of MOFs composites have been developed, which include electrochemical, microwave, mechanochemical, spray drying, and flow chemistry routes [93].

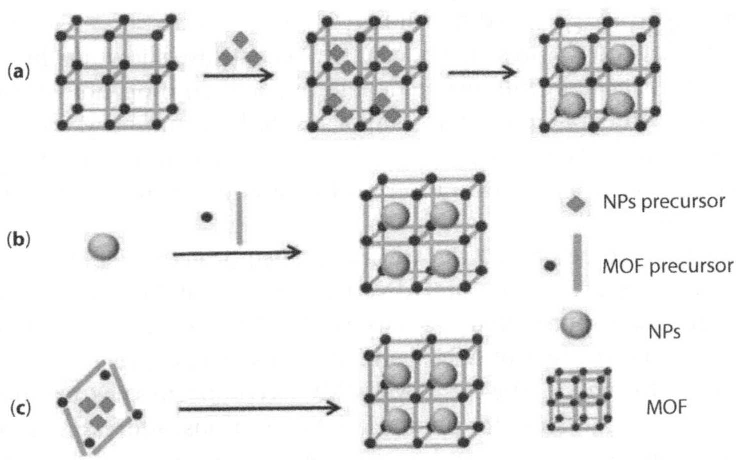

Figure 14.20 Different synthetic approaches for the preparation of MN/MOF composites: (a) the "ship-in-a-bottle" approach, (b) the "bottle-around-ship" approach, and (c) the "one-pot" approach. Reprinted with permission from Ref. [87].

Concerning the characterization techniques employed to determine the bulk and surface properties of MNs/MOFs composites, the following can be listed [83, 94]: inductively coupled plasma (ICP) spectroscopy, elemental analysis, infrared (IR) spectroscopy, UV-vis spectroscopy, nuclear magnetic resonance (NMR) spectroscopy, powder X-ray diffraction, nitrogen physisorption, thermogravimetric analysis (TGA), X-ray photoelectron spectroscopy (XPS), Electron paramagnetic resonance (EPR), Scanning electron microscopy (SEM), Energy-dispersive X-ray spectroscopy (EDS), Transmission electron microscopy (TEM), Electron energy loss spectroscopy (EELS), High-angle annular dark-field imaging-scanning transmission electron microscopy (HAADF-STEM), Selected area electron diffraction (SAED), X-ray absorption spectroscopy, Time-resolved microwave conductivity (TRMC), among others. Undoubtedly, the characterization of the composites using a combination of the techniques mentioned above, as well as the monitoring of their evolution during the different stages of the synthesis and evaluation enriches the deep knowledge of these complex materials [95].

On the other hand, it has been shown that MOFs are good hosts for MNs, mainly due to their high porosity and tunable cavity dimensions providing a confined space avoiding NPs growth, as well as a good dispersion throughout the MOFs. This feature, together with the intimate contact between the MOFs and the encapsulated MNs and the fixed spatial arrangement favor the charge transfer, which leads to good photocatalytic behavior [96]. The first successful report of the MNs /MOFs synergy was reported in 2012, with Pt NPs loaded into de cavities of UiO MOFs, built from Ir-phosphor-derived linear dicarboxylate ligands and $Zr_6(\mu_3\text{-}O)_4(\mu_3\text{-}OH)_4(\text{carboxylate})_{12}$ SBUs [97]. This composite was an efficient photocatalyst for hydrogen evolution reaction, where it was evidenced by a facile electron transfer from the photoreduced Ir phosphor to the entrapped Pt NPs [97]. This fact marked the starting point for continuing the study of noble and non-noble metals deposited in MOFs, to obtain a high photocatalytic activity.

It should be noted that each MN/MOF photocatalytic system behaves differently depending on the type of reaction and the metal (Ag, Au, Cu, Ni, Pd, Pt, etc.) location on the MOF [96]. Some recent examples of MN/MOF systems applied in photocatalytic reactions are described and analyzed below.

14.3.1 Ag–MOF Composites

Ag nanoparticles were deposited on hollow Co-MOF-74, which was obtained by reacting Co-ZIF-67 with the ligand of MOF-74,

2,5-dihydroxyterephthalic acid, and tested in the CO_2 photocatalytic reduction with $Ru(bpy)_3Cl_2{\cdot}6H_2O$ as photosensitizer and TEOA as sacrificial electron donor irradiated with visible light. Authors confirmed and proposed a reaction mechanism to convert Co-ZIF-67 in hollow Co-MOF-74. In addition, they confirmed a higher photocatalytic activity toward gaseous products ($CO + H_2$) with the presence of Ag NPs as cocatalyst on Co-MOF-74 (Figure 14.21) [98].

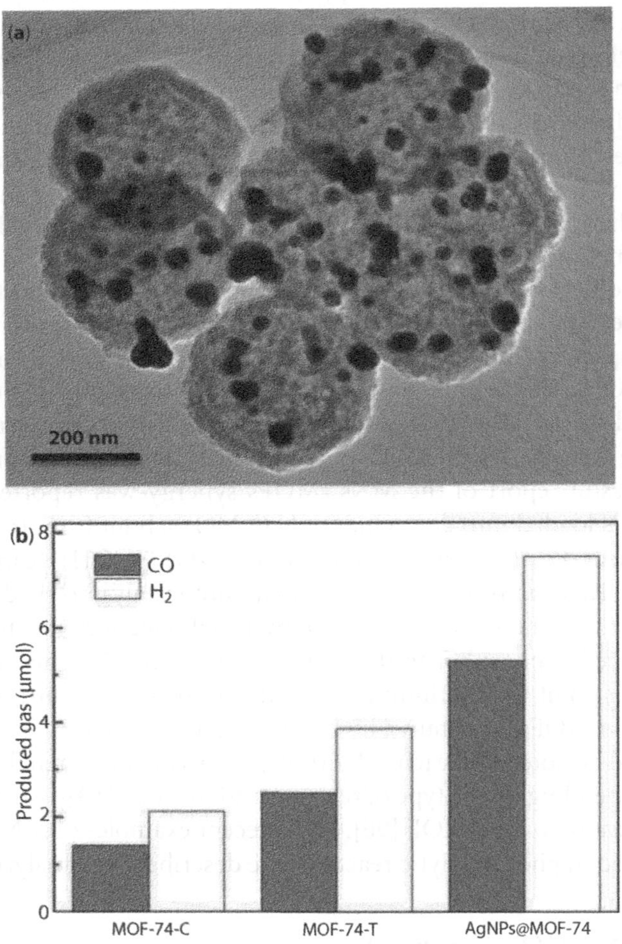

Figure 14.21 (a) TEM image of AgNPs@MOF-74 used in photocatalysis; (b) Comparison of photocatalytic activities by utilizing MOF-74-C, MOF-74-T, and AgNPs@MOF-74 as cocatalysts. Reprinted with permission from Ref. [98].

14.3.2 Au–MOF Composites

Thin nanosheets of pillared layered PPF-3 (paddle-wheel frameworks) were synthesized by a PVP assisted solvothermal route as a platform of Au nanoparticles to improve charge transfer and mass transport in the photocatalytic CO_2 reduction under visible light [99]. According to the results, Au/PPF-3 photocatalysts presented a high activity and selective production of formic acid, ~fourfold higher than that over pristine PPF-3 MOF under visible light irradiation. These results were explained in terms of an enhanced electron transfer from Au plasmonic nanoparticles to the PPF-3 MOF. However, a critical factor to maintain high electron and energy transfer into the hybrid system was the PPF-3 nanosheets thickness. A reaction mechanism was proposed (Figure 14.22) in which the excited electrons were transferred to Co^{2+} ions, leading to formation of Co^{+} ions and then CO_2 molecules adsorbed on cobalt catalytic sites are reduced into HCOOH by a proton coupled electron transfer process.

14.3.3 Cu–MOF Composites

Noble metal-free/MOF composites are not common for applications in catalysis or photocatalysis due to the instability of certain metals (e.g., Cu or Ni) to remain as reduced metal species under reaction conditions or thermal treatments. However, by selecting an appropriate synthesis strategy, it is possible to obtain active, selective, and stable photocatalytic systems. For instance, Xiao *et al.* [100] prepared a Cu/Cu@UiO-66 photocatalyst, which metallic copper was deposited on the surface, in addition to being encapsulated in the MOF, respectively. The above was planned to have copper plasmonic nanoparticles (on the MOF surface) and Cu quantum dots in close contact with the MOF forming a

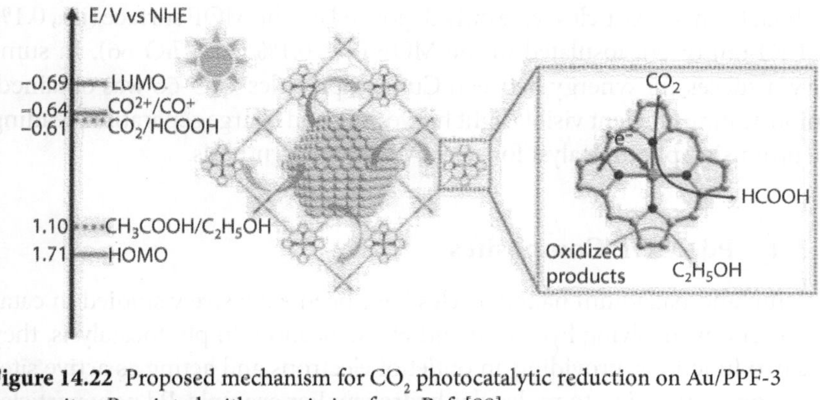

Figure 14.22 Proposed mechanism for CO_2 photocatalytic reduction on Au/PPF-3 composites. Reprinted with permission from Ref. [99].

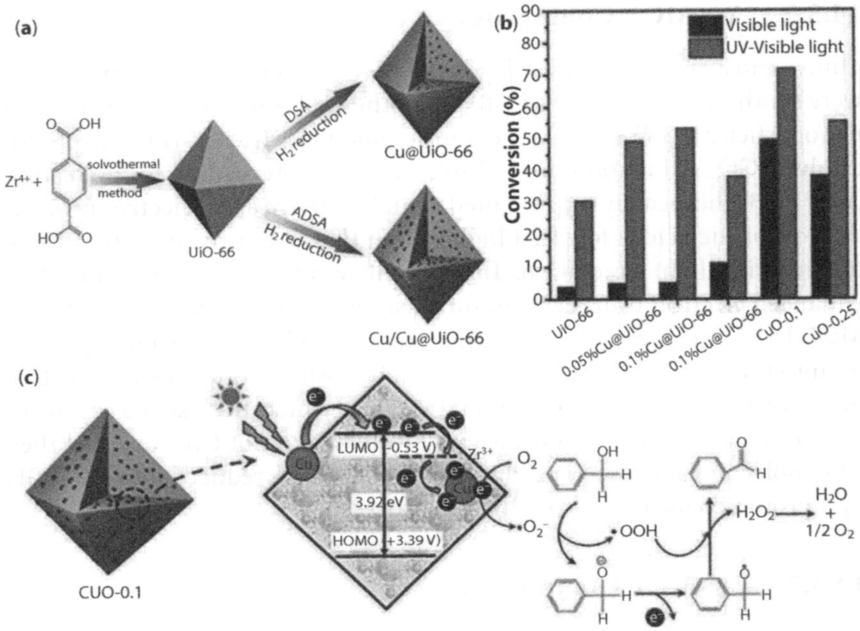

Figure 14.23 Scheme showing the synthesis of Cu–UiO-66 composites (a); a comparison of photocatalytic activity with the different photocatalysts (b); proposed reaction mechanism for the selective oxidation of benzyl alcohol (c). Reprinted with permission from Ref. [100].

Schottky barrier, as shown in Figure 14.23. Authors used an advanced double-solvent approach followed by a careful reduction to prepare their composites, which were evaluated in the partial oxidation of aromatic alcohols under visible light. Authors found that the 0.1% Cu/Cu@UiO-66 photocatalyst (CUO-0.1) showed the higher activity and selectivity compared with other photocatalysts in which Cu nanoparticles were only deposited on the MOF surface (i.e., 0.1% Cu/UiO-66) or encapsulated in the MOF (i.e., 0.1% Cu@UiO-66). In summary, a successful synergy between Cu nanoparticles-UiO-66 was obtained, demonstrating efficient visible light harvesting and charge separation, leading to a promising photocatalyst for selective organic synthesis.

14.3.4 Pd–MOF Composites

Platinum and palladium nanoparticles have been extensively studied in catalytic reactions involving hydrogen and H_2 adsorption. In photocatalysis, they have two functions, providing an outlet of electrons and acting as active sites for protons conversion to molecular hydrogen. For example, Pd nanoparticles

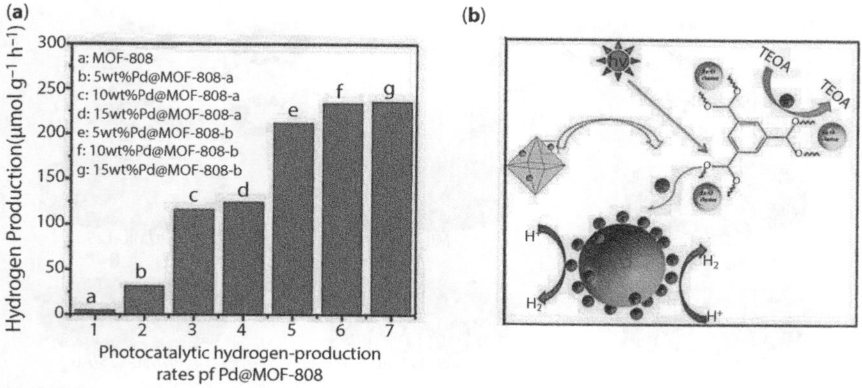

Figure 14.24 Photocatalytic hydrogen production on different Pd@MOF-808 composites (a); photocatalytic water hydrogen production mechanism of Pd@MOF-808 series materials with TEOA as a sacrificial donor under visible light irradiation (b). Reprinted with permission from Ref. [102].

linked with a MOF (e.g., Pd/CeMIL-101), have shown a superb photocatalytic H_2 production, close to 500 μmol under visible light irradiation [101]. On the other hand, Pd–MOFs composites, apart from presenting an enhanced photocatalytic activity, they have a great capacity for hydrogen adsorption. Even though this effect is already well known, it is magnified when Pd-nanoparticles interact with a MOF. For instance, Xu *et al.* [102] synthesized Pd@MOF-808 series materials by a bottle-around-ship approach by a simple solution method. Core–shell and penetrated Pd particles on MOF-808 were the two procedures for the preparation of the composites. According to the results, the higher photocatalytic activity in the hydrogen evolution reaction (256 μmol/g-h) in presence of TEOA was obtained with the penetrated Pd nanoparticles (Figure 14.24) showing also a H_2 storage capacity of 8.2% at 4 MPa y 77K.

14.3.5 Pt–MOF Composites

Pt-MOF, Au-MOF composites and bimetallic–MOF composites have been the most studied in applications in catalysis and photocatalysis [103]. It is worth noting, a recent article, where the integration of plasmonic effects (Au) and Schottky junctions (Pt) into MOF composites are demonstrated [104]. Two strategic routes for the preparation of Pt@MIL-125/Au and Pt/MIL-125/Au and MIL-125/Au were followed by using preformed Pt nanoparticles and Au nanorods, as shown in Figure 14.24. The higher photocatalytic hydrogen generation (1743 μmol/g h) in the presence of TEOA/

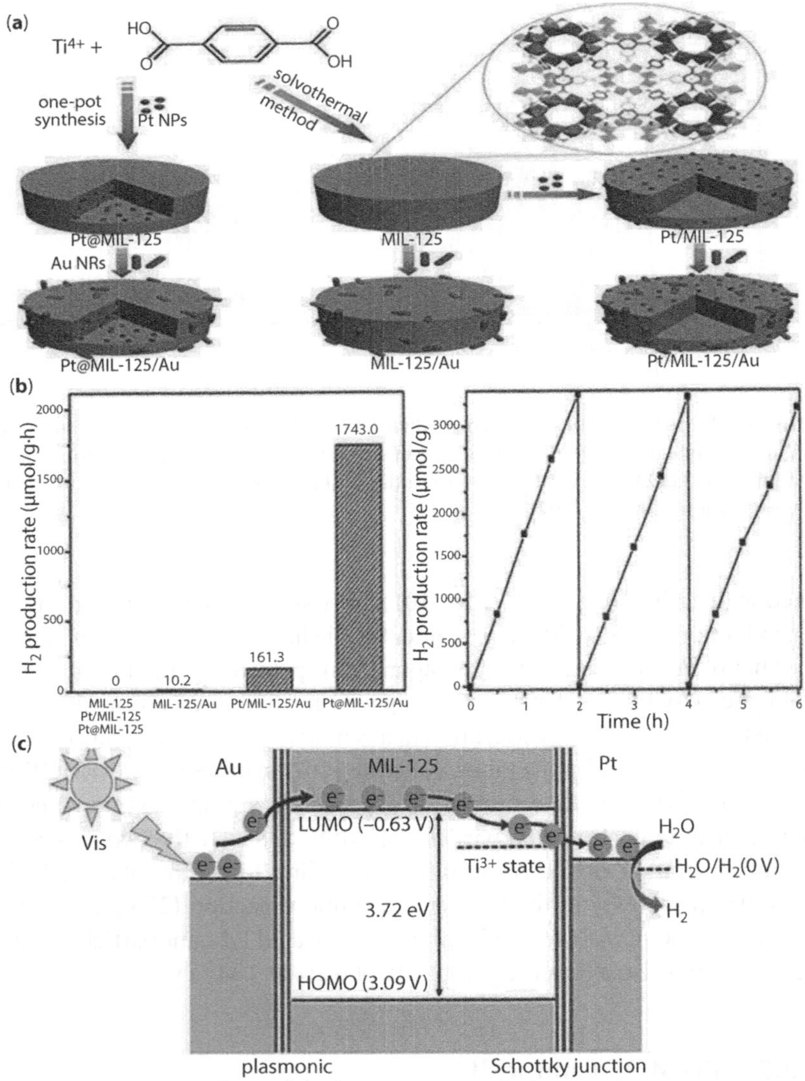

Figure 14.25 (a) Representation of synthesis method for Pt@MIL-125/Au and Pt/MIL-125/Au materials. (b) Photocatalytic H2 production rates of different catalysts, (c) Schematic illustration showing the electron migration at the two metal–MOF interfaces based on the energy levels. Reprinted with permission from Ref. [104].

MeCN aqueous solution was obtained with the Pt@MIL-125/Au composite (Figure 14.25). Authors explained their magnificent results according to the crucial point played by the MOF, which having a porous semiconductor-like structure can spatially separate Au and Pt to steer the charge flow for high catalytic efficiency. In other words, this work paves the way to evaluate the behavior of wide-bandgap semiconductors, not necessarily MOFS, to effect photocatalytic reactions under visible light with the accurate inclusion of Au and Pt nanoparticles.

14.4 Semiconductor–MOF Composites and Their Application in Photocatalysis

Many semiconductors have been used as photocatalysts since the pioneering work of Fujishima and Honda in 1972 [105]. Many have shown promising results; but TiO_2 is the most employed due to its outstanding properties, such as high chemical stability, wide availability, low cost, and high photocatalytic activity in the decomposition of organic and inorganic pollutants; however, TiO_2 shows a poor performance in the water-splitting reaction or the photocatalytic production of H_2. Other disadvantages of TiO_2 are that it is active in the UV region of the electromagnetic spectrum; therefore, it can only use about 5% of the solar energy; also, the recombination of the photogenerated electron-hole pairs is high. These problems are not exclusive to TiO_2 since most of the employed photocatalysts show similar behavior. Various strategies have been employed to overcome these disadvantages: for example, to improve the use of the solar energy, many researchers have used organic molecules as photosensitizers or semiconductors that absorb visible light; also, noble metals (Au, Ag, Pt) or semiconductors (heterojunctions) have been used to improve the electron–hole separation [106].

Recently, it has been proved that pristine MOFs behave as semiconducting materials, and they are used in several photocatalytic applications, as described earlier. Such as, MOFs can be used along with other semiconductors to obtain composite materials or heterojunctions (hereafter semiconductor–MOF composites). In these materials, MOFs can act as support due to its high surface area, as a photosensitizer to enhance the harvesting of visible light, depending on the chemical structure of the ligand, or improve the electron-hole separation. The employed semiconductors include metal oxides (TiO_2, ZnO, Ag_2O, Cu_2O, MoO_3, manganese oxides, Fe_3O_4, SnO_2), chalcogenides (Bi_2S_3, CdS, CdTe, copper sulfides, In_2S_3), and others (BiOCl, $BiVO_4$, GR, rGO).

Semiconductor–MOF composites have been used in several photocatalytic applications, such as the degradation of water or air pollutants, production of H_2, and the reduction of CO_2, where the degradation of pollutants is the most-studied application. The application of these materials has been extensively covered in the literature [107–109] and, therefore, in the next sections, only the most recent works are highlighted.

14.4.1 TiO_2–MOF Composites

As the most employed photocatalyst, the TiO_2–MOF composites are the most studied. As can be seen in Table 14.5, many MOFs and MOF families have been used recently in the synthesis of TiO_2–MOF composites; for example, UiO-66, the MIL family, Cu–BTC, and ZIF-8. The application of pristine MOFs to water decontamination has been widely studied [110–112], and with a similar trend, it can be observed in Table 14.5 that the most reported application is the degradation of water pollutants. Among the toxic compounds used in these works are Rhodamine B [113, 114], Methylene blue [113–118], Tetracycline [119], and Reactive red 198 [120]. Also, these composites have been employed in the degradation of air pollutants [121] or in the photo-oxidative desulfurization of model compounds [122].

The degradation of organic pollutants is often enhanced using MOFs in these composites due to its large surface area. It is common to observe that the adsorption of organic pollutants is higher in the semiconductor–MOF composite than in the pristine semiconductor; however, the presence of the semiconductor reduces the adsorption capacity of the MOF, compared to that of the pristine MOF. In several of the reviewed works, the contribution of the adsorption to the removal of the organic contaminant was higher than the contribution of the photocatalytic reaction. For example, a TiO_2–UiO-66 composite was used to degrade RhB and MB under basic and acid conditions [113]; in this work, it was observed that the adsorption was as high as the 90% of the total removal in the case of MB and 50% in the case of RhB. Also, a TiO_2–MIL-100(Fe) composite was used to eliminate both pollutants from aqueous solutions, and similar behavior was observed [114]. However, these works did not present a characterization of the spent catalyst; this characterization is desirable to elucidate if the pollutant is effectively degraded or just is adsorbed on the surface of the photocatalyst.

On the other hand, the scarcity of works related to the photocatalytic production of H_2 and CO_2 reduction could be attributed to the low performance of bare TiO_2 in both reactions [125, 126] and, generally, the

Table 14.5 Recent applications of TiO_2–MOF composites.

MOF	Method of synthesis	Application/ degraded compound	Highlighted results	References
UiO-66	Solvothermal	Rhodamine B and Methylene blue	The adsorption of RhB and MB was heavily increased by the MOF	[113]
MIL-100(Fe)	Solvothermal	Rhodamine B and Methylene blue	TOC removal was above 80% under visible light irradiation	[114]
MIL-101 (Cr)-NH_2	Solvothermal	Methylene blue	H_2O_2 was used to enhance the degradation rate of MB	[116]
MIL-88B (Fe)-NH_2	Solvothermal	Methylene blue	The composites presented a high stability after 5 reaction cycles	[117]
MIL-125 (Ti)-NH_2	Solvothermal	Tetracycline	The composites showed a good performance despite the presence of amorphous TiO_2	[119]
Cu-BTC	Solvothermal	Degradation of air pollutants	The degradation of isopropyl alcohol was performed in the gas phase	[121]
Cu-BTC	Wet route**	Photo-oxidative desulfurization	Benzothiophene and dibenzothiophene were completely degraded in 60 min	[122]

(*Continued*)

Table 14.5 Recent applications of TiO_2–MOF composites. (*Continued*)

MOF	Method of synthesis	Application/ degraded compound	Highlighted results	References
MIL–125 (Ti)-NH_2	Solvothermal	Hydrogen production	The composite presented an enhanced production, 70 times higher than pristine MIL-125(Ti)–NH_2	[123]
ZIF-8	Ultrasound	Hydrogen production	The presence of ZIF-8 in the composites enhanced up to 3 times the production rate, compared to bare TiO_2	[124]
Cu–BTC	Wet route**	Hydrogen production	The H_2 production depended on the preparation method of the composites	[125]
Cu–BTC	Aerosol method	CO_2 reduction	The CO_2 reduction proceeded via the formation of carbonates	[126]
CPO-27(Mg)	Solvothermal	CO_2 reduction	The presence of alkaline sites enhanced the adsorption of CO_2	[127]
UiO-66–NH_2	Thermal*	CO_2 reduction	CO was reported as the main product of CO_2 photocatalytic reduction	[13]

deposition of noble-metal nanoparticles is used to enhance the performance of TiO_2 [128, 129]. However, the use of TiO_2–MOF composites can improve the reaction rate without the use of noble metals, using UV or visible light; e.g., a TiO_2–MIL-125–NH_2 composite was used in the photocatalytic production of H_2 using visible light [123]. This material showed a performance higher than the activity presented by both bare TiO_2 and MIL-125(Ti)–NH_2, without the use of any other co-catalyst. It is worth to note that this composite was obtained by using a solvothermal treatment of MIL-125–NH_2, during which the latter was decomposed to form TiO_2. Nowadays, the use of MOFs as precursors of metal oxides is extensively used, and many composites with an interesting structure and properties can be obtained by this approach [130].

Also, TiO_2–Cu–BTC composites have been used in the photocatalytic production of hydrogen, and it has also been observed an increment in the performance, compared to the pure semiconductors. For example, a TiO_2–Cu–BTC composite was used in the photocatalytic photo-reforming of glycerol to produce hydrogen [125]. In these materials, the preparation method and the mass ratio between TiO_2 and Cu–BTC played a crucial role in enhancement of the photocatalytic activity; it was found that the interaction between both compounds was enhanced by using a preformed TiO_2 material and growing the MOF on top of it, instead of growing TiO_2 on a preformed MOF. This result was explained in terms of the quality of the MOF structure in the final composite, i.e., the MOF structure partially collapsed when TiO_2 was synthesized by using a solvothermal method. This collapse of the Cu-BTC structure caused a weak interaction with the TiO_2 and, probably, a deficient charge carrier transport between the semiconductors.

Besides the preparation method, another important aspect that affects the photocatalytic performance of these composites is the mass or molar ratio between the semiconductor and the MOF. For example, in the work described earlier, it was found that the best TiO_2/Cu-BTC mass ratio was 0.5. Also, a similar trend has been observed in the photocatalytic reduction of CO_2. For example, TiO_2-Cu-BTC composites prepared by an aerosol method [126] were used in this reaction and the observed photocatalytic activity changed in function of the TiO_2/Cu-BTC molar ratio, been 1/3.3 the best.

It is worth to note that in several works, CO was reported as the main product of the photocatalytic reduction of CO_2. As it is well known, the reduction of CO_2 is a difficult reaction, according to its thermodynamics and, in this sense, TiO_2–MOF composites seem to be not an adequate photocatalyst for this reaction, because noncomplex organic compounds are not reported in the mentioned works. This behavior seems to be a general trend in the literature because, in general, CO is reported as the

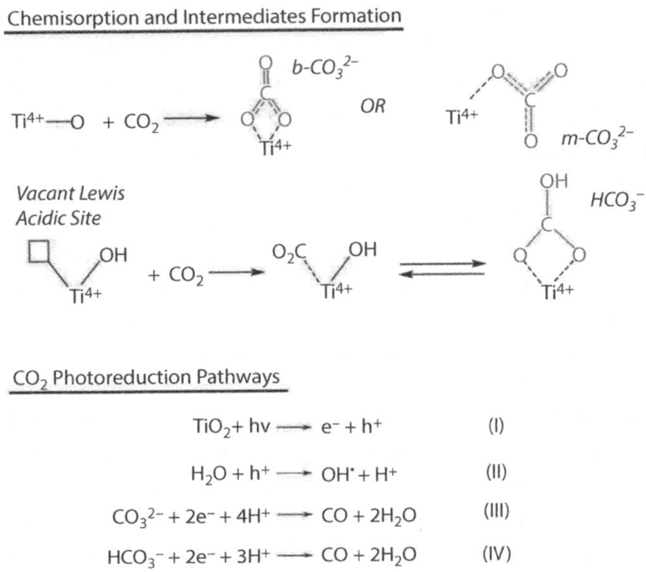

Figure 14.26 Proposed mechanism of the photocatalytic reduction of CO_2. Reprinted with permission from Ref. [126].

main product of the CO_2 photocatalytic reduction using these composites; however, there are some exceptions where the formation of CH_4 is also reported [131].

Figure 14.26 presents one of the proposed mechanisms of CO formation in this reaction [126]. As it can be observed, the CO production is through the formation of a carbonate intermediary on the surface of the photocatalyst. One solution to enhance the production of organic compounds in the photocatalytic reduction of CO_2 could be to introduce another co-catalyst, to obtain more complex MOF-based composites, as described in the following sections.

14.4.2 Graphitic Carbon Nitride–MOF Composites

Carbon nitride is a polymeric material recently used as photocatalyst. This compound has seven crystalline phases [132]; however, among all the allotropes, the tri-*s*-triazine-based phase is the most energetically favored and the most stable under a wide range of chemical or thermal conditions [133]. Thus, tri-*s*-triazine-based phase is generally recognized as the basic unit for the formation of graphitic carbon nitride (g-C_3N_4). g-C_3N_4 is highly stable in aqueous solutions in all the pH range, it is thermally stable up to 550°C, and it does not suffer from corrosion under light irradiation. g-C_3N_4 can be activated by visible light due to its bandgap (2.7–2.8 eV) and the position of its valence band (VB) (ca. 1.6 eV vs. NHE) and its conduction band (CB) (ca. −1.1 eV vs. NHE) allows to perform several photocatalytic reactions such as H_2 production, CO_2 reduction, water oxidation, and degradation of organic or inorganic pollutants [133].

Despite all the advantages described, g-C_3N_4 has some shortcomings. As many inorganic semiconductors, it presents a high recombination rate of e^-–h^+ pairs, and it has low electrical conductivity. Several approaches have been employed to overcome these disadvantages, e.g., deposition of noble metal nanoparticles or to couple with other semiconductors to form heterojunctions. Several MOFs and MOF families have been used to form such heterojunctions or g-C_3N_4–MOF composites, including the MIL family, UiO-66, and ZIF-8, as shown in Table 14.6.

The g-C_3N_4–MOF composites have been recently used in the photocatalytic degradation of organic compounds, such as Rhodamine B [134, 135, 140], Methylene blue [136, 140], and the reduction inorganic pollutants such as Cr(VI) [136]. However, the reduction of CO_2 and H_2 production has been less studied. In the reviewed works, the enhancement of the photocatalytic activity in these composites has been attributed to specific surface area and the decrease of the recombination of charge carriers.

As mentioned earlier, the adsorption of organic contaminants is an essential step toward its degradation and, hence, the specific surface area is an important property of the photocatalyst. In general, g-C_3N_4 shows a low surface area, as many inorganic semiconductors; however, g-C_3N_4–MOF composites could have a high surface area, even as that of the bare MOFs, depending on its morphology. For example, Hong *et al.* [134] obtained several g-C_3N_4- MIL-100(Fe) composites, where the g-C_3N_4 showed a nanosheet morphology (CNNs, 85 $m^2\ g^{-1}$), with a surface area as high as 1122 $m^2\ g^{-1}$ (See Figure 14.27). As a comparison, the pristine MIL-100(Fe) showed an area value of 1225 $m^2\ g^{-1}$.

g-C_3N_4 presents a high recombination of charge carriers, as demonstrated by the PL spectra of the material. Figure 14.28 shows the PL spectra of bulk g-C_3N_4, g-C_3N_4 nanosheets functionalized with benzoic acid (CFB), and several CFB-MIL-125 (Ti)–NH_2 composites (CFBM) [137]. As it can be seen, the incorporation of MIL-125 (Ti)–NH_2 in the composites caused a decrease in the PL signal, attributed to a lower recombination of the charge carriers, proportional to the content of CFB in the composites (5 wt% and 10 wt%). The dependence of the PL signal on the composition of the g-C_3N_4–MOF composite has been observed in several works. For example, Lei *et al.* [136] reported that g-C_3N_4–MIL-88B(Fe) composite showed

Table 14.6 Recent applications of g-C_3N_4–MOF composites.

MOF	Method of synthesis	Application/degraded compound	Highlighted results	References
MIL-100(Fe)	Thermal*	Rhodamine B	The composites showed areas similar to that of the MIL-100(Fe)	[134]
MIL-125 (Ti)	Solvothermal	Rhodamine B	The composites showed a good photocatalytic activity under visible light irradiation	[135]
MIL-88B(Fe)	Solvothermal	Methylene blue and Cr(VI)	The performance of the composites depended on the g-C_3N_4 content, showing a maximum at 6 wt%	[136]
MIL-125 (Ti)–NH_2	Solvothermal	H_2 production	The composites showed a better performance than that of the physical mixture of the components	[137]
UiO-66	Solvothermal	H_2 production	The PL of the composites depended on the content of g-C_3N_4	[138]
ZIF-8	Wet route**	CO_2 reduction	The composites produced methanol as product of the CO_2 reduction	[139]

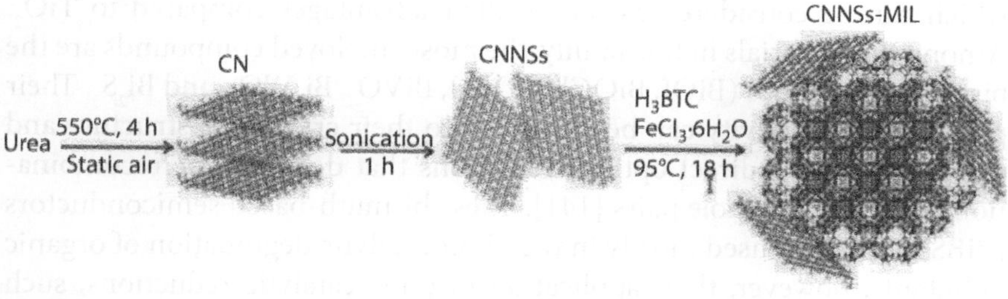

Figure 14.27 g-C_3N_4–TiO_2 composite with a high surface area. Reprinted with permission from Ref. [134].

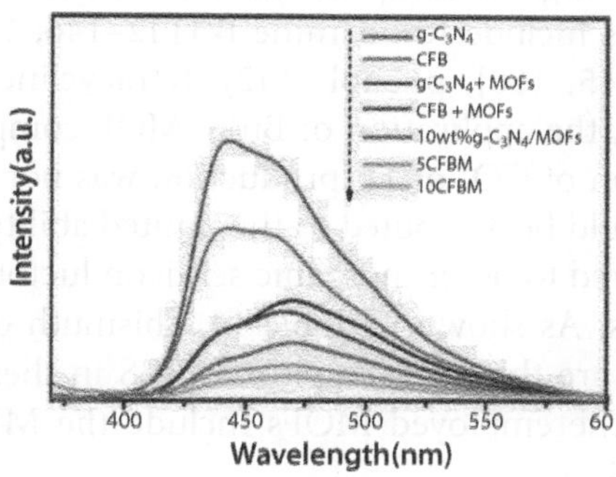

Figure 14.28 PL spectra of the several CFB-MIL-125 (Ti)–NH_2 composites at an excitation wavelength of 325 nm. Reprinted with permission from Ref. [137].

similar behavior, where it was found that the highest decrease in the PL signal corresponded to a 6 wt% of g-C_3N_4. The composites with a lower or higher quantity of g-C_3N_4 showed a higher recombination of charge carriers.

In the CO_2 photocatalytic reduction, some g-C_3N_4–MOF composites have shown a better selectivity to useful products, such as methanol. For example, a g-C_3N_4–ZIF-8 composite [139] showed a good production of methanol when the photocatalyst where irradiated with an unfiltered 300 W Xe lamp. In this case the formation of methanol was explained by a mechanism where several carbonate species were formed, like the mechanism presented in Figure 14.26.

14.4.3 Bismuth-Based Semiconductors

In recent years, bismuth-based semiconductors have been used in several photocatalytic applications, as possible candidates to replace TiO_2. Due to their bandgap, these semiconductors can be activated by visible light, which can be considered as one of their advantages compared to TiO_2. Among the materials in this family, the most employed compounds are the bismuth oxyhalides (BiOI, BiOCl, BiOBr), $BiVO_4$, Bi_2WO_6, and Bi_2S_3. Their photocatalytic activity can be attributed to their crystalline structure and the presence of indirect optical transitions that decrease the recombination rate of electro-hole pairs [141]. These bismuth-based semiconductors (BiBS) have been used mostly in the photocatalytic degradation of organic pollutants; however, their application in photocatalytic reductions, such as H_2 production, CO_2 reduction or N_2 fixation, is limited due to the low reduction potential of their conduction band [141].

In this sense, BiBS–MOF composites have been used in the photocatalytic degradation of organic compounds, as can be seen in Table 14.7. The degraded pollutants include Rhodamine B [142–146, 148, 149, 151, 152], Methylene blue [145, 150], phenol [142], tetracycline and bisphenol A [147]. As expected, the application of BiBS–MOF composites in the photocatalytic reduction of CO_2 or H_2 production was not reported in recent works. This fact could be attributed to the limited ability of BiBS and pristine MOFs, compared to other inorganic semiconductors, to perform such reduction reactions. As shown in Table 14.7 bismuth oxyhalides, such as BiOBr and BiOCl, are the most-employed BiBS in these composites, followed by $BiVO_4$. The employed MOFs include the MIL family, UiO-66, CAU-17, and ZIF-8.

In a similar manner as in the case of the composites described in previous sections, the enhancement of the photocatalytic activity of the BiBS–MOF composites has been attributed to a better charge-carrier separation and an increase in the superficial area. Besides the PL experiments, photocurrent analysis has also been employed to measure this separation. For example, Zhu *et al.* [142] employed photocurrent measurements to demonstrate an effective charge-carrier separation and a faster interfacial charge transfer, as show in Figure 14.29. As can be observed, the response of the BNM-7 composite (BiOBr–MIL-125(Ti)–NH_2 composite, with a MOF content of 7 wt%) increased almost two-fold compared to pristine BiOBr. Also, it is interesting to note that unmodified MIL-125(Ti)–NH_2 showed a photocurrent response, indicating the semiconducting behavior of the MOF, as observed in other works [144].

14.4.4 Reduced Graphene Oxide–MOF Composites

Since its discovery, graphene (GR), and recently graphene oxide (GO), has been used in electronics, catalysis and energy applications owing to its excellent electrical and thermal conductivity, and high surface area.

Table 14.7 Recent applications of BiBS-MOF composites.

MOF	Semiconductor	Method of synthesis	Degraded compound	Highlighted results	References
MIL-125 (Ti)–NH_2	BiOBr	Solvothermal	Rhodamine B and phenol	Photocurrent measurements demonstrated a better interfacial charge transfer	[142]
UiO-66	BiOBr	Solvothermal	Rhodamine B	The composites showed a higher superficial area, compared to BiOBr	[143]
UiO-66–NH_2	BiOBr	Solvothermal	Rhodamine B	The physical mixture of the components showed was less active than the synthetized composites	[144]
CAU-17	BiOBr	Solvothermal	Rhodamine B, Methylene blue, and Methyl orange	–	[145]
MIL-88B (Fe)	BiOBr	Thermal*	Rhodamine B	The composites showed a BiOBr@MOF core–shell structure	[146]
MIL-125 (Ti)–NH_2	BiOCl	Solvothermal	Tetracycline and bisphenol A	The activity of the composites depended on the mass content of the MOF, been 10 wt% the best	[147]

(Continued)

Table 14.7 Recent applications of BiBS-MOF composites. *(Continued)*

MOF	Semiconductor	Method of synthesis	Degraded compound	Highlighted results	References
UiO-66	BiOCl	Solvothermal	Rhodamine B	The composite showed an enhanced adsorption due to BiOCl, compared to the MOF	[148]
MIL-101 (Cr)	$BiVO_4$	Solvothermal	Rhodamine B	–	[149]
ZIF-8	$BiVO_4$	Wet route**	Methylene blue	The composites were more stable than the physical mixture of the components, after 3 reaction cycles	[150]
MIL-125(Ti)	$BiVO_4$	Solvothermal	Rhodamine B	–	[151]
MIL-100 (Fe)	Bi_2S_3	Solvothermal	Rhodamine B	The charge-carrier separation in the composites were demonstrated by electrochemical impedance spectroscopy	[152]

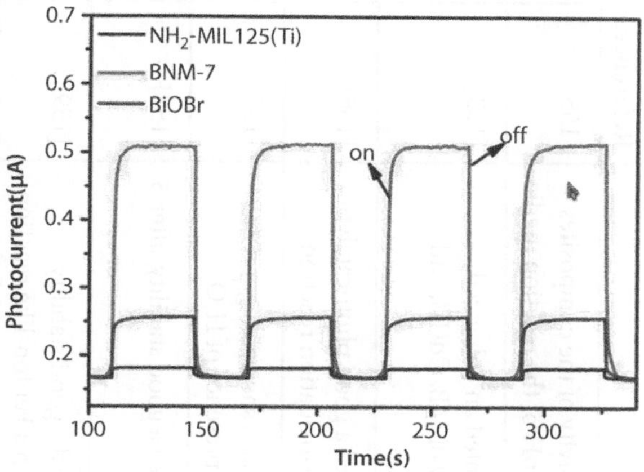

Figure 14.29 Photocurrent response of a BiOBr-MIL-125(Ti)–NH_2 composite. Reprinted with permission from Ref. [142].

GR is composed of a layer of interconnected hexagonal carbon rings, and it is considered as the basic construction unit of other nanocarbons, such as fullerenes, carbon nanotubes, or graphite. Due to its properties, GR has been one the most promising carbon-based material in the synthesis of semiconductor heterojunctions or composites for photocatalytic applications. For example, GR is an excellent electron pool or conduit due its high electrical conductivity, which enhances the electron-hole separation in the composites; also, the high surface area of GR enhances the adsorption of pollutants or provides a higher number of reactive sites.

Graphene is commonly obtained by the reduction of GO, so in the literature is more frequent to found references to reduced graphene oxide (rGO), instead of graphene. rGO share almost the same properties of GR; however, these properties can be tuned by controlling the reduction degree of GO, which represents one the advantages of rGO. Graphene oxide is hydrophilic and rGO is partially hydrophilic, due to presence of several oxygenated functional groups in their structure, such as the carboxylic groups. These groups allow the chemical attaching of several semiconductors to form more efficient heterojunctions, which normally present an enhanced separation of charge carriers and absorption of visible light.

Among the used semiconductors in the synthesis of such heterojunctions or composites, MOFs represent a good alternative, due to the properties described earlier. As can be seen in Table 14.8, several MOFs have been used in the synthesis of the rGO-MOF or GO-MOF composites, such as ZIF-8 [153], UiO-66 [154], and several members of the MIL family [155–159]. In these composites, the predominant application is the

Table 14.8 Recent applications of rGO–MOF or GO–MOF composites.

MOF	Method of synthesis	Application/degraded compound	Highlighted results	References
MIL-68 (In)–NH_2	Solvothermal	Amoxicillin	The photocatalytic activity of the composites was affected by the pH of the reaction media	[155]
ZIF-8	Wet route**	Methylene blue	The composites presented an enhanced adsorption capacity of MB, Cd(II), and Pb(II)	[153]
MIL-LIC-1(Eu)	Solvothermal	Benzyl alcohol and water oxidation	The composites showed a good photocatalytic activity in the water-oxidation reaction	[156]
MIL-53(Fe)	Solvothermal	Rhodamine B	The evaluated composites showed a good activity only in the presence of H_2O_2	[157]
MIL-125 (Ti)–NH_2	Solvothermal	Methylene blue	The composites showed a good stability, after 5 reaction cycles	[158]
MIL-88B(Fe)	Solvothermal	Reactive red 195	The composites showed a good stability (3 reaction cycles) in a Fenton-like reaction	[159]
UiO-66–NH_2	Microwave	CO_2 reduction	The composites produced CH_4 and formic acid, but H_2 was also produced	[154]

degradation of water pollutants; e.g., amoxicillin [155], Methylene blue [153, 158], Rhodamine B [157], Reactive red 195 [159], among others. The application of these composites in the CO_2 photocatalytic reduction or the H_2 production has been scarce; however, it is worth to note the application of the GO–MIL–LIC-1(Eu) composite in the water-oxidation reaction [156].

The GO–MIL-LIC-1(Eu) composite was tested in the presence of a ruthenium complex as sensitizer ($[Ru(bpy)_3]Cl_2$) and $Na_2S_2O_8$ as electron acceptor (sacrificial agent) in a nitrogen atmosphere, under visible-light irradiation [156]. The GO content in the composites varied between 1 and 10 wt% been the composite with 1 wt% of GO (MIL@GO-1%) the most active, as it can be seen in Figure 14.30. The pristine MIL-LIC-1(Eu) produced almost 50 μmol of O_2 after 3 h of reaction time, but the addition of 1 wt% of GO increased the production of O_2 to almost 125 μmol in the same time.

In other interesting application, several Fe-based MOFs–rGO composites have been used in Fenton-like or photo-Fenton-like reactions. For example, rGO–MIL-53(Fe) composites [157] were used in the degradation of RhB as model organic pollutant. These composites showed almost no activity when they were evaluated in the absence of H_2O_2, even under light irradiation; however, in the presence of this oxidant agent, the degradation of RhB achieved 100% after 60 min of reaction time. In a similar manner, rGO–MIL-88(Fe) composites were employed to degrade Reactive red 195 in a photo-Fenton-like and Fenton-like reaction [159]. In this case, it was observed that the rGO–MIL-88(Fe) composites presented a good stability after three reaction cycles, losing just 5% of its original conversion.

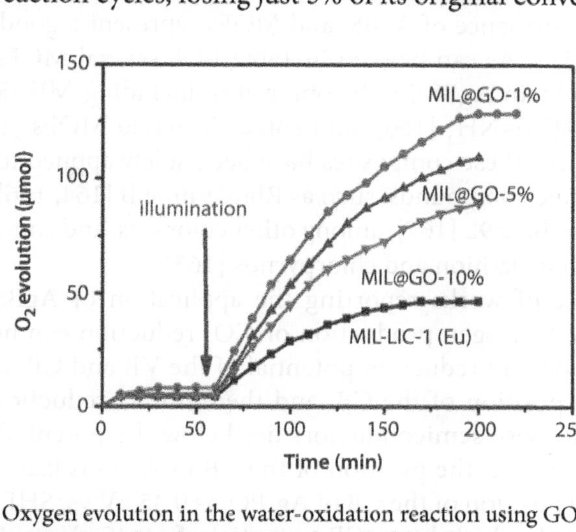

Figure 14.30 Oxygen evolution in the water-oxidation reaction using GO-MIL-LIC-1(Eu) composites. Reprinted with permission from Ref. [156].

14.4.5 Silver-Based Semiconductors

Many silver-based semiconductors (AgBS) have been applied in photocatalytic reactions because they represent a good alternative photocatalysts to harvest visible light. Virtually all these AgBS have a band gap energy bellow 3.0 eV, or even less than 2.0 eV, which make them active under visible light irradiation [160], and an alternative to TiO_2 in solar-driven photocatalysis. There are several AgBS reported in the literature, such as: A_2S, Ag_3OsO_4, silver halides (AgCl, AgBr, AgI), silver oxyhalides, among others [160], and they have been primarily applied in the degradation of organic compounds [161].

Since the times of the analogic photography, it is well known that silver halides can decompose under light irradiation, and it is not different in photocatalysis. All AgBS are prone to suffer photocorrosion when they are irradiated, yielding metallic silver due to low reduction potential of the Ag^+/Ag^0 pair [161]. Actually, it is a common practice to employ some inorganic or organic silver precursors to produce metallic silver nanoparticles (AgNPs) by photochemical or photocatalytic reactions [162, 163]. Although, the formation of AgNPs is not at all a disadvantage because, as it is well known, metallic NPs can enhance the photocatalytic performance when they are deposited on the surface of a semiconductor. This enhancement is because metallic AgNPs can act as a trap for the photogenerated electrons, increasing the electron-hole separation.

Besides the *in situ* formation metallic AgNPs, the synthesis of hetero-junctions or composites is another approach to enhance the photocatalytic performance of AgBS, and MOFs represent a good alternative in such materials. As can be seen in Table 14.9, several MOFs have been employed to obtain AgBS-MOF composites, including MIL-53(Fe) [164, 165], MIL-125(Ti)–NH_2 [166], and some Cu-based MOFs [167–169]. It can be noted that these composites have been solely applied to the degradation of organic compounds, such as Rhodamine B [164, 166], Orange G [168], and Acid blue 92 [167], among other colorants, and some herbicides such as methyl malathion and chlorpyrifos [165].

The absence of works reporting the application of AgBS–MOFs in photocatalytic hydrogen production or CO_2 reduction can be explained by the oxidation and reduction potential of the VB and CB, respectively, of AgBS. The position of the CB, and therefore its reduction potential, of virtually all these semiconductors lies below the potential of the H^+/H_2 pair; for example, the position of the CB of Ag_2O is 0.20 eV vs. SHE [170], and the position of the CB of Ag_3PO_4 is 0.45 eV vs. SHE [160]. One exception is AgI, which has a CB potential of –0.45 eV vs. NHE [164].

Table 14.9 Recent applications of AgBS–MOF composites.

MOF	Semiconductor	Method of synthesis	Degraded compound	Highlighted results	References
MIL-53(Fe)	AgI	Solvothermal	Rhodamine B	The performance of the composites depended on the AgI content	[164]
MIL-125 (Ti)–NH_2	Ag_3PO_4	Solvothermal	Rhodamine B and Methylene blue	The performance showed by the composites was higher than that of TiO_2 P25	[166]
MIL-53(Fe)	$AgIO_3$	Solvothermal	Methyl malathion and chlorpyrifos	The mixture of both herbicides was completely degraded in 60 min, in the presence of $Na_2S_2O_8$	[165]
Cu–TPA	Ag_2O	Thermal*	Acid blue 92	The composites showed a good stability, losing 8% of its original activity, after 5 reaction cycles	[167]
Cu–BDC	Ag_2O	Solvothermal	Orange G	The Ag_2O–Cu–BDC composite showed better performance, compared to the pristine components	[168]
CU–BTC	Ag_2CrO_4	Solvothermal	Congo Red and Ponceau BS	The decrease in the photocatalytic activity was attributed to the formation of metallic Ag	[169]

Consequently, the photogenerated electrons are more likely to reduce Ag^+ ion of the silver-based semiconductor that participate in other reduction reaction. In the other hand, the VB potential of AgBS is comparable to that of other common semiconductor, such as ZnO and TiO_2 [160], which make them good photocatalysts for oxidation or degradation of organic pollutants.

14.4.6 Other Semiconductors

Several semiconductors have been successfully used in photocatalytic applications, searching for an alternative to TiO_2. These semiconductors include several metallic oxides, such as ZnO, Fe_2O_3, WO_3, and SnO_2, among others; chalcogenides like CdS, CdTe, ZnS; and others such as titanates, vanadates, and tungstates, to name a few. These semiconductors have some advantages, compared to TiO_2; for example, several chalcogenides, like CdS or CdTe, absorb visible light owing its relatively small bandgap energy; others have a potential reduction higher than TiO_2, hence they can be applied in photocatalytic production of hydrogen or CO_2 reduction. However, almost all these semiconductors have a major drawback as they suffer of photocorrosion, being unstable under light irradiation, releasing inorganic cations, which, sometimes, are extremely toxic, like Cd^{2+}.

One approximation to avoid the photocorrosion is, as mentioned earlier, use these semiconductors together with other materials, specially semiconductors, to obtain diverse heterojunctions or composites. MOFs have been successfully applied to synthetize semiconductor–MOF (SC–MOF) composites, providing enhanced properties like a higher surface area or better separation of charge carriers. Among the recently used MOFs in these composites are UiO-66 [171], ZIF-8 [172], several members of the MIL family (MIL-101(Cr) [173], MIL-125(Ti) [174], MIL-88B(Fe) [175]), TMU-5 [176], NU-1000 [177], among others (see Tables 14.10 and 14.11).

The application of the SC-MOF composites is as diverse as the properties of the involved semiconductors; for example, photocatalytic degradation of organic pollutants, photocatalytic hydrogen production, and CO2 reduction [178]. In Table 14.11 are shown several SC–MOF composites employed in the photocatalytic degradation of organic compounds. Metallic oxides (MoO_3 [176], MnO_x [173], SnO_2 [172], ZnO [179, 180]), and chalcogenides (Cu_xS [171], In_2S_3 [174], CdTe [181]) are the most-used semiconductors in this application. With a similar trend, as observed in the last sections, these composites are mainly used in the degradation of

Table 14.10 Recent applications of SC–MOF composites in the photocatalytic degradation of water pollutants.

MOF	Semiconductor	Method of synthesis	Application/degraded compound	Highlighted results	References
UiO-66	Cu_xS	Solvothermal	Rhodamine B	The presence of Cu_xS enhanced the adsorption of RhB, compared to bare UiO-66	[171]
MIL-101(Cr)	MnO_x	Solvothermal	Rhodamine B and Methylene blue	MnO_x was obtained by photo-deposition	[173]
MIL-88B(Fe)	Fe_3O_4	Solvothermal	Rhodamine B and Methylene blue	Fe_3O_4 was synthetized by decomposition of the MOF	[175]
ZIF-8	ZnO	Thermal*	Methylene blue	ZnO was used as Zn^{2+} precursor in the MOF synthesis	[179]
MOF-46	ZnO	Wet route**	Methylene blue	The PL emission of the composites depended on the solvent	[180]
ZIF-8	SnO_2	Ultrasound	Methylene blue	The composite was stable during 10 reaction cycles	[172]
$[CoNi(\mu_3\text{-tp})_2(\mu_2\text{-pyz})_2]$ MOF	$CuWO_4$	Solvothermal	Methylene blue and 4-nitrophenol	The activity of the SC-MOF composite depended on its mass ratio	[182]
TMU-5	MoO_3	Ultrasound	Photo-oxidative desulfurization	The composites lose 3% of the original MoO_3	[176]
MIL-125 (Ti)	In_2S_3	Solvothermal	Tetracycline	In_2S_3 enhanced the absorption of tetracycline	[174]
NTU-9	CdTe	Thermal*	Rhodamine 6G	The composites showed a good performance with light between 300 and 800 nm	[181]

Table 14.11 Recent applications of diverse SC–MOF composites in the photocatalytic H_2 production.

MOF	Semiconductor	Method of synthesis	Highlighted results	References
NH_2–MIL-125 (Ti)	$ZnIn_2S_4$	Solvothermal	The best performance was achieved with a 40 wt% of MOF, and the composite was stable for 5 reaction cycles	[186]
NH_2–UiO-66	$Cd_{0.2}Zn_{0.8}S$	Solvothermal	The composites showed a good activity in both CO_2 reduction and hydrogen production	[178]
ZIF-8	CdS	Wet route**	The CO formation was reduced by the presence of the MOF in the composite	[183]
UiO-66	CdS	Thermal*	The electron transfer between the MOF and the SC was evidenced using femtosecond transient absorption spectroscopy	[184]
NH_2–MIL-125 (Ti)	Ni_2P	Ultrasound-assisted Thermal*	The SC–MOF composite showed a production rate twice than that shown by a Ni_2P–TiO_2 composite	[187]
NH_2–MIL-125 (Ti)	Mo_XS_X	Thermal*	$Mo_3S_{13}^{2-}$–NH_2–MIL–125 composite showed an enhanced H_2 production, compared to a MoS_2–NH_2–MIL–125 composite	[185]
NH_2–MIL-125 (Ti)	Cu_2O	Solvothermal	Ti^{3+} sites were formed during the composite synthesis, which contributed to reduce the electron–hole recombination	[188]
NU-1000	NiS_x	Thermal*	–	[177]

organic dyes, probably because these compounds are used as model compounds in the photocatalytic degradation of water pollutants. Among the degraded dyes are Rhodamine B [171, 173, 175], Methylene blue [172, 173, 175, 179, 180, 182], and Rhodamine 6G [181].

Table 14.11 shows the SC–MOF composites recently used in the photocatalytic production of hydrogen. It should be noted that just one work reported the application of $Cd_{0.2}Zn_{0.8}S$–UiO-66–NH_2 composite in the photocatalytic reduction of CO_2 [178]. In this case, chalcogenides are the predominant semiconductors; for example, CdS [183, 184], MoS_2 [185], NiS_x [177], and other more complex sulfur-based chalcogenides such as $ZnIn_2S_4$ [186], and $Cd_{0.2}Zn_{0.8}S$. The used MOFs comprise ZIF-8, Ui-O66, NU-1000 [177], and NH_2-modified MIL-125(Ti) or UiO-66.

As in the case of the previous SC-MOF composites described earlier, the enhancement in the photocatalytic activity of the composites listed in Tables 14.10 and 14.11 has been also explained in terms of the improved surface area of the composite, compared to that of the pristine semiconductor, and an enhanced charge-carrier separation. The latter is normally evidenced by steady-state PL studies; however, time-resolved techniques are a better solution to estimate the electron-hole recombination velocity. For example, Xu *et al.* [184] femtosecond transient absorption spectroscopy (fsTAS) to analyze the recombination of charge carriers in several CdS–UiO-66 composites as a function of the mass content of CdS (10, 20, and 40 wt%), as showed in Figure 14.31. It can be observed that the excitation bleach recovery, and therefore the electron transfer rate between CdS and UiO-66, decreased as follows: CdS–UiO-66(10) > CdS–UiO-66(20) > CdS–UiO-66(40) ≈CdS. According to the authors, this result was like the observed photocatalytic activity of the composites.

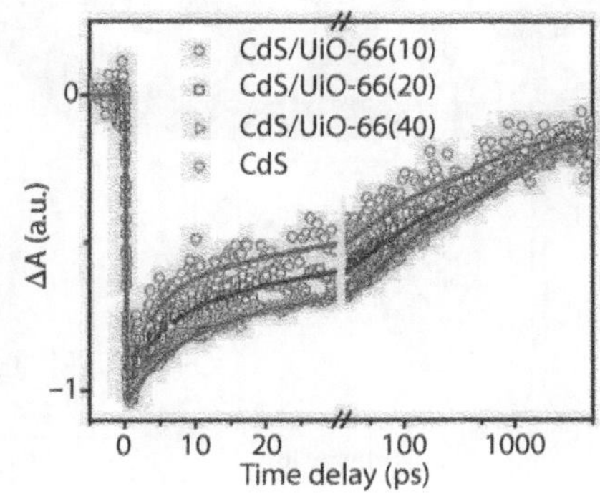

Figure 14.31 fsTAS kinetic traces at 650 nm of CdS–UiO-66 composites as a function of the CdS mass percent. Reprinted with permission from Ref. [184].

14.5 MOF-Based Multicomponent Composites and Their Application in Photocatalysis

MOFs have been modified by adding two compounds to increase their catalytic activity. Among these are semiconductors, metal nanoparticles, carbon compounds. These individual cases were already revised in previous sections. Modified MOFs have been used in various reactions, mainly in two cases, fuels generation and pollutants degradation. Even though, some review articles have already been published addressing this type of multi-component MOFs in the aforementioned reactions [18, 189], the most recent investigations in these topics will be analyzed.

14.5.1 Semiconductor–Semiconductor–MOF Composites

Coupling a semiconductor with MOFs is a way to improve the optical properties and increase the photocatalytic activity, which has been reported, mainly, for hydrogen production reactions. For example, in the case of the TiO_2–Ti_3C_2–UiO-66–NH_2 system, the MOF was first modified with amino groups to confer particular properties on its surface and thus, be active under

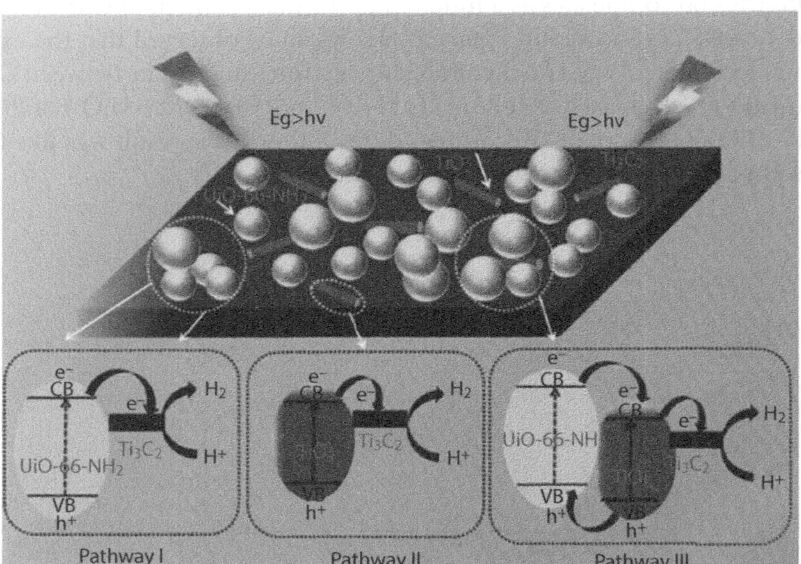

Figure 14.32 Schematic illustration of the charge-transfer pathways for Ti_3C_2/TiO_2/UiO-66-NH_2. Reprinted with permission from Ref. [190].

visible light and take advantage of its porosity to stimulate the exchange of photogenerated charge carriers. Second, after depositing a thin film of the UiO-66–NH_2 on glass, TiO_2 was added randomly, leaving space available for the addition of Ti_3C_2. Subsequently, Ti_3C_2 particles were deposited on the surface of UiO-66–NH_2, as well as in TiO_2, leading to an intimate contact between the three components [190]. Each arrangement was activated individually, and there were also particles of the three components interacting each other. It was corroborated, by the characterization techniques employed, that the global composite had the highest efficiency in terms of electron transfer compared to that obtained using each component individually. It was shown that the use of two semiconductors in a MOF substantially increased the photocatalytic activity, without the use of a noble metal. A schematic illustration of the charge transfer mechanism and the involved reactions is shown in Figure 14.32. Several combinations, semiconductor1–semiconductor2–MOF, have been recently reported to have new active and stable materials for various photocatalytic reactions, such as those shown in the Table 14.12.

14.5.2 Semiconductor–Metal–MOF Composites

MOFs modification by coupling a semiconductor with metal nanoparticles has been used lately, to ensure a good charge transfer between the components and prevent recombination therefore during irradiation. These new photocatalysts have been applied, mostly, in hydrogen production reactions. An example was MIL-101 modified with cadmium sulfide (CdS), and then doped with gold particles to increase its photocatalytic activity, which was enhanced by the interaction of three components, generating a synergistic effect between them. Notably, gold nanoparticles facilitated the electron transfer from MIL-101 to CdS, giving rise to a higher stability of the photocatalytic system [191].

In another case study, MIL-125 was modified with amino groups as a first step, and then g-C_3N_4 particles were supported, and finally deposit Ni and Pd nanoparticles on the MOF-semiconductor support [192]. To complete this complex photocatalytic system, Eritrosin B was added as a photosensitizer in the reaction medium and starts the charge transfer process by irradiation with visible light. The reaction mechanism to generate H_2 was explained by the electron transfer from the excited photosensitizer to g-C_3N_4 and MIL-125, and finally to Ni and Pd nanoparticles to convert protons in molecular hydrogen. An illustration of the mentioned reaction mechanism is shown in Figure 14.33.

An interesting approach to modify MOFs type MIL-53 and MIL-88A was done by using AgCl and Ag nanoparticles approach with applications

Table 14.12 Recent investigations related to the use of modified MOFs with two components and their applications in photocatalytic reactions.

MOF	Semiconductor or/and metal	Method of synthesis	Application/degraded compound	Reference
ZIF-67(Co)	g-C_3N_4-MoS_2	Solvothermal	Hydrogen generation	[199]
Cu–BTC	ZnO–GO	Solvothermal	Hydrogen generation	[200]
MIL-101(Cr)	Carbon Dots–CdS	Solvothermal	Hydrogen generation	[201]
MOF-74 (Ni)	CdS–Co_3O_4	Solvothermal	Hydrogen generation	[202]
UiO-66 (Zr)	Ag–AgBr	Hydrothermal	Methyl orange	[203]
UiO-66 (Zr)	g-C_3N_4–Ag	Hydrothermal	Rhodamine B	[204]

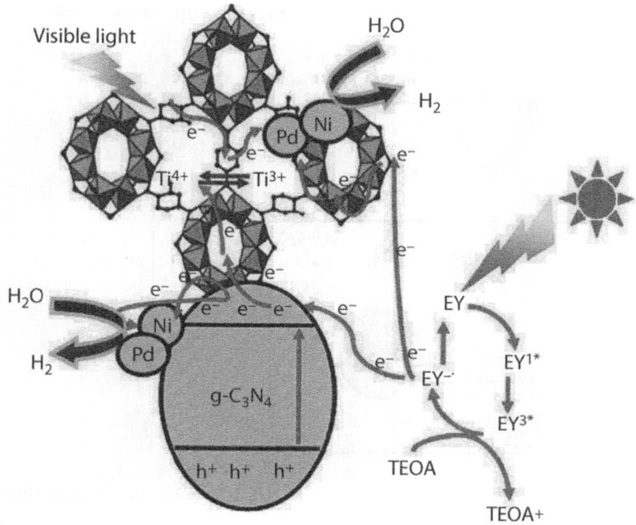

Figure 14.33 Mechanism of H_2 evolution reaction over EY-sensitized NH_2–MIL-125(Ti)/CN/NiPd under visible illumination. Reprinted with permission from Ref. [192].

dyes degradation [193, 194]. Particularly, these cases include Rhodamine B and Chromium (VI) reduction [195], and Ibopruphene [196], providing valuable information on the simultaneous and individual degradation of the mentioned pollutants. On the other hand, Rhodamine B was degraded by using MIL-101, which was doped with anatase, and then gold nanoparticles were added (i.e., MIL-101-core–Au/anatase–shell composite). There was a synergistic effect between the TiO_2 and Au nanoparticles favoring the electron transfer to anatase and MIL-101, increasing significantly the degradation of Rhodamine B [197]. A possible explanation for the reaction mechanism for the dye degradation under visible light is shown in Figure 14.34.

Finally, another interest application using modified MOFs with two components, was the reduction of CO_2 to CO. UiO-67 was doped with CdS and Co nanoparticles. The benefits of occupying a modified MOF is to take advantage of the organic chains of this compound, which have the ability to adsorb CO_2 from the environment and once in the MOF, the particles of CdS and Co particles were activated with visible light, finding that Co was a key point, to carry out the reduction of CO_2 [198]. Table 14.12. summarizes some outstanding examples of modified MOFs with two components and their applications in photocatalytic reactions.

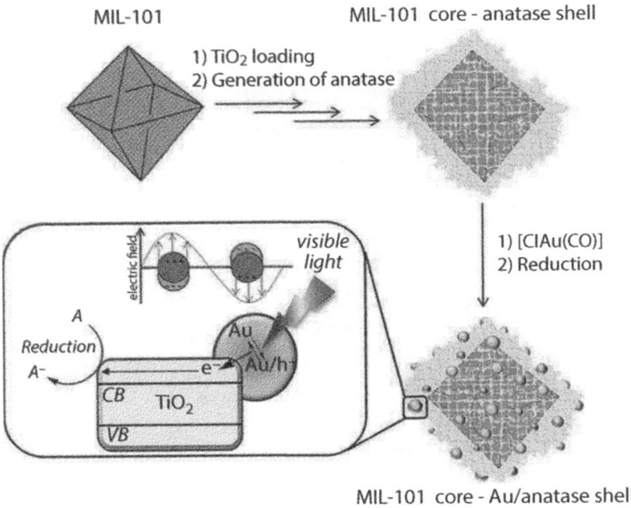

Figure 14.34 Scheme of synthesis procedure of the MIL-101-core–Au/anatase–shell material and a schematic illustration of the generation of photogenerated pairs and is plasmonic excitation of gold particles on the surface of the anatase shell under visible-light illumination. Reprinted with permission from Ref. [197].

14.6 Conclusions

In the last decade, there has been a notable increase in the development of new advanced materials with the intention of solving the current energy and environmental problems, among other applications. MOFs belong to this category, as they have relevant properties, such as, high specific surface area, regular and tunable pore structure, diversity of the organic linkers and the metal nodes [205]. In addition, MOFs are easily prepared with the possibility of using various organic precursors and metal clusters and incorporating various components in their structure. In particular, MOFs have found applications in photocatalysis because they carry out similar processes as those performed by traditional semiconductors. Due to the versatility to tune the organic linkers and metal clusters species in MOFs, it is possible to harvest light and generate electron-hole pairs required in photocatalytic reactions. As is well-known, the charge transfer mechanisms of photocatalytic reactions are governed by three processes: localized metal-to-ligand charge transfer (MLCT), a ligand-to-metal charge transfer (LMCT), or a π–π^* transition of the aromatic ligand [107].

Consequently, in the last 3 years a great variety of pristine and modified MOFs (i.e., composites) presenting different structures, morphologies and photocatalytic properties, have been synthesized. In this chapter, we have addressed recent advances in the synthesis and applications of pristine MOFs and MOFs-based composites in photocatalytic reactions concerning topics of energy, environment, and organic synthesis. With respect to pristine MOF, the following aspects can be highlighted: (1) Several strategies were followed to obtain pristine MOFs of group 4 periodic table (PT), as efficient photocatalysts under irradiation with visible light, e.g., the introduction electron-donor groups on MOF UiO-66 (Zr)–X (X = –OH, $-NH_2$, –COOH, $-NO_2$, and –H) and the cation-exchange (Ti incorporation on substitution of Zr in MOF UiO-66). (2) The exchange strategy was also used to obtain core@shell MOFs, type ZIF-67 @Co-MOF-74 or other combinations with metal clusters belonging to groups 8 and 9 (PD). (3) Group 11 metallic clusters were poorly studied, except for Cu MOFs. The most relevant is Cu–BTC (HKUST-1), however, due its low stability in water, different organic ligands were analyzed with to Cu clusters in order to improve the photocatalytic activity and stability. It should be noted that the metal clusters mostly used for this purpose were those of Fe, Co, and Ni. (4) With Group 12 metallic clusters (Zn and Cd), different synthesis routes were explored to obtain interpenetrated and pillared MOFs, which gave rise to significant differences in the band gap, thermal stability, pore size, and BET surface area, compared to the starting compound ZIF-8.

In general, MOF-composites showed improved properties (i.e., higher conductivity, stability, etc.), which presented a higher photocatalytic activity compared to the pristine MOF. Concerning metal nanoparticles–MOF composites, significant advances were shown with the use of Au, Ag, or Cu nanoparticles, which can perform simultaneously, two substantive functions in photocatalysis, i.e., visible light absorption (plasmonic effect) and low electron-hole recombination (Schottky barrier effect).

Multicomponent photocatalysts comprising MOF–semiconductor1–semiconductor2 (MOF–sem1–sem2) and MOF–semiconductor–metal (MOF–sem–metal) were particularly useful for hydrogen generation and pollutant degradation. The three components interaction favored light absorption, electron–hole recombination, reactants adsorption and active sites suitable for the surface reactions. A key point for promoting the studied photocatalytic reactions was the presence of metal nanoparticles, which acted, either as electron reservoirs and plasmonic species for transferring electrons to the semiconductor and MOF.

References

1. Batten, S.R., Champness, N.R. *et al.*, Terminology of metal–organic frameworks and coordination polymers (IUPAC Recommendations 2013). *Pure Appl. Chem.*, 85, 1715, 2013.
2. Nasalevich, M.A., van der Veen, M. *et al.*, Metal–organic frameworks as heterogeneous photocatalysts: Advantages and challenges. *CrystEngComm*, 16, 4919, 2014.
3. Khan, N.A. and Jhung, S.H., Synthesis of metal–organic frameworks (MOFs) with microwave or ultrasound: Rapid reaction, phase-selectivity, and size reduction. *Coord. Chem. Rev.*, 285, 11, 2015.
4. Li, H., Eddaoudi, M. *et al.*, Design and synthesis of an exceptionally stable and highly porous metal–organic framework. *Nature*, 402, 276, 1999.
5. Zhou, H.-C.J. and Kitagawa, S., Metal–Organic Frameworks (MOFs). *Chem. Soc. Rev.*, 43, 5415, 2014.
6. Rowsell, J.L.C. and Yaghi, O.M., Metal–organic frameworks: A new class of porous materials. *Microporous Mesoporous Mater.*, 73, 3, 2004.
7. Fang, Y., Ma, Y. *et al.*, Metal–organic frameworks for solar energy conversion by photoredox catalysis. *Coord. Chem. Rev.*, 373, 83, 2018.
8. Gao, Q., Xu, J. *et al.*, Recent advances about metal–organic frameworks in the removal of pollutants from wastewater. *Coord. Chem. Rev.*, 378, 17, 2019.
9. Kobielska, P.A., Howarth, A.J. *et al.*, Metal–organic frameworks for heavy metal removal from water. *Coord. Chem. Rev.*, 358, 92, 2018.
10. Guan, H.-Y., LeBlanc, R.J. *et al.*, Recent progress in the syntheses of mesoporous metal–organic framework materials. *Coord. Chem. Rev.*, 369, 76, 2018.
11. Horike, S., Shimomura, S. *et al.*, Soft porous crystals. *Nat. Chem.*, 1, 695, 2009.
12. Jiao, L., Seow, J.Y.R. *et al.*, Metal–organic frameworks: Structures and functional applications. *Mater. Today*, 27, 43, 2019.
13. Crake, A., Christoforidis, K.C. *et al.*, CO_2 capture and photocatalytic reduction using bifunctional TiO_2/MOF nanocomposites under UV–vis irradiation. *Appl. Catal. B Environ.*, 210, 131, 2017.
14. Wang, C.-C., Li, J.-R. *et al.*, Photocatalytic organic pollutants degradation in metal–organic frameworks. *Energy Environ. Sci.*, 7, 2831, 2014.
15. Zhu, B., Zou, R. *et al.*, Metal–Organic Framework Based Catalysts for Hydrogen Evolution. *Adv. Energy Mater.*, 8, 1801193, 2018.
16. Lei, Z., Xue, Y. *et al.*, MOFs-Based Heterogeneous Catalysts: New Opportunities for Energy-Related CO_2 Conversion. *Adv. Energy Mater.*, 8, 1801587, 2018.
17. Shen, L., Liang, R. *et al.*, Strategies for engineering metal–organic frameworks as efficient photocatalysts. *Chin. J. Catal.*, 36, 2071, 2015.
18. Qiu, J., Zhang, X. *et al.*, Modified metal–organic frameworks as photocatalysts. *Appl. Catal. B Environ.*, 231, 317, 2018.

19. Li, S., Sun, S. *et al.*, Effects of electron-donating groups on the photocatalytic reaction of MOFs. *Catal. Sci. Technol.*, 8, 1696, 2018.
20. Ghalei, B., Sakurai, K. *et al.*, Enhanced selectivity in mixed matrix membranes for CO_2 capture through efficient dispersion of amine-functionalized MOF nanoparticles. *Nat. Energy*, 2, 17086, 2017.
21. Misra, R. and Gautam, P., Tuning of the HOMO–LUMO gap of donor-substituted symmetrical and unsymmetrical benzothiadiazoles. *Org. Biomol. Chem.*, 12, 5448, 2014.
22. Zhang, T. and Lin, W., Metal–organic frameworks for artificial photosynthesis and photocatalysis. *Chem. Soc. Rev.*, 43, 5982, 2014.
23. Laurier, K.G.M., Vermoortele, F. *et al.*, Iron(III)-Based Metal–Organic Frameworks As Visible Light Photocatalysts. *J. Am. Chem. Soc.*, 135, 14488, 2013.
24. Chen, X., Shen, S. *et al.*, Semiconductor-based Photocatalytic Hydrogen Generation. *Chem. Rev.*, 110, 6503, 2010.
25. de Miguel, M., Ragon, F. *et al.*, Evidence of Photo induced Charge Separation in the Metal–Organic Framework MIL-125(Ti)–NH_2. *ChemPhysChem*, 13, 3651, 2012.
26. Kuila, A., Surib, N.A. *et al.*, Metal Organic Frameworks: A New Generation Coordination Polymers for Visible Light Photocatalysis. *ChemistrySelect*, 2, 6163, 2017.
27. Fu, Y., Sun, D. *et al.*, An Amine-Functionalized Titanium Metal–Organic Framework Photocatalyst with Visible-Light-Induced Activity for CO_2 Reduction. *Angew. Chem. Int. Ed.*, 51, 3364, 2012.
28. Chen, D., Xing, H. *et al.*, Highly efficient visible-light-driven CO2 reduction to formate by a new anthracene-based zirconium MOF via dual catalytic routes. *J. Mater. Chem. A*, 4, 2657, 2016.
29. Wang, X., Lu, W. *et al.*, Topology-guided design of an anionic bor-network for photocatalytic [Ru(bpy)3]2+ encapsulation. *Chem. Commun.*, 52, 1926, 2016.
30. Brozek, C.K. and Dincă, M., Ti^{3+}-, $V^{2+/3+}$-, $Cr^{2+/3+}$-, Mn^{2+}-, and Fe^{2+}-Substituted MOF-5 and Redox Reactivity in Cr- and Fe–MOF-5. *J. Am. Chem. Soc.*, 135, 12886, 2013.
31. Wang, D., Huang, R. *et al.*, Fe-Based MOFs for Photocatalytic CO_2 Reduction: Role of Coordination Unsaturated Sites and Dual Excitation Pathways. *ACS Catal.*, 4, 4254, 2014.
32. Subudhi, S., Rath, D. *et al.*, A mechanistic approach towards the photocatalytic organic transformations over functionalised metal organic frameworks: A review. *Catal. Sci. Technol.*, 8, 679, 2018.
33. Deria, P., Mondloch, J.E. *et al.*, Beyond post-synthesis modification: Evolution of metal–organic frameworks via building block replacement. *Chem. Soc. Rev.*, 43, 5896, 2014.
34. Karagiaridi, O., Bury, W. *et al.*, Synthesis and characterization of isostructural cadmium zeolitic imidazolate frameworks via solvent-assisted linker exchange. *Chem. Sci.*, 3, 3256, 2012.

35. Guillerm, V., Kim, D. *et al.*, A supermolecular building approach for the design and construction of metal–organic frameworks. *Chem. Soc. Rev.*, 43, 6141, 2014.
36. Zhao, H., Xia, Q. *et al.*, Construction of Pillared-Layer MOF as Efficient Visible-Light Photocatalysts for Aqueous Cr(VI) Reduction and Dye Degradation. *ACS Sustain. Chem. Eng.*, 5, 4449, 2017.
37. Tu, J., Zeng, X. *et al.*, Microwave-induced fast incorporation of titanium into UiO-66 metal–organic frameworks for enhanced photocatalytic properties. *Chem. Commun.*, 53, 3361, 2017.
38. Nasalevich, M.A., Hendon, C.H. *et al.*, Electronic origins of photocatalytic activity in d^0 metal organic frameworks. *Sci. Rep.*, 6, 23676, 2016.
39. Doan, T.L.H., Nguyen, H.L. *et al.*, Tailoring the Optical Absorption of Water-Stable Zr^{IV}- and Hf^{IV}-Based Metal–Organic Framework Photocatalysts. *Chem. – Asian J.*, 10, 2660, 2015.
40. Yuan, S., Qin, J.-S. *et al.*, Thermodynamically Guided Synthesis of Mixed-Linker Zr-MOFs with Enhanced Tunability. *J. Am. Chem. Soc.*, 138, 6636, 2016.
41. Sun, M., Yan, S. *et al.*, Enhancement of visible-light-driven CO_2 reduction performance using an amine-functionalized zirconium metal–organic framework. *Dalton Trans.*, 47, 909, 2018.
42. Chen, E.-X., Qiu, M. *et al.*, Acid and Base Resistant Zirconium Polyphenolate-Metalloporphyrin Scaffolds for Efficient CO_2 Photoreduction. *Adv. Mater.*, 30, 1704388, 2018.
43. Zhu, W., Zhang, C. *et al.*, Selective reduction of CO_2 by conductive MOF nanosheets as an efficient co-catalyst under visible light illumination. *Appl. Catal. B Environ.*, 238, 339, 2018.
44. Guo, C., Guo, J. *et al.*, Synthesis of core–shell ZIF-67@Co–MOF-74 catalyst with controllable shell thickness and enhanced photocatalytic activity for visible light-driven water oxidation. *CrystEngComm*, 20, 7659, 2018.
45. Ai, L., Zhang, C. *et al.*, Iron terephthalate metal–organic framework: Revealing the effective activation of hydrogen peroxide for the degradation of organic dye under visible light irradiation. *Appl. Catal. B Environ.*, 148–149, 191, 2014.
46. Xu, W.-T., Ma, L. *et al.*, Metal–organic frameworks MIL-88A hexagonal microrods as a new photocatalyst for efficient decolorization of methylene blue dye. *Dalton Trans.*, 43, 3792, 2014.
47. Zhang, Z., Li, X. *et al.*, Hexagonal microspindle of NH_2–MIL-101(Fe) metal–organic frameworks with visible-light-induced photocatalytic activity for the degradation of toluene. *RSC Adv.*, 6, 4289, 2016.
48. Xu, J.-Y., Zhai, X.-P. *et al.*, *In situ* preparation of a MOF-derived magnetic carbonaceous catalyst for visible-light-driven hydrogen evolution. *RSC Adv.*, 6, 2011, 2015.
49. Li, Y., Hou, G. *et al.*, Facile synthesis of MOF 235 and its superior photocatalytic capability under visible light irradiation. *RSC Adv.*, 6, 16395, 2016.

50. Qu, L.-L., Wang, J. *et al.*, Iron(III)-based metal–organic frameworks as oxygen-evolving photocatalysts for water oxidation. *Sustain. Energy Fuels*, 2, 2109, 2018.
51. Liu, N., Jing, C. *et al.*, Effect of synthesis conditions on the photocatalytic degradation of Rhodamine B of MIL-53(Fe). *Mater. Lett.*, 237, 92, 2019.
52. Zhao, J., Wang, Q. *et al.*, A hexanuclear cobalt metal–organic framework for efficient CO2 reduction under visible light. *J. Mater. Chem. A*, 5, 12498, 2017.
53. Zhang, M., Wang, L. *et al.*, Two pure MOF-photocatalysts readily prepared for the degradation of methylene blue dye under visible light. *Dalton Trans.*, 47, 4251, 2018.
54. Qin, L., Chen, H.-Z. *et al.*, Photodegradation of Some Organic Dyes over Two Metal–Organic Frameworks with Especially High Efficiency for Safranine T. *Cryst. Growth Des.*, 17, 1293, 2017.
55. Sanram, S., Boonmak, J. *et al.*, Ni(II)–metal–organic frameworks based on 1,4-phenylenedipropionic acid: Solvothermal syntheses, structures, and photocatalytic properties. *Polyhedron*, 119, 151, 2016.
56. Peterson, G.W., Wagner, G.W. *et al.*, Ammonia Vapor Removal by $Cu_3(BTC)_2$ and Its Characterization by MAS NMR. *J. Phys. Chem. C*, 113, 13906, 2009.
57. DeCoste, J.B., Peterson, G.W. *et al.*, The effect of water adsorption on the structure of the carboxylate containing metal–organic frameworks Cu–BTC, Mg–MOF-74, and UiO-66. *J. Mater. Chem. A*, 1, 11922, 2013.
58. Zhou, E.H., Jia, J.L. *et al.*, Multifunctional Gas Adsorption and Photocatalytic Degradation Based on a Porous Metal–Organic Framework Material. *Russ. J. Coord. Chem.*, 44, 173, 2018.
59. Wang, L., Zeng, T. *et al.*, Syntheses, structures and catalytic mechanisms of three new MOFs for aqueous Cr(VI) reduction and dye degradation under UV light. *Polyhedron*, 157, 152, 2019.
60. Chang, H.-N., Liu, L.-W. *et al.*, A 3D Ag(I) metal–organic framework for sensing luminescence and photocatalytic activities. *J. Mol. Struct.*, 1155, 496, 2018.
61. Jing, H.-P., Wang, C.-C. *et al.*, Photocatalytic degradation of methylene blue in ZIF-8. *RSC Adv.*, 4, 54454, 2014.
62. Masoomi, M.Y., Bagheri, M. *et al.*, High efficiency of mechanosynthesized Zn-based metal–organic frameworks in photodegradation of Congo red under UV and visible light. *RSC Adv.*, 6, 13272, 2016.
63. Masoomi, M.Y., Bagheri, M. *et al.*, Enhancement of photocatalytic performance in two zinc-based metal–organic frameworks by solvent assisted linker exchange. *CrystEngComm*, 19, 5749, 2017.
64. Zhang, S., Li, L. *et al.*, Hierarchical metal–organic framework nanoflowers for effective CO_2 transformation driven by visible light. *J. Mater. Chem. A*, 3, 15764, 2015.
65. Xu, Q.-Q., Fan, H.-J. *et al.*, Zn (Ce,Mn)-MOFs with (3,4,5)-connected 3-D topology network and test of photocatalysis on the reduction of Cr^{6+} by Zn (Ce)-MOFs. *Polyhedron*, 92, 60, 2015.

66. Jin, J.-C., Wu, J. *et al.*, A 3D luminescent Zn(II) MOF for the detection of high explosives and the degradation of organic dyes: An experimental and computational study. *CrystEngComm*, 19, 6464, 2017.
67. Wu, P., Jiang, M. *et al.*, Highly efficient photocatalytic hydrogen production from pure water via a photoactive metal–organic framework and its PDMS@ MOF. *J. Mater. Chem. A*, 5, 7833, 2017.
68. Tang, Y.-H., Wang, F. *et al.*, Construction of unprecedented pillar-layered metal organic frameworks via a dual-ligand strategy for dye degradation. *Dalton Trans.*, 47, 4032, 2018.
69. Zhang, C.Y., Ma, W.X. *et al.*, Structure, photoluminescent properties and photocatalytic activities of a new Cd(II) metal–organic framework. *Spectrochim. Acta A. Mol. Biomol. Spectrosc.*, 118, 657, 2014.
70. Zhang, S., Li, L. *et al.*, Construction of Interpenetrated Ruthenium Metal–Organic Frameworks as Stable Photocatalysts for CO_2 Reduction. *Inorg. Chem.*, 54, 8375, 2015.
71. Zhang, C., Ma, D. *et al.*, Preparation, structure and photocatalysis of metal–organic frameworks derived from aromatic carboxylate and imidazole-based ligands. *J. Coord. Chem.*, 69, 985, 2016.
72. Ji, W.-J., Hao, R.-Q. *et al.*, Design of two isoreticular Cd-biphenyltetracarboxylate frameworks for dye adsorption, separation and photocatalytic degradation. *Dalton Trans.*, 47, 700, 2018.
73. Gu, J.-Z., Cai, Y. *et al.*, A new series of Cd(II) metal–organic architectures driven by soft ether-bridged tricarboxylate spacers: Synthesis, structural and topological versatility, and photocatalytic properties. *Dalton Trans.*, 47, 14327, 2018.
74. Xia, L., Ni, J. *et al.*, Photoactive metal–organic framework as a bifunctional material for 4-hydroxy-4′-nitrobiphenyl detection and photodegradation of methylene blue. *Dalton Trans.*, 47, 16551, 2018.
75. Zuo, L.-Q., Zhang, T.-F. *et al.*, A 3D binuclear salen-based multifunctional MOF: Degradation of MO dye and highly selective sensing of Fe3+. *Inorg. Chem. Commun.*, 99, 113, 2019.
76. Elemike, E.E., Onwudiwe, D.C. *et al.*, Noble metal–semiconductor nanocomposites for optical, energy and electronics applications. *Sol. Energy Mater. Sol. Cells*, 201, 110106, 2019.
77. Ma, L., Chen, S. *et al.*, Recent Progress in Constructing Plasmonic Metal/ Semiconductor Hetero-Nanostructures for Improved Photocatalysis. *Catalysts*, 8, 634, 2018.
78. Fu, Y., Li, J. *et al.*, Metal/Semiconductor Nanocomposites for Photocatalysis: Fundamentals, Structures, Applications and Properties. *Nanomaterials*, 9, 359, 2019.
79. Wen, Y., Ding, H. *et al.*, Preparation and visible light photocatalytic activity of Ag/TiO2/graphene nanocomposite. *Nanoscale*, 3, 4411, 2011.
80. Ray, C. and Pal, T., Recent advances of metal–metal oxide nanocomposites and their tailored nanostructures in numerous catalytic applications. *J. Mater. Chem. A*, 5, 9465, 2017.

81. Rycenga, M., Cobley, C.M. *et al.*, Controlling the Synthesis and Assembly of Silver Nanostructures for Plasmonic Applications. *Chem. Rev.*, 111, 3669, 2011.
82. Wang, D., Pillai, S.C. *et al.*, Plasmonic-based nanomaterials for environmental remediation. *Appl. Catal. B Environ.*, 237, 721, 2018.
83. Yang, Q., Xu, Q. *et al.*, Metal–organic frameworks meet metal nanoparticles: Synergistic effect for enhanced catalysis. *Chem. Soc. Rev.*, 46, 4774, 2017.
84. Chen, L., Luque, R. *et al.*, Encapsulation of metal nanostructures into metal–organic frameworks. *Dalton Trans.*, 47, 3663, 2018.
85. Jiao, L., Wang, Y. *et al.*, Metal–Organic Frameworks as Platforms for Catalytic Applications. *Adv. Mater.*, 30, 1703663, 2018.
86. Zhu, Q.-L. and Xu, Q., Metal–organic framework composites. *Chem. Soc. Rev.*, 43, 5468, 2014.
87. Xiang, W., Zhang, Y. *et al.*, Nanoparticle/Metal–Organic Framework Composites for Catalytic Applications: Current Status and Perspective. *Molecules*, 22, 2103, 2017.
88. Hu, P., Morabito, J.V. *et al.*, Core–Shell Catalysts of Metal Nanoparticle Core and Metal–Organic Framework Shell. *ACS Catal.*, 4, 4409, 2014.
89. Li, G., Zhao, S. *et al.*, Metal–Organic Frameworks Encapsulating Active Nanoparticles as Emerging Composites for Catalysis: Recent Progress and Perspectives. *Adv. Mater.*, 30, 1800702, 2018.
90. Chen, L., Luque, R. *et al.*, Controllable design of tunable nanostructures inside metal–organic frameworks. *Chem. Soc. Rev.*, 46, 4614, 2017.
91. Liu, Y., Liu, Z. *et al.*, Metal or metal-containing nanoparticle@MOF nanocomposites as a promising type of photocatalyst. *Coord. Chem. Rev.*, 388, 63, 2019.
92. Kirchon, A., Feng, L. *et al.*, From fundamentals to applications: A toolbox for robust and multifunctional MOF materials. *Chem. Soc. Rev.*, 47, 8611, 2018.
93. Rubio-Martinez, M., Avci-Camur, C. *et al.*, New synthetic routes towards MOF production at scale. *Chem. Soc. Rev.*, 46, 3453, 2017.
94. Butova, V.V., Soldatov, M.A. *et al.*, Metal–organic frameworks: Structure, properties, methods of synthesis and characterization. *Russ. Chem. Rev.*, 85, 280, 2016.
95. Howarth, A.J., Peters, A.W. *et al.*, Best Practices for the Synthesis, Activation, and Characterization of Metal–Organic Frameworks. *Chem. Mater.*, 29, 26, 2017.
96. Dhakshinamoorthy, A., Li, Z. *et al.*, Catalysis and photocatalysis by metal organic frameworks. *Chem. Soc. Rev.*, 47, 8134, 2018.
97. Wang, C., deKrafft, K.E. *et al.*, Pt Nanoparticles@Photoactive Metal–Organic Frameworks: Efficient Hydrogen Evolution via Synergistic Photoexcitation and Electron Injection. *J. Am. Chem. Soc.*, 134, 7211, 2012.
98. Deng, X., Yang, L. *et al.*, Shape-Defined Hollow Structural Co–MOF-74 and Metal Nanoparticles@Co–MOF-74 Composite through a Transformation Strategy for Enhanced Photocatalysis Performance. *Small*, 15, 1902287, 2019.

99. Chen, L., Wang, Y. *et al.*, A simple strategy for engineering heterostructures of Au nanoparticle-loaded metal–organic framework nanosheets to achieve plasmon-enhanced photocatalytic CO2 conversion under visible light. *J. Mater. Chem. A*, 7, 11355, 2019.
100. Xiao, L., Zhang, Q. *et al.*, Copper-mediated metal–organic framework as efficient photocatalyst for the partial oxidation of aromatic alcohols under visible-light irradiation: Synergism of plasmonic effect and Schottky junction. *Appl. Catal. B Environ.*, 248, 380, 2019.
101. Wen, M., Mori, K. *et al.*, Design and architecture of metal organic frameworks for visible light enhanced hydrogen production. *Appl. Catal. B Environ.*, 218, 555, 2017.
102. Xu, J., Liu, J. *et al.*, Synthesis, structure and properties of Pd@MOF-808. *J. Mater. Sci.*, 54, 12911, 2019.
103. Wang, Q. and Astruc, D., State of the Art and Prospects in Metal–Organic Framework (MOF)-Based and MOF-Derived Nanocatalysis. *Chem. Rev.*, 2019.
104. Xiao, J.-D., Han, L. *et al.*, Integration of Plasmonic Effects and Schottky Junctions into Metal–Organic Framework Composites: Steering Charge Flow for Enhanced Visible-Light Photocatalysis. *Angew. Chem. Int. Ed.*, 57, 1103, 2018.
105. Fujishima, A. and Honda, K., Electrochemical Photolysis of Water at a Semiconductor Electrode. *Nature*, 238, 37, 1972.
106. Park, H., Park, Y. *et al.*, Surface modification of TiO_2 photocatalyst for environmental applications. *J. Photochem. Photobiol. C Photochem. Rev.*, 15, 1, 2013.
107. Zhang, L. and Zhang, J., Metal–organic frameworks for CO_2 photoreduction. *Front. Energy*, 13, 221, 2019.
108. Shi, Y., Yang, A.-F. *et al.*, Applications of MOFs: Recent advances in photocatalytic hydrogen production from water. *Coord. Chem. Rev.*, 390, 50, 2019.
109. Bedia, J., Muelas-Ramos, V. *et al.*, A Review on the Synthesis and Characterization of Metal Organic Frameworks for Photocatalytic Water Purification. *Catalysts*, 9, 52, 2019.
110. Wu, Z., Yuan, X. *et al.*, Photocatalytic Decontamination of Wastewater Containing Organic Dyes by Metal–Organic Frameworks and their Derivatives. *ChemCatChem*, 9, 41, 2017.
111. Wang, C.-C., Du, X.-D. *et al.*, Photocatalytic Cr(VI) reduction in metal–organic frameworks: A mini-review. *Appl. Catal. B Environ.*, 193, 198, 2016.
112. Dias, E.M. and Petit, C., Towards the use of metal–organic frameworks for water reuse: A review of the recent advances in the field of organic pollutants removal and degradation and the next steps in the field. *J. Mater. Chem. A*, 3, 22484, 2015.
113. Wang, Y., Liu, H. *et al.*, A dual-functional UiO-66/TiO_2 composite for water treatment and CO_2 capture. *RSC Adv.*, 7, 16232, 2017.

114. Huang, J., Song, H. *et al.*, Facile synthesis of N-doped TiO_2 nanoparticles caged in MIL-100(Fe) for photocatalytic degradation of organic dyes under visible light irradiation. *J. Environ. Chem. Eng.*, 5, 2579, 2017.
115. Liu, X., Dang, R. *et al.*, A sandwich-like heterostructure of TiO_2 nanosheets with MIL-100(Fe): A platform for efficient visible-light-driven photocatalysis. *Appl. Catal. B Environ.*, 209, 506, 2017.
116. Li, X., Pi, Y. *et al.*, TiO_2 encapsulated in Salicylaldehyde–NH_2–MIL-101(Cr) for enhanced visible light-driven photodegradation of MB. *Appl. Catal. B Environ.*, 191, 192, 2016.
117. Li, Y., Jiang, J. *et al.*, TiO_2 Nanoparticles Anchored onto the Metal–Organic Framework NH2–MIL-88B(Fe) as an Adsorptive Photocatalyst with Enhanced Fenton-like Degradation of Organic Pollutants under Visible Light Irradiation. *ACS Sustain. Chem. Eng.*, 6, 16186, 2018.
118. Oveisi, M., Mahmoodi, N.M. *et al.*, Facile and green synthesis of metal–organic framework/inorganic nanofiber using electrospinning for recyclable visible-light photocatalysis. *J. Clean. Prod.*, 222, 669, 2019.
119. Li, X., Pi, Y. *et al.*, Amorphous TiO_2@NH_2–MIL-125(Ti) homologous MOF-encapsulated heterostructures with enhanced photocatalytic activity. *Chem. Commun.*, 54, 1917, 2018.
120. Khajavian, R., Ghani, K. *et al.*, Decolorization of reactive red 198 using preferentially-oriented metal–organic framework thin films: A structure/property correlation. *Inorg. Chem. Commun.*, 79, 65, 2017.
121. Wang, H., Yu, T. *et al.*, Enhanced Photocatalytic Oxidation of Isopropanol by HKUST-1@TiO_2 Core–Shell Structure with Ultrathin Anatase Porous Shell: Toxic Intermediate Control. *Ind. Eng. Chem. Res.*, 55, 8096, 2016.
122. Liu, J., Li, X.-M. *et al.*, Combining the Photocatalysis and Absorption Properties of Core–Shell Cu–BTC@TiO_2 Microspheres: Highly Efficient Desulfurization of Thiophenic Compounds from Fuel. *Materials*, 11, 2209, 2018.
123. Zhang, B., Zhang, J. *et al.*, MIL-125–NH_2@TiO_2 Core–Shell Particles Produced by a Post-Solvothermal Route for High-Performance Photocatalytic H_2 Production. *ACS Appl. Mater. Interfaces*, 10, 16418, 2018.
124. Zhang, M., Shang, Q. *et al.*, Self-template synthesis of double-shell TiO_2@ZIF-8 hollow nanospheres via sonocrystallization with enhanced photocatalytic activities in hydrogen generation. *Appl. Catal. B Environ.*, 241, 149, 2019.
125. Martínez, F.M., Albiter, E. *et al.*, Hydrogen Production from Glycerol Photoreforming on TiO_2/HKUST-1 Composites: Effect of Preparation Method. *Catalysts*, 9, 1, 2019.
126. He, X., Gan, Z. *et al.*, Rapid Formation of Metal–Organic Frameworks (MOFs) Based Nanocomposites in Microdroplets and Their Applications for CO_2 Photoreduction. *ACS Appl. Mater. Interfaces*, 9, 9688, 2017.

127. Wang, M., Wang, D. *et al.*, Self-assembly of CPO-27–Mg/TiO_2 nanocomposite with enhanced performance for photocatalytic CO_2 reduction. *Appl. Catal. B Environ.*, 183, 47, 2016.
128. Barrios, C.E., Albiter, E. *et al.*, Photocatalytic hydrogen production over titania modified by gold – Metal (palladium, nickel and cobalt) catalysts. *Int. J. Hydrog. Energy*, 41, 23287, 2016.
129. Luna, A.L., Dragoe, D. *et al.*, Photocatalytic Hydrogen Evolution Using Ni–Pd/TiO2: Correlation of Light Absorption, Charge-Carrier Dynamics, and Quantum Efficiency. *J. Phys. Chem. C*, 121, 14302, 2017.
130. Zhan, W., Sun, L. *et al.*, Recent Progress on Engineering Highly Efficient Porous Semiconductor Photocatalysts Derived from Metal–Organic Frameworks. *Nano-Micro Lett.*, 11, 1, 2019.
131. Li, R., Zhang, W. *et al.*, Metal–Organic-Framework-Based Catalysts for Photoreduction of CO_2. *Adv. Mater.*, 30, 1705512, 2018.
132. Teter, D.M. and Hemley, R.J., Low-Compressibility Carbon Nitrides. *Science*, 271, 53, 1996.
133. Ong, W.-J., Tan, L.-L. *et al.*, Graphitic Carbon Nitride (g-C3N4)-Based Photocatalysts for Artificial Photosynthesis and Environmental Remediation: Are We a Step Closer To Achieving Sustainability? *Chem. Rev.*, 116, 7159, 2016.
134. Hong, J., Chen, C. *et al.*, Carbon nitride nanosheet/metal–organic framework nanocomposites with synergistic photocatalytic activities. *Catal. Sci. Technol.*, 6, 5042, 2016.
135. Wang, H., Yuan, X. *et al.*, Synthesis and applications of novel graphitic carbon nitride/metal–organic frameworks mesoporous photocatalyst for dyes removal. *Appl. Catal. B Environ.*, 174, 175–445, 2015.
136. Lei, Z.-D., Xue, Y.-C. *et al.*, The Influence of Carbon Nitride Nanosheets Doping on the Crystalline Formation of MIL-88B(Fe) and the Photocatalytic Activities. *Small*, 14, 1802045, 2018.
137. Zhou, G., Wu, M.-F. *et al.*, Synthesis and characterizations of metal-free Semiconductor/MOFs with good stability and high photocatalytic activity for H_2 evolution: A novel Z-Scheme heterostructured photocatalyst formed by covalent bonds. *Appl. Catal. B Environ.*, 220, 607, 2018.
138. Wang, R., Gu, L. *et al.*, Quasi-Polymeric Metal–Organic Framework UiO-66/g-C_3N_4 Heterojunctions for Enhanced Photocatalytic Hydrogen Evolution under Visible Light Irradiation. *Adv. Mater. Interfaces*, 2, 1500037, 2015.
139. Liu, S., Chen, F. *et al.*, Enhanced photocatalytic conversion of greenhouse gas CO_2 into solar fuels over g-C_3N_4 nanotubes with decorated transparent ZIF-8 nanoclusters. *Appl. Catal. B Environ.*, 211, 1, 2017.
140. Huang, J., Zhang, X. *et al.*, Protonated graphitic carbon nitride coated metal–organic frameworks with enhanced visible-light photocatalytic activity for contaminants degradation. *Appl. Surf. Sci.*, 441, 85, 2018.
141. Ye, L., Deng, Y. *et al.*, Bismuth-Based Photocatalysts for Solar Photocatalytic Carbon Dioxide Conversion. *ChemSusChem*, 12, 3671, 2019.

142. Zhu, S.-R., Liu, P.-F. *et al.*, Enhanced photocatalytic performance of BiOBr/NH_2–MIL-125(Ti) composite for dye degradation under visible light. *Dalton Trans.*, 45, 17521, 2016.
143. Sha, Z. and Wu, J., Enhanced visible-light photocatalytic performance of BiOBr/UiO-66(Zr) composite for dye degradation with the assistance of UiO-66. *RSC Adv.*, 5, 39592, 2015.
144. Bibi, R., Shen, Q. *et al.*, Hybrid BiOBr/UiO-66–NH_2 composite with enhanced visible-light driven photocatalytic activity toward RhB dye degradation. *RSC Adv.*, 8, 2048, 2018.
145. Zhu, S.-R., Wu, M.-K. *et al.*, *In Situ* Growth of Metal–Organic Framework on BiOBr 2D Material with Excellent Photocatalytic Activity for Dye Degradation. *Cryst. Growth Des.*, 17, 2309, 2017.
146. Yue, X., Guo, W. *et al.*, Preparation of Efficient BiOBr/MIL-88B(Fe) Composites with Enhanced Photocatalytic Activities. *Water Environ. Res.*, 89, 614, 2017.
147. Hu, Q., Di, J. *et al.*, *In-situ* preparation of NH_2–MIL-125(Ti)/BiOCl composite with accelerating charge carriers for boosting visible light photocatalytic activity. *Appl. Surf. Sci.*, 466, 525, 2019.
148. Tong, X., Yang, Z. *et al.*, BiOCl/UiO-66 composite with enhanced performance for photo-assisted degradation of dye from water. *Appl. Organomet. Chem.*, 32, e4049, 2018.
149. Xu, Y., Lv, M. *et al.*, $BiVO_4$/MIL-101 composite having the synergistically enhanced visible light photocatalytic activity. *RSC Adv.*, 5, 43473, 2015.
150. Si, Y., Li, Y. *et al.*, Fabrication of Novel ZIF-8@$BiVO_4$ Composite with Enhanced Photocatalytic Performance. *Crystals*, 8, 432, 2018.
151. Yang, Z., Ding, J. *et al.*, Preparation of $BiVO_4$/MIL-125(Ti) composite with enhanced visible-light photocatalytic activity for dye degradation. *Appl. Organomet. Chem.*, 32, e4285, 2018.
152. Abdpour, S., Kowsari, E. *et al.*, MIL-100(Fe) nanoparticles supported on urchin like Bi_2S_3 structure for improving photocatalytic degradation of rhodamine-B dye under visible light irradiation. *J. Solid State Chem.*, 266, 54, 2018.
153. Mao, J., Ge, M. *et al.*, Constructing multifunctional MOF@rGO hydro-/aerogels by the self-assembly process for customized water remediation. *J. Mater. Chem. A*, 5, 11873, 2017.
154. Wang, X., Zhao, X. *et al.*, Microwave irradiation induced UIO-66–NH_2 anchored on graphene with high activity for photocatalytic reduction of CO_2. *Appl. Catal. B Environ.*, 228, 47, 2018.
155. Yang, C., You, X. *et al.*, A novel visible-light-driven In-based MOF/graphene oxide composite photocatalyst with enhanced photocatalytic activity toward the degradation of amoxicillin. *Appl. Catal. B Environ.*, 200, 673, 2017.
156. Cai, J., Lu, J.-Y. *et al.*, Eu-Based MOF/graphene oxide composite: A novel photocatalyst for the oxidation of benzyl alcohol using water as oxygen source. *New J. Chem.*, 41, 3882, 2017.

157. Zhang, C., Ai, L. *et al.*, Graphene Hybridized Photoactive Iron Terephthalate with Enhanced Photocatalytic Activity for the Degradation of Rhodamine B under Visible Light. *Ind. Eng. Chem. Res.*, 54, 153, 2015.
158. Huang, L. and Liu, B., Synthesis of a novel and stable reduced graphene oxide/MOF hybrid nanocomposite and photocatalytic performance for the degradation of dyes. *RSC Adv.*, 6, 17873, 2016.
159. Vu, T.A., Le, G.H. *et al.*, Highly photocatalytic activity of novel Fe-MIL-88B/GO nanocomposite in the degradation of reactive dye from aqueous solution. *Mater. Res. Express*, 4, 035038, 2017.
160. Li, J., Fang, W. *et al.*, Ag-based semiconductor photocatalysts in environmental purification. *Appl. Surf. Sci.*, 358, 46, 2015.
161. Li, G., Wang, Y. *et al.*, Recent progress in highly efficient Ag-based visible-light photocatalysts. *RSC Adv.*, 4, 53649, 2014.
162. Albiter, E., Hai, Z. *et al.*, A Comparative Study of Photo-Assisted Deposition of Silver Nanoparticles on TiO_2. *J. Nanosci. Nanotechnol.*, 13, 4943, 2013.
163. Albiter, E., Valenzuela, M.A. *et al.*, Photocatalytic deposition of Ag nanoparticles on TiO_2: Metal precursor effect on the structural and photoactivity properties. *J. Saudi Chem. Soc.*, 19, 563, 2015.
164. Han, Y., Bai, C. *et al.*, A facile strategy for fabricating AgI–MIL-53(Fe) composites: Superior interfacial contact and enhanced visible light photocatalytic performance. *New J. Chem.*, 42, 3799, 2018.
165. Oladipo, A.A., Vaziri, R. *et al.*, Highly robust $AgIO_3$/MIL-53 (Fe) nanohybrid composites for degradation of organophosphorus pesticides in single and binary systems: Application of artificial neural networks modelling. *J. Taiwan Inst. Chem. Eng.*, 83, 133, 2018.
166. Abdelhameed, R.M., Tobaldi, D.M. *et al.*, Engineering highly effective and stable nanocomposite photocatalyst based on NH_2–MIL-125 encirclement with Ag_3PO_4 nanoparticles. *J. Photochem. Photobiol. Chem.*, 351, 50, 2018.
167. Mohaghegh, N., Kamrani, S. *et al.*, Nanoporous Ag_2O photocatalysts based on copper terephthalate metal–organic frameworks. *J. Mater. Sci.*, 50, 4536, 2015.
168. Tabatabaei, N., Dashtian, K. *et al.*, Novel visible light-driven Cu-based MOFs/Ag_2O composite photocatalysts with enhanced photocatalytic activity toward the degradation of orange G: Their photocatalytic mechanism and optimization study. *New J. Chem.*, 42, 9720, 2018.
169. Mehraj, O., Sofi, F.A. *et al.*, Synthesis of novel silver chromate incorporated copper–metal–organic framework composites with exceptionally high photocatalytic activity and stability. *J. Mater. Sci. Mater. Electron.*, 29, 3358, 2018.
170. Wang, G., Ma, X. *et al.*, Controlled synthesis of Ag_2O microcrystals with facet-dependent photocatalytic activities. *J. Mater. Chem.*, 22, 21189, 2012.
171. Ma, X., Lou, Y. *et al.*, New UiO-66/Cu_xS Heterostructures: Surface Functionalization Synthesis and Their Application in Photocatalytic Degradation of RhB. *Bull. Chem. Soc. Jpn.*, 91, 515, 2018.

172. Mehta, J.P., Tian, T. *et al.*, Sol–Gel Synthesis of Robust Metal–Organic Frameworks for Nanoparticle Encapsulation. *Adv. Funct. Mater.*, 28, 1705588, 2018.
173. Ding, D., Jiang, Z. *et al.*, Enhanced photocatalytic activity and mechanism insight of MnO_x/MIL-101. *J. Taiwan Inst. Chem. Eng.*, 82, 226, 2018.
174. Wang, H., Yuan, X. *et al.*, *In situ* synthesis of In_2S_3@MIL-125(Ti) core–shell microparticle for the removal of tetracycline from wastewater by integrated adsorption and visible-light-driven photocatalysis. *Appl. Catal. B Environ.*, 186, 19, 2016.
175. Jin, Z., Dong, W. *et al.*, One-Pot Preparation of Hierarchical Nanosheet-Constructed Fe_3O_4/MIL-88B(Fe) Magnetic Microspheres with High Efficiency Photocatalytic Degradation of Dye. *ChemCatChem*, 8, 3510, 2016.
176. Bagheri, M., Masoomi, M.Y. *et al.*, A MoO_3–Metal–Organic Framework Composite as a Simultaneous Photocatalyst and Catalyst in the PODS Process of Light Oil. *ACS Catal.*, 7, 6949, 2017.
177. Peters, A.W., Li, Z. *et al.*, Toward Inexpensive Photocatalytic Hydrogen Evolution: A Nickel Sulfide Catalyst Supported on a High-Stability Metal–Organic Framework. *ACS Appl. Mater. Interfaces*, 8, 20675, 2016.
178. Su, Y., Zhang, Z. *et al.*, $Cd_{0.2}Zn_{0.8}S$@UiO-66–NH_2 nanocomposites as efficient and stable visible-light-driven photocatalyst for H_2 evolution and CO_2 reduction. *Appl. Catal. B Environ.*, 200, 448, 2017.
179. Yu, B., Wang, F. *et al.*, Self-template synthesis of core–shell ZnO@ZIF-8 nanospheres and the photocatalysis under UV irradiation. *Mater. Lett.*, 156, 50, 2015.
180. Rad, M. and Dehghanpour, S., ZnO as an efficient nucleating agent and morphology template for rapid, facile and scalable synthesis of MOF-46 and ZnO@MOF-46 with selective sensing properties and enhanced photocatalytic ability. *RSC Adv.*, 6, 61784, 2016.
181. Kaur, R., Rana, A. *et al.*, Efficient photocatalytic and photovoltaic applications with nanocomposites between CdTe QDs and an NTU-9 MOF. *RSC Adv.*, 7, 29015, 2017.
182. Ramezanalizadeh, H. and Manteghi, F., Synthesis of a novel MOF/$CuWO_4$ heterostructure for efficient photocatalytic degradation and removal of water pollutants. *J. Clean. Prod.*, 172, 2655, 2018.
183. Zeng, M., Chai, Z. *et al.*, Core–shell CdS@ZIF-8 structures for improved selectivity in photocatalytic H_2 generation from formic acid. *Nano Res.*, 9, 2729, 2016.
184. Xu, H.-Q., Yang, S. *et al.*, Unveiling Charge-Separation Dynamics in CdS/Metal–Organic Framework Composites for Enhanced Photocatalysis. *ACS Catal.*, 8, 11615, 2018.
185. Nguyen, T.N., Kampouri, S. *et al.*, Photocatalytic Hydrogen Generation from a Visible-Light-Responsive Metal–Organic Framework System: Stability versus Activity of Molybdenum Sulfide Cocatalysts. *ACS Appl. Mater. Interfaces*, 10, 30035, 2018.

186. Liu, H., Zhang, J. *et al.*, Construction of heterostructured $ZnIn_2S_4$@NH_2–MIL-125(Ti) nanocomposites for visible-light-driven H_2 production. *Appl. Catal. B Environ.*, 221, 433, 2018.
187. Kampouri, S., Nguyen, T.N. *et al.*, Photocatalytic hydrogen generation from a visible-light responsive metal–organic framework system: The impact of nickel phosphide nanoparticles. *J. Mater. Chem. A*, 6, 2476, 2018.
188. Karthik, P., Balaraman, E. *et al.*, Efficient solar light-driven H_2 production: Post-synthetic encapsulation of a Cu_2O co-catalyst in a metal–organic framework (MOF) for boosting the effective charge carrier separation. *Catal. Sci. Technol.*, 8, 3286, 2018.
189. Jiang, D., Xu, P. *et al.*, Strategies to improve metal organic frameworks photocatalyst's performance for degradation of organic pollutants. *Coord. Chem. Rev.*, 376, 449, 2018.
190. Tian, P., He, X. *et al.*, Enhanced charge transfer for efficient photocatalytic H2 evolution over UiO-66–NH2 with annealed Ti3C2Tx MXenes. *Int. J. Hydrog. Energy*, 2018.
191. Wang, Y., Zhang, Y. *et al.*, Controlled fabrication and enhanced visible-light photocatalytic hydrogen production of Au@CdS/MIL-101 heterostructure. *Appl. Catal. B Environ.*, 185, 307, 2016.
192. Xu, J., Gao, J. *et al.*, NH2–MIL-125(Ti)/graphitic carbon nitride heterostructure decorated with NiPd co-catalysts for efficient photocatalytic hydrogen production. *Appl. Catal. B Environ.*, 219, 101, 2017.
193. Sofi, F.A., Majid, K. *et al.*, The visible light driven copper based metal–organic-framework heterojunction: HKUST-1@Ag–Ag_3PO_4 for plasmon enhanced visible light photocatalysis. *J. Alloys Compd.*, 737, 798, 2018.
194. Zhou, T., Zhang, G. *et al.*, Highly efficient visible-light-driven photocatalytic degradation of rhodamine B by a novel Z-scheme Ag_3PO_4/MIL-101/$NiFe_2O_4$ composite. *Catal. Sci. Technol.*, 8, 2402, 2018.
195. Liu, Q., Zeng, C. *et al.*, Boosting visible light photoreactivity of photoactive metal–organic framework: Designed plasmonic Z-scheme Ag/AgCl@MIL-53–Fe. *Appl. Catal. B Environ.*, 224, 38, 2018.
196. Huang, W., Jing, C. *et al.*, Integration of plasmonic effect into spindle-shaped MIL-88A(Fe): Steering charge flow for enhanced visible-light photocatalytic degradation of ibuprofen. *Chem. Eng. J.*, 349, 603, 2018.
197. Tilgner, D. and Kempe, R., A Plasmonic Colloidal Photocatalyst Composed of a Metal–Organic Framework Core and a Gold/Anatase Shell for Visible-Light-Driven Wastewater Purification from Antibiotics and Hydrogen Evolution. *Chem. - Eur. J.*, 23, 3184, 2017.
198. Chen, C., Wu, T. *et al.*, Highly effective photoreduction of CO_2 to CO promoted by integration of CdS with molecular redox catalysts through metal–organic frameworks. *Chem. Sci.*, 9, 8890, 2018.

199. Wang, Z., Jin, Z. *et al.*, Efficient hydrogen production over MOFs (ZIF-67) and g-C_3N_4 boosted with MoS_2 nanoparticles. *Int. J. Hydrog. Energy*, 43, 13039, 2018.
200. Shi, X., Zhang, J. *et al.*, Photocatalytic H_2 evolution improvement for H free-radical stabilization by electrostatic interaction of a Cu–BTC MOF with ZnO/GO. *Nano Res.*, 11, 979, 2018.
201. Meng, X.-B., Sheng, J.-L. *et al.*, Metal–organic framework as nanoreactors to co-incorporate carbon nanodots and CdS quantum dots into the pores for improved H_2 evolution without noble-metal cocatalyst. *Appl. Catal. B Environ.*, 244, 340, 2019.
202. Zhang, Y., Wang, G. *et al.*, CdS p–n heterojunction co-boosting with Co_3O_4 and Ni–MOF-74 for photocatalytic hydrogen evolution. *Dalton Trans.*, 47, 11176, 2018.
203. Cui, S., Ye, Z. *et al.*, Construction of ternary Ag/AgBr@UIO-66(NH_2) heterojunctions with enhanced photocatalytic performance for the degradation of methyl orange. *J. Mater. Sci. Mater. Electron.*, 29, 15138, 2018.
204. Feng, S., Wang, R. *et al.*, Synthesis of Zr-based MOF nanocomposites for efficient visible-light photocatalytic degradation of contaminants. *Res. Chem. Intermed.*, 45, 1263, 2019.
205. Liang, Z., Qu, C. *et al.*, Pristine Metal–Organic Frameworks and their Composites for Energy Storage and Conversion. *Adv. Mater.*, 30, 1702891, 2018.

Index

Also of Interest

Check out these other forthcoming and published titles from Scrivener Publishing

By the same editors:

Potassium-Ion Batteries: Materials and Applications, Edited by Inamuddin, Rajender Boddula, and Abdullah M. Asiri, ISBN 9781119661399. Edited by one of the most well-respected and prolific engineers in the world and his team, this is the most thorough, up-to-date, and comprehensive volume on potassium-ion batteries available today. *COMING IN APRIL 2020*

Zinc Batteries: Basics, Developments, and Applications, Edited by Rajender Boddula, Inamuddin, and Abdullah M. Asiri, ISBN 9781119661894. Edited by one of the most well-respected and prolific engineers in the world and his team, this is the most thorough, up-to-date, and comprehensive volume on zinc batteries available today. *COMING IN APRIL 2020*

Rechargeable Batteries: History, Progress, and Applications, edited by Rajender Boddula, Inamuddin, Ramyakrishna Pothu, and Abdullah M. Asiri, ISBN 9781119661191. Edited by one of the most well-respected and prolific engineers in the world and his team, this is the most thorough, up-to-date, and comprehensive volume on rechargeable batteries available today. *COMING IN APRIL 2020*

Other Books of Interest:

Advanced Battery Materials, edited by Chunwen Sun, ISBN 9781119407553. This book details the latest R&D in electrochemical energy storagetechnologies for portable electronics and electric vehicle applications. *NOW AVAILABLE!*

Nanomaterials for Electrochemical Energy Storage Devices, edited by Poulomi Roy and Suneel Kumar Srivastava, ISBN 9781119510031. Presents state-of-the-art review chapters by subject matter specialists in electrochemical energy storage systems. *NOW AVAILABLE!*

Polymers for Energy Storage and Conversion, edited by Vikas Mittal, ISBN 9781118344545. One of the first comprehensive books to focus on the role polymers is playing in the burgeoning energy materials market. *NOW AVAILABLE!*

Energy Storage: A New Approach 2nd Edition, by Ralph Zito and Haleh Ardebili, ISBN 9781119083597. A revision of the groundbreaking study of methods for storing energy on a massive scale to be used in wind, solar, and other renewable energy systems. *NOW AVAILABLE!*